18th International Summer School on Vacuum, Electron and Ion Technologies

(VEIT 2013)

Journal of Physics: Conference Series Volume 514

Sozopol, Bulgaria
7-11 October 2013

Editors:

Richard van de Sanden
Miglena Dimitrova
Chavdar Ghelev

ISBN: 978-1-63439-295-2
ISSN: 1742-6588

Printed from e-media with permission by:

Curran Associates, Inc.
57 Morehouse Lane
Red Hook, NY 12571

Printed by Curran Associates, Inc. (2014)

Institute of Physics
Dirac House, Temple Back
Bristol BS1 6BE UK

Phone: 44 1 17 929 7481
Fax: 44 1 17 920 0979

techtracking@iop.org

Additional copies of this publication are available from:

Curran Associates, Inc.
57 Morehouse Lane
Red Hook, NY 12571 USA
Phone: 845-758-0400
Fax: 845-758-2634
Email: curran@proceedings.com
Web: www.proceedings.com

Table Of Contents

THIN FILMS DEPOSITION. COATINGS FOR ADVANCED APPLICATIONS

18th International Summer School on Vacuum, Electron and Ion Technologies

(VEIT 2013)

Journal of Physics: Conference Series Volume 514

Sozopol, Bulgaria
7-11 October 2013

ISBN: 978-1-63439-295-2
ISSN: 1742-6588

IOP Institute of Physics

SURFACES AND THIN FILMS PROCESSING AND ANALYSIS

PLASMA–SURFACE INTERACTION, PLASMA DIAGNOSTICS. MODELING AND COMPUTER SIMULATION

Journal of Physics: Conference Series **514** (2014) 011001 doi:10.1088/1742-6596/514/1/011001

Preface

The International Summer School on Vacuum, Electron and Ion Technologies (VEIT) has been organized biennially since 1977, when the series of VEIT School was launched by the Institute of Electronics, Bulgarian Academy of Sciences with the aim to act as a forum for exchange and dissemination of knowledge and ideas on the latest developments in electron-, ion-, and plasma-assisted technologies. The organizers of the 2013 edition of the event were the Institute of Electronics, Bulgarian Academy of Sciences, Sofia, Bulgaria and the the Dutch Institute for Fundamental Energy Research, Nieuwegein, The Netherlands. While the school has initially been providing a meeting place for researchers mainly from Eastern and Central European countries, its importance has grown issue by issue. The school is now a major scientific event and a meeting place for young scientists from Eastern and Western Europe involved in research and development associated with high-tech industries. Many former school participants have gone on to become leading scientists in research establishments and companies throughout the world. Leading international companies, such as High Voltage Engineering, Balzers, Varian, and Hauzer have used the VEIT forum to present their products through oral presentations, poster contributions or exhibits. The School Proceedings have been published in special issues of the international journals *Vacuum, Plasma Processes and Polymers*, and *Journal of Physics: Conference Series*.

The eighteenth edition of VEIT was held in the Black Sea resort Sozopol, Bulgaria, on 7–11 October 2013. It was attended by 91 participants from 15 countries: Bulgaria, Czech Republic, France, Germany, Greece, The Netherlands, Portugal, Romania, Russia, Serbia, Slovakia, Slovenia, Sweden, Ukraine and the UK.

Following the tradition of publishing the VEIT Proceedings, a selection of papers presented at the event is published in this special issue of *Journal of Physics: Conference Series*, under the originality and quality criteria of acceptance by the journal, including peer reviewing.

The school comprised of fifteen plenary and three poster sessions. At the plenary sessions, 21 invited talks of general interest were presented by highly renowned international experts in the field, as well as 10 progress reports by young scientists. In total, 66 contributed papers were presented during the poster sessions. There were several scientific highlights covering fundamentals of interaction of fast particles with solids and challenging practical applications ranging from novel techniques for creating hard coatings, optical/protective layers, biocompatible materials to nanosized structures produced by evaporation, sputtering or external irradiation. Latest results were presented on ion-beam synthesis and modification in both low-energy (deposition and film growth) and high-energy (sputtering, implantation) regimes, and processing of solid materials aiming at patterning the surface or at creating nanophase systems for electronic or tribological/wear resistant applications.

Despite the busy scientific program, the atmosphere was relaxed and informal. The early afternoons of most conference days were free to stimulate both scientific and social interaction between participants, which often took place on the beach. The social program included a welcome party, an official dinner, and an outing to historical landmarks in old Sozopol. VEIT 2013 owes its success to many people. The International Advisory Committee shaped the scientific program and ensured high-quality plenary

Journal of Physics: Conference Series **514** (2014) 011001 doi:10.1088/1742-6596/514/1/011001

presentations by careful selection of invited speakers. The Local Committee bore the brunt of the organization both at the conference site and in dealing with correspondence, abstracts, and manuscripts for these proceedings. We are grateful to our sponsors, the Dutch Institute for Fundamental Energy Research, Nieuwegein, The Netherlands, the companies AllData and Astel, and the Ministry of Education and Science of Republic of Bulgaria for their generosity, which enabled us to support the attendance of students and provided support to deal with mailing, printing, renting the conference site, etc. We are also grateful to the members of the International Advisory Committee — Th Czerwiec, M Dimitrova, G Dinescu, A Ehiasarian, N Guerassimov, I Katardjiev, H Kersten, D Mataras, W Möller, I Petrov, Z Petrovic, B Rauschenbach, M C M (Richard) van de Sanden and M Ürgen — for their continuous concern about the high scientific level and the future of the event, as well as for their contribution to the formation of a new generation of scientists and experts in the field. And, last but not least, we would also like to thank all authors for their valuable contributions to these proceedings and to the school, and all reviewers for their hard and tedious, but very important, work. The next conference in the series will be held in September 2015.

M C M van de Sanden, Miglena Dimitrova and Chavdar Ghelev
Guest Editors

Peer review statement

All papers published in this volume of *Journal of Physics: Conference Series* have been peer reviewed through processes administered by the proceedings Editors. Reviews were conducted by expert referees to the professional and scientific standards expected of a proceedings journal published by IOP Publishing.

VIS/IR spectroscopy of thin AlN films grown by pulsed laser deposition at 400°C and 800°C and various N_2 pressures

S Bakalova[1,4], **A Szekeres**[1], **M Anastasescu**[2], **M Gartner**[2], **L Duta**[3], **G Socol**[3], **C Ristoscu**[3] **and I N Mihailescu**[3]

[1]Acad. Georgy Nadjakov Institute of Solid State Physics, Bulgarian Academy of Sciences, 72 Tsarigradsko Chaussee, 1784 Sofia, Bulgaria
[2]Institute of Physical Chemistry, Romanian Academy of Sciences, 202 Spl. Independentei, 060021 Bucharest, Romania
[3]Lasers Department, National Institute for Lasers, Plasma, and Radiation Physics, PO Box MG-54, 77125 Bucharest-Magurele, Romania

E-mail: sbakalova@issp.bas.bg

Abstract. The optical properties of pulsed-laser-deposited (PLD) AlN films on silicon are reported in the wide visible and infrared spectral range from $5x10^4$ cm^{-1} to 350 cm^{-1}. The films were deposited at 400 °C or 800 °C and in vacuum or nitrogen gas ambient at pressures from 0.1 Pa to 10 Pa. The optical constants of the films were determined by modelling the spectroscopic ellipsometry data recorded in both the visible and infrared spectral ranges. The PLD AlN films deposited in vacuum have a bandgap energy of 6.2 eV in a good agreement with the crystalline AlN material, while an effective bandgap narrowing was observed for films deposited in N_2 ambient. This effect was most pronounced in films deposited at 400°C, where the optical absorption below the fundamental edge was the largest due to a larger degree of disorder in the films. Fourier transform infrared spectroscopy was used to identify AlN phonon modes and stress in the films. The shift and broadening of the E_1(TO) peak towards the higher wave numbers indicates that an increased compressive stress is developed in the deposited films as the nitrogen pressure is increased.

1. Introduction

Aluminum nitride (AlN) films have great potential for various applications due to their excellent electronic, optical and thermal properties, which can be tailored by varying the deposition methods and the technological conditions. AlN and other thin III-nitride films are finding applications in high-power and high-frequency devices, high-power switches and blue and ultraviolet light-emitting devices and photo-detectors. The successful fabrication of devices based on thin AlN films requires a better understanding of the optical and electronic properties as related to the growth condition. Various techniques are employed for the deposition of thin AlN films, such as metal organic chemical vapor deposition (MOCVD) [1], reactive magnetron sputtering [2,3] and PLD [4-8]. Compared to MOCVD, PLD has the main advantage of ensuring the growth of thin films with good crystallinity and stoichiometry at relatively low temperatures.

[4] To whom any correspondence should be addressed.

Journal of Physics: Conference Series **514** (2014) 012001 doi:10.1088/1742-6596/514/1/012001

AlN with a stable wurtzite structure has a direct bandgap of 6.2 eV [9,11]. Cases where the AlN bandgap has been determined to be considerably below 6.2 eV have been attributed to oxygen contamination. The index of refraction reported in the visible spectrum is ranging from 1.99 to 2.25 dependening on the AlN crystallinity. The phonon structure of AlN has also been subject to several studies, where the LO and TO modes have been reported at about 905 cm^{-1} and 667 cm^{-1}, respectively, while the E_2 mode has been observed at 665 cm^{-1} [9].

In our recent studies, PLD AlN films on silicon were optimized by using different nitrogen pressures [5,6] at relatively high temperatures. The crystalline structure of the AlN was found to be promoted by using higher repetition rates for the ablation and deposition at lower nitrogen pressures. In this study, the focus is on the optical properties of the optimized PLD films by complementary VIS/IR ellipsometry and Fourier transform infrared spectroscopy (FTIR). The optical constants and vibrational spectra are related to the film structure and deposition conditions.

2. Experimental details

AlN films were deposited on (100)Si substrates by laser ablation of a polycrystalline AlN target by a 248-nm COMPEXPro 205 pulsed KrF* excimer laser source. The incident laser fluence was 4 J/cm^2 with a pulse duration of 25 ns and a pulse repetition rate of 40 Hz. The substrate-target distance was 5 cm. The Si substrate was heated to either 800°C or 400°C. The depositions at 800°C were carried out in a vacuum ambient at 10^{-5} Pa residual pressure or in nitrogen gas with a dynamic pressure of 0.1 Pa and 10 Pa. For deposition at 400°C the nitrogen pressure was 0.1 Pa. The target was rotated at 0.3 Hz to avoid piercing and to ensure a clean surface during the multi-pulse laser irradiation. A shutter was placed between the target and the substrate during the first 1 000 laser pulses. Thus, the impurities and defects still present on the target surface after cleaning could be collected and removed before reaching the deposition area. For the synthesis of each structure, 20 000 consecutive laser pulses were applied which resulted in different film thicknesses (from 500 nm to 700 nm) depending on the nitrogen pressure and the laser pulse repetition rate.

The optical response of the AlN films was measured in the wide spectral region from 190 nm to 28 μm by UV/visible and infrared spectroscopic ellipsometry (VIS/IR SE) and Fourier transform infrared (FTIR) transmission/absorption measurements. The VIS SE measurements were performed using a Woollam VASE rotating compensator ellipsometer at multiple angles of incidence ($\Phi = 55°$, 60°, 65°, 70°, 75°). Ellipsometric data in the IR range 400 – 4000 cm^{-1} were recorded with 16 cm^{-1} spectral resolution at angles of incidence $\Phi = 60°$ and $\Phi = 70°$. The FTIR spectroscopic measurements were carried on a Shimadzu FTIR-8400s spectrometer in absorption mode. The spectra were collected over the range of 7.8×10^3 to 350 cm^{-1} with a resolution of 4 cm^{-1}.

3. Results

The optical constants of the AlN films were obtained by modelling the measured ellipsometric parameters Ψ and Δ. For the case of a multilayered structure consisting of the substrate, the film and a surface roughness layer, a parameterized optical model was exploited to extract the complex refractive index of interest from the measured dispersion of Ψ and Δ. The parameters in the model were the AlN film thickness d, the surface roughness layer thickness d_r, the AlN optical response (n, k) and the established Si optical response. Using an effective medium approximation, the surface roughness layer was modelled as commonly assumed as containing 50% air and 50% AlN material. The optical constants of the AlN films were properly modelled using the Cauchy equation combined with an Urbach absorption tail. For illustration of the optical model fitting accuracy, the Ψ and Δ spectra measured at $\Phi = 60°$ and the corresponding model-generated ones are presented in figure 1 for an AlN film deposited at 800°C and 0.1 Pa nitrogen gas ambient.

In figure 2, the refractive index and extinction coefficient of AlN films deposited at different ambient pressures and temperatures are presented in the region of normal dispersion from 180 to 1200 nm. The index of refraction, n, of PLD AlN films is 2.0-2.1 for wavelengths between 550 and 1200 nm, increasing to about 2.8 at 180 nm. The high-frequency dielectric constant $\varepsilon_{r\infty}$ was calculated

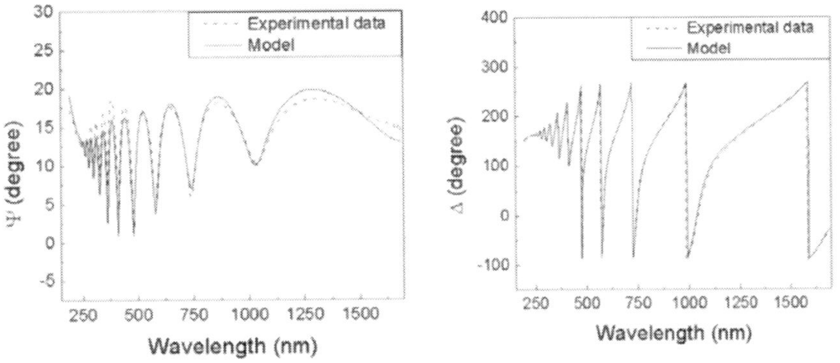

Figure 1. Experimental and model ellipsometric data of the AlN/Si structure deposited at 800°C and 0.1 Pa N_2 pressure; model data is given from the best fitting. The experimental data is recorded for an angle of incidence of 60° in the spectral range from 190 to 1680 nm. The mean-square error value of the fit is 31, which implies good approximation in the case of thick films. The fringes are induced by the finite film thickness of 672 nm. The surface roughness layer obtained in the model is 2.92 nm, in a very good agreement with the AFM data, where the root-mean-square roughness parameter is estimated at 2.6 nm.

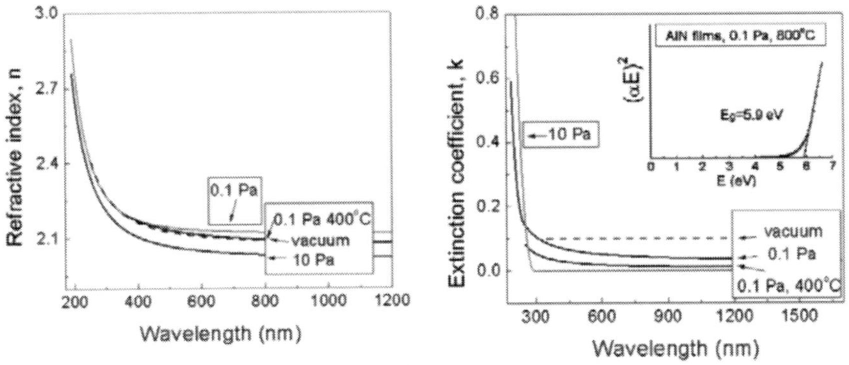

Figure 2. Optical response of AlN films deposited at different ambient pressures and temperatures in the region of normal dispersion from 180 to 1200 nm.

from the long-wavelength refractive index using the dispersion theory: $\varepsilon(\omega) = \varepsilon_r(\omega) + i\varepsilon_i(\omega) = [n(\omega)+ik(\omega)]^2$ yielding $\varepsilon_{r\infty} = n^2$ for long wavelengths. The $\varepsilon_{r\infty}$ values for different deposition conditions are presented in table 1 and are close to the high-frequency dielectric constant of AlN $\varepsilon_{r\infty} = 4.84$ reported by Collins at al. [10]. The AlN films are mostly transparent in the 400 – 900 nm spectral range with absorption registered below 400 nm (figure 2). The refractive index values were higher for the films deposited at low N_2 pressure and in a vacuum ambient due to the optimized crystallization under these conditions. Using a low substrate temperature and increased nitrogen pressure led to deterioration of the film's crystalline quality, so that the effective refractive index decreased. Our previous structural investigations [6] did also show that the crystallization is promoted at lower pressures, confirming our present results.

The extinction coefficient k at 400 – 1000 nm is almost zero when nitrogen ambient is used during deposition and becomes about 0.1 for AlN films synthesized in vacuum. Using the extinction coefficient, k, we determined the optical absorption of the films, α, ($\alpha = 4\pi k/\lambda$). The width of the optical bandgap E_g is estimated from the Tauc plot (figure 2, inset) represented as $(\alpha E)^2$ vs. E (photon energy) for direct bandgap materials. The linear part in the Tauc plot indicates the onset in the absorption due to the bandgap transitions and, when extrapolated, the optical band gap of the material was approximated (e.g. figure 2, inset). The bandgap energies of the AlN films deposited at different ambient pressures are presented in figure 3. Single crystalline AlN has a direct bandgap transition width of 6.2 eV [9,11]. The PLD AlN films deposited in vacuum have a bandgap energy of 6.2 eV in a good agreement with the crystalline AlN material. The slightly lower bandgap energy for other films deposited at 800°C and N_2 ambient is due to optical absorption below the fundamental edge, or the so-called Urbach tail. This effect is explained by the exponential distribution of the density of localized states in the tails of the allowed bands due to disorder and defects in the crystalline structure. One can observe the characteristic exponential absorption edge of Urbach tails resulting in linear approximation of ln(α) vs. photon energy, as illustrated in figure 3, inset, for AlN films deposited at 800°C and 0.1 Pa N_2. The narrowing of the bandgap due to this effect is most clearly observable in the AlN films deposited at 400°C, where E_g ~4.1 eV is considerably lower (figure 3). This is consistent with the fact that a lower temperature deposition of AlN favors the formation of crystalline defects with energy levels extending deep in the AlN bandgap thus reducing the effective forbidden gap.

The optical response of the AlN films was extended to the infrared range using IR ellipsometry. The ellipsometric parameters Ψ and Δ were measured in the range 2.5 – 25 μm. The optical model included the AlN layer thickness, roughness and transitions that were modelled using TOLO and Gauss oscillators. The fit of the experimental data with the optical model yielded a very low mean-square error MSE ~ 8-11. The dispersion features appearing in the spectra can be analyzed in the dispersion of the imaginary, $\varepsilon_2(\omega)$, part of the dielectric function of PLD AlN films derived from the spectroscopic ellipsometry. The $\varepsilon_2(\omega)$ dispersion showed a pronounced feature in the photon wavenumber range 400 – 800 cm^{-1} arising from absorption due to phonon modes in the AlN films (figure 4). The phonon band is centered at about 650 cm^{-1} for most of the films and is shifted to the lower wavenumbers (~610 cm^{-1}) for the AlN deposited in vacuum. This is consistent with the findings that vacuum laser deposition results in a slightly different stoichiometry of AlN films [12].

Table 1. High-frequency dielectric constant $\varepsilon_{r\infty}$ of PLD AlN films, synthesized at different nitrogen ambient pressures.

N_2 pressure	$\varepsilon_{r\infty}$
vacuum	4.34
0.1 Pa	4.49 (4.35 for 400°C)
10 Pa	4.08

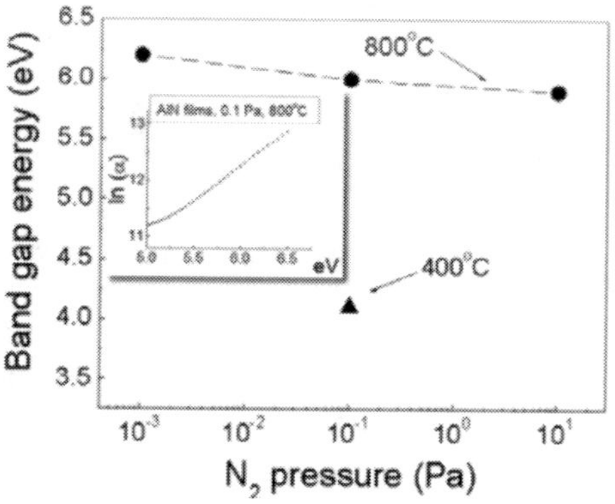

Figure 3. Optical band gap of AlN films deposited at different ambient pressures and temperatures as estimated from the absorption onset.

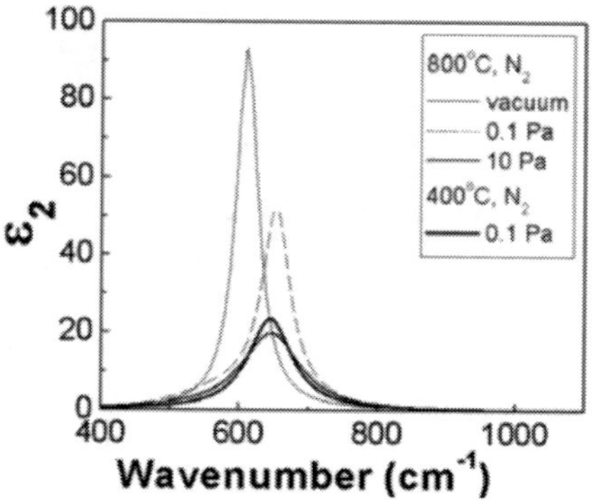

Figure 4. IR dispersion of the imaginary part $\varepsilon_2(\omega)$ of the complex dielectric function of AlN films deposited at different ambient pressures and temperatures.

The dispersions of $\varepsilon_2(\omega)$ shown in figure 4 are the result of the modelling; thus they lack details on the separate phonon modes contributing to the absorption. To have details on the phonon modes in this region, Fourier transform infrared spectroscopy was performed providing a better spectral resolution (figure 5). The broad absorption band appearing in the modelled $\varepsilon_2(\omega)$ dispersion was registered in the FTIR in the region 750 – 550 cm^{-1} with a distinct peak at about 686 cm^{-1} arising from the $E_1(TO)$ phonon mode of AlN. The shift and broadening of the $E_1(TO)$ peak towards the higher wavenumber indicates that an increased compressive stress is developed in the deposited films as the nitrogen pressure is increased. The peak observed at about 620 cm^{-1} originates from the vibrational modes in the Si substrate. Additional modes were observed at 669 cm^{-1} and 706 cm^{-1} for the films deposited at 800°C and at low pressure ambient (vacuum and 0.1 Pa N_2).

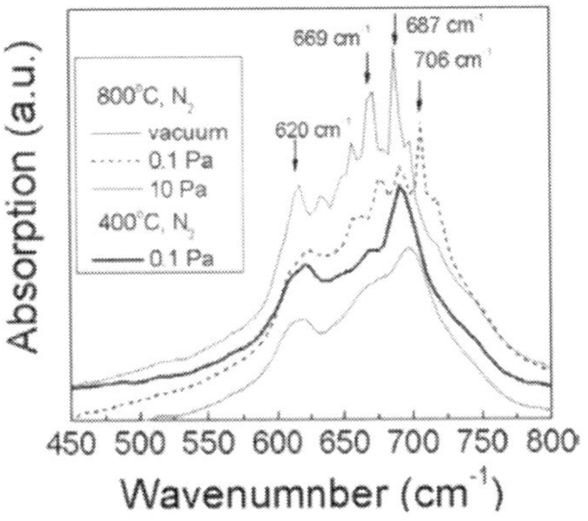

Figure 5. FTIR absorption spectra of AlN films.

4. Conclusions

Optical constants of PLD AlN films on Si are reported in a wide region from the visible to the IR spectra (5×10^4 cm^{-1} to 350 cm^{-1}) as measured by spectroscopic ellipsometry and FTIR spectroscopy. The effects of the nitrogen pressure and the deposition temperature are related to the bandgap width and the optical constants dispersion. The vibrational spectra analysis shows the presence of the characteristic AlN $E_1(TO)$ phonon mode. AlN films deposited at 0.1 Pa N_2 pressure and a higher temperature have the widest bandgap and exhibit a small broadening in the dispersions due to crystalline disorder.

References

[1] Kung P, Saxler A, Zhang X, Walker D, Wang T C, Furguson I and Razeghi M 1995 *Appl. Phys Lett.* **66** 2958
[2] Stan G E, Pasuk I, Galca A C and Dinescu A 2010 *Digest J. Nanomat. & Biostructures* **5**/4 1041
[3] Auger M A, Vázquez L, Jergel M, Sánchez O and Albella J M 2004 *Surf. Coat. Technol.* **180–181** 140
[4] Shukla G and Khare A 2008 *Appl. Surf. Sci.* **255** 2057
[5] Ristoscu C, Ducu C, Socol G, Craciunoiu F and Mihailescu I N 2005 *Appl. Surf. Sci.* **248** 411
[6] Bakalova S, Szekeres A, Cziraki A, Lungu C P, Grigorescu S, Socol G, Axente E and Mihailescu I N 2007 *Appl. Surf. Sci.* **253**/19 8215
[7] Vispute R D, Narayan J, Wu H and Jagannadham K 1995 *J. Appl. Phys.* **77** 4724
[8] Ren Z M, Lu Y F, Ni H Q, Liew T Y F, Cheong B A, Chow S K, Ng M L and Wang J P 2000 *J. Appl. Phys.* **88** 7346
[9] Strite S and Morkoç H 1992 *J. Vac. Sci. Technol.* B **10** 1237
[10] Collins A T, Lightowlers E C and Dean P J 1967 *Phys. Rev.* **158** 833
[11] Christensen N E and Gorczyca I 1994 *Phys. Rev.* B **50** 4397
[12] Gyorgy E, Ristoscu C, Mihailescu I N, Klini A, Vainos N, Fotakis C, Ghica C, Schmerber G and Faerber J 2001 *J. Appl. Phys.* **90** 456

PECVD preparation of silicon and germanium with different isotopic composition via their tetrafluorides

P G Sennikov[1,2,3], R A Kornev[1], L A Mochalov[1] and S V Golubev[2]

[1]Laboratory of Plasma Chemistry, G. G. Devyatykh Institute of Chemistry of High-Purity Substances, Russian Academy of Sciences, 603950 Nizhny Novgorod, Russia
[2]Laboratory of Plasma Physics, Institute of Applied Physics, Russian Academy of Sciences, 603950 Nizhny Novgorod, Russia

E-mail: pgsen@rambler.ru

Abstract. Plasma-chemical reduction of silicon and germanium fluorides with different isotopic composition by hydrogen at low pressure was studied experimentally. Samples of silicon of natural isotopic composition and germanium polycrystalline "flakes" with isotopic numbers 72 and 74 were obtained and used to grow poly- and single crystals by the Czochralski method. The contamination by the most important impurities in each type of silicon and germanium was determined. A chemical mechanism for the reduction process was also proposed. It was shown that the direct method of reduction of isotopically-enriched fluorides is suitable for obtaining small amounts of high-purity isotopes of those elements to be used in unique physical experiments.

1. Introduction

Gaseous tetrafluorides of silicon and germanium are widely used in the PECVD technologies for preparation of thin-layer structures for multilayer cell elements, photodetectors, uncooled micro-bolometers [1-3]. Silicon tetrafluoride is of interest from the point of view of preparing bulk silicon samples with the purity grade required for photovoltaics taking into account its potentially high stocks as a byproduct of fluoride production [4]. In fact, there are no alternatives to the fluorides of silicon and germanium as initial substances for preparation of isotopic species of these IV-group elements, as there exist well-developed centrifugal technologies for enrichment of their isotopes [5]. The isotopes of silicon and germanium both in the form of thin layers and bulk crystals are used for the development of nano-electronic devices of new generation and in metrology [6,7] as well as for manufacturing the unique instruments in high-energy physics [8]. The PECVD technologies simplify considerably the process of conversion of fluorides into the target product, namely, films and polycrystalline powders (granulate) used for growth of crystals.

The goal of the present work is to systemize our recent results on preparation of bulk samples of silicon and germanium with different isotopic composition by PECVD, as well as to estimate the promising aspects of this approach for the solution of some practical problems.

[3] To whom any correspondence should be addressed.

2. Experimental

The process of plasma-chemical hydrogen reduction of SiF_4 and GeF_4 was carried out in a set-up whose functional diagram is shown in figure 1.

The power of the RF generator (13.56 MHz) was 1 kW. The power applied to the area of plasma discharge was measured by a calibrated coupler in the range 300 – 700 W. The plasma density in this source can reach 5×10^{11} cm^{-3}.

The pressure during the experiment was in the range 0.1 – 0.3 Torr. The flow rate of the plasma gas H_2+SiF_4 was set by mass flow controllers in the range 35 – 125 cm^3/min. The molar ratio H_2/SiF_4 was set in the range 2 – 6. The flow rate of the GeF_4+H_2 plasma gas was varied in the range 70 – 120 cm^3/min, with the molar ratio H_2/GeF_4 being in the range 4 – 6.

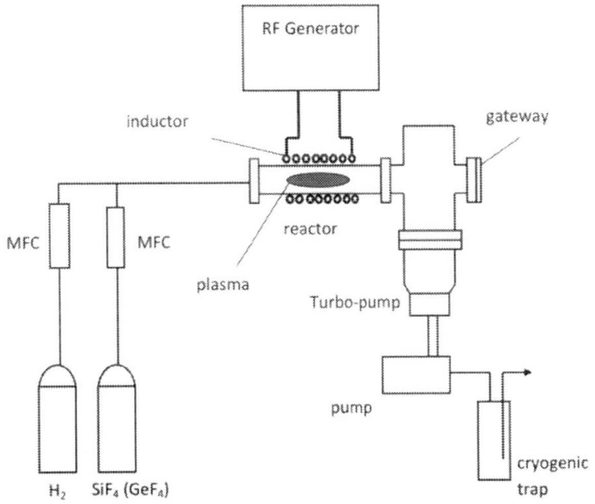

Figure 1. The set-up for plasma reduction of silicon tetrafluoride in RF inductively-coupled discharge.

The plasma chemical reactor consisted of a quartz glass tube equipped with an inductor coupled to the RF generator via a matching device. The reactor comprised an additional tube-substrate upon which the elemental silicon and germanium were deposited in the form of polycrystals or "flakes" with an average size of several millimeters and thickness up to 100 μm. The duration of a single process with continuous deposition was four hours.

The emission spectra of H_2+SiF_4-based plasma were studied by a Jobin Yvon HR 640 spectrometer in the range 190 – 750 nm (with a spectral resolution of 0.1 E).

The crystals of silicon and germanium were grown at the Leibnitz Institute of Crystal Growth (Berlin, Germany) by Dr. N. Abrosimov by the so-called "mini-Czochralski method" developed especially for growing crystals of isotopically pure silicon at extremely low loads of the crucible [9].

Secondary ion mass-spectrometry (SIMS), laser mass spectrometry (LMS), IR and Raman spectroscopy, and X-ray phase analysis were used to determine the elemental composition. The electro-physical parameters of silicon and germanium crystals were determined by standard techniques.

The experiments were conducted using silicon tetrafluoride of 99.99% purity with natural isotopic abundance (Astor Company, Sankt-Petersburg, Russia) and fluorides of germanium of 99,9% purity, namely, $^{74}GeF_4$ (enriched up to 86%) and $^{72}GeF_4$ (enriched up to 52%) (Electrochemical Plant JSC PA, Zelenogorsk, Russia).

3. Results and discussions

3.1. PECVD of silicon

The binding energy of Si-F is 5.6 eV (as a comparison, the binding energy of Si-H is 3.1 eV). This is why thermal decomposition or reduction to elemental silicon with the use of the known reducing agents is actually impossible [10]. A large number of papers have appeared (see e.g. [11–13]) devoted to the preparation of thin films of amorphous and micro-crystalline silicon from low-temperature non-equilibrium plasma based both on pure SiF_4 and on its mixture with silane and maintained by discharges of different types. However, it was only recently that this approach was used for the first time to prepare polycrystalline silicon powder [14]. A finely-dispersed powder of Si was prepared in a high-frequency induction discharge at 13.56 MHz, an input power of 4 kW and pressure of the gas mixture SiF_4:H_2:Ar from 0.1 to 30 Torr; the process was organized in such a way as the silicon particles grew larger during the fluoride decomposition. The deposition rate was in the range from 0.5 to 5 g/hour. The authors of the cited patent [14] believe that, with the aim to increase the yield, the deposition process can be carried out at powers of up to 50 kW, pressure of 100 – 1000 Torr and fluoride-hydrogen ratio of 1:1. However, data on the implementation of this technology, as well as quantitative data on the degree of purity of the prepared silicon are not given.

Our experiments were conducted at the lower pressure of 0.1 – 0.3 Torr. The flow rate of the plasma gas $H_2 + SiF_4$ was set in the range 35 – 125 cm^3/min. The molar ratio of H_2/SiF_4 was varied in the range 2 – 6. The deposition rate of silicon under these conditions was 0.2 g/hour. We studied the dependence of the silicon yield on the specific energy input to the RF inductively-coupled discharge. The specific energy input (*P*) was defined as the ratio of the power (*W*) applied to the discharge from the source of RF oscillations to the plasma gas flow (*Q*):

$$P = W/Q. \quad (1)$$

The yield of silicon increased monotonically from 13 to 56 % as the energy input was raised from 7 to 25 MJ/mol. The further increase of the energy input of 25 – 37.3 MJ/mol did not change the yield of silicon. Probably, the yield of silicon, equal to 56 % under these conditions, is of a quasi-equilibrium nature and the energy input value of 25 MJ/mol is the optimum one. The minimum energy input (the ratio of energy input to the degree of fluoride conversion) is 10.2 kWh per mol of silicon and corresponds to the "energy cost" of the product.

According to [15], the energy of free electrons in the plasma of a RF discharge is 7 – 10 eV at a pressure of 0.1 – 0.3 Torr. As it is indicated in [16], this energy is sufficient for the dissociation of the H_2 molecule and formation of atomic hydrogen. This is why we can assume with high probability that the reduction of SiF_4 occurs with the participation of $H^\bullet$. Other intermediate particles assisting the reduction of silicon tetrafluoride were identified from the emission spectra of plasma. The set of our and literature data [17-21] makes it possible to assume the following mechanism of solid-phase formation of silicon under the conditions of our experiment.

$$H_2 \rightarrow 2H, \quad (2)$$
$$SiF_4 + H^\bullet \rightarrow SiF_3^\bullet + HF, \quad (3)$$
$$SiF_3^\bullet + H_2 \rightarrow SiF^\bullet + 2HF, \quad (4)$$
$$SiF^\bullet \rightarrow SiF_{ads} + H^\bullet \rightarrow \mathbf{Si}\downarrow + HF. \quad (5)$$

In addition, since the line corresponding to the $SiH^\bullet$ radical is present in the plasma emission spectrum, it is reasonable to assume that the following reactions take place:

$$SiF^\bullet + H_2 \rightarrow SiH^\bullet + HF, \quad (6)$$
$$SiH \rightarrow \mathbf{Si}\downarrow + H. \quad (7)$$

The silicon deposited on the inner surface of the reactor consists of "flakes" with a size of about 3 – 5 mm and a polycrystalline structure, as evidenced by X-ray phase analysis. A bulk polycrystal of p-type with a mass of about 10 g, a concentration of carriers of 10^{17} cm^{-3} and a resistivity of $\rho = 0.3$ Ohm cm was grown from the prepared silicon by the Czochralski method.

Table 1 presents data on the content of impurities in the initial tetrafluoride, in the polycrystalline "flakes" immediately after deposition and in the bulk polycrystal.

The main conclusion from the data presented in table 1 is that the polycrystalline silicon "flakes" prepared were contaminated by oxygen, carbon, hydrogen, and, probably, by fluorine during the plasma-chemical deposition. The reason for the increased content of the latter two impurities is understandable. The sharp increase in the concentration of oxygen is connected with the process of oxidation of silicon powder with a developed surface during the contact with the environment. The increase in the concentration of carbon is less clear and can be connected with the use of gaskets made of synthetic materials subjected to a relatively high temperature. The content of the cited gas-forming impurities in the crystal grown by the Czochralski method is decreased by more than an order of magnitude. It should be noted that the content of oxygen in the crystal, according to the data of IR spectroscopy and SIMS, is at the level of 10^{18} cm^{-3} (which is typical for the Czochralski method where oxygen enters the melt from the quartz crucible); that of carbon is at the level of 5×10^{17} cm^{-3} (which exceeds by an order of magnitude the content of this impurity in "Czochralski" silicon prepared by the known technologies). The content of electro-active impurities (P, As, B) does not increase during transition from gas to crystal.

Table 1. Comparison of the content of impurities (ppm wt) in the initial SiF_4, in polycrystalline silicon "flakes" after plasma chemical deposition and in the polycrystal prepared by Czochralski (Cz) method.

Impurity	SiF_4 (GC^a and AAS^b data according to certificates of manufacturer)	PECVD Si "flakes" (SIMS data)	Cz bulk silicon polycrystal	
			SIMS and IR data	LMS^c data
H	0.1	3	<0.3	-
O	0,5	130 – 700	13	-
C	1.2	20 – 200	5	-
F	-	16 – 160	<16	<2
P	0.5	<2.5	<0.25	<0.6
As	0.5	<6	<0.6	0.3
B	0.5	<1	0.5-1	0.04

[a] GC - gas chromatography, [b] AAS - atomic absorption spectroscopy, [c] LMS - laser mass spectrometry.

Thus, the PECVD method makes it possible to prepare silicon polycrystals of satisfactory quality. However, to implement it even in an enlarged laboratory version it is necessary to solve a number of problems connected with increasing the process rate and the silicon yield, as well as with the extracting the prepared powder from the reactor without contact with the environment. If these problems are successfully solved, the technology proposed can be used, first of all, for production of silicon enriched with the silicon-28 isotope. The demand for this expensive material for metrology can amount to tens of kilograms per year [6].

3.2. PECVD of germanium

The binding energy of Ge-F is somewhat less than the binding energy of Si-F, namely, 4.8 eV. However, achieving a sufficiently high yield of germanium as a result of thermal decomposition or chemical reduction of the fluoride is quite problematic, as evidenced by thermodynamic calculations. Moreover, in contrast with silicon, during the plasma-chemical reduction of germanium tetrafluoride, apart from deposition of particles of polycrystalline germanium, formation of polyfluorogermanes of different compositions, easily hydrolyzed in air, was observed; this was proved by X-ray phase and mass spectrometric analyses. The yield of germanium is constant in the range of energy input from 27 to 50 MJ/mol and is 85 %. The minimum energy consumption is 8.6 kWh per mol of germanium. Meanwhile, the yield of polyfluorogermanes reaches 15 %. However, it was found that raising the energy to more than 70 MJ/mol results in increasing the yield of germanium up to 90 %, which is accompanied by a sharp drop in the efficiency of polyfluorogermanes formation. The deposition rate of germanium under our conditions was 5 g/h.

The reduction of germanium tetrafluoride by atomic hydrogen at a small energy input via the stages of fluorogermanes formation can be represented in the following way:

$$H_2 \rightarrow 2H^\bullet, \quad (8)$$
$$GeF_4 + H^\bullet \rightarrow GeHF_3 + F^\bullet, \quad (9)$$
$$GeHF_3 + H^\bullet \rightarrow GeH_2F_2 + F^\bullet, \quad (10)$$
$$GeH_2F_2 + H^\bullet \rightarrow GeH_3F + F^\bullet, \quad (11)$$
$$GeH_3F + H^\bullet \rightarrow GeH_4 + F^\bullet, \quad (12)$$
$$GeH_4 \rightarrow \mathbf{Ge}\downarrow + 2H_2, \quad (13)$$
$$F^\bullet + H^\bullet \rightarrow HF, \quad (14)$$

with subsequent polymerization of the fluorogermanes molecules. According to the literature data [22], in a certain temperature range in an electric discharge, formation of radicals of germanium fluorides $GeF^\bullet$ and $GeF_2^\bullet$ is possible followed by polymerization:

$$GeF_4 + 2H \rightarrow GeF_2^\bullet + 2HF, \quad (15)$$
$$nGeF_2^\bullet \rightarrow (GeF_2^\bullet)n, \text{ where } n = 2\text{-}8 \quad (16)$$
or
$$nGeF_2^\bullet + mGeF^\bullet \rightarrow (GeF_2^\bullet)_n(GeF^\bullet)_m, \quad (17)$$

where n and m denote the mole number of the corresponding species.

This was proved by the initial increase in the yield of polyfluorogermanes with the increase of the energy input. However, as the energy input is raised further, the following processes can, probably, be the determinative ones in providing a high yield of germanium:

$$GeF_4 + H^\bullet \rightarrow GeF_3^\bullet + HF, \quad (18)$$
$$GeF_3^\bullet + H^\bullet \rightarrow GeF_2^\bullet + HF, \quad (19)$$
$$GeF^\bullet_2 + H^\bullet \rightarrow GeF^\bullet + HF, \quad (20)$$
$$GeF^\bullet + H^\bullet \rightarrow \mathbf{Ge}\downarrow + HF. \quad (21)$$

As was already mentioned, we were provided with samples of GeF_4 with different degrees of enrichment by the isotopes ^{74}Ge and ^{72}Ge.

Table 2 presents the content of impurities in ^{74}Ge in the form of polycrystalline “flakes” immediately after deposition and in the single crystal prepared on its basis (with a mass of about 20 g). The charge carriers concentration in the single n-type crystal was $n = 9\times10^{14}$ cm^{-3}, with the resistivity being 1.9 Ohm cm. A slight dilution of the germanium-74 isotope was observed during growth (83 % against 85.9 %) connected with the use of a seed of natural germanium. As in the case of silicon,

Table 2. Comparison of the content of impurities (at/cm^3) in polycrystalline ^{74}Ge “flakes” after plasma chemical deposition and in the single crystal grown on its basis by the Czochralski method.

Impurity	Prepared by PECVD ^{74}Ge "flakes" (SIMS data)	^{74}Ge single crystal	
		SIMS and IR data	LMS data
H	$10^{17} - 10^{18}$	$10^{15} - 10^{16}$	-
O	$10^{19} - 10^{20}$	$10^{16} - 5\times10^{16}$ (1×10^{17} by IR-spectroscopy data)	-
C	$10^{18} - 10^{19}$	$10^{15} \times 10^{16}$	$<2\times5\cdot10^{19}$
F	$10^{18} - 10^{19}$	$10^{16} - 5\times10^{16}$	$<1.5\times10^{16}$
P	$<5\times10^{15} - 10^{16}$	$<5\times10^{15} - 10^{16}$	$<5\times10^{15}$
As	$<5\times10^{15} - 10^{16}$	$<5\times10^{15} - 10^{16}$	$<2.5\times10^{19}$
B	$<5\times10^{15} - 10^{16}$	$<5\times10^{15} - 10^{16}$	$<8\times10^{15}$

during the transition to the bulk crystal the content of the gas-forming impurities decreased and the content of the electro-active impurities was below the limit of detection by SIMS and LMS. While comparing the data of the tables 1 and 2, one can see that the germanium prepared by PECVD is characterized, as a whole, by a lower content of impurities as compared with silicon.

The mass of the prepared polycrystalline ^{72}Ge was about 50 g. Four single crystals were grown sequentially from this material. The mass of the last (the purest) crystal was 20.6 g. The concentration of charge carriers in its initial ("pure") part was $n = 3.5\times10^{12}$ cm^{-3} ($\rho > 40$ Ohm cm) (note the difference with the data for the ^{74}Ge crystal given above). The content of phosphorus and arsenic in the "pure" part was determined by IR spectroscopy at helium temperature. According to this data, the content of phosphorus was 8×10^{13} and 9×10^{12} cm^{-3}, the content of arsenic was 4×10^{14} and 7×10^{12} cm^{-3}, respectively. The content of oxygen in the samples investigated was determined by the same method at room temperature; it was less than 1.0×10^{15} cm^{-3}. In what concerns the characteristics quoted, the single crystal of ^{72}Ge corresponds to the best samples of semiconductor germanium. This is why the direct PECVD method proposed seems quite promising for the preparation of ^{76}Ge crystals with a degree of enrichment of about 8 0% which are planned to be used in the experiments on determining the neutrino mass within the framework of the GERDA project [8].

4. Conclusions

The experiments conducted indicate that the PECVD method based on a RF discharge with fluorides as the initial plasma-forming substances can be successfully used for the preparation of bulk samples of silicon and germanium. The substance deposited is in the form of "flakes", the bulk poly- or single crystals being subsequently grown from them. The main technological parameters are still inferior to those of germanium. The approach proposed is considered to be promising, first of all, for preparation of comparatively small (tens of kilograms) amounts of expensive isotopically pure crystals of silicon and germanium since their tetrafluorides enriched by centrifuging are commercially available.

Acknowledgments

The authors wish to thank Dr. Andreev, M. Drozdov, Dr. A. Potapov for determining the impurities in PECVD silicon and germanium by spectroscopic and mass-spectroscopic methods, Dr. Abrosimov for growing Si and Ge crystals and Prof. A. Bulanov for assisting in SiF_4 and GeF_4 samples handling.

References

[1] Haller E E 2006 *Mater. Sci. Semicond. Processing* **9** 408
[2] Torres A, Kosarev A, .Garcia Cruz V L and Ambrosio L 2003 *J. Non-Cryst. Solids* **329** 179
[3] Moreno M, Torres A, Ambrosio R, Rosales P, Heredia A, Kosarev A, Torres E, Zuniga C, Reyes-Betanzo C and Dominigues M 2012 *J. Non-Cryst. Solids* **358** 2099
[4] Loginov A V and Garbar A M 1989 *High-purity Substances* **3** 747 (in Russian)
[5] Tarbeyev Y V, Kaliteyevski A K, Sergeyev V I, Smirnov R D and Godisov O N 1994 *Metrology* **31** 269 (in Russian)
[6] Becker P, Schiel D, Pohl H-J, Kaliteevski A K, Godisov O N, Churbanov M F, Devyatykh G G, Gusev A V, Bulanov A D, Adamchik S A, Gavva V A, Kovalev I D, Abrosimov N V, Hallmann-Seiffert B, Riemann H, Valkiers S, Taylor P, De Bièvre P and Dianov E M 2006 *Meas. Sci. Technol.* **17** 1854
[7] Becker P, Pohl H-J, Riemann H and Abrosimov N 2010 *Phys. Status Solidi* A **207** 49
[8] Bellotti E 2006 *J. Phys.: Conf. Series* **39** 318
[9] Abrosimov N V, Riemann H, Schröder W, Pohl H-J, Kaliteevski A K, Godisov O N, Korolyov V A and Zhilnikov A Ju 2003 *Cryst. Res. Technol.* **38** 654
[10] Kut'in A M, Polyakov V S, Sennikov P G 2008 *Abstract Book 5th Int. Conf. Silicon 2008* (Chernogolovka Russia) p 178 (in Russian)
[11] Kumar S R, Brenot R, Kalache B, Tripathi V, Vanderhaghen R, Drevillon B and Roca i Cabarrocas P 2001 *Solid State Phenom.* **80-81** 237

[12] Bruno G, Capezzuto P and Cicala G 1991 *J. Appl. Phys.* **69**/10 7256
[13] Vodopyanov A V, Golubev S V, Mansfeld D A, Sennikov P G and Drozdov Yu N 2011 *Rev. Sci. Instr.* **82** 063503
[14] Takahashi N, Yatsurugi Y and Yusa A 1979 *Method and apparatus for production of high purity silicon* US Patent 4150168 A
[15] Fridman A A 2008 *Plasma Chemistry* (Cambridge University Press New York) 978
[16] Polak L S, Ovsyannikov A A, Slovetskii D I and Vursel' F B 1975 *Theoretic and Applied Plasma Chemistry* (Nauka Moskva) 304 (in Russian)
[17] Madan A 1984 In: *Topics in Appl. Phys.* (Springer Berlin) **55** 245
[18] Cicala G, Bruno G and Capezzuto P 1991 *Pure and Appl. Chem.* **68** 1143-49
[19] Mutsukura N, Ohuchi M, Satoh S and Machi Y 1983 *Thin Solid Films* **109** 47-57
[20] Ling-Fai Wang J, Krishnan C N and Margrave J L 1973 *J. Molecul. Spectr.* **48** 346
[21] Yuliang He, Yin C, Cheng G, Wang L and Xiangna L 1994 *J. Appl. Phys.***75** 797–803
[22] Tananaev I V and Shpirt M Ya 1967 *Chemistry of germanium* (Khimiya Moskva) 452 (in Russian)

Journal of Physics: Conference Series **514** (2014) 012003 doi:10.1088/1742-6596/514/1/012003

Optical and electrical properties of very thin chromium films for optoelectronic devices

V Lozanova, A Lalova, L Soserov and R Todorov[1]

Acad. J. Malinowski Institute of Optical Materials and Technologies,
Bulgarian Academy of Sciences, Acad. G. Bonchev Str., Bl. 109, 1113 Sofia, Bulgaria

E-mail: rossen@iomt.bas.bg

Abstract. Establishing the optimal experimental conditions for the development of transparent metal contacts to be used in optoelectronic devices, such as organic light-emitting diodes and solar cells, is an important task. In this paper we present an overview of the development of very thin e-beam-deposited chromium films with high optical transparency. The surface morphology is investigated by scanning electron microscopy. The variation is examined of the films' electrical and optical properties (transmittance and complex refractive index) with the variation of the thickness and deposition rate. We observed that, for a given thickness of the chromium films, the absorption coefficient increases when the deposition rate is decreased. We also found that the thin films with a thickness of less than 10 nm show an average transmittance exceeding 60 % in the spectral range 400 – 1500 nm. The films' resistivity, ρ, is determined by the four-point probe method. The value of ρ varies in the range of 10^{-3} – 10^{-4} Ω cm for chromium coatings in the thickness interval 5 – 100 nm. The results obtained show that very thin metal films could be an alternative to the transparent conductive oxides.

1. Introduction

Thin chromium films have found many applications, for example as layers for improvement of the adhesion to transparent substrates [1], in mask production [2] and as transparent electrodes [3]. Recently, it has been demonstrated that the ultrathin transparent metal contacts provide a number of advantages over the more commonly used conductive transparent metal oxides, such as indium tin oxide (ITO) [4]. According to [5], chromium and nickel thin films possess optical transparency comparable to that of the ITO in the visible and near-infrared range (0.4 – 2.5 μm), while it can be significantly higher in the ultraviolet (175 – 400 nm) and mid-infrared (2.5 – 25 μm) regions.

The conditions of deposition, as deposition rate, temperature of substrate, etc., influence very strongly the properties of the metal coating [6]. It has been shown that the deposition rate and argon partial pressure in the case of cathode sputtering strongly affect the microstructure and electrical properties of thin chromium films [7, 8]. Metal films deposited by ion-beam sputtering display exceptionally low surface roughness, the films being as thin as about 20 Å [3].

The goal of the present work is the investigation of the optical and electrical properties of very thin chromium films deposited by an electron-beam technique.

[1] To whom any correspondence should be addressed.

Journal of Physics: Conference Series **514** (2014) 012003 doi:10.1088/1742-6596/514/1/012003

2. Experimental details

Thin films of chromium were deposited by the e-beam technique in a Leybold-Heraeus A 702 Q thin films deposition system. The e-beam evaporation was performed in vacuum $p \sim 3\times10^{-3}$ Pa at a deposition rate of 0.2 nm/s and controlled by a quartz monitoring technique. The thickness of the thin films was also cross-checked ex-situ by a Talystep profilometer.

The surface of the thin films was observed by a Joel Superprobe 733 (Japan) scanning electron microscope. The film's resistivity, ρ, was determined by the four-point probe method. The transmittance (T) and reflectance (R) were measured by a Cary 5E (Australia) UV–VIS–NIR spectrophotometer in the range 350 – 2000 nm to an accuracy of $\Delta T = \pm 0{,}1\%$ and $\Delta R = \pm 0{,}5\%$.

3. Results and discussion

The SEM image of a 15-nm thick chromium film is presented in figure 1. A smooth surface was observed for all thin films.

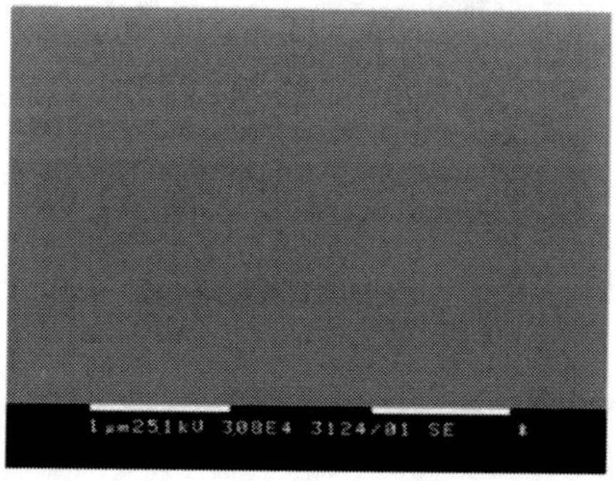

Figure 1. SEM image of a thin chromium film deposited by e-beam evaporation.

The transmittance spectra of thin chromium films deposited by e-beam evaporation are shown in figure 2. The values of T are lower at the shorter wavelengths and increase at higher λ. It is seen that the films' transmittance is above 65 % for thin films with a thickness of 10 nm. The thin film with a thickness of 5 nm exhibited transmittance in the range of 67 – 80 % in the visible spectral range and transmittance higher than 80 % in the near infrared range.

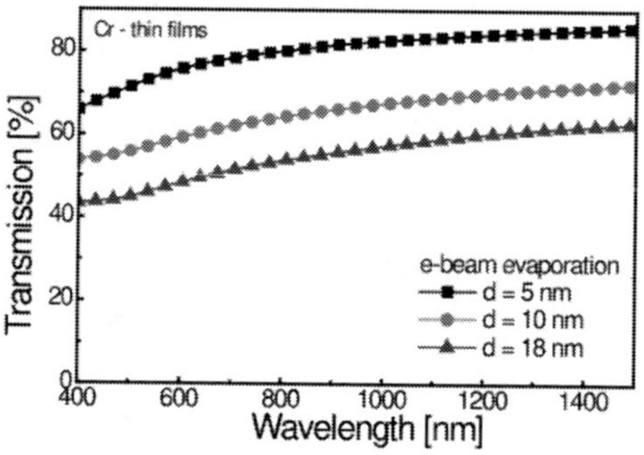

Figure 2. Transmittance spectra of thin Cr films deposited by e-beam evaporation.

The refractive index, n, the extinction coefficient, k, and the film thickness, d, were calculated from the transmittance or reflectance spectra as described in [9]. The method is based on minimization of the discrepancy between the theoretically calculated ($T(n(\lambda), k(\lambda), d)$ and $R(n(\lambda), k(\lambda), d)$ and the experimentally measured values $T_{exp.}$ and $R_{exp.}$ of the transmittance and reflectance at normal incidence until the difference becomes lower than the accuracy of the measurements, ΔT and ΔR [10-11].

$$\begin{aligned} T(n(\lambda), k(\lambda), d) - T_{exp.} &= \Delta T \\ R(n(\lambda), k(\lambda), d) - R_{exp.} &= \Delta R \end{aligned} \qquad (1)$$

Discontinuities appear in the solution for n and k due to the requirement for an accurate film thickness value, while there is loss of solutions at the critical points [12]. The thickness d is not computed from set (1), but is introduced as a parameter. For an initial approximation for the thickness we used the value measured by a profilometer with an accuracy of ±1 nm. Hence, by varying the thickness slightly around the approximate one, we chose the value yielding the smallest discontinuities in the solution for n and k. The accuracy in the determination of the refractive index, n, was better than ±0.005 and that for the absorption coefficient, k, around the absorption edge was about ±0.01 [11, 13].

The real and imaginary parts of the complex permittivity, $\varepsilon = \varepsilon' + i\varepsilon''$ can be calculated from the refractive index and extinction coefficient by the following equations:

$$\varepsilon' = n^2 - k^2 \quad \text{and} \quad \varepsilon'' = 2nk \qquad (2)$$

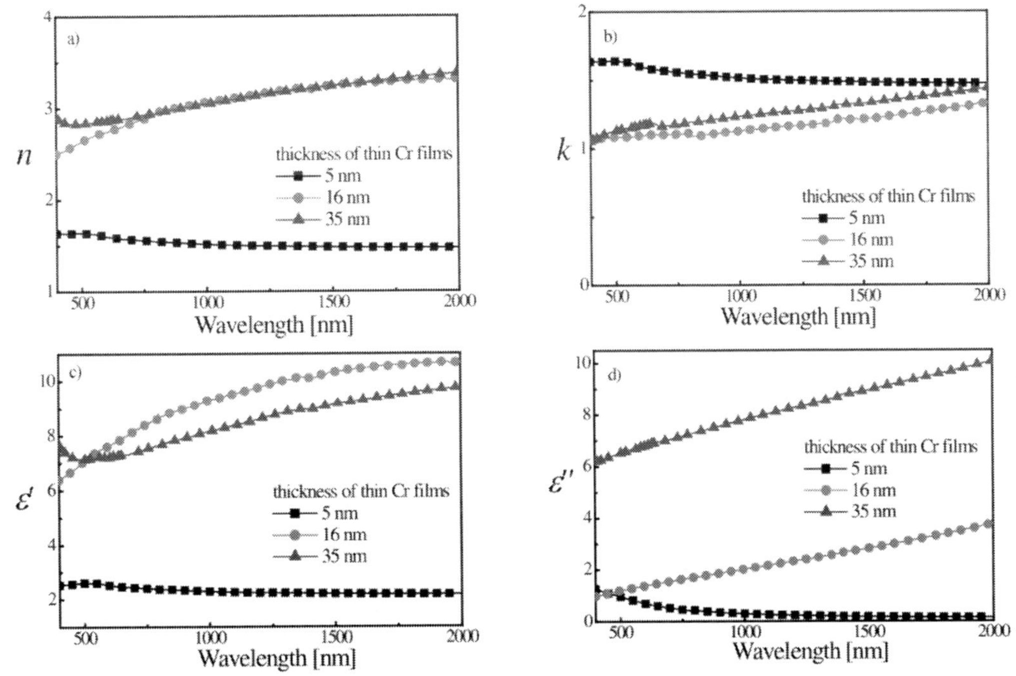

Figure 3. Dispersions of the refractive index, the extinction coefficient, and the real and imaginary parts of the complex permittivity of thin chromium films.

The results for the refractive index, the extinction coefficient, and the real and imaginary parts of permittivity are given in figure 3. The decrease of the films' thickness leads to a drop of the refractive index. The values of *n* are in the range of 2.5 – 3.5 for thin films with thickness 16 and 35 nm, while *n* varies from 1.47 to 1.64 for the 5-nm coating in the 400 – 1500 nm spectral range.

The trends in the variation of the imaginary and real part of the permittivity relative to the films' thickness are similar to those of the refractive index. As is seen in figure 3d, the imaginary part of the complex permittivity, ε", decreases as the films' thickness is decreased.

The variation of the film's resistivity, ρ, is given in table 1. It can be observed that the value of the resistivity decreases for thicker films. A sharp increase of the values of the resistivity is observed for films thinner than 22 nm. The results for ρ are in the same order of magnitude as for the thin films deposited by cathode magnetron sputtering [14]. According to [15], the sheet resistance, R_s, of ITO films with thickness 50 – 100 nm is in the range 2 – 6 kΩ/sq.

Table 1. Resistivity of thin chromium films, ρ, sheet resistance, R_s and their thickness, *d*.

Film thickness, *d* [nm]	ρ [Ω cm]	R_s kΩ/sq
5	6.3×10^{-3}	12.6
10	2.25×10^{-3}	2.25
18	2.70×10^{-3}	1.5
22	2.33×10^{-4}	0.106
92	2.11×10^{-4}	0.02293
94	2.16×10^{-4}	0.02298
110	2.83×10^{-4}	0.03093

Taking into account that the relation between the real part of the conductivity, σ' and ε" is

$$\sigma' = \omega\varepsilon''/4\pi, \tag{3}$$

where σ'=1/ρ is the real part of complex conductivity and ω is the angular frequency. It is seen from equation (3) that the imaginary part of the complex permittivity, ε", is proportional to the conductivity. Consequently, the increase of ε" with the thin films' thickness (see Figure 3d) is related with the increase of σ' (or the decrease of ρ). This observation indicates that in the interval of thicknesses 10 – 20 nm we can expect a transition of the film structure from discrete islands to a continuous network [14].

4. Conclusions

In the work presented, the optical and electrical properties of thin chromium films are investigated. It is determined that thin films thinner than 10 nm possess a relatively high transmittance ($T > 65$ %). The optical constants and the complex permittivity are calculated. It is found that the imaginary part of the permittivity, ε", decreases with the decrease of the films' thickness. It is further established that the resistivity varies in the range of $6.3\times10^{-3} - 2.83\times10^{-4}$ Ω cm and is comparable with that of ITO films.

References

[1] Vial A and Laroche T 2007 *J. Phys.* D: *Appl. Phys.* **40** 7152
[2] Shy S-L, Lei T F, Chu C H, Chang C Y, Lee S H and Loong W A 1994 *Proc. SPIE* **2194** 287
[3] Rajani K V, Daniels S, McNally P J, Olabanji Lucas F and Alam M M 2010 *Phys. Status Solidi* A **207** 1586
[4] Klauk H, Huang J-R, Nichols J A and Jackson T N 2000 *Thin Solid Films* **366** 272
[5] Ghosh D S, Martinez L, Giurgola S, Vergani P and Pruneri V 2009 *Opt. Lett.* **34**/3 325
[6] Kawagoe T and Mizoguchi T 1993 *Jpn. J. Appl. Phys.***32** 935
[7] Li H-T, Jiang B-L and Yang B 2011 *Appl. Surf. Sci.* **258** 935
[8] Foroughi-Abari A, Xu C and Cadien K C 2012 *Thin Solid Films* **520** 1762
[9] Todorov R, Lalova A, Petkov K and Tasseva J 2012 *Semicond. Sci. Technol.* **27**/11 115014
[10] Laaziz Y, Bennouna A, Chahboun N, Outzourhit A and Ameziane E L 2000 *Thin Solid Films* **372** 149
[11] Todorov R, Tasseva J, Babeva Tz and Petkov K 2010 *J. Phys.* D: *Appl. Phys.* **43** 505103
[12] Del Pozo J M and Diaz L 1992 *Thin Solid Films* **209** 137
[13] Todorov R and Petkov K 2001 *J. Optoelectron. Adv. Mater.***3** 311
[14] Lin J-P, Lin L-M, Guan G-Q, Wu Y-W and Lai Fa-Ch 2012 *Acta Photonica Sinica* **41** 8
[15] Kulkarni A K, Schulz K H, Lim T-S and Khan M 1997 *Thin Solid Films* **308-309** 1

Journal of Physics: Conference Series **514** (2014) 012004 doi:10.1088/1742-6596/514/1/012004

Improvement of the organic solar cells functional parameters

D Dimov[1,3], I Spassova[1], G Danev[1], I Zhivkov[1,2] and J Assa[1]

[1]Acad. J. Malinowski Institute of Optical Materials and Technologies, Bulgarian Academy of Sciences, Acad. G. Bonchev Str., Bl. 109, 1113 Sofia, Bulgaria
[2]Faculty of Chemistry, Center for Materials Research, Brno University of Technology, 118 Purkynova, 612 00 Brno, The Czech Republic

E-mail: dean@iomt.bas.bg

Abstract: By changing the concentration ratio (zinc-phthalocyanine-ZnPhTc:Fullerene-C_{60} from ZnPhTc:C_{60} = 1:0 through ZnPhTc:C_{60} = 1:1, to 1 C_{60}: ZnPhTc = 1:0) in the active layer, as well as the architecture of the organic solar cells (OSC), a substanial increase of their short-circuit current is achieved. We suggest that the blurred interlayer boundaries lead to an enhanced motion of the holes and the electrons to the electrodes – ITO and Al respectively. Regardless of the shorter distances between the places of excitons creation, the excitons have enough "life-time" for diffusion to the suitable places where their decomposition to the charge carriers – holes and electrons, takes place. This is one of the well-known approaches to achieve a higher efficiency of the OSCs. We thus suggest that the possible interactions and processes discussed are obviously essential prerequisites for improving the functional parameters of the OSCs.

1. Introduction

The possibilities for improving the functionality and the production facilities of the OSCs have increased in the last few years due to the combination of theoretical and material advances and the particularly improved understanding of the processes occurring in the interaction of light with matter. This allowed the introduction of a variety of new materials (e.g., dyes, organic low-molecular weight compounds and polymers) in the production of organic solar cells (OSC) or organic light emitting diodes. The organic materials are attractive not only because of the almost unlimited number of compounds, but also due to the combination of their influence on the OSCs functional parameters and the low production costs [1]. At present, the efficiency achieved of the OSCs is about 10 %, which is too low. One of the possible ways of improving these devices' efficiency is to develop contacts of a heterojunction layer or an add-layer which would affect the velocity of holes and electrons in order to reach close values. The aim of the present work is to study this assumption by following the changes of the electrical parameters and structure of vacuum-deposited (VD) donor – zinc phthalocyanine (ZnPhTc) and acceptor – Fullerene (C_{60}) layers as active films in the OSCs.

[3] To whom any correspondence should be addressed.

2. Experimental

Most of the experiments were carried out on VD films consisting of condensed ZnPhTc and C_{60}. In the basic SC configurations, the upper layer is located between the donor – ZnPhTc and the cathode – a 100 – 120-nm Al film. The substrate used was a flexible polymer band – polyethylene-terephtalat (PET) covered by electroconductive films of indium thin oxide (ITO), 30 Ω/sq., fabricated by Kintec Ltd., Hong Kong. It is well known [2] that the characteristics – roughness, defectlessnes and granularity of the final condensed films, depend on the growth, the microstructure and the size of the microcrystallites of each preceding deposited layer. Therefore, one of the goals of the experiments was to ensure evaporation conditions for producing a one- or multi-layer organic SC with the desired reproducible composition and microstructure. The dyes studied were evaporated from three independently heated vessel sources [3], thus overcoming the difficulties in evaporating at controlled rates because of the low and insufficient thermal stability and thermal conductivity of the substances used. Deposition rates of 0.1 – 0.4 nm/sec were achieved at 550 °C for ZnPhTc, 700 °C for C_{60} and 130 – 160 °C for bathocuproine (BCP) at a basic pressure of 4×10^{-3} Pa. The composition and thermal stability of VD ZnPhTc layers on KBr substrates were studied by Fourier transform infrared spectroscopy (FTIR). The FTIR absorbance spectra were recorded by a Bruker Tensor 27 spectrometer in the range 4400 – 450 cm^{-1} with a resolution of 2 cm^{-1}. The surface morphology of the layers was studied using scanning electron microscopy (SEM Philips 515). The films' structure and crystallinity were studied by TEM (HRTEM JEOL JEM 2100). The electrical characteristics were measured by a Keithley 617 (USA) high-precision electrometer.

3. Results and discussion

3.1. FTIR investigations

This part of the study aimed to follow the possible changes of the composition from the precursor ZnPhTc bulk material during the VD of the layers. The results from the FTIR studies are presented in figure 1, namely, the FTIR spectra of initial ZnPhTc dispersed on a KBr pellet. The spectra are normalized with respect to the band at 720 cm^{-1}. The peaks at 720 cm^{-1}, 1120 cm^{-1}, 1165 cm^{-1} and 1333 – 1335 cm^{-1} in figure 1 – curves 1 and 2, are typical and characteristic of the existence of the ZnPhTc compound. It is obvious that these essential peaks are present in both the bulk and the evaporated materials, which confirms the similarity of the composition.

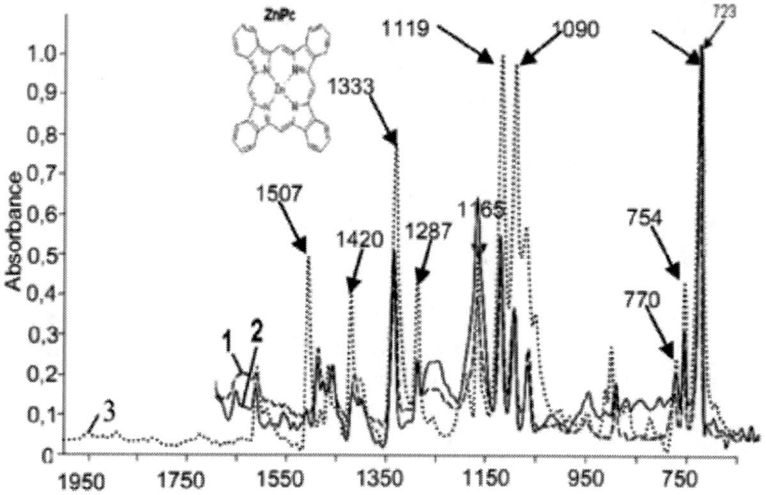

Figure 1. FTIR spectra of vacuum deposited ZnPhTc; 1 – untreated, 2 – thermaly treated at 200 °C, 3 – FTIR spectrum of initial ZnPhTc on a KBr pellet.

Journal of Physics: Conference Series **514** (2014) 012004 doi:10.1088/1742-6596/514/1/012004

Moreover, it can be concluded that the composition of the vacuum-deposited dyes used is identical to the initial one and the thermal treatment up to 200 °C (figure 1, curve 2) for both kinds of films has only a negligible effect on their compositions. This allows thermal annealing of the deposited films as a way of reducing their structural defects.

3.2. SEM and TEM investigations

Electron-microscopy observations of the surface morphology of the various substrates (figure 2 a, c) types were carried out prior to and after a deposition of ZnPhTc films with different thickness, given in relative units (figure 3). It was thus proved that the PET/ITO–60 Ω/sq. flexible substrates (figure 2 a, c and 2 b, d) are the most suitable ones from the point of view of the fine surface morphology and defectlessness of the active layer.

On the other hand, the use of 30 Ω/sq. ITO layers is more suitable for OSCs because of the low sheet resistance providing a better OSC performance. To reduce the influence of the substrate on the structure of the active film, an additional buffer layer must be added. Thus, a pyromelythic diahnhydride (PMDA) blocking layer was chosen [4] as a way of reducing the transport of holes towards the ITO anode.

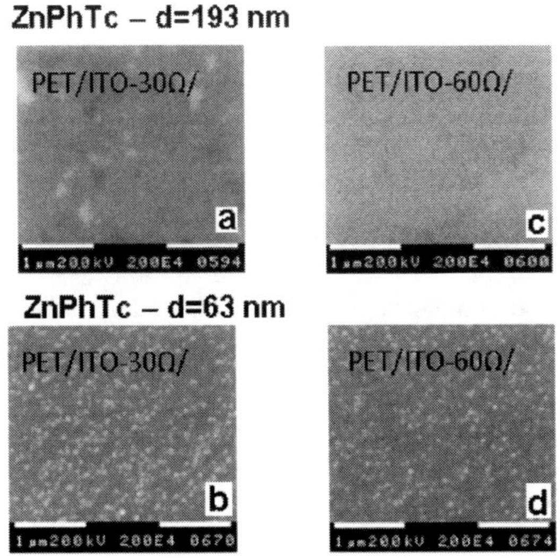

Figure 2. SEM images of ZnPhTc evaporated on PET/ITO substrates:
a, b – 30 Ω/sq.; c,d – 60Ω/sq.,
a ,c – thickness 193 nm; b, d – thickness 63 nm.

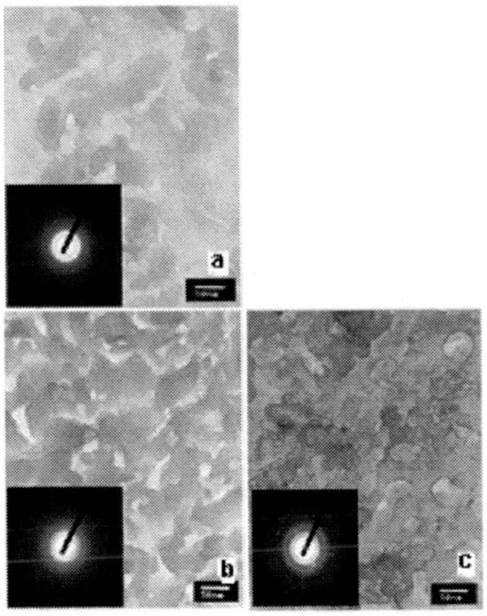

Figure 3. TEM images taken from a composite of ratio 1:1; the thicknesses are measured in relative units (ru), where 5000 ru ≈ 100 nm:
a) (ZnPhTc+Fullerene) ≈ 5000 ru,
b) (ZnPhTc+Fullerene) /BCP ≈ 6000 ru,
c) (ZnPhTc+Fullerene) /BCP ≈ 7500 ru.

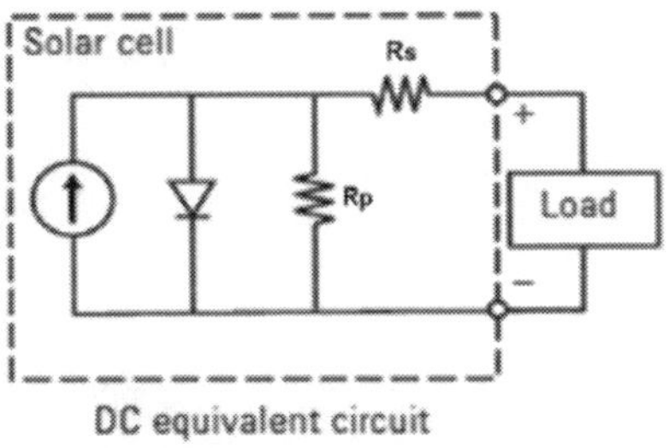

Figure 4. Simplified equivalent circuit diagram for modelling the DC behavior of a solar cell.

3.3. Electrical measurements

Figure 4 presents a simplified equivalent circuit diagram of an OSC. The important cell parameters are the parallel (R_p) and serial (R_s) resistance. The values of R_p (table 1) influence mainly the slope of the curve in the range around 0 V (3rd and 4th quadrant). At a lower R_p, the open-circuit voltage (V_{oc}) drops, while the short-circuit current (I_{sc}) remains unchanged. R_s defines the slope of the *J/V* curve (figure 5) in the 1st quadrant, i.e. in the range above the V_{oc} (transmitting diode direction). The fill factor (*FF*) and the I_{sc} decrease, while the V_{oc} remains unchanged when R_s increases [5].

Analyzing these results (figure 5 and table 1), one can see that the deposition of C_{60} leads to increased R_s, which is associated with the contact barrier. This barrier creates an electric field which is opposite to the field of the cell. This is confirmed by the decrease of the I_{sc} and V_{oc}. However, the same result may be due to the difference between the donor and acceptor LUMO levels, as well as to the not very good quality of the electrical contacts.

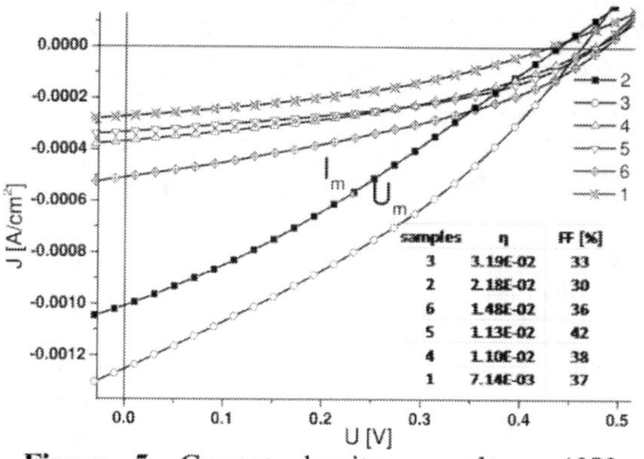

Figure 5. Current density – voltage (*J/V*) characteristics of organic photovoltaic devices with various layer structures illuminated by white light at 6 mW cm^{-2}.

As already discussed, an approach to create OSCs with good functional characteristics is to provide an effective separation of excitons in the entire volume of the active film. Therefore, our study aimed to establish conditions for creating "blurred boundaries" of the active composite (figure 6, 2 and 3) layer with a ratio of 1:1. This assumption was proved by the best parameters of the structures studied (figure 6, 2 and 3). These samples are composed of a film of donor ZnPhTc and acceptor C_{60} with blurred boundaries at the molecule-molecule contact in the largest part of the volume of the active layer. It is also seen in figure 6, 2 and 3 that increasing the thickness of the active layer leads to an increase of the I_{sc}, V_{oc}, FF, and the energy conversion efficiency (η).

Table 1. Parallel (R_p) and serial (R_s) resistance values.

Ω cm^{-2}/samples	1	2	3	4	5	6
R_s	758	337	199	513	377	280
R_p	3585	680	559	2796	3541	1742

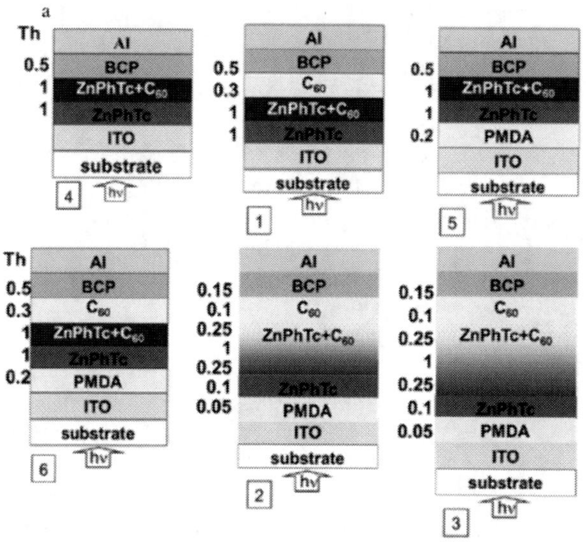

Figure 6. Different structures of OSC – 1, 4, 5, 6 develop the idea of a PMDA buffer layer; 2, 3 – layers with "blurred " boundaries.

[a] Th- thicknesses in relative units.

4. Conclusions

We discussed and experimentally confirmed a novel idea for improvement of some functional OSC parameters, namely, the enhanced exciton decay and subsequent charge separation on the grain boundary of the ZnPhTc-C_{60} donor – acceptor composite layer. Complementary PMDA and C_{60} layers were added to the OSC structure in view of facilitating the extraction of the photogenerated charge carriers through the electrodes.

Acknowledgment

The present work was financially supported by the National Science Fund at the Ministry of Education and Science of Republic of Bulgaria under contract DO-02/254-2009 and by the Grant Agency of the Czech Republic under contract: No. P205/10/2280.

References

[1] Brabec C, Dyakonov V, Parizi J, Sariciftchi N 2003 *Organic Photovoltaics: Concepts and Realization* (Springer, Berlin, Germany) p 60
[2] Leamy J H, Gilmar G H and Dirks D 1980 *Science* **6** 1
[3] Karashanova D, Dimov D, Spassova E, Assa J, Danev G 2012 *Nanosci. & Nanotechnol.* **12** 106
[4] Eunkyoung N, Seungsik O, Donggeun J, Hyoungsub K, Heeyeop C and Junsin Y 2012 *Semicond. Sci. Technol.* **27** 105004
[5] Johnev B 2005 *Doctor rerum naturalium* Department of Biology, Chemistry and Pharmacy of Freie Universität Berlin, Berlin

Vacuum-deposited diphenyl-diketo-pyrrolopyrroles structures with photoelectrical applications

Y Georgiev[1,3], I Zhivkov[1,2,4], T Takov[3], G Angelov[3], R Prikryl[1], S Stritesky[1] J Honova[1] and M Weiter[1]

[1]Faculty of Chemistry, Center for Materials Research, Brno University of Technology, 118 Purkynova, 612 00 Brno, The Czech Republic
[2]Acad. J. Malinowski Institute of Optical Materials and Technologies, Bulgarian Academy of Sciences, Acad. G. Bonchev str., Bl. 101/109, 1113 Sofia, Bulgaria
[3]Faculty of Electronic Engineering and Technologies, Department of Microelectronics, Technical University of Sofia, 8 Kl. Ochridski Blvd., 1756 Sofia, Bulgaria

E-mail: zhivkov@fch.vutbr.cz

Abstract. Modern organic semiconductor devices are designed as stacked organic/inorganic thin films. In such multilayer structures, strong requirements are set forth to the properties of the specific films, e.g., film homogeneity, thickness uniformity and roughness. The physical vacuum deposition (PVD) technique fulfils all of these requirements having even the capability of producing the whole multilayer structure in the same vacuum cycle. Diphenyl-diketo-pyrrolopyrroles (DPP) are low-molecular-weight materials with promising luminescence and photoelectrical properties. This study deals with PVD of thin DPP films and measurement of their properties. The organic films and inorganic electrodes of the multilayer structures were prepared in the same vacuum cycle, and then the structures were encapsulated in a glove box avoiding exposure to oxygen. ITO|composite|Al structures were prepared with active DPP-C_{60} composite films (ratio 60:40 mass %) of about 100 nm thicknesses. The surface morphology of the films was studied by SEM imaging, which revealed uniform films containing separate spherical structures. The dark current and the photoconductivity were measured and the photoelectrical behavior of the structures was estimated. It was found that the optimized preparation of the structures under the aforementioned conditions leads to an improvement of their photoelectrical properties.

1. Introduction

Diphenyl-diketo-pyrrolopyrroles (DPP) are low-molecular-weight materials with promising luminescence and photoelectrical properties. The first known derivative, 1,4-diketo-3,6-difenyl-pyrrolo-[3,4-c]-pyrrole (DPP) was described for the first time in 1974, having been synthesized by mistake.

In 1986, the first derivative was introduced to the market and, since then, many other derivatives have fulfilled their potential, especially in the industry of colorants. The interest in these pigments has grown recently and they have become subjects of patents in many areas of applications.

[4] To whom any correspondence should be addressed.

Journal of Physics: Conference Series **514** (2014) 012005 doi:10.1088/1742-6596/514/1/012005

These compounds have been the object of intensive research for pigment applications; they exhibit a variety of shades in the solid state and, especially, chemical, light and thermal stability [1]. DPP itself has a high molar absorption coefficient, as well as a high fluorescence quantum yield; therefore, the low-molecular-weight derivatives of DPP have been extensively studied concerning their optical and photophysical properties [2–4]. Potential applications have been reported of DPP derivatives as luminescent media in a polymer matrices [5], solid-state dye lasers [6], OLED devices [7] and organic field-effect transistors [8].

Thin films of low-molecular-weight semiconductors are usually prepared by means of a variety of complex techniques, including physical or chemical vapor deposition, organic molecular-beam epitaxy or solution-based deposition techniques. The performance of small molecular organic devices has been shown to be highly sensitive to the films' morphology and processing conditions. Often, the solution-deposited active layers of devices (e.g. spin cast films) exhibit a high portion of microcrystallites and aggregates, whereas the vapor deposition techniques provide high-quality crystalline films characterized by improved charge-transport properties compared with those of solution-deposited films. The relationship between the organic thin film morphology and the device performance is nowadays subject of comprehensive research activities.

DPP based polymers, as well as small molecules, have found their place in the organic electronics. This paper deals with 3,6-bis(5-(benzofuran-2-yl)thiophen-2-yl)-2,5-bis(2-ethylhexyl)pyrrolo[3,4-c]pyrrole-1,4(2H,5H)-dione, denoted hereinafter as DPP (figure 1). Using this material as a donor for bulk-heterojunction-based solar cells leads to a photovoltaic conversion efficiencies up to 4,4 % [9]. Conjugated systems provide high charge-carrier mobility across the structure.

In the present paper, DPP-C_{60} donor-acceptor composite films were prepared by vacuum co-deposition and the photoelectrical properties of ITO|composite| films were investigated.

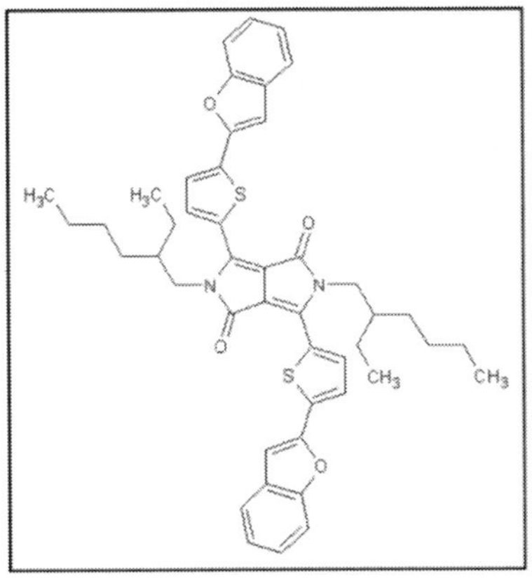

Figure 1. Chemical structure of the active material DPP.

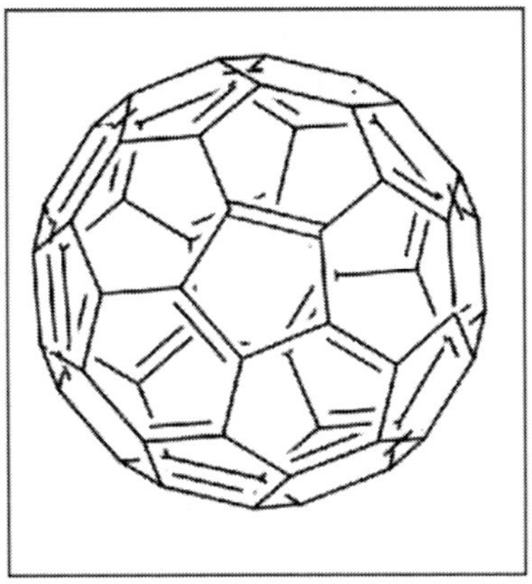

Figure 2. Chemical structure of the fullerene C_{60}.

2. Experimental

3,6-bis(5-(benzofuran-2-yl)thiophen-2-yl)-2,5-bis(2-ethylhexyl)pyrrolo[3,4-c]pyrrole-1,4(2H,5H)-dione was provided by our partner Center for Organic Chemistry Ltd. (COC), who synthetized it according to the route published in 2009 by Walker et. al. [9]. Fullerene C60 PCBM ([6,6]-Phenyl-C61-butyric acid methyl ester) (figure 2) was purchased from Ossila Ltd.

To fabricate our samples we used the Ossila OLED/OPV Pixelated Anode Substrate System and the Ossila Encapsulation System, which consist of a pre-patterned ITO-covered glass substrate with a standard size of 20×15 mm, an active area deposition mask, a cathode deposition mask, encapsulation epoxy, encapsulation glass slides and electrical connection legs with a standard 0.1-inch (2.54 mm) pitch. The pre-patterned ITO structure consists of one cathode strip along the long side of the substrate and six fingers, which, combined with the deposition masks, form six active pixels with dimensions 4×1.5 mm and six anodes connected to them.

The composite thin film was deposited onto Ossila substrates in an MBraun vacuum system from two heat sources simultaneously at evaporation temperatures of 180 °C for DPP and 420 °C for C_{60} and deposition rates of 4.2 Å/s and 2.5 Å/s, respectively, through the active area deposition mask. After deposition of the active composite layer, the samples were taken out in nitrogen atmosphere, the mask was changed with the cathode deposition mask and the samples were returned back into the evaporation chamber. When the vacuum in the evaporation chamber reached sufficient levels, aluminum electrodes were deposited.

Finally, the samples were encapsulated according to a standard procedure.

An optical micrograph image of the sample prepared was taken and the overlapping of the layers was inspected.

Selected samples were studied by a Philips 515 scanning electron microscope (SEM). The SEM morphology characterization of the DPP-C_{60} composite films was carried out on the Ossila substrates without implementing the final encapsulation procedure and the result is shown in figure 3.

Figure 3. SEM images of the deposited composite thin films: a) magnification of 10 000; b) magnification of 40 000.

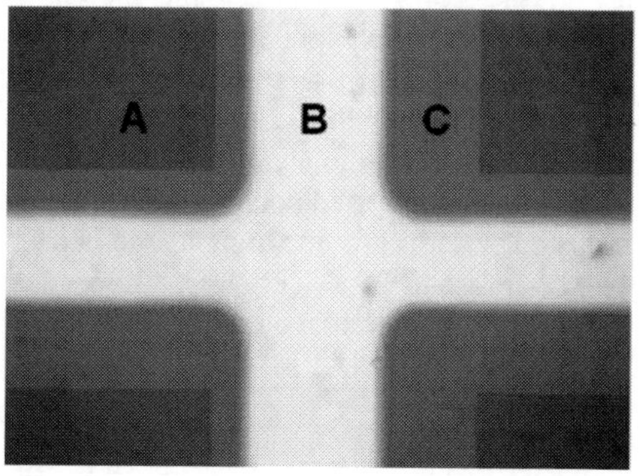

Figure 4. Optical micrograph in polarized light; the device structure is as follows: *A* is ITO, *B* is Al and *C* is the active composite layer.

The setup for photoelectrical measurements consisted of a Keithley 6517A electrometer and voltage source; monochromatic light was produced by an LSH502 LOT-Oriel halogen lamp and an MSH101 LOT-Oriel monochromator. The light power was controlled by using gray filters and measured by a S120VC standard Si photodiode power sensor and a Keithley 485 picoampermeter.

The photoelectrical measurements started with the spectral dependence of the photocurrent at zero applied voltage; the *I-V* characteristics were then measured in both directions of the voltage scale in dark and when exposed to monochromatic light. Finally, the dependence of the photocurrent on the incidental light power (irradiance) was measured. The sample was exposed to monochromatic light at the wavelength of 533 nm.

3. Results and discussion

3.1. Sample characterization

An optical micrograph image of the sample prepared for electrical characterization is presented in figure 4. As is seen, uniform films covering entirely the ITO electrode area were prepared.

A more precise film-surface characterization of the composite films was performed by SEM. SEM images taken at two different magnifications from the same sample without applying the encapsulation procedure are presented in figure 3. The image taken at the lower magnification of 10 000 (figure 3a) confirms that smooth and uniform DPP-C_{60} films without a presence of pinholes were produced. The

SEM image taken at a magnification of 40 000 (figure 3b) reveals the formation of spheres with a size of about 100 - 200 nm. The spheres observed could be related to the formation of a single phase of one of the components. To clarify this problem, further studies of samples with different DPP-C_{60} ratios should be carried out. In general, one can conclude that the DPP-C_{60} films deposited are suitable for electrical measurements in "sandwich" type samples.

3.2. Photoelectrical measurements

Figure 5 presents the spectral dependence of the photocurrent measured at zero applied voltage between the electrodes of ITO|composite|Al structures with active DPP-C_{60} composite films. The maximum of the spectrum was found at 533 nm. A similar spectral dependence of the photocurrent taken from samples with the same DPP-C_{60} ratio of a composite film prepared by spin-coating technique was published in the literature [9]. To excite this peak during the following experiments, a monochromatic light (λ = 533 nm) was used.

I-V characteristics measured in dark and under illumination of ITO|composite|Al structures with active DPP-C_{60} composite films of about 100 nm thickness are presented in figure 6. To estimate the electrical parameters of the samples, the graphs are plotted on a semi-logarithmic scale (the negative values of the current are multiplied by −1).

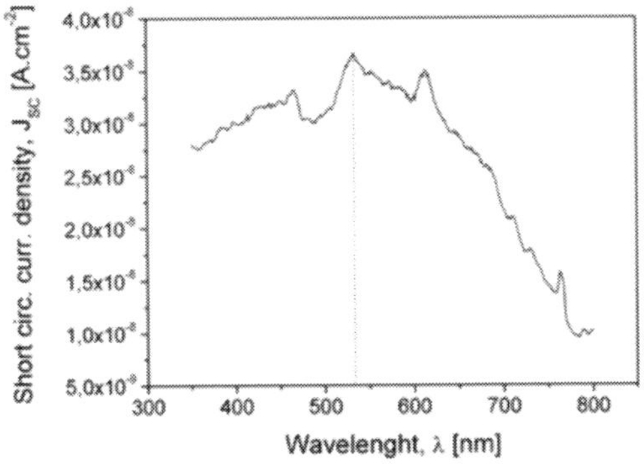

Figure 5. Spectral dependence of the photocurrent at zero applied voltage.

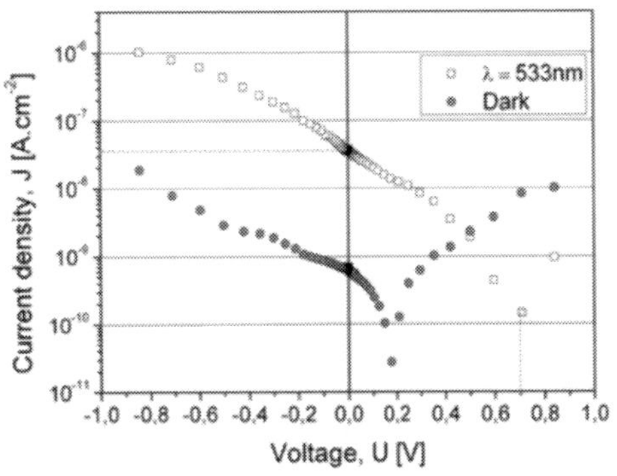

Figure 6. *I-V* characteristics measured in the dark and under monochromatic light with irradiance 0.35 mW cm^{-2}

The dark current measurements yielded non-linear and almost symmetrical characteristics with a contact barrier of about 0.2 V. These effects could be related to the imperfections of the electrodes.

It should be mentioned that no additional intermediate layers serving to reduce the contact barrier were used. The optimization of the stacked structures will be a subject of a future study. Under illumination by monochromatic light, the non-optimized samples produced photocurrent of almost two orders of magnitude higher than what was observed in the dark measurements. By processing the data for the DPP-C_{60} structure, a short-circuit current of $J_{SC} = 36\times10^{-9}$ A cm^{-2} and an open-circuit voltage of U_{OC} = 0.71 V were determined.

Using the area confined within J_{OS} and U_{OC}, the dependence of the electrical power on the voltage is plotted in figure 7, right *y* axis. The maximum electrical power found was $P_{MAX} = 1{,}2\times10^{-8}$ mW cm^{-2} at U_{MP} = 0.25 V and $J_{MP} = 5{,}24\times10^{-9}$ A cm^{-2}.

The dependence of the photocurrent on the incident light power (irradiance) for samples with DPP-C_{60} films under illumination by monochromatic light is plotted in figure 8. It was found that the photocurrent is proportional to G^{γ}, where G is the photogeneration rate and γ = 1.15 is the slope of the graph presented. This result could be connected to a monomolecular photogeneration mechanism.

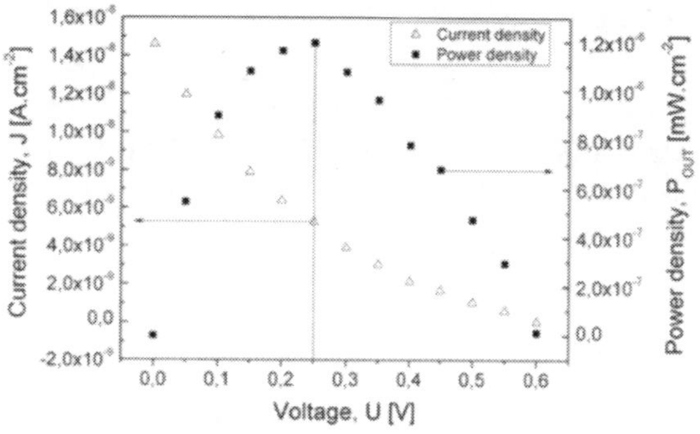

Figure 7. Dependence of the power density (right y axis) and current density (left y axis) on the voltage applied, as calculated from figure 6, light curve.

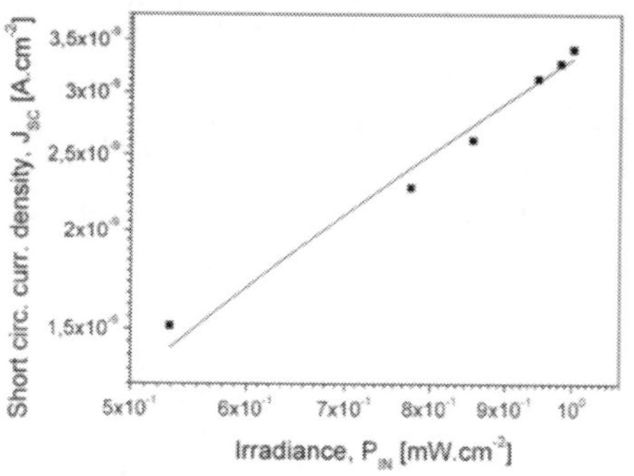

Figure 8. Dependence of the photocurrent on the light intensity at λ = 533 nm and applied voltage U_{MP} = 0.25 V.

4. Conclusions

ITO|composite|Al structures with active DPP-C_{60} composite films of about 100-nm thickness were prepared by vacuum deposition and encapsulated in an inert atmosphere.

The surface morphology of the films prepared was studied by SEM imaging, which revealed formation of spheres with a size of about 100 - 200 nm.

Photoelectrical measurements were carried out. It could be concluded that the vacuum-deposited DPP-C_{60} composite active layers have promising properties for incorporation in the future organic solar cells. Further optimization should be conducted in view of reducing the contact barrier and facilitating the excitation decay and charge carrier extraction through addition of intermediate interface layers.

Acknowledgements

This work was supported by the Ministry of Industry and Trade of the The Czech Republic under project No FR-TI1/144, by the Grant Agency of the Czech Republic under projects No. P205/10/2280 and 13-29358S and by the Center for Materials Research at FCH BUT project No. CZ.1.05/2.1.00/01.0012 supported by ERDF. The work was also funded by contract No. 132PD0048-03 supporting PhD students at the Technical University of Sofia. Jana Honova is Brno PhD Talent Scholarship Holder – Funded by the Brno City Municipality.

References

[1] Wallquist O 2002 *VCH publishers* (Weinheim)
[2] Fukuda M, Kodama K, Yamamoto H and Mito K 2002 *Dyes Pigments* **53** 67
[3] Mizuguchi J and Rochat A C 1988 *J. Imag. Sci.* **32** 135
[4] Mizuguchi J 2000 *J. Phys. Chem.* A **104** 1817
[5] Smet M, Metten B and Dehaen W 2001 *Tetrahedron Lett.* **42** 6527
[6] Fukuda M, Kodama K, Yamamoto H and Mito K 2004 *Dyes Pigments* **63** 115
[7] Patents W O 2004/090046, US 2005/0008892
[8] Yanagisawa H, Mizuguchi J, Aramakil S and Sakai Y 2008 *Jpn. J. Appl. Phys.* **47**/6 4728
[9] Walker B, Tamayo A B, Dang X D, Zalar P, Seo J H and Garcia A 2009 *Adv. Funct. Mater.* **19** 3063

Aerosol flow homogenization in the spray polyphenylene vinylene thin film deposition

V Milenkov[1], J Honova[1], I Zhivkov[1,2,4], R Yordanov[3], M Vala[1], D Mladenova[1,2] and M Weiter[1]

[1]Faculty of Chemistry, Center for Materials Research, Brno University of Technology, 118 Purkynova, 612 00 Brno, The Czech Republic
[2]Acad. J. Malinowski Institute of Optical Materials and Technologies, Bulgarian Academy of Sciences, Acad. G. Bonchev Str., Bl. 109, 1113 Sofia, Bulgaria
[3]Faculty of Electronic Engineering and Technologies, Department of Microelectronics, Technical University of Sofia, 8 Kl. Ochridski Blvd., 1756 Sofia, Bulgaria

Email: zhivkov@fch.vutbr.cz

Abstract. Two different spray deposition (SD) techniques – SD and homogenized SD (h-SD), were compared in the case of poly[2-methoxy-5-(3,7-dimethyloctyloxy)-1,4-phenylene vinylene] (MDMO-PPV) used as a model material. The h-SD setup includes a homogenizer (a chamber) for homogenization of the spray flow. The homogenizer allows droplet size control and spraying at a low air pressure. Films of about 1μm thickness were prepared by SD and h-SD from a toluene solution with a concentration of 0.5 g l^{-1}. Optical micrograph images show that using h-SD results in obtaining uniform films without a distinguishable structure. Imaging by polarized light reveals the typical discs observed often in the spray deposition experiments. The spectral dependence of the photocurrent consists of two peaks. This property makes the structures with h-SD films more attractive for a photovoltaic application since this enhances the photovoltaic conversion in a wider visible light range. The *I-V* characteristics measured on samples with h-SD films show that exposure to white light increases the current by two orders of magnitude. One can thus conclude that the ITO|MDMO-PPV|Al structures with h-SD MDMO-PPV films behave as a photovoltaic cell.

1. Introduction

Organic semiconductor thin films are potential candidates for applications in microelectronic devices due to their easy processing and low cost [1].

Poly[2-methoxy-5-(3,7-dimethyloctyloxy)-1,4-phenylene vinylene] (MDMO-PPV) is a widely investigated organic semiconductor (HOMO and LUMO orbital positions at 5.4 and 3.2 eV, respectively, [2]) with a potential application as photoconductive or electroluminescent layers.

A variety of methods are being used to produce PPV films – spin-coating [3], electrophoretic deposition [4], Langmuir-Blodgett [5] and ink-jet printing [6].

[4] To whom any correspondence should be addressed.

Recently, spray-deposition (SD) through a conventional airbrush was introduced as an efficient large-area deposition technique [7, 8].

The SD method produces films on a large covered area at relatively low substrate temperatures (20 – 100 °C). The main problems impeding the wide usage of the spray deposition arises from the non-uniformity of the sprayed films. As a result, films with voids or considerably high grain size are usually produced.

In our previous experiments we found that MDMO-PPV films sprayed at room temperature of the substrate tend to cross-link in network fibers separated by voids [9]. This effect was reduced by increasing the substrate temperature during the spray deposition which decreases the void size and increases the covered part of the substrate [10]. In this paper, a way of increasing further the film uniformity is suggested by homogenizing the sprayed flow. Usually, the homogenizer consists of a heated chamber [11] where the solution is sprayed and heated in contact with the chamber walls. Our construction suggests a chamber of approximately 1 l volume where a more precise heating control is achieved by mixing directly the sprayed flow with hot air. The air is heated outside the chamber in an air convector to temperatures of 30 – 70 °C. SD using homogenization of the sprayed flow by mixing with hot air is denoted hereinafter as h-SD.

The paper presents a comparative study of the properties of SD and h-SD poly[2-methoxy-5-(3,7-dimethyloctyloxy)-1,4-phenylene vinylene] (MDMO-PPV) films. The study seeks potential applications for these films in solar energy conversion technologies and other branches of organic electronics.

2. Experimental

A base solution (concentration of 1 g l^{-1}) was prepared by dissolving 20 mg MDMO-PPV (Sigma-Aldrich, catalogue number 546461) in 20 ml toluene (Sigma-Aldrich Toluene, catalogue number PLC22C11X). The solution was heated for 5 min at 60 °C, followed by ultrasonic treatment for 15 min and filtering by a filter of 0.35 μm hole size. A work solution without undissolved particles with a concentration of 0.5 g.l^{-1} was prepared by diluting 20 ml of the base solution by 20 ml toluene.

The aerosol flow was produced by a Paasche EZ-STARTER Single Action Airbrush Kit with a nozzle diameter of 0.5 mm. The parameters used for the SD process were: substrate temperature 30°C; distance from the spray nozzle to the substrate 10 cm; air pressure used for the formation of the sprayed flow 1 kPa; work solution amount 20 ml; deposition time 4 min; substrate motion velocity 1 mm/s.

The parameters used for h-SD process were: substrate temperature 30 °C; distance from the homogenizer's outlet to the substrate 2 cm; air pressure for the formation of the sprayed flow 1 kPa; work solution amount 100 ml; deposition time 20 min; substrate motion velocity 1 mm/s.

Thin films for surface morphology characterization were spray deposited on microscope-grade glass. The substrates were cleaned with detergent and deionized water followed by a plasma treatment.

Selected samples were studied by scanning electron microscope (SEM) Philips 515.

The samples for photoelectrical characterization were deposited on Ossila substrates with preliminary prepared ITO electrodes after cleaning according to a standard procedure.

Aluminum electrodes were deposited in oil-free vacuum of 8×10^{-4} Pa.

The photoelectrical measurements were carried out in an oil-free vacuum of 2.2×10^{-5} Pa by a Keithley 6517A Electrometer. Monochromatic light was produced by an LSH502 LOT-Oriel halogen lamp and an MSH101 LOT-Oriel monochromator. The light power was controlled by adding grey filters and measured by a Gigahertz-Optik - X97 Irradiance Radiometer.

The photoelectrical measurements started by taking the spectral dependence of the photocurrent at zero applied voltage; the *I-V* characteristics were then measured in both directions of the voltage scale in dark and under exposure to a white light. Finally, the dependence of the photocurrent on the incident light power (irradiance) was measured.

3. Results and discussion

3.1. Sample characterization

Figure 1 presents an optical micrograph of a SD film deposited under the aforementioned conditions. It confirms that the SD film tends to cross-link in a network of fibers separated by voids. Comparing the color in the scratched line (in the middle of the picture), one can clearly see that voids (areas not covered by the material) have been formed. This observation is supported by electrical measurements of sandwich samples with SD films prepared under the same conditions; the measurements resulted in a short circuit between the electrodes.

A precise film surface characterization of the h-SD films was performed by SEM.

SEM images of a h-SD sample at two different magnifications are presented in figure 2. At the lower magnification of 10000 (figure 2a), spherical or disc-shaped particles of size of about 1 µm could be observed. They are probably related to the presence of dust or to non-dissolved MDMO-PPV particles. In figure 2b, a SEM image of the same film taken at magnification of 40000 is presented. Spheres or discs with a size of about 200 nm could still be observed and considered as being dust particles. The other, more uniform part of the film, could be related to the individual structure of the film. The SEM images of h-SD films confirmed that uniform and relatively smooth films free of pinholes have been obtained. These films are suitable for electrical measurements in a "sandwich" type samples.

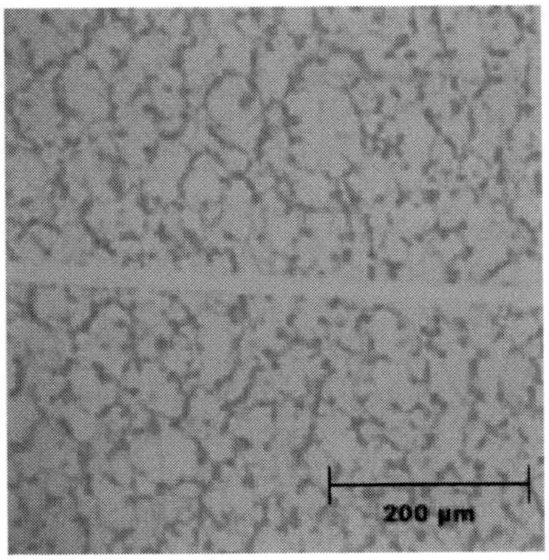

Figure 1. Optical micrograph of a SD MDMO-PPV film.

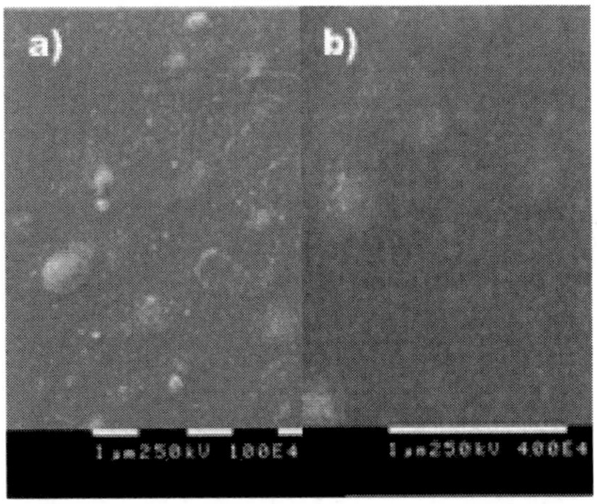

Figure 2. SEM images of a h-SD film: a) magnification 10000; b) magnification 40000.

3.2. Photoelectrical measurements

Figure 3 presents the spectral dependence of the photocurrent measured at zero voltage applied between the electrodes of structures with h-SD films. It consists of two peaks. The peak at 560 nm is known from photovoltaic structures with spin-coated and electrophoretically deposited MDMO-PPV films [12]. Another stronger and wider peak appears at 370 nm. The existence of these two peaks makes the structures with h-SD films more attractive for photovoltaic applications as it allows photovoltaic conversion in a wider visible light range. White light were used in the subsequent experiments on excitation of these two peaks.

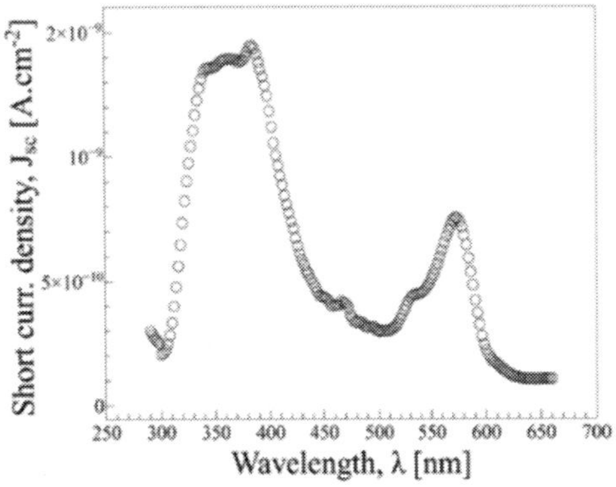

Figure 3. Spectral dependence of the photocurrent at zero applied voltage.

The *I-V* characteristics measured in dark and under white light illumination of a ITO|MDMO-PPV|Al structure with h-SD MDMO-PPV films of about 1 μm thickness are presented in figure 4. To estimate the electrical parameters of the samples, the data are plotted in a semilogarithmic scale (the negative values of the current are multiplied by −1).

The dark-current measurements (figure 4, curve *a*) resulted in non-linear and almost symmetrical characteristics with a contact barrier of about 0.3 V. These effects could be related to the bulk material properties, which seem to dominate in the thick MDMO-PPV film. An optimization of the spray parameters to obtain films of a lower thickness is, therefore, necessary. Despite the non-optimized conditions of forming a thick film where charge-carriers recombination is possible to occur, a photocurrent higher by two orders of magnitude (marked by the vertical line in figure 4) was obtained under illumination by white light. This result demonstrates the potential of the h-SD method for deposition of thin films for photovoltaic applications.

The dependence of the photocurrent on the incident light power (irradiance) for samples with h-SD MDMO-PPV films under illumination by white light is plotted in figure 5. It is seen that the photocurrent is proportional to G^{γ}, where G is the photogeneration rate and $\gamma = 1.7$ is the slope of the graph presented.

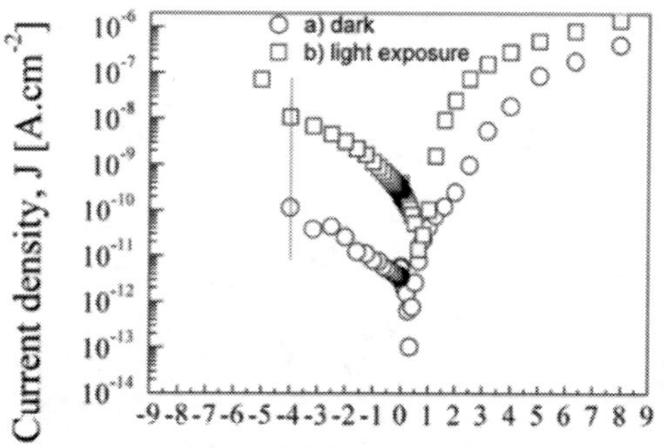

Figure 4. *I-V* characteristics measured: a) in the dark and b) under illumination by white light, irradiance of 0.59 mW cm^{-2}) of ITO|MDMO-PPV|Al structures with h-SD MDMO-PPV films.

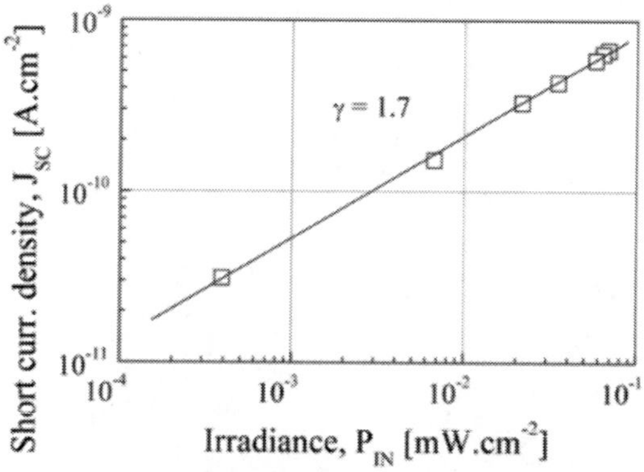

Figure 5. Dependence of the photocurrent on the light intensity under illumination by white light, of ITO|MDMO-PPV|Al structures with h-SD MDMO-PPV films.

To summarize, it could be pointed out that the spray deposition parameters have a considerable influence on the uniformity, coverage and grain size of the SD MDMO-PPV films. The homogenizer introduced allows one to control the droplet size by using low nozzle air pressure and achieve a very low air pressure by controlling the temperature and pressure of the additional flow that is mixed with the first one.

4. Conclusions

Thin films of thickness of about 1 μm were prepared by SD and h-SD from a solution with a concentration of 0.5 g.l^{-1}.

It was found that the homogenization of the sprayed aerosol flow by mixing it with hot air reduces the droplet size thus leading to deposition of more uniform films.

Optical micrograph images show that using h-SD results in forming uniform films without distinguishable structure.

The spectral dependence of the photocurrent consists of two peaks. This property makes the structures with h-SD films more attractive for photovoltaic applications as photovoltaic conversion takes place in a wider visible light region. The *I-V* characteristics measured on samples with h-SD films show that exposure to white light increases the current by two orders of magnitude.

More experiments connected with the determination of the droplet size and size distribution should be carried out to achieve a full control over the process parameters. The application of the h-SD thin films in the development of organic electronics necessitates further studies on the mechanism and kinetics of the spray process, and on the structure and properties of the films.

Acknowledgements
This work was supported by the South Moravian Region and the Seventh Framework Programme for Research and Development (grant SIGA 885), by the IGA Brno University of Technology via the FCH/FEKT-S-11-2 project, by the Grant Agency of The Czech Republic project No. P205/10/2280, the project "Centre for Materials Research at FCH BUT" No. CZ.1.05/2.1.00/01.0012 supported by ERDF. Jana Honova is holder of Brno Ph.D. Talent Scholarship funded by the Brno City Municipality.

References

[1] Tada K and Onoda M 2009 *Molecular Crystals and Liquid Crystals* **505** 124
[2] Thompson B C and Frechet J M J 2008 *Angew. Chem. Int. Ed.* **47** 58
[3] Gómez R and López J 2011 *Lat. Am. J. Phys. Educ.* **5** 2
[4] Tada K and Onoda M 2005 *Current Appl. Phys.* **5** 5
[5] Fitrilawati F, Renu R, Baskar C, Xu L G, Chan H S, Valiyaveettil S, Tamada K and Knoll W 2005 *Langmuir* **21** 12146
[6] Shah V and Wallace D 2004 *Proc. IMAPS 37-th Annual Int. Symp. Microelectronics* **1**
[7] Vak D, Kim S, Jo J, Oh S, Na S, Kim J and Kim D 2007 *Appl. Phys. Lett.* **91** 081102
[8] Na S, Yu B, Kim S, Vak D, Kim T, Yeo J, Kim D 2010 *Sol. Energy Mater. Sol. Cells* **94** 1333
[9] Aleksandrova M, Dobrikov G, Zhivkov I, Kyosev B and Dobreva T 2009 *J. Optoelectr. Adv. Mater.* **11** 1400
[10] Zhivkov I, Milenkov V, Stoimenov R, Mladenova D and Weiter M 2012 *EDS IMAPS CS Int. Con. Proc.* 120
[11] Fujita K, Ishikawa T and Tsutsui T 2002 *Jpn. J. Appl. Phys.* **41** L70
[12] Mladenova D, Zhivkov I, Ouzzane I, Vala M, Heinrichová P, Budurova D and Weiter M 2012 *J. Phys.: Conf. Series* **398** 012040

Journal of Physics: Conference Series **514** (2014) 012007 doi:10.1088/1742-6596/514/1/012007

Substrate temperature control in a system fordeposition of carbon nanostructures at atmospheric pressure

Zh Kiss'ovski[1], N Djermanova, D Mitev and V Vachkov

Faculty of Physics, St. Kl. Ohridski University of Sofia,
5 J. Bourchier Blvd., 1164 Sofia, Bulgaria

E-mail: kissov@phys.uni-sofia.bg

Abstract. The paper presents a plasma-enhanced chemical vapor deposition system for synthesis of carbon nanostructures on a metal substrate at atmospheric pressure with a system for substrate temperature control. It is realized by using a miniature microwave plasma source with a surface-wave discharge. An important part of the system is the movable target holder which allows adjustment of the distance between the plasma flame and the substrate. The substrate temperature increases during the deposition process due to the heating by the plasma flame. A constant value of the substrate temperature (400 – 700 °C) is essential for the synthesis of various carbon structures. An electronic system with sensors positioned in the holder is designed for measurement and control of the substrate temperature. The substrate temperature is measured in continuous and pulsed modes of the discharge, with the temperature value being controlled by the amplitude, repetition rate and duration of the microwave pulses.

1. Introduction

The unique properties of single-layer carbon nanostructures (graphene), as electrical conductivity, hardness and transparency, are attractive in view of a large number of applications in microelectronics. There exists a wide variety number of carbon nanostructures (nanowalls, nanoflakes, etc.) with multiple graphene layers, which are characterized by high reactivity, and also have great potential for various applications [1]. One of the methods for deposition of carbon nanostructures is the plasma-enhanced chemical vapor deposition (PECVD). Most of the systems for deposition of various carbon nanostructures operate at a low pressure of the neutral gas. In microwave plasma-enhanced chemical vapor deposition systems, the substrates used are sputter-coated with transition metals as catalysts. The conditions are similar to those used for the PECVD growth of carbon nanotubes but with pressure in the processing chamber lower than 1 Torr, when heating of the substrate is necessary. We present a design of a plasma-enhanced chemical vapor deposition system for synthesis of carbon nanostructures at atmospheric pressure with a system for substrate temperature control. It is realized by using a miniature microwave plasma source with a surface-wave discharge. This low-power source produces dense plasma with high neutral-gas temperature [2], which are the main factors for effective decomposition of complex molecules and generation of highly reactive species. The process is carried out in a chamber at a controlled gas mixture of argon, hydrogen and methane. The main part of the

[1] To whom any correspondence should be addressed.

system is the movable target holder which allows adjustment of the distance between the plasma flame and the substrate, thus determining the values of the plasma density and the gas temperature at the substrate surface. The holder contains also a K-type thermocouple in contact with the substrate for temperature measurement. The design of the holder includes also wires for biasing the substrate by a DC voltage. The substrate temperature increases due to heating by the plasma flame. A constant value of the substrate temperature is essential for the synthesis of various carbon structures: nanotubes (400 – 600 °C), nanoflakes (600 – 700 °C), nanowalls [3-4]. We designed an electronic system for precise measurement of the substrate temperature by a K-type thermocouple whereby the temperature can be controlled by a small-size Peltier element. Another manner to control the substrate temperature is used at a pulsed mode of operation of the discharge. The substrate temperature is measured by an electronic system and its value is controlled by the amplitude, repetition rate and duration of the microwave pulses at a fixed distance from the discharge tube end.

2. Experimental set-up and electronic system for measuring the substrate temperature

The experiments were carried out on the set-up presented in figure 1. A miniature microwave plasma source with a surface-wave discharge produces a plasma flame in a dielectric deposition chamber with two metal flanges (top and bottom). A microwave signal at a frequency f = 2.45 GHz from an MPG-4M generator (0 – 20 W) is fed to the coaxial surface-wave exciter. The forward and reflected powers are measured from a Pasternack PE2219-30 directional coupler by a HP 437B power meter. The exciter of the surface waves has an outer diameter of 2 mm; its inner conductor is a steel capillary with a diameter of 1 mm. A thin alumina tube is used as a discharge tube because the ceramics can withstand high gas temperatures and has high thermal conductivity. The initial tuning of the source at a frequency of 2.45 GHz is carried out by a Maury 1878C triple stub. The neutral gas-mixture ($Ar/H_2/CH_4$) is fed trough the steel capillary to the discharge tube. The gas flows are measured by APEX-AX-MC mass-flow controllers.

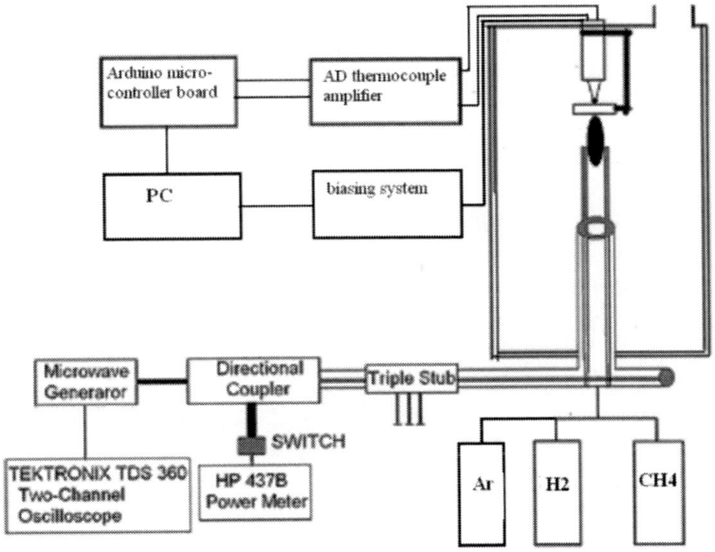

Figure 1. Experimental set-up and system for substrate temperature measurement.

The electron temperature and density in the argon gas are estimated by double-probe and optical diagnostics. The results for the plasma parameters at different gas flows and applied power obtained by both diagnostics in the middle of the discharge tube are as follows: plasma density in the range $n_e = 4\times10^{20} - 8\times10^{20}$ m^{-3} and electron temperature $T_e = 1.6 \pm 0.1$ eV [5, 6]. In the plasma flame, the electron temperature and density and the gas temperature decrease rapidly at a distance of 5 mm. This is why the copper substrate with dimensions 1.5 mm×1.5 mm is exposed to the plasma flame at 2 mm from the ceramic tube end. The substrate is tightly glued to the thermocouple and fixed to the movable holder. To measure accurately the substrate temperature by a thermocouple, one needs to compensate the error voltage by a reference (cold) junction. The electronic circuit must compensate the changes in the temperature at the reference junction of the thermocouple in order for the output voltage to be an accurate representation of the hot junction measurement. Converting the voltage generated by a thermocouple into an accurate temperature reading is not easy for many reasons: a small voltage signal, a nonlinear temperature-voltage relationship, an error due to the reference junction; also, the thermocouples may pose grounding problems. The small signal requires the use of a high-gain amplifier stage before the analog-to-digital conversion. To measure the temperature of a substrate for deposition of carbon nanostructures in the range of 400 – 700 °C, we used a K-type thermocouple (chromel-alumel). This type of thermocouple allows measurement from –200 to 1000°C with an almost invariable Seebeck coefficient – 41,84μV in this wide range of temperature measurement. An appropriate measurement circuit solution when using type a K thermocouple is the Analog Devices thermocouple amplifier with cold junction compensation – AD8495ARMZ, which converts the temperature of the thermocouple to a signal voltage with a transfer coefficient of 5mV/°C. This voltage is measured by an Arduino-microcontroller board and fed to a PC; the substrate temperature is thus monitored and displayed. AD8495ARMZ is a fixed gain instrumentation amplifier with thermocouple cold junction compensation on an integrated circuit. The circuit is laser trimmed to match the characteristics of K type (chromel-alumel) thermocouples and amplifies the small thermocouple voltage to provide a 5 mV/°C output. The high common-mode rejection of the amplifier blocks common-mode noise that the long thermocouple leads can pick up.

3. Results for the substrate temperature

The temperature of the substrate is measured in continuous and pulsed modes of operation of the discharge at a gas flow-rate of 250 sccm in a gas mixture $Ar/H_2/CH_4$. The result for the substrate temperature dependence on the input microwave power for plasma production is presented in figure 2. These temperatures can be considered as steady-state values, as they are measured after an initial rapid increase with time.

The temperature obtained is in the range 500 – 900 °C, which is necessary for deposition carbon nanostructures in gas-mixtures of argon, hydrogen and methane. In the continuous mode, the heating of the plasma source in the chamber is significant, so that it can only operate for a limited period of time without cooling. In a pulsed mode of operation, the source can function for a long period of time without cooling. The temperature dependence on the repetition rate of microwave pulses with an amplitude $P = 15$ W and a pulse duration $\tau = 0.435$ ms is presented in figure 3. These temperature values are not sufficiently high for effective deposition of carbon nanostructures. The plasma system operation is very stable at a comparatively low pulse repetition rates; the dependence of the substrate temperature on the pulse duration at 100 Hz was studied (figure 4) at an input power $P = 20$ W. This mode allows one to fix precisely the substrate temperature for a long

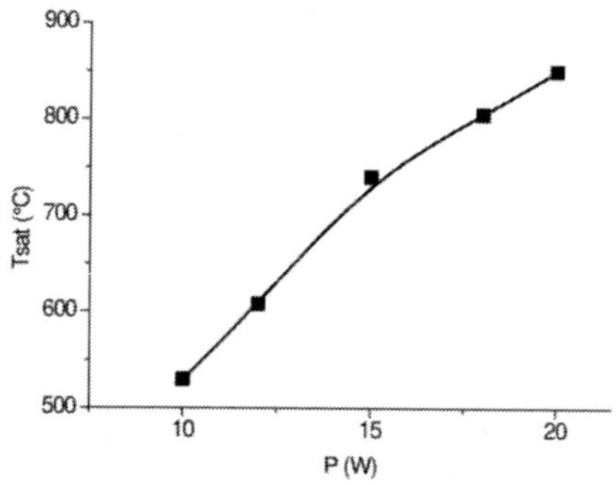

Figure 2. Dependence of the substrate temperature on the input power in continuous mode.

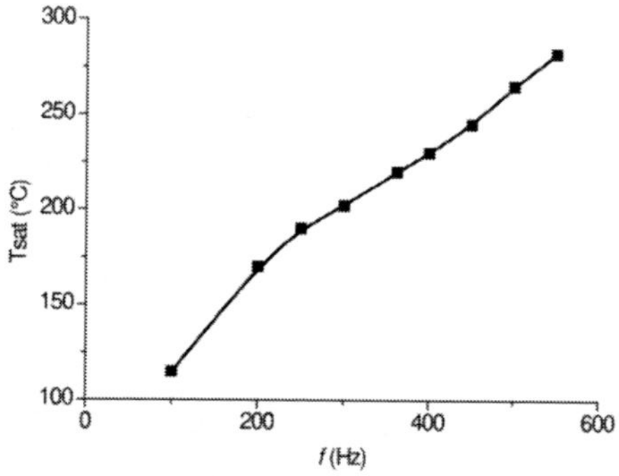

Figure 3. Dependence of the substrate temperature on the microwave pulse repetition rate.

period of time by using a temperature stabilized deposition system.

4. Conclusions

A system with a miniature microwave plasma source is designed for deposition of carbon nanostructures at atmospheric pressure. An electronic system is developed for precise measurement of the substrate temperature. The results show that, at a fixed distance of the substrate to the discharge tube, the substrate temperature can be controlled precisely in a pulsed mode of the discharge by varying the pulse amplitude and duration. The distance between the substrate and the plasma flame is optimized in order to obtain temperatures in the range 400 – 800 °C.

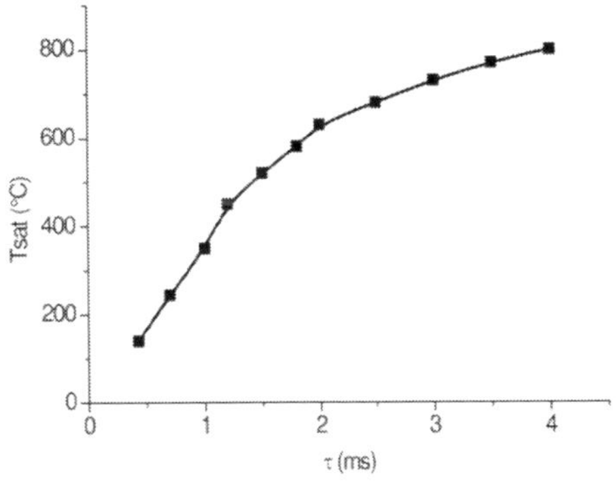

Figure 4. Dependence of the substrate temperature on the microwave pulse duration at $f = 100$ Hz.

Acknowledgements

This work was financially supported by thee Science Fund of University of Sofia, project N84/2013.

References

[1] Ostrikov K, Cvelbarand U and Murphy A 2011 *J. Phys. D: Appl. Phys.* **44** 174001
[2] Kiss'ovski Zh, Kolev M, Ivanov A, Lishev St and Koleva I 2009 *J. Phys. D: Appl. Phys.* **42** 182
[3] Ohashi T, Kato R, Takumi Ochiai T, Tokun T and Kawarada H 2012 *Diamond & Related Materials* **24** 184–7
[4] Malesevic A,Vitchev R, Schouteden K,Volodin A, Zhang L,van Tendeloo G, Vanhulsel A and van Haesendonck C 2008 *Nanotechnol.* **19** 305604
[5] Kiss'ovski Zh, Ivanov A, Iordanova S and Koleva I 2011 *J Phys. D: Appl. Phys.* **44** 205203
[6] Kiss'ovski Zh., Ivanov A and Kolev St 2012 *Proc. ESCAMPIG XXI* (July 2012 Viana do Castelo Portugal) Topic 5 pp 10-4

Study of sol-gel Cu-doped Al_2O_3 thin films: structural and optical properties

T Ivanova[1,4], A Harizanova[1], T Koutzarova[2] and B Vertruyen[3]

[1]Central Laboratory of Solar Energy and New Energy Sources,
Bulgarian Academy of Sciences, 72 Tsarigradsko Chaussee, 1784 Sofia, Bulgaria
[2]Acad. E. Djakov Institute of Electronics,
Bulgarian Academy of Sciences, 72 Tsarigradsko Chaussee, 1784 Sofia, Bulgaria
[3]LCIS/SUPRATECS, Institute of Chemistry B6, University of Liege,
Sart-Tilman, B-4000 Liege, Belgium

E-mail: tativan@phys.bas.bg

Abstract. We present a study of of copper-doped Al_2O_3 films prepared by sol-gel deposition. The films were spin-coated on Si and quartz substrates and a high-temperature treatment was conducted in the range 500 – 800 °C in oxygen or nitrogen. The impact was followed of the annealing procedures on the $Al_xCu_{1-x}O_3$ films's properties. XRD was used to determine the films' structure; it revealed a mixture of amorphous and crystalline phases. Optical characterization was performed by UV-VIS spectroscopy. The sol-gel films prepared are very transparent. The band gaps of the Al-Cu-O films were estimated from the optical data, with the values ranging from 3.4 to 4.8 eV depending on the Cu content, the annealing and the gas ambients.

1. Introduction

Aluminum oxide is a technologically important material due to its high transparency in a wide spectral range (from UV to mid-IR), and its stability with good electrical and mechanical properties. Al_2O_3 can be used as dielectric and sensing layers and optical films [1]. Various physical and chemical procedures have been employed in the synthesis of mixed metal oxide films, including pulsed laser deposition, sputtering, e-beam evaporation. [2, 3]. The sol-gel method proposed for preparation of $Al_xCu_{1-x}O_3$ films is a cost-effective technology, with advantages such as control of the composition and homogeneity, and a possibility of coating substrates of complex shapes [4].

In this paper we report a study of Cu–Al–O thin films prepared by a sol-gel technology from sols with different concentration of the Cu precursor followed by an annealing treatment. The XRD analysis revealed the structure of and the phases in the films studied. The vibrational properties were investigated by FTIR spectroscopy, while the optical transmittance and reflectance were measured by UV-VIS spectrophotometry. The dependence of the optical band gaps on the annealing temperature was also determined.

[4] To whom any correspondence should be addressed.

Journal of Physics: Conference Series **514** (2014) 012008 doi:10.1088/1742-6596/514/1/012008

2. Experimental

The sol solution for Al_2O_3 deposition was synthesized using alumina-sec-butoxide and isopropyl alcohol. Acetylacetone was used as a chelating agent. The resulting solution 0.4M $Al(C_4H_9O)_3$ was aged by stirring for 1h at 50 °C. The Cu sol was prepared by dissolving Cu acetate in ethanol (0,2 M concentration) with monoethanolamine as a complexing agent. The two solutions were mixed as described in table 1; the films prepared from the corresponding sols are referred to in the text by the sol name. The mixed sols were stirred by a magnetic stirrer at 50°C/2 hours.

Table 1. Preparation of mixed sols for deposition of $Al_xCu_{1-x}O_3$ films.

Sol name	**Al**	**AC0,2**	**AC0,4**	**AC0,6**	**AC5**	**AC10**
Al sol (ml)	20	19,8	19,6	19,4	15	10
Cu sol (ml)	-	0,2	0,4	0,6	5	10

The layers were deposited by spinning at 2000 rpm on Si and quartz substrates. The preheating treatment was conducted at 300°C/30 min; the annealing was carried out at temperatures from 500 to 800°C for 1 hour in O_2 or N_2 ambients. The XRD spectra of the films were recorded by a Bruker D8XRD diffractometer at a grazing angle of 2° and a step of 0.1° for 8 s per step. The optical measurements were performed by a UV-3600 Shimadzu spectrophotometer in the range 185 – 900 nm.

3. Results and discussion

The phase composition of the samples was identified using the XRD patterns shown in figure 1. The undoped Al_2O_3 films and the $Al_xCu_{1-x}O_3$ films with low Cu content manifested featureless spectra, an

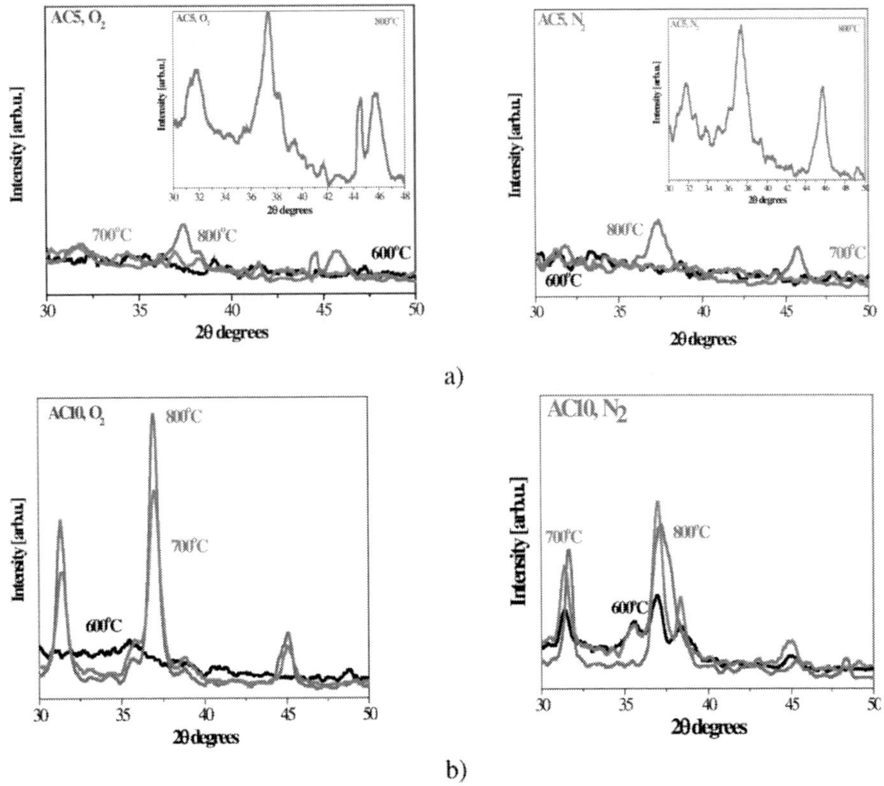

Figure 1. XRD spectra of $Al_xCu_{1-x}O_3$ films annealed at 600, 700 and 800°C in O_2 and N_2 ambient; a) films obtained from sol AC5 and b) from AC10.

indication of an amorphous structure, independent of the annealing temperature and ambient. The XRD analysis of $Al_xCu_{1-x}O_3$ films obtained from sols AC3 and AC4 after annealing at 500 °C in O_2 and N_2 revealed spectra without any XRD peaks, a sign of an amorphous structure. Broad XRD lines began to appear as the annealing temperature was raised, revealing some amorphous fraction still present even after a 800 °C treatment. A considerable difference was seen for the films obtained from sols AC5 and AC10. The AC5 films remained amorphous up to 700 °C regardless of the gas ambient (figure 1a). The inset shows the enlarged XRD lines of AC5 films treated at 800°C. The two lines (31.9° and 37.3°) are due to reflections (006) and (101) of $CuAlO_2$ (JCPDS 00-009-0185), the line (46°) is assigned to γ-Al_2O_3 [5]. The peak at 44.5°, appearing only for a film annealed in oxygen, is related to $CuAl_2O_4$ (JCPDS 01-078-0556). $CuAlO_2$ is the predominant phase in $Al_xCu_{1-x}O_3$ films (sol AC5) with no CuO or Cu_2O phases.

The AC10 films exhibit a different behavior (figure 1b), as clear XRD lines for films annealed at 600 °C in N_2 are seen; for the films treated in O_2, crystallization starts after annealing at 700 °C. Three XRD lines (31.9°, 35.6°, 37.3°) are due to $CuAlO_2$ (JCPDS 00-009-0185; JCPDS 04-007-5000), and a weak line, to $CuAl_2O_4$. A feature at 38.3° corresponds to tenorite CuO (JCPDS 00-003-0867). Rise of the temperature to 700°C leads to the same XRD lines, but with a greater intensity. Annealing at 800 °C in O_2 results in a well-crystallized AC10 film with no $CuAl_2O_4$ phase. The AC10 film treated in N_2 shows a lower intensity of the lines, together with additional peaks related to the (104) and (009) orientations of $CuAlO_2$ and a line at 48.3° assigned to cuprite Cu_2O (JCPDS 00-005-0067). The XRD study further reveals that $CuAlO_2$ is the phase found predominantly in the AC5 and AC10 films. Similar XRD results for Al-Cu-O films were reported in [4]. It must be noted that the $CuAlO_2$ films are studied due to their low electrical resistivity, high optical transparency in the visible region, the existence of a photovoltaic effect and a field emission effect; it is thus known as a multi-functional p-type semiconductor [3].

Table 2 presents the crystallite sizes as determined by the Debay-Scherrer equation from the XRD line of $CuAlO_2$. As is seen, for AC5 films the peak is relatively weak and the crystallites are very small. In the AC10 films, the crystallites become bigger as the annealing temperatures is raised; the crystallites are larger for samples annealed in N_2 compared to those annealed in O_2.

The sol-gel $Al_xCu_{1-x}O_3$ films were compared to pure Al_2O_3 by FTIR spectroscopy. The absorption bands around 3400 – 3700 cm^{-1} are related to O-H stretching vibrations; their intensity decrease with the annealing. The line at 2350 cm^{-1} is due to CO_2 from atmospheric contamination, while the weak band at 1585 cm^{-1} is assigned to the COO stretching vibration. The FTIR analysis was focused on the spectral range 350 – 1000 cm^{-1}, where the characteristic metal-oxygen vibrations can be detected. The detailed FTIR studies of the films from the sols AC0,2-0,6 (small Cu concentrations) reveal that the Cu additive affects the shapes and the intensity of the IR lines without the appearance of new absorption bands connected to Cu-O bonds. Thus, the FTIR spectra of sol-gel films with a higher copper content (sols AC5 and AC10) are given in figure 2. The spectra of the preheated samples show a difference in the absorption bands shapes and intensity for all lines attributed to Al-O bonds.

Table 2. Crystallite size as estimated from XRD data ($CuAlO_2$).

Material	Annealing	2θ, (006)	Crystallite size, [nm]
AC5	600°C, N_2	31.73	18.1
	700°C, O_2	31.37	18.4
	700°C, N_2	31.37	25.9
	800°C, O_2	31.45	15.3
	800°C, N_2	31.60	38.8
AC10	800°C, O_2	31.90	5.8
	800°C, N_2	31.83	8.3

The annealing at 500 and 600°C in O_2 and N_2 (not shown here) led to the detection of new weak bands around 500 cm^{-1}, which can be due to Cu-O bonds. The high temperature treatments at 700 and 800 °C resulted in FTIR spectra with stronger differences from pure Al_2O_3. The absorption bands at 660 and 440 cm^{-1} in all spectra are due to Al-O stretching vibrations of AlO6 and to bending vibrations of AlO4. The main absorption peaks at 370 cm^{-1} become stronger for AC5 and AC10 films compared to the Al_2O_3 film. A clear absorption line at 490 cm^{-1} appears in the spectra of AC10 films treated at 700 and 800 °C and for AC5 films at 800 °C; this band is related to Cu-O bonds, an indication of a CuO phase. The presence of a small CuO fraction in these films was detected by XRD. Other interesting features are a weak line at 615 cm^{-1} and a clear band at 640 cm^{-1} for AC5 films and AC10 films annealed at 800 °C in O_2 and for AC10 in N_2. These bands are assigned to Cu(I)-O bonds in Cu_2O. The cuprite was not seen in the XRD spectra, but it might be in an amorphous state or below the XRD sensitivity. Similar FTIR results for $CuAlO_2$ films were reported in [6]. The FTIR bands are broad, indicating an amorphous and disordered structure, thus confirming the XRD results. The small Cu contents lead to negligible changes in the films' structure, similarly to the case of Al_2O_3 films. Higher amounts of the copper additive results in the formation of a new phase ($CuAlO_2$).

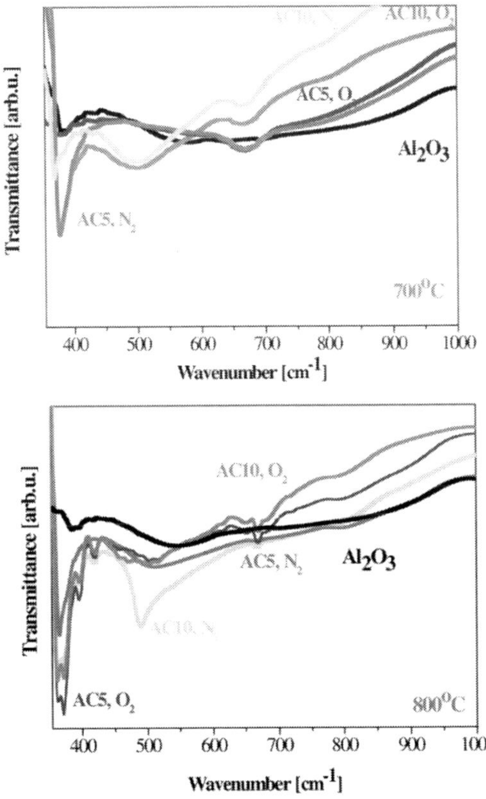

Figure 2. FTIR spectra of sol-gel Al_2O_3 and $Al_xCu_{1-x}O_3$ films annealed at different temperatures in oxygen and nitrogen ambient.

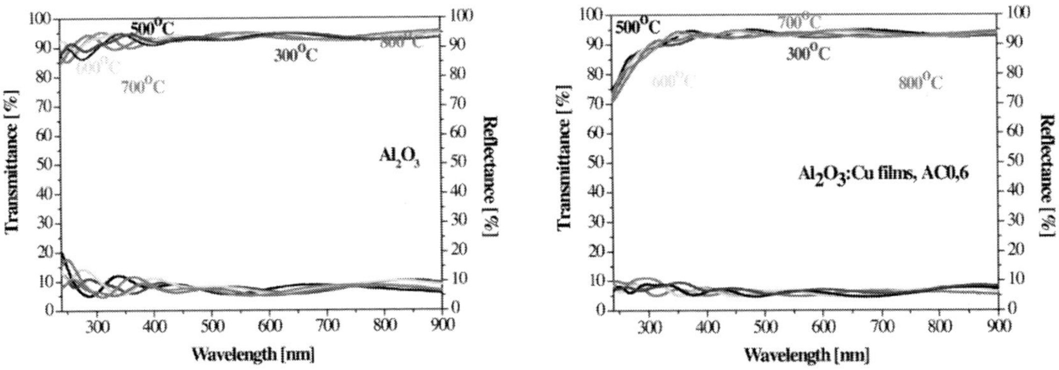

a) b)

Figure 3. UV-VIS transmittance and reflectance spectra of; a) Al_2O_3 films and b) $Al_xCu_{1-x}O_3$ films obtained from the sol AC0,6.

The optical transmittance and reflectance spectra were recorded in the 240 – 900 nm spectral range for sol-gel Al_2O_3 and $Al_xCu_{1-x}O_3$ films. The Al_2O_3 films exhibit a very high transmittance (~94 %) (figure 3a). The $Al_xCu_{1-x}O_3$ films with a smaller Cu content manifest an optical behavior very similar to that of Al_2O_3 with a slightly lower transmittance (figure 3b). For these films, no absorption edge is observed in the spectral range studied. These results are supported by the structural studies, where the characteristics of the films with a small Cu content differ slightly from those of Al_2O_3 films. Raising the Cu content results in quite different optical characteristics (figure 4). The change in the optical properties is explained by the existence of a new phase, namely $CuAlO_2$, as revealed by the XRD analysis. The AC5 films show a transparency around 87 % (at 550 nm) after an annealing up to 700 °C in both gas ambient. After a high temperature treatment at 800 °C, the transparency decreases and absorption bands appear at 450 nm (O_2 annealed films) and at 440 nm (N_2).

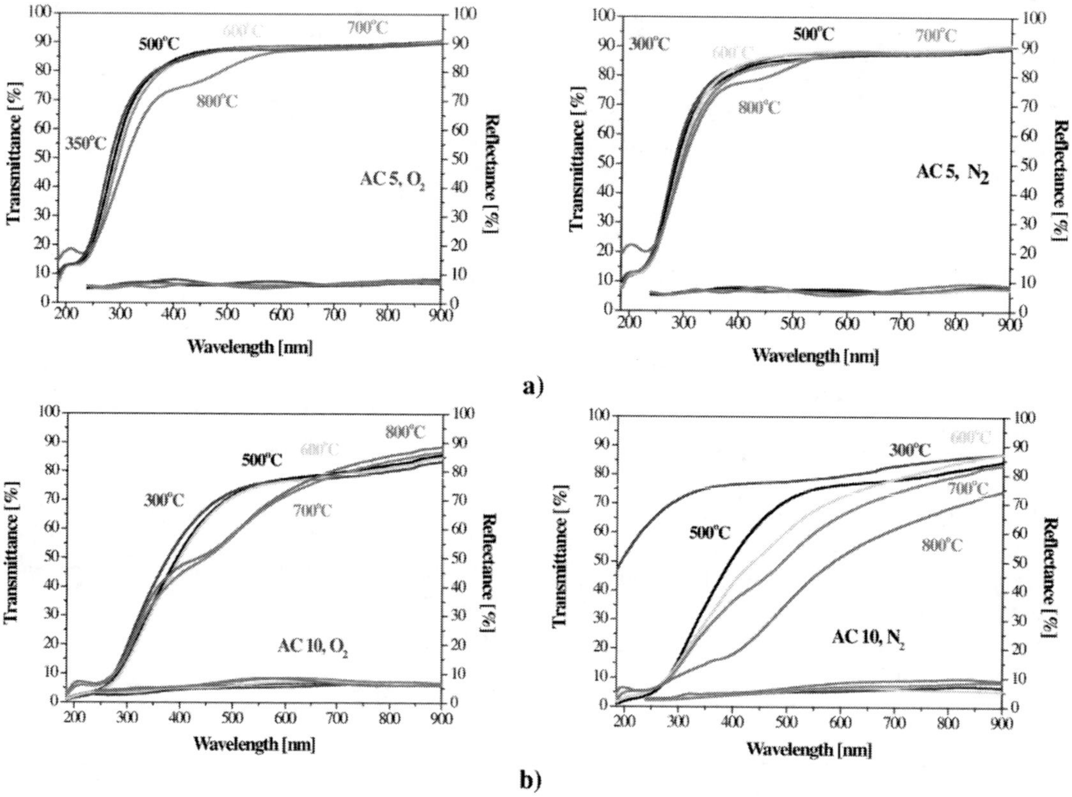

Figure 4. Transmittance and reflectance spectra of $Al_xCu_{1-x}O_3$ films; a) films obtained from the sol AC5 and b) AC10 films annealed at different temperatures.

The AC10 films display a lower transparency; different optical properties are obtained depending on the gas ambient. For the O_2 ambient, these films have a transmittance around 77 % (550 nm) after annealing in the 300 – 600 °C range. The films treated at 700 and 800 °C show absorption features at 490 nm. The AC10 films in N_2 manifest a clear tendency of transparency reduction from 79 % (300 °C, 550 nm) to 45 % (800 °C). The reflectance is between 9 – 10% in the entire spectral range studied. The absorption observed at 450 – 490 nm in the transmittance spectra of some $Al_xCu_{1-x}O_3$ films is attributed to the presence of Cu_2O nanoparticles [7]. The presence of the Cu_2O phase was not detected by the XRD measurements, but was found by the FTIR study.

Journal of Physics: Conference Series **514** (2014) 012008 doi:10.1088/1742-6596/514/1/012008

The optical band gaps estimated from the spectrophotometric data are given in figure 5. The optical band gap values for the preheated samples are 4,63 eV for the AC5 films and 4,06 eV for the AC10 films. It can be seen that in the AC5 films the optical band gap slightly decreases with the annealing temperature, with the films treated in nitrogen possessing higher values. The AC10 films are characterized by significantly lower values. The values reported for $CuAlO_2$ films are in the range 3,38 – 3,80 eV [8,9], some authors [3] quote the value of 4.7 eV. In our case, the films' structures are with mixed crystalline and amorphous fractions, the $CuAlO_2$ phase being predominant according to the XRD results. The broadening observed of the band gap of the AC5 films can be related to the smaller particle sizes (table 2), i.e., the quantum confinement effect of a size dependency of the band gap that is often found in semiconductors.

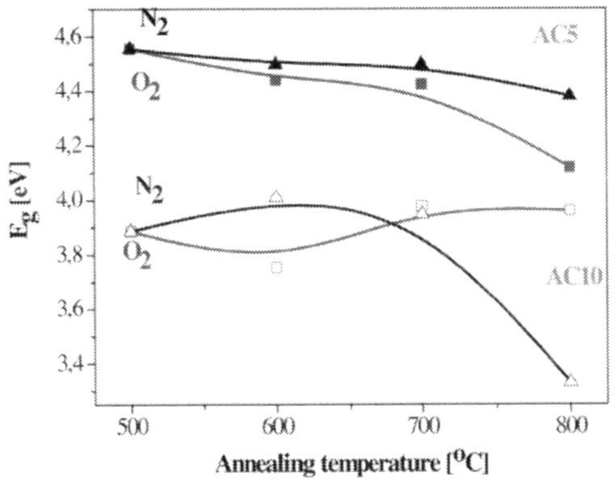

Figure 5. Optical band gaps of $Al_xCu_{1-x}O_3$ films.

4. Conclusions

The Al_2O_3 and Cu doped Al_2O_3 films were successfully prepared by a sol-gel technology and spin coating. The influence was studied of the amount of the copper additive. Increasing the copper incorporation leads to $Al_xCu_{1-x}O_3$ films with a predominant $CuAlO_2$ crystalline phase with small inclusions of CuO and A_2O_3. The optical properties depend on the copper amount, the annealing temperature and the gas ambient.

References

[1] Li X, Chen Q, Sang L, Yang L, Liu Z and Wang Z 2011 *Phys. Procedia* **18** 100

[2] Prasanna S, Mohan R, Jayakumar S, Kannan M and Ganesan V 2012 *Thin Solid Films* **520** 2689

[3] Zhang Y, Liu Zh, Zang D and Feng L. 2014 *Vacuum* **99** 160

[4] Banerjee A, Chattopadhyay K K and Joo S W 2011 *Physica* B: *Cond. Matter.* **406** 220

[5] Musil J, Blažek J, Zeman P, Prokšová Š, Šašek M and Čerstvý R 2010 *Appl. Surf. Sci.* **257** 1058

[6] Saha B, Thapa R and Chattopadhyay K K 2009 *Mater. Lett.* **63** 394

[7] Massera J, Choi J, Petit L, Richardson M, Obeng Y and Richardson K 2008 *Mater. Res. Bulletin* **43** 3130

[8] Lu Y M, He Y B, Yang B, Polity A, Volbers N, Neumann C, Hasselkamp D and Meyer B K 2006 *Phys. Status Solidi* C **8** 2895

[9] Zhang Y, Liu Z, Zang D Y, Che X S, Feng L P and Bai XX 2013 *J. Phys. Chem. Solids* **74** 1672

Journal of Physics: Conference Series **514** (2014) 012009 doi:10.1088/1742-6596/514/1/012009

Effect of different technological approaches on the optical properties of ZnO sol-gel thin films

T Ivanova[1,4], A Harizanova[1], T Koutzarova[2], B Vertruyen[3] and Z Nenova[1]

[1]Central Laboratory of Solar Energy and New Energy Sources, Bulgarian Academy of Sciences, 72 Tsarigradsko Chaussee, 1784 Sofia, Bulgaria
[2]Acad. E. Djakov Institute of Electronics, Bulgarian Academy of Sciences, 72 Tsarigradsko Chaussee, 1784 Sofia, Bulgaria
[3]LCIS/SUPRATECS, Institute of Chemistry B6, University of Liege, Sart-Tilman, B-4000 Liege, Belgium

E-mail: tativan@phys.bas.bg

Abstract. The work presents a sol-gel approach with two types of solvents for ZnO films deposition in view of varying the films' structural and optical properties. The ZnO films were characterized by X-Ray diffraction (XRD) and found to have a polycrystalline wurtzite phase structure. The grain sizes and the lattice parameters were determined; it was established that the ZnO crystallites' size increases with the annealing temperatures from 25 nm to 36 nm. The ZnO films possess high transmittance in the visible spectral range. Further, the optical band gap values were estimated. The material's vibration properties were analyzed by Fourier Transform Infrared (FTIR) spectroscopy. The two technological sol-gel approaches for deposition of ZnO films, using two different solvents, proved to be successful in producing structures with different structural and optical properties. The thin films fabricated were very smooth and uniform and exhibited a high transparency in the visible spectral range.

1. Introduction

ZnO has been the object of intensive scientific research due to its interesting properties, such as a wide band gap (3.37 eV at room temperature) with large exciton binding energy (60 meV), a high chemical stability, a high refractive index, a high thermal conductivity, as well as binding, antibacterial and UV protection properties [1]. Based on these attractive features, ZnO has found numerous applications as gas sensors, thin-film transistors, surface acoustic wave devices and solar cell windows [2, 3]. Another interesting application is as dilute magnetic semiconductors, which have been lately emerging as materials for spintronic applications [5]. Recently, ZnO films have been prepared by RF magnetron sputtering, chemical vapor deposition, electrochemical deposition, spray pyrolysis [6, 7] etc. The optical properties of ZnO films are of particular interest from both fundamental and technological viewpoints. The advantages of the sol-gel method include excellent control of the stoichiometry, possibilities to modify the composition, etc. Among the variety of factors affecting the quality of the ZnO sol-gel films is the sol concentration, the chelating agent, the pre-heating temperatures, the subsequent annealing treatments, the aging and a dip or spin-coating deposition [8].

[4] To whom any correspondence should be addressed.

This work describes two sol solutions for spin coating of ZnO films in view of modifying their optical properties. The effect of using two preheating temperatures was also studied. The films' structural and optical properties were investigated by XRD, FTIR and UV-VIS spectrophotometry.

2. Experimental

Two types of sol solutions were developed using zinc acetate dehydrate as a zinc precursor with a concentration of 0,4 M.

Solution 1) The zinc acetate was dissolved in isopropanol. The complexing agent and stabilizer used was trietanolamine (TEA). The molar ratio TEA/Zn was fixed at 1.

Solution 2) the solvent used was absolute ethanol and the corresponding complexing agent was monoethanolamine (MEA). The molar ratio MEA/Zn was fixed at 1.

The films were deposited by spin coating at 4000 rpm on quartz and Si substrates. The preheating temperatures used (the firing between layers) were 250 °C/10 minutes and 300 °C/10 minutes. The films were annealed at temperatures of 500 °C, 600 °C, 700 °C and 800 °C for one hour.

The XRD spectra of the sol-gel films were recorded by means of a Bruker D8XRD diffractometer at a grazing of angle 2° and a step of 0.1° for 8 s per step. The FTIR measurements were performed in the spectral region 350 – 4000 cm^{-1} on an IR Prestige-21 Shimadzu FTIR Spectrophotometer. The optical measurements were conducted by a Shimadzu 3600 UV-VIS-NIR spectrophotometer.

3. Results and discussions

The film thickness of the ZnO sol-gel films prepared did not differ significantly, which allowed us to compare their structural and optical properties. Depending on the annealing treatment, the ZnO films of solution 1 had thicknesses of 150 – 140 nm in the case of a preheating temperature of 250 °C; the corresponding values for a preheating at 300 °C were in the range 140 – 160 nm. The film thicknesses for the films of solution 2 were 155 – 160 nm (for a preheating at 250 °C) and 160 nm (300 °C).

Figure 1 presents the XRD spectra of the ZnO films of solution 1 treated at the two preheating temperatures and subsequently annealed at high temperatures. The XRD data shows that the films have crystallized in wurtzite ZnO, as the main five reflections correspond to this crystal phase (JCPDS 01-07-8070). The films formed at the lower preheating temperature (250 °C) possess weaker XRD lines compared to the films preheated at 300 °C for all additional annealings from 500 °C to 800 °C. Thus, the preheating temperature influences the crystallization behavior, so that increasing the preheating temperature results in ZnO films of higher crystallinity.

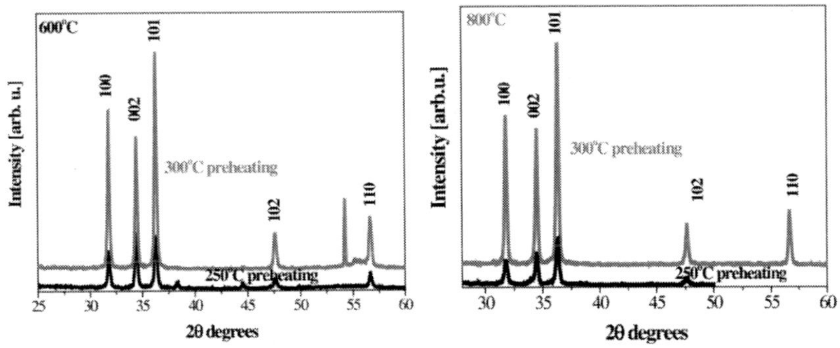

Figure 1. XRD spectra of ZnO sol-gel films prepared from solution 1 as a function of the preheating temperatures (annealing at 600 °C and 800 °C).

The structure of the films deposited from solution 2 was studied for a preheating temperature of 300 °C. Figure 2 shows XRD spectra of films prepared from the two solutions. The degree of crystallization of the films deposited from solution 1 is higher than that of the corresponding samples deposited from solution 2. The films are well crystallized without a preferential orientation.

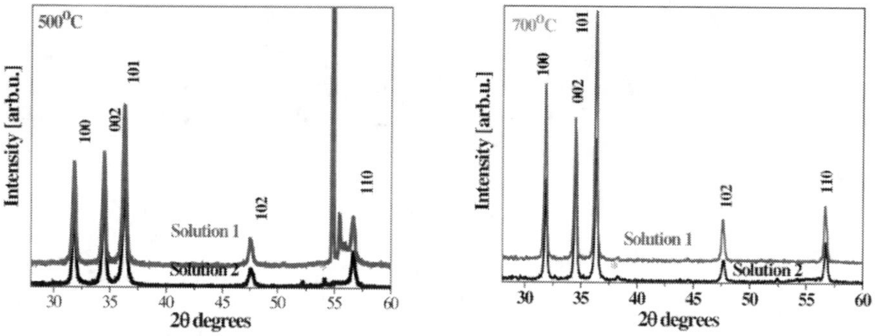

Figure 2. XRD spectra of ZnO films deposited from the two solutions and preheated at 300°C. The asterisk marks cubic ZnO_2.

The average crystallite sizes were estimated from the XRD patterns according to Scherrer's formula; the results are shown in figure 3. In the case of ZnO films of solution 1, the crystallites of the samples preheated at 250 °C are smaller than those of the films preheated at 300 °C. Another interesting feature is that the crystallites' size decreases with the annealing temperatures for the films treated at the lower preheating temperature, while the crystallites grow larger in the ZnO films preheated at 300 °C for for both solutions. When comparing the films of the two solutions, one can see that, for the first set of samples, the crystallites are slightly larger, with the highest value found for the annealing temperature of 700 °C. For the films of solution 2, the rise of the crystallites' size is almost linear when the annealing temperature is raised.

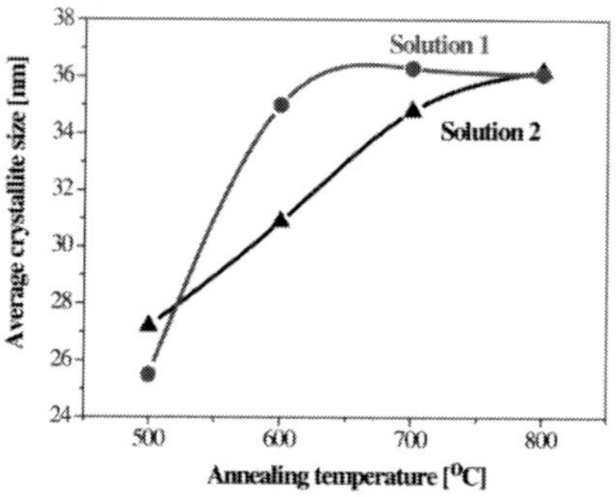

Figure 3. Average crystallites' sizes of the sol-gel ZnO (solution 1 – blue line) and ZnO (solution 2 – black) films, as a function of the annealing temperatures.

Thus, the XRD investigation reveals that both the solvent and the MEA used influence the crystallization behavior of the ZnO films. A difference of 50 °C in the preheating temperature affects the size of the crystallites and the degree of crystallinity.

The vibrational properties of the ZnO sol-gel films were investigated as a function of the preheating and annealing temperatures. Figure 4 shows FTIR spectra of sol-gel films annealed at 500 °C and 800 °C; the effect of the other annealing temperatures was also analyzed. The FTIR spectra indicate that the solution type and the preheating temperatures (not given here) affect the shapes and

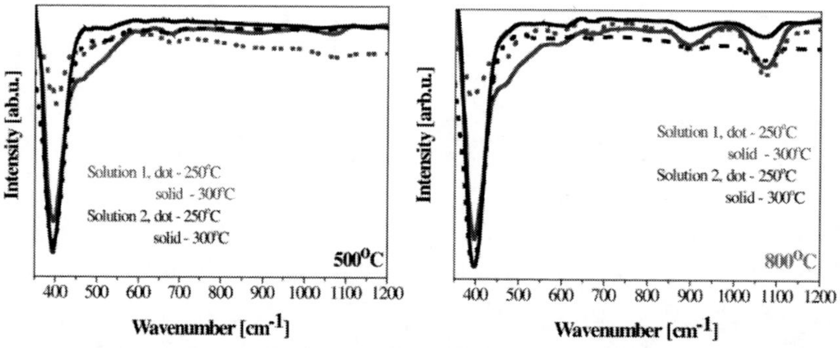

Figure 4. FTIR spectra of sol-gel ZnO films, annealed at 500 and 800°C.

positions of the bands. Intensive absorption bands at 3300 – 3500 cm^{-1} for the preheated samples due to the stretching vibrations of OH groups [9] can be seen in the spectra of the films of solution 1, while the corresponding bands for the films of solution 2 are weaker. The films of solution 1 (250 °C) exhibit a broad band of a lower intensity with two clear maxima at 463.0 cm^{-1} and at 392.1 cm^{-1}. Increasing the preheating temperature to 300 °C results in a stronger broad band centered at 433.2 cm^{-1} with two additional peaks at 406.3 cm^{-1} and 386.4 cm^{-1}. The films from solution 2 display stronger absorption bands at 404 cm^{-1} (preheating 250 °C) and 390 cm^{-1} (300 °C). These absorption bands are typical for ZnO [10]. The high temperature treatments lead to a similar tendency in the spectra. The main absorption band for the samples of solution 1 appeares at 402 cm^{-1} with two distinctive peaks at 464 $cm^{-1}$and 383 cm^{-1} (preheating 250°C) and, respectively, one strong line at 394.5 cm^{-1} (preheating 300 °C). In contrast, the films of solution 2 show a strong absorption line at 394 cm^{-1} with a greater intensity for the film preheated at the higher temperature. In all films studied, raising the temperature above 600 °C results in a main band at 396 cm^{-1} with slight differences in the position and and intensity. In summary, the FTIR analysis reveals that the ZnO films possess different vibrational properties depending on the solution used and the heat treatment.

The optical properties were studied for ZnO films deposited on quartz substrates under similar technological conditions. The differences in the films' properties are presented in figure 5. For the films of solution 1, preheating at 250 °C results in more transparent coatings compared with those preheated at 300 °C; it is independent of the heat treatment. For the ZnO films of solution 2, the tendency is reverse, namely, the films preheated at 300 °C are always more transparent than those preheated at 250 °C. The high-temperature annealing (500 – 800 °C) lowers the films' transmittance.

The specific bands observed in the optical spectra below the band gap wavelength are due to the excitonic absorption of ZnO (see table 1). The excitonic feature of bulk ZnO is located near 373 nm. The excitonic absorption of ZnO films from solution 2 for the two preheating temperatures are weaker compared to the other ZnO samples (solution 1).

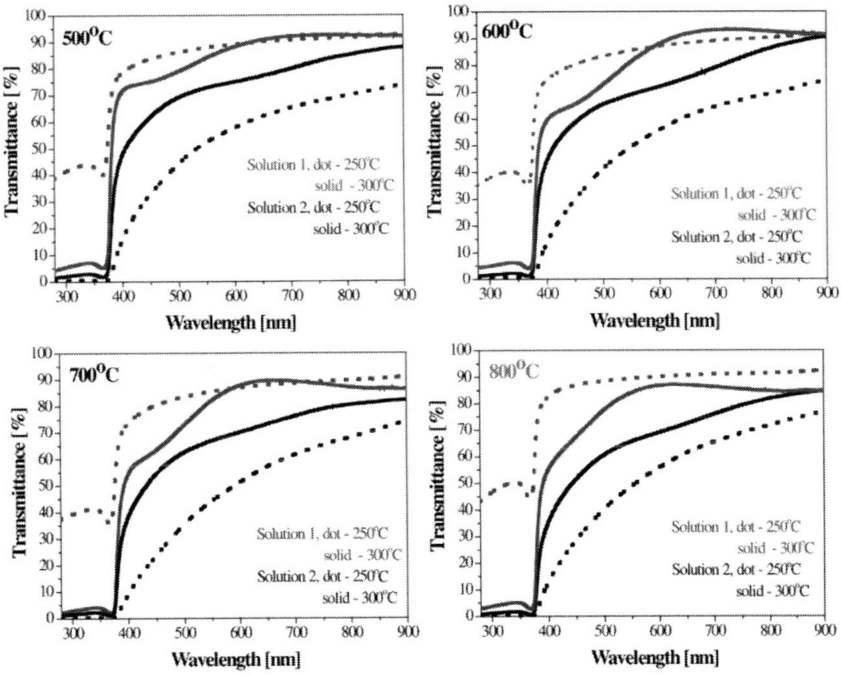

Figure 5. Transmittance spectra of ZnO films depending on the preheating and annealing temperatures.

Table 1. Excitonic absorption and optical band gap energy, estimated from optical data.

Preheating	Annealing	excitonic absorption [nm]		E_g [eV]	
		Solution 1	Solution 2	Solution 1	Solution 2
250 °C	-	301	-	3,52	3,28
	500 °C	335	347	3,33	3,27
	600 °C	339	347	3.31	3,25
	700 °C	341	347	3,31	3,25
	800 °C	344	347	3.30	3,27
300 °C	-	-	331	3,36	3,30
	500 °C	344	337	3,31	3,28
	600 °C	345	337	3,29	3,27
	700 °C	347	343	3,29	3,28
	800 °C	347	347	3,29	3,28

On the other hand, the excitonic absorption in the case of the preheating temperature of 300 °C is better expressed. The values of the optical band gap are close to the reported value (3,37 eV) for bulk ZnO [11] and to those cited in the literature for ZnO films [12]. The films prepared from solution 1 at the preheating temperature of 250 °C have optical band gaps decreasing from 3,33 eV to 3,30 eV depending on the annealing, while the films of the same solution preheated at 300 °C exhibit E_g = 3.29 eV regardless of the annealing temperatures with the exception of 500 °C. No clear tendency is seen for the ZnO films of solution 2 with the values ranging from 3,25 eV to 3,28 eV.

4. Conclusions

Nanosized ZnO thin films were prepared by the sol-gel technique using two types of sol solutions. The sols prepared remained stable for more than three months. The ZnO films had the polycrystalline structure of hexagonal wurtzite. The ZnO grain size varied from 25 nm to 37 nm depending on the solution type and the heat treatments. The transmittance in the visible range was 77 – 90 %, while the optical band gaps ranged from 3.25 eV to 3.52 eV. The sol-gel technology was thus proved to be promising for deposition of ZnO films intended for applications in different optoelectronic devices or in solar cells.

References

[1] Goyal A and Kachhwaha S 2012 *Mater. Let.s* **68** 354–6
[2] Djurisic A B, Ng A M C and Chen X Y 2010 *Progress in Quantum Electr.* **34** 191–259
[3] Lee J H, Ko K H and Park B O 2003 *J. Crystal Growth* **247**/1-2 119–25
[4] Riaz S, Naseem S and Xu Y B 2011 *J Sol-Gel Sci. Technol.* **59** 584–90
[5] Hongsingthong A, Yunaz I and Konagai M 2011 *Solar Energy Mater. Solar Cells* **95** 171–4
[6] Vimalkumar T, Poornima N, Kartha C and Vijayakumar K P 2011 *Appl. Surf. Sci.* **257** 8334–40
[7] Houng B, Huang Ch-L and Tsai S-Y 2007 *J. Crystal Growth* **307** 328-33
[8] Caglar Y, Illican S, Yakuphanoglu F, Gao K, Lu P and Xue D 2009 *J. Alloys Comp.* **448** 885-8
[9] Ning G, Zhao X P, Li Y and Zhang Ch Q 2006 *Optical Mater.* **28**/4 385-90
[10] Li C P, Jang, B H, Wang X C, Li M J, Su L and Li X W 2011 *Appl. Surf. Sci.* **257** 5998-6003
[11] Prajapati C S and Sahay P P 2012 *Appl. Surf. Sci.* **258** 2823-8
[12] Chattopadhyay S, Neogi S K, Pandit P, Dutta S, Rakshit T, Jana D, Chattopadhyay S, Sarkar A and Ray S K 2012 *J. Luminescence* **132** 6-11

Journal of Physics: Conference Series **514** (2014) 012010 doi:10.1088/1742-6596/514/1/012010

Low-temperature deposition of ultrathin SiO_2 films on Si substrates

P Vitanov[1], A Harizanova, T Ivanova and H Dikov

Central Laboratory of Solar Energy and New Energy Sources,
Bulgarian Academy of Sciences, 72 Tsarigradsko Chaussee, 1784 Sofia, Bulgaria

E mail: vitanov@phys.bas.bg

Abstract. We present a detailed investigation of the properties of silicon dioxide deposited at a low temperature. The advantages of this process include its low thermal requirements (about 200 °C), the absence of corrosive by-products and the lack of need of vacuum equipment. Sol solutions were prepared for the deposition of ultrathin SiO_2 films by spin-coating at the low annealing temperature of 200 °C. The layers' thickness was 24 nm and 5 nm. We describe in detail the material properties of this novel low-temperature SiO_2 layers obtained by extensive characterization using Fourier transform infrared spectroscopy (FTIR), atomic force microscopy (AFM), XPS spectroscopy, capacitance-voltage (*C-V*) and current-voltage (*I-V*) measurements. The ultrathin oxide layers on Si substrates show good dielectric properties.

1. Introduction

Silicon oxide is widely used for optical and electronic applications. SiO_2 of the highest quality can be formed by thermal oxidation of Si at temperatures over 800 °C in dry O_2. Thermal oxide is only grown on Si substrates at high temperatures, which limits its applications, as many of those require SiO_2 film deposition on different substrates [1]. SiO_2 nanolayers have been produced by plasma oxidation, atomic layer deposition, deposition from sol-solutions and oxidation by wet-chemical methods.

In this work we propose a technological approach consisting of liquid chemical deposition from a sol solution that has the following advantages over other deposition techniques: it is cost effective, no vacuum equipment is needed, control of the film' coverage and thickness is possible. The preparation procedure is also presented of the sol solutions for ultrathin SiO_2 layers of different thickness at 200 °C. Two sets of samples were studied, namely, SiO_2 films with thickness of 24 nm and 5 nm. The film morphology was studied by AFM. the FTIR investigation confirmed the formation of SiO_2. The chemical analysis was performed by XPS. The dielectric properties of the MOS structures were studied by capacitance-voltage(*C-V*) and current-voltage (*I-V*) measurements.

2. Experimental

Thev SiO_2 layers were prepared by the sol-gel technique using tetraethyl orthosilicate (TEOS) as a precursor. The introduction of glacial acetic acid at the molar ratio TEOS:acetic acid 1:15 caused acetate modification, resulting in an exothermic reaction. Further, the solution was modified by a small amount of water. Acetylacetone acts as a stabilizer, so that it was added in the molar ratio

[1] To whom any correspondence should be addressed.

TEOS:acetylacetone 1:1. Two solutions were prepared with different molar content of TEOS in order to deposit SiO_2 layers of different thickness. The films were obtained by spin coating at 8000 rpm for 30 s on Si substrates, the latter being cleaned beforehand by the RCA procedure, which is a standard one for wafer cleaning involving removal of the organic contaminants, of the oxide layer and of the ionic contamination (for details see [2]). The samples were prepared on Si wafers: (n- and p-type Si) with different surfaces: one side polished and one side etched (p/e) and both sides polished (p/p). The annealing procedure included heating at 200 °C for 30 min.

The AFM studies were conducted on a scanning probe microscope DiMultimode V (Veeco). The FTIR measurements were performed by an IRPrestige-21 Shimadzu FTIR Spectrophotometer. The layers were deposited on Si substrates, with bare Si wafer used as a background. The electrical properties of the layers were studied using MOS structures. The XPS studies were performed by a VG Escalab II system using AlK_α radiation with an energy of 1486.6 eV. The chamber pressure was 10^{-7} Pa. The binding energies (BE) were determined utilizing the C1s line (from adventitious carbon) as a reference with an energy of 285.0 eV. The accuracy of the BE measured was ± 0.2 eV. The SiO_2 films thickness was determined by ellipsometry, thus preparing two sets of films film with thicknesses of 24 and 5 nm.

3. Results and discussions

AFM images of ultrathin SiO_2 films (5 nm thick) are shown in figures 1 and 2. The R_a is the arithmetic mean of the absolute values of the height of the surface profile $Z(x)$, where $Z(x)$ is a function describing the surface profile analyzed in terms of the height (Z) and position (x) of the sample over the evaluation length L. The root-mean-square roughness R_q of a surface is similar to the roughness average, the only difference being the mean squared absolute values of the surface roughness profile. The values obtained for the surface roughness were $R_a = 0.74$ nm and $R_q = 0.94$ nm for $L = 500$ nm and, respectively, $R_a = 0.85$ nm, $R_q = 1.07$ nm for $L = 1000$ nm.

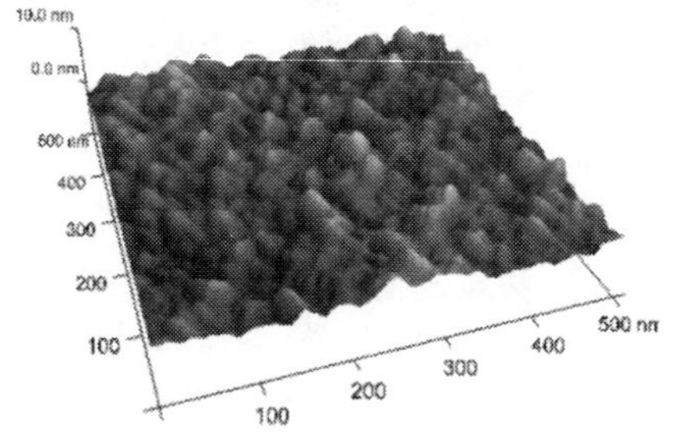

Figure 1. 3-D image of a sample with a 5 nm SiO_2 layer.

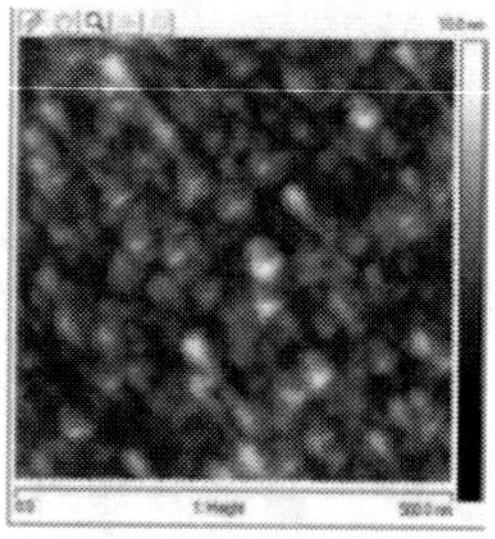

Figure 2. AFM image of a 5-nm-thick SiO_2 layer.

For a Gaussian distribution of asperity height, the statistical theory yields that the ratio of R_q to R_a should be 1.25. Some authors note that the asperity height distribution of most engineering surfaces (tribology) may be approximated by a Gaussian distribution with R_q/R_a values of up to 1.31. For our sample, the values of R_q/R_a using data collected from AFM imaging were 1.26 and 1.27, reasonably close to the value of 1.25 predicted by the theory. This result is significant since it indicates that, at the imaging scale, the asperity height distribution of these surfaces are approximately Gaussian and that the statistical relationships for the surface roughness are applicable. The height of the surface profile can be observed from the section analysis. The spectrum of the profile in one section is illustrated in figure 3. The surface smoothness of the layer is assessed by the parameters R_p, R_v and RT. The maximum profile peak height (R_p) denotes the highest peak around the surface profile with respect to the baseline. Likewise, the maximum profile valley depth (R_v) is the measure of the deepest valley across the surface profile analyzed from the baseline. Thus, the maximum height of the profile (RT) is defined as the vertical distance between the deepest valley and highest peak: $RT = R_p + R_v$. For the samples studied these parameters were as follows: $R_p = 2.5$ nm, $R_v = 1.2$ nm, and $RT = 3.7$ nm.

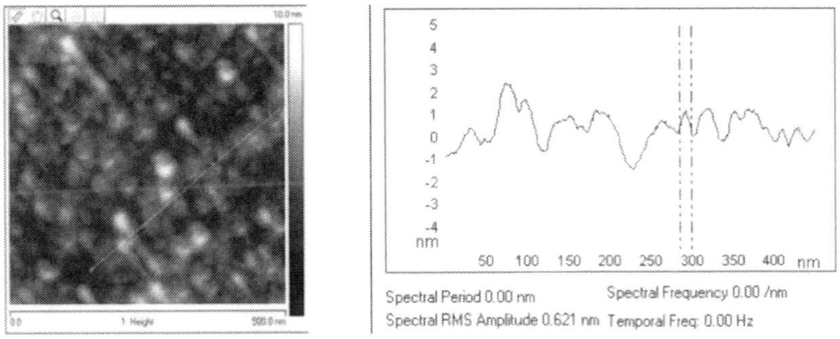

Figure 3. Spectrum of the surface profile.

The results of the AFM measurements demonstrate that the technique used results in homogeneous layers with good a surface coverage of the silicon substrate. The surface layer roughness was in the order of several nanometers and could only be observed by AFM.

XPS spectroscopy was used for determining if SiO_2 was formed (figure 4). The presence of carbon (C1s peak at 285.0 eV) was only registered on the film surfaces; it disappeared after a short-term sputtering (1 minute).

The binding energy of O1s is 532.8 eV, which corresponds to oxide. The Si2p signal is split in two peaks located at 99.31 eV and 103.5 eV. The first peak at 99.31 eV corresponds to a Si-Si bond; it originates from the Si substrate as the sol-gel film is very thin (5 nm). The second peak at 103.5 eV is Si2p3/2, which is an indication of SiO_2. Thus, the XPS analysis proved that we were able to deposit very thin SiO_2 films at a relatively low temperature.

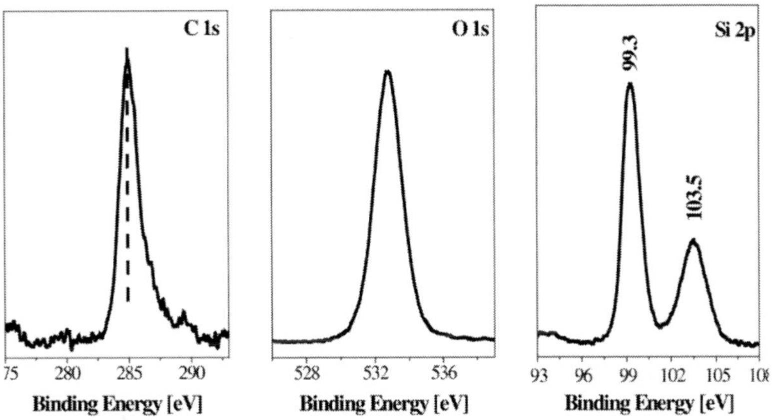

Figure 4. XPS spectra of C1s, O1s and Si2p of a sol-gel SiO_2 film.

FTIR spectra of the films are shown on figure 5. The SiO_2 films annealed at 200 °C and 300 °C manifested an absorption band in the range of 3400 – 3800 cm^{-1} assigned to the OH stretching vibration. The absorption bands due to Si-O-Si stretching, bending and rocking modes were observed at 1070,8 cm^{-1}, 805,6 cm^{-1} and 480 cm^{-1} in SiO_2 [3]. The absorption at 670.4 cm^{-1} can be attributed to Si-Si bonds due to oxygen vacancies [4, 5]. The FTIR spectra reveal that ultrathin SiO_2 films have been formed at these low annealing temperatures. These results are in agreement with the results obtained for low temperature thermal-ALD SiO_2 [6].

The electrical properties of the sol-gel SiO_2 films were investigated by measuring their capacitance-voltage (*C-V*) and current-voltage (*I-V*) characteristics. For the capacitance- voltage measurements we used a probe head with an area of the Hg dot of $3{,}4\times10^{-3}$ cm^2 and 2×10^{-4} cm^2. The MOS structure samples investigated are presented in table 1. Figure 6 presents the volt-capacitance characteristics of the two groups of samples.

The dielectric permeability k of sample 5 with thickness 24 nm as determined from the capacitance measurements was 3.89; it was lower for the ultrathin layers (5 nm). The hysteresis of the *C-V* curves was also studied. The measurement started at 0 V, the voltage was then swept to inversion, to accumulation and was looped back. The flat band voltage V_{fb1} was determined from the first *C-V* curve swept from inversion to accumulation, while V_{fb2} was determined from the *C-V* curve swept from accumulation to inversion. Using the difference $V_{fb1} - V_{fb2}$, the density N_t of the trapped charge was calculated using the formula $N_t = \dfrac{C_{ox}(V_{fb1} - V_{fb2})}{q}$, where C_{ox} is the capacitance of the layer per cm^2.

Table 2 presents the results for N_t calculated from the measurements at three points of the samples described in table 1. For structures with a p-Si substrate, the hysteresis character of the *C-V* curves indicates trapping of holes in the dielectric. In the

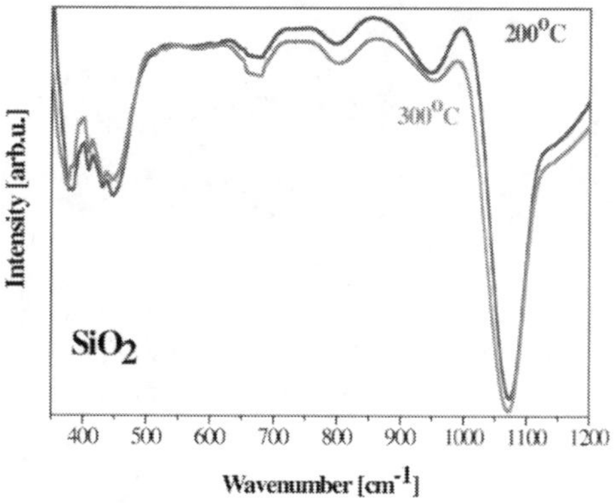

Figure 5. FTIR spectra of SiO_2 films annealed at 200 °C and 300 °C for 30 min.

Table 1. Description of the sol-gel samples.

Sample No.	Substrate type	Film thickness [nm]
5	p- Si , 3-5 Ω cm	24
10	n- Si , 6-9 Ω cm	24
15	p- Si , 3-5 Ω cm	5
20	n- Si , 6-9 Ω cm	5

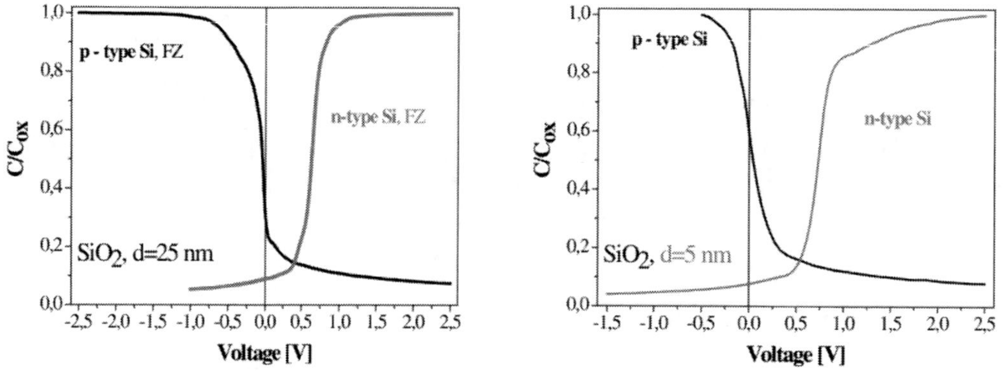

Figure 6. *C-V* curves of SiO_2 films with different thickness deposited on p- and n-type Si.

ultrathin SiO_2 layer (sample 15), the trapped charge is positive and 2.4 times greater than that for sample 5. For structures with an n-Si substrate, the hysteresis character of the *C-V* curves indicates trapping of electrons in the dielectric (negative trapped charge). In the ultathin SiO_2 layer (sample 20), the trapped charge is negative and 4 times greater than that for sample 10 (d = 24 nm) [7].

The *I-V* measurements were conducted at a negative voltage polarity with respect to the Hg dot for the p-Si substrate, and a positive one for the n-Si substrate, to ensure that the Si substrate is in accumulation mode (avoiding the voltage drop across the depletion layer in the Si substrate). The *I-V* data are presented in figure 7. These characteristics demonstrate the good dielectric properties of the ultrathin SiO_2 layers.

Table 2. Estimated results for the trapped charge density N_t.

No	Substrate type	Oxide thickness [nm]	V_{fb1}	V_{fb2}	$V_{fb1} - V_{fb2}$	N_t [cm^{-2}]
5	p-Si	24	0.096	-1.280	1.376	3.70E+12
5	p-Si	24	0.037	-1.278	1.315	3.70E+12
5	p-Si	24	0.006	-1.217	1.223	3.29E+12
10	n-Si	24	0.117	0.546	-0.429	-1.05E+12
10	n-Si	24	0.051	0.526	-0.475	-1.18E+12
10	n-Si	24	0.128	0.501	-0.374	-9.87E+11
15	p-Si	5	-0.022	-1.232	1.210	8.61E+12
15	p-Si	5	-0.036	-1.242	1.206	8.76E+12
15	p-Si	5	0.170	-1.160	1.330	7.96E+12
20	n-Si	5	0.070	0.667	-0.597	-4.42E+12
20	n-Si	5	0.103	0.661	-0.558	-4.06E+12
20	n-Si	5	0.107	0.665	-0.558	-4.03E+12

The sign "−" denotes a negative trapped charge, "+", a positive one.

4. Conclusions

The technological process proposed allows one to successfully deposit ultrathin films (5 and 24 nm) of SiO_2 on Si substrates at low temperatures. We present detailed material properties of the SiO_2 layers obtained by extensive characterization using FTIR, AFM, XPS spectroscopy, capacitance-voltage (*C-V*) and current-voltage (*I-V*) measurements. Thus, we proved that the SiO_2 films deposited at low-temperaturse are uniform, smooth and possess good electrical properties. The results obtained indicate that these films can be applied as tunneling-type layers and passivation layers in advanced solar cells or in optoelectronics.

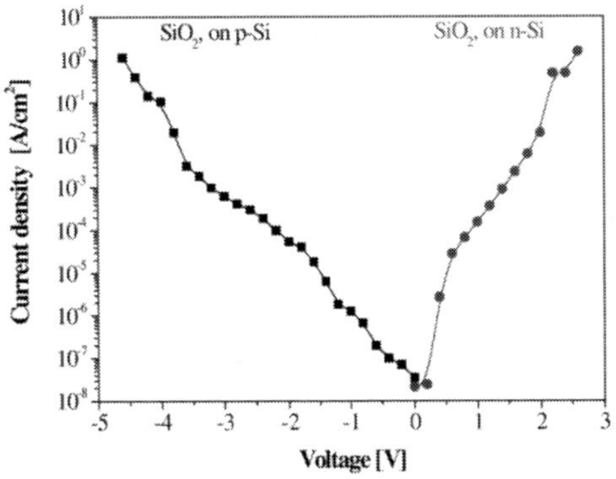

Figure 7. *I-V* characteristics of SiO_2 films deposited on p- and n- type Si substrates. The film thickness is 5 nm.

Acknowledgments

This works was supported in part under contract NANOPV "Nanomaterials and Nanotechnology for Advanced Photovoltaics", Reference 246331. Irina Bineva (ISSP-BAS) is acknowledged for performing the AFM measurements.

References

[1] Moreno L, Codinachs I, Birkenstock C, Garma T, Zierold R, Bachmann J, Nielsch K, Schöning M J and Morral A F I 2009 *Phys. Status Solidi* A **206** 435
[2] Kern W 1990 *J. Electrochem. Soc.* **137** 1887
[3] Ay F and Aydinly A 2004 *Optical Mater.* **26** 33
[4] Morales-Sánchez A, Barreto J, Dominguez C, Aceves M, Yu Z and Luna López J A 2008 *Phys. Status Solidi* C **5** 3651
[5] Dingemans G, van Helvoirt C A A, van de Sanden M C M and Kessels W M M 2011 *ECS Transactions* **35**/4 19
[6] Jurecka S, Kobayashi H, Takahashi M, Matsumoto T and Pincik E 2012 *Appl. Surf. Sci.* **258** 840
[7] Mitani Y, Satake H, Nakasaki Y and Toriumi A 2003 *IEEE Trans. Electron Devices* **50** 2221

Journal of Physics: Conference Series **514** (2014) 012011 doi:10.1088/1742-6596/514/1/012011

Characterization of ZrO_2 and $(ZrO_2)_X(Al_2O_3)_{1-X}$ thin films on Si substrates: effect of the Al_2O_3 component

P Vitanov[1], A Harizanova and T Ivanova

Central Laboratory of Solar Energy and New Energy Sources,
Bulgarian Academy of Sciences, 72 Tsarigradsko Chaussee, 1784 Sofia, Bulgaria

E mail: vitanov@phys.bas.bg

Abstract. ZrO_2 and $(ZrO_2)_x(Al_2O_3)_{1-x}$ films were deposited by the sol-gel technique on Si substrates. The effect of the Al_2O_3 additive on the film surface morphology was studied by atomic force microscopy (AFM). The mixed oxide films showed a smoother morphology and lower values of the root-mean-square (RMS) roughness compared to ZrO_2. Further, FTIR spectra indicated that ZrO_2 underwent crystallization. The electrical measurements of the MIS structure revealed that the presence of Al_2O_3 and the amorphization affects its dielectric properties. The MIS structure with $(ZrO_2)_x(Al_2O_3)_{1-x}$ showed a lower fixed charge ($\sim 6\times10^{10}$ cm^{-2}) and an interface state density in the middle of the band gap of 6×10^{11} eV^{-1} cm^{-2}). The dielectric constant measured was 22, with the leakage current density decreasing to 2×10^{-8} A cm^{-2} at 1×10^{6} V cm^{-1}.

1. Introduction

A large number of dielectrics have a higher permittivity than SiO_2, some of them possessing a low leakage current, a high thermal stability and good interface properties comparable to the Si/SiO_2 interface. Binary metal oxides such as Al_2O_3, TiO_2, ZrO_2 and HfO_2 are being intensively studied as they are predicted to be thermodynamically stable on Si and have a reasonably large band gap and, consequently, a good barrier height for electrons and holes in the semiconductor. Amorphous films are recommended for numerous applications due to their improved dielectric properties. To stabilize the amorphous phase of ZrO_2, SiO_2, and Al_2O_3 can be added to form a pseudobinary alloy [1]. It has been reported that sputtered $(ZrO_2)_x(Al_2O_3)_{1-x}$ [2] and co-evaporated $(Zr_{0.6}Al_{0.4})O_{1.8}$ films [3] showed increased stability of the amorphous phase. Recently, films of pseudobinary alloys as $(Al_2O_3)_x(TiO_2)_{1-x}$ and $(ZrO_2)_x(Al_2O_3)_{1-x}$ obtained on Si by chemical solution deposition (CSD) have been studied in view of different possible applications [4]. We synthesized mixed oxide films by the sol-gel technique because of the several advantages it offers, such as low processing temperature, homogeneity and cost effectiveness. Unlike other deposition technologies, the sol-gel process does not require high-vacuum equipment; the films prepared have been used as functional materials in optical, microelectronic, and opto-electronic applications [5].

We report that the addition of Al_2O_3 changes the properties of the dielectric films, such as degree of crystallinity, roughness, thermal stability on Si, and the films' dielectric properties as well.

[1] .To whom any correspondence should be addressed.

2. Experimental

Zirconium (IV) propoxide was used as a Zr precursor to form the ZrO_2 films. A more detailed description can be found in [6]. For the aluminum component, $AlCl_3$ anhydrous dissolved in absolute ethanol was used. The Zr/Al molar ratio was 1:1.5. The thin films were deposited by spin coating at 8000 rpm. Firing was carried out at 350 °C/30 min to pyrolize the organic components; this was followed by a high-temperature annealing at 750°C/60 min in N_2. This deposition method requires individual processing; under the same conditions (sol density and spinning rate), the same thickness is reproduced. In the single-layer deposition procedure, the layer thickness was around 30 nm for ZrO_2 and 50 nm for $(ZrO_2)_x(Al_2O_3)_{1-x}$. After repeating the deposition, the thicknesses were 60 nm for ZrO_2 and 102 nm for the mixed system. ZrO_2 and $(ZrO_2)_x(Al_2O_3)_{1-x}$ layers were deposited on 3`` p-type polished Si with resistivity of 5-7 Ω □cm.

The films' thickness and refractive index were measured by a He-Ne laser ellipsometer at 632.8 nm; the results are presented in table 1. The thickness uniformity across the wafer was 2 – 5 %.

Table 1. Refractive index and thickness of ZrO_2 and $(ZrO_2)_X(Al_2O_3)_{1-X}$ layers annealed at 750 °C as estimated from the ellipsometric data.

Layer type	Refractive index	Thickness [nm]
$(ZrO_2)_x(Al_2O_3)_{1-x}$	1.68	50 – 100
ZrO_2	1.88	30 – 60

The FTIR measurements were performed in the spectral region 350 – 1600 cm^{-1} by an IRPrestige-21Shimadzu FTIR Spectrophotometer in view of studying the chemical bonding configuration. The XRD measurements were performed on a SIEMENS D500 diffractometer in a Bragg-Brentano geometry with secondary graphite monochromator and CuKα radiation.

3. Results and discussions

A typical top-view AFM image showing the surface morphology of the ZrO_2 films (60-nm thick) is presented in figure 1. Using these images, the root-mean-square (RMS) roughness and average roughness were estimated. The microstructure observed of the sol-gel film is dense and fine grained. The average roughness values vary from 0.9 nm to 0.8 nm for different wafer areas. The RMS values are in the range 1.18 nm to 1.16 nm. Figure 2 shows an AFM micrograph of a $(ZrO_2)_x(Al_2O_3)_{1-x}$ film. The surface is smooth and uniform. The RMS values vary from 0.17 to 0.21 nm, while the average roughness ranges from 0.16 nm and 0.14 nm,. The values for the mixed system are lower by about a factor of ten compared with those of ZrO_2, suggesting a very high degree of smoothness.

Figure 1. AFM micrograph of ZrO_2 films deposited on Si, annealed at 750°C.

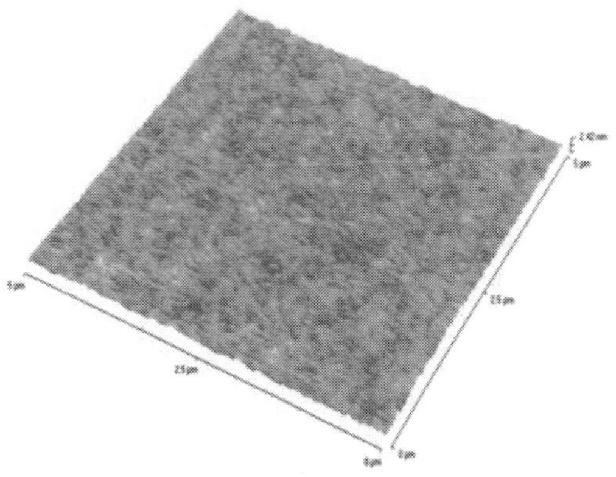

Figure 2. AFM image of $(Al_2O_3)_x(ZrO_2)_{1-x}$ layers deposited on Si, annealed at 750°C.

These values are close to the reported RMS of ALD $Pr_xAl_{1-x}O_3$ [7]. The higher values for the ZrO_2 layers can be due to their polycrystalline structure [8]. The same tendency has been previously observed by SEM investigation [6] for thicker films. The ZrO_2 films show a columnar structure and a rougher surface compared to the mixed films, the latter revealing a much smoother surface with a small-grain structure. Both microscopic methods demonstrate that adding Al_2O_3 results in smoother and amorphous-like films.

The ZrO_2 and mixed $(ZrO_2)_x(Al_2O_3)_{1-x}$ films were studied by FTIR spectroscopy to characterize the chemical bonding and molecules vibrations; the measured spectra are presented in figure 3. The ZrO_2 film's FTIR spectrum exhibits an intense band at 408.9 cm^{-1} and two weaker bands at 391.6 cm^{-1} and 435.9 cm^{-1}. The IR lines at 408.9 and 572.9 cm^{-1} can be assigned to Zr-O bond vibrations in monoclinic ZrO_2 [9].

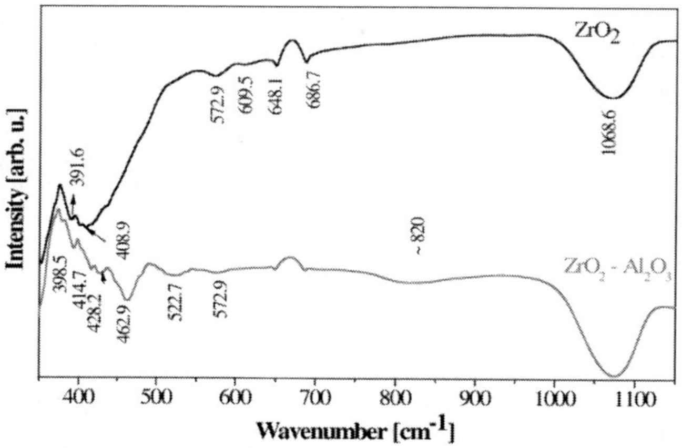

Figure 3. FTIR spectra of ZrO_2 and $(ZrO_2)_x(Al_2O_3)_{1-x}$ films.

The absorption bands at 609.5 and 686.7 are connected to the Zr-O stretching mode. The absorption peak at 648.1 cm^{-1} may be due to the stretching vibrations of tetragonal ZrO_2. The intense line at 1068.6 cm^{-1} is due either to Si-O bonds of silicon dioxide or to the zirconyl bonds (Zr=O) [10]. Some overlapping of these two bands might exist. The spectrum of $(ZrO_2)_x(Al_2O_3)_{1-x}$ presents some new additional bands, while the other bands are slightly shifted with respect to those of ZrO_2. The weak broad band at 522.7 cm^{-1} has been reported to be related with crystalline ZrO_2. The absorption band at 648.1 cm^{-1} can be related to tetragonal ZrO_2. The presence of the Al_2O_3 component is manifested by a broad absorption band centred at 823.6 cm^{-1}, which can be assigned to the stretching vibrations of Al-O bonds [11].

We also recorded XRD spectra of ZrO_2 and $(ZrO_2)_x(Al_2O_3)_{1-x}$ layers annealed at 750 °C for 1 hour (not shown here) [6]. The spectra revealed that ZrO_2 is well crystallized, exhibiting four XRD peaks of the tetragonal crystal phase (JPCDS card 50-1089) and two diffraction features due to monoclinic ZrO_2 [6]. The XRD spectrum of $(ZrO_2)_x(Al_2O_3)_{1-x}$ layer is without characteristic features, indicating that the film's structure is amorphous, despite of the high-temperature treatment [6].

Based on the FTIR analysis, one can conclude that ZrO_2 film undergoes crystallization. The spectra of the $(ZrO_2)_x(Al_2O_3)_{1-x}$ layers exhibit new absorption bands, which suggests a low degree of crystallization of the ZrO_2 component in the mixed system. This confirms the results from the microscope studies, namely, that the addition of Al_2O_3 induces amorphization of the film structure.

The dielectric properties of the layers were studied using an upper contact of Al dots, thus forming a MIS (metal-insulator-semiconductor) structure. The layers were deposited on chemically-cleaned Si surfaces, where the thickness of native SiO_2 did not exceed 1 nm. Figure 4 presents *C-V* curves of ZrO_2 and $(ZrO_2)_x(Al_2O_3)_{1-x}$ layers at 1 MHz with Al dots and a contact area of 2.5×10^{-3} cm^2. The ZrO_2 film showed a positive fixed charge with a density of 5.42×10^{11} cm^{-2}. The fixed charge density of the $(ZrO_2)_x(Al_2O_3)_{1-x}$ film was smaller: 6.46×10^{10} cm^{-2}, which is due to the Al_2O_3 content.

The frequency dependence of the dielectric constant *k* was measured in a parallel circuit mode by an LCR HP 4274A device (see table 2). The dielectric constant was determined from the capacitance in strong accumulation and the physical film thickness. The estimated *k* for ZrO_2 films was smaller than that reported by other authors ($k = 27$) [12], which can be due to the presence of a monoclinic phase that has a lower permittivity thus decreasing the overall permittivity [12].

Moreover, the dielectric constants of the $(ZrO_2)_x(Al_2O_3)_{1-x}$ films were smaller (10 - 17.2) due to the Al_2O_3 additive. The sol-gel Al_2O_3 films annealed at 750 °C possessed a dielectric constant of 7.4 (1 MHz) [13].

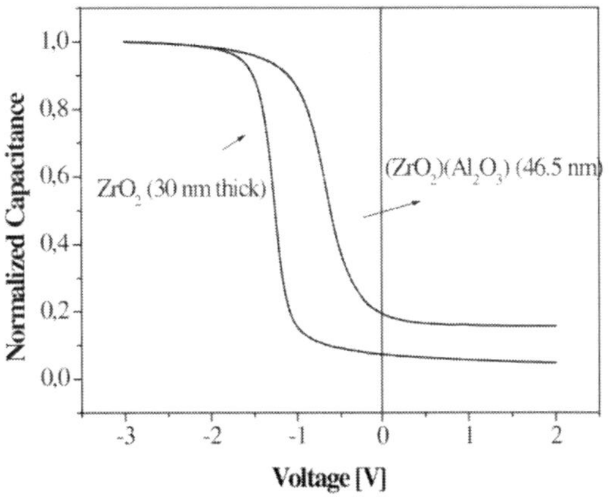

Figure 4. *C-V* curves of MIS structures with $(ZrO_2)_x(Al_2O_3)_{1-x}$ and ZrO_2 films. Al contacts were treated at 410°C/15 min/N_2.

Table 2. Dielectric constant *k* for ZrO_2 and $(ZrO_2)_X(Al_2O_3)_{1-X}$ films at different frequencies.

Oxide type	1 MHz	100 kHz	10 kHz	4 kHz	1 kHz
$(ZrO_2)_x(Al_2O_3)_{1-x}$	10.7	12.2	14.5	15.5	17.2
ZrO_2	11.5	15.3	20.2	21.8	23.9

The quasistatic pulse method (*CV* map 92B) was used to study the dielectric/Si interface. This technique allows one to avoid the dielectric loss and the error in the phase-angle capacitance measurement. The results for the dielectric constant *k* and the density of the interface states, D_{it}, in the middle of the band gap are summarized in table 3. In [14], the effective dielectric constant *k* of $(ZrO_2)_x(Al_2O_3)_{1-x}$ was determined to be k =22.1. This is a proof ofr the thermal stability of the layer on a Si substrate after a high-temperature treatment.

Table 3. Dielectric constant *k* and interface state density in midband gap, as measured by the quasistatic pulse technique.

Oxide type	Dielectric constant *k* (QS method)	D_{it} interface state density in midband gap (cm^{-2} eV^{-1})
$(ZrO_2)_x(Al_2O_3)_{1-x}$	22.3	6.00×10^{11}
ZrO_2	24.1	1.45×10^{12}

The values of D_{it} registered for the pseudobinary alloy $(ZrO_2)_x(Al_2O_3)_{1-x}$ are considerably smaller that those of the ZrO_2 layer, which can be connected with the influence of the Al_2O_3 component and the amorphization of the layer.

Figure 5 illustrates the corresponding leakage current (*J-E*) characteristics of MIS structures with $(ZrO_2)_x(Al_2O_3)_{1-x}$ and ZrO_2 layers, where *J* is the current density and *E*, the electric field.

The *C-V* measurements of the MIS structures with a $(ZrO_2)_x(Al_2O_3)_{1-x}$ film showed that the addition of aluminum oxide leads to a very small decrease of the dielectric constant. At high electric fields, the leakage currents of the pseudobinary layers are lower by several orders of T magnitude compared to those of pure ZrO_2. One can presume that the ZrO_2 films have higher leakage currents caused by grain boundaries, which serve as leakage paths. The composite system $(ZrO_2)_x(Al_2O_3)_{1-x}$ possesses an amorphous structure, which decreases the possibilities for a similar type of leakage. The other possibility is to assume that the presence of the Al_2O_3 component changes the conductivity as a result of the larger equivalent band gap of the pseudobinary dielectric film. It is known that two transport mechanisms are important for the leakage in dielectrics: Schottky effect and Poele-Frenkel emission. They result from the lowering of the Coulomb potential barrier by a high electric field. The Schottky effect is associated with the barrier at the interface metal electrode-insulator and therefore, the current is electrode-limited. The Poele-Frenkel emission is related to barriers of the traps in the bulk of the material. It should be expected that the larger equivalent band gap of $(ZrO_2)_x(Al_2O_3)_{1-x}$ layer will lead to a decrease in the dielectric conductivity. This has been observed experimentally for $(ZrO_2)_x(SiO_2)_{1-x}$ films [15].

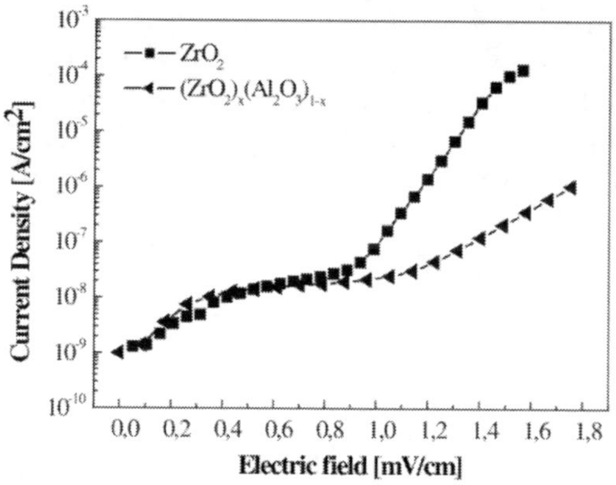

Figure 5. Leakage current characteristics of MIS structures with $(ZrO_2)_x(Al_2O_3)_{1-x}$ and ZrO_2 layers.

4. Conclusions

Thin films of ZrO_2 and $(ZrO_2)_x(Al_2O_3)_{1-x}$ were deposited by similar sol-gel technological procedures on Si wafers. The surface and electrical characteristics of the layers was thus improved as the presence of Al_2O_3 leads to amorphization of the ZrO_2 structure and to a smoother surface. This results in lowering the refractive index and the leakage current, while maintaining a rather high value of the dielectric constant. The system – $(ZrO_2)_x(Al_2O_3)_{1-x}$ films on silicon – shows a good thermodynamic stability at high-temperature annealing (750 °C for 60 minutes), as no degradation was observed of the electrical properties. The results are encouraging and determine the pseudobinary alloy films $(ZrO_2)_x(Al_2O_3)_{1-x}$ as a promising material with good dielectric properties.

References

[1] Sarkar D K, Brassard D, El Khakani M A and Ouellet L 2007 *Thin Solid Films* **515** 4788
[2] Van Dover R B, Lang D V, Green M L, Manchanda L 2001 *J. Vac. Sci. Techn.* A **19** 2778
[3] Di Z, Zhang M, Liu W, Luo S, An Z, Zhang Z and Liu C 2005 *J. Non-Crystalline Solids* **351** 93
[4] Vitanov P, Harizanova A, Ivanova T, Alexieva Z and Agostineli G 2006 *Jpn. J. Appl. Phys.* **45** 5894
[5] Sharma S K, Vishwas M, Rao K N, Mohan S, Reddy D S and Gowda K V A 2009 *J. Alloys Compounds* **471** 244
[6] Vitanov P, Harizanova A and Ivanova T 2009 *Materials Sci. Engin.* B **165** 178
[7] de Rouffignac R and Gordon R G 2006 *Chemical Vapor Deposition* **12** 152
[8] Triyoso D H 2005 *J. Appl. Phys.* **98** 054104
[9] Ortiz A, Alonso J C and Haro-Poniatowski E 2006 *J. Electronic Materials* **34** 150
[10] Guo G-Y and Chen Y-L 2006 *Appl. Phys.* A **84** 431
[11] Padmaja P, Warrier K G K, Padmanabhan M, Wunderlich W, Berry F Y, Mortimer M and Creamer N J 2006 *Mater. Chem. Phys.* **95** 56
[12] Brunet M, Mafhoz Kotb H, Bouscayrol L, Scheid E, Andrieux M, Legros C and Schamm-Chardon S 2011 *Thin Solid Films* **519** 5638
[13] Vitanov P, Harizanova A, Ivanova T and Dimitrova T 2009 *Thin Solid Films* **517** 6327
[14] Li Y, Zhu J, Liu H and Liu Z 2006 *Microelectr. Engin.* **83** 1905
[15] Wilk G D, Wallace R M, Anthony J M 2000 *J. Appl. Phys.* **87** 484

Influence of annealing on the optical, structural and electrical properties of ZnO:Al/Ag/ZnO:Al multilayer stack structures

K Lovchinov[1], M Petrov[1,3], O Angelov[1], H Nichev[1], D Karashanova[2] and D Dimova-Malinovska[1]

[1]Central Laboratory of Solar Energy and New Energy Sources, Bulgarian Academy of Sciences, 72 Tsarigradsko Chaussee, 1784 Sofia, Bulgaria
[2]Acad. J. Malinowski Institute of Optical Materials and Technologies, Bulgarian Academy of Sciences, Acad. G. Bonchev Str. Bl. 109, 1113 Sofia, Bulgaria

E mail: petrov80@abv.bg

Abstract. Multilayer stack structures ZnO:Al(20 nm)/Ag(x)/ZnO:Al(20 nm) with different thickness of the middle Ag(x) layer – x = 6 nm, 10 nm, 12 nm, 16 nm and 20 nm, were prepared by r.f. magnetron sputtering. The ZnO:Al and Ag layers were deposited by magnetron sputtering on glass substrates without heating. The two-layer structures ZnO:Al(20)/Ag(x) were annealed in $N_2 + H_2$ at 180 °C, 50 min., and then capped by the top ZnO:Al(20 nm) layer. The optical, structural and electrical properties were studied of the as-deposited and of the annealed three-layer stacks. The TEM, SEM and AFM analyses demonstrated changes of the stack structure with the Ag film thickness and after annealing. The transmittance and reflectance spectra revealed bands of Ag electrons plasma oscillations and inter-band d-shell Ag electrons which were red-shifted after annealing. The as-deposited structures had low resistivity (1×10^{-3} to 4×10^{-5} Ω cm), which decreased with the thickness of the Ag film due to the change in the carrier transport mechanisms. The value of the resistivity increased after annealing.

1. Introduction

Systems consising of layered structures, namely, transparent conductive oxide (TCO)/noble metal (Cu, Au, Ag) have been recently studied intensively because of the possibility for application in thin film solar cells [1]. The difference in the work functions of the TCO and the noble metals offers an opportunity to increase the conductivity of the TCO as a rear electrode [2]. The main tasks in the development and fabrication of thin-film solar cells are increasing the light absorption, minimizing the reflection losses and increasing the reflection of the light transmitted through the solar cell.

This work addresses the optical, structural and electrical properties of three-layer stack ZnO:Al(20 nm)/Ag(x)/ZnO:Al(20 nm) structures with different thicknesses of the middle Ag(x) layer, x = 6 nm, 10 nm, 12 nm, 16 nm and 20 nm. The influence of the annealing in $N_2 + H_2$ at 180 °C, 50 min., is studied in the case of annealing the first two-layers ZnO:Al(20 nm)/Ag(x) structures before depositing the top ZnO:Al film.

[3] To whom any correspondence should be addressed.

Journal of Physics: Conference Series **514** (2014) 012012 doi:10.1088/1742-6596/514/1/012012

2. Experimental

ZnO:Al(20 nm)/Ag(x)/ZnO:Al(20 nm) multilayer structures were deposited on glass substrates by successive r.f. magnetron sputtering of ZnO:Al and Ag targets without intentional heating. The Ag layers had thicknesses of 6 nm, 10 nm, 12 nm, 16 nm and 20 nm. The ZnO:Al layers were deposited in an atmosphere of Ar (2 Pa) + H_2 (0.02 Pa), and the Ag layers, in Ar (0.2 Pa) at a r.f. power of 50 W and 20 W, respectively. The two-layer structures ZnO:Al(20)/Ag(x) were annealed in $N_2 + H_2$ at 180 °C, 50 min., and then capped with the top ZnO:Al(20 nm) layer. The TEM images were obtained by a JEOL JEM 2100 microscope, and the SEM micrographs, on a Philips 515 apparatus. The AFM images were taken by an NT-MDT Solver 47 Pro system, with the root-mean-square (rms) surface roughness being calculated by the Gwyddion software. The optical properties of the stacks were measured by a Shimadsu UV.3600 spectrophotometer in the wavelength range 350 – 1200 nm. The feature at about 800 nm in the optical spectra is an artifact due to the change of the detectors for the different spectral ranges. The electrical properties were measured by the four-point probe method using a VEECO instrument.

3. Results and discussion

TEM and SEM micrographs of the multilayer stack structures ZnO:Al(20 nm)/Ag(x)/ZnO:Al(20 nm), as-deposited and after annealing of the two-layer structures, are shown in figure 1 and figure 2. The phase composition of the samples was identified from the SAED patterns presented as insets in figure 1, as follows: hexagonal ZnO (PDF 89-1397), hexagonal Ag (PDF 87-0598), monoclinic Al_2O_3 (PDF 86-1410) and orthorhombic Al_2O_3 (PDF 88-0107) [3]. The TEM and SEM micrographs of the as-deposited structures reveal an increase in the Ag grains size as the thickness of the middle Ag layer is increased from 6 nm to 12 nm. The Ag grains aggregate into chains in the stacks with $x = 16$ nm (figure 1c), while the Ag film's structure is transformed from a discontinuous one to a semi-continuous one ($x = 20$ nm, figure 1d). Annealing the two-layer ZnO:Al(20 nm)/Ag(x) structure results in an increase of the Ag grain size due to coalescence (figure 1e, f, g , h and figure 2e, f, g and h). In the stacks with $x = 16$ nm and $x = 20$ nm, the structure of the Ag films is transformed from a semi-continuous island-like one into a structure with separate grains of spheroidal and ellipsoidal shape (figure 1g and h and figure 2g and h). Similar results have been reported by other authors [4].

The surface roughness of the samples was studied by AFM scanning. The AFM images and the rms roughness of the as-deposited stacks with $x = 6$ nm, 16 nm and 20 nm and the annealed ones with x = 16 nm and 20 nm are presented in figure 3. The rms roughness decreases from 6 nm to 4 nm as the thickness of the middle Ag layer is increased from 6 nm to 16 nm (figure 3a and b) due to the increased degree of substrate covering by Ag nano-grains. A further increase of the Ag film thickness to 20 nm

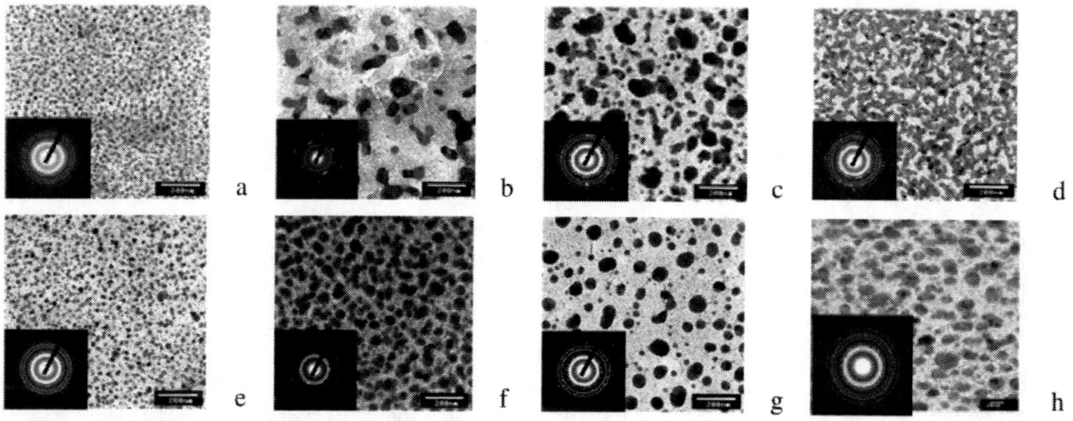

Figure 1. TEM micrographs of as-deposited (a, b, c and d) and annealed (e, f, g and h) multilayer stacks ZnO:Al(20 nm)/Ag(x)/ZnO:Al(20 nm) with different thickness of the Ag film for $x = 6$ nm (a and e), 12 nm (b and f), 16 nm (c and g) and 20 nm (d and h). The markers correspond to 200 nm.

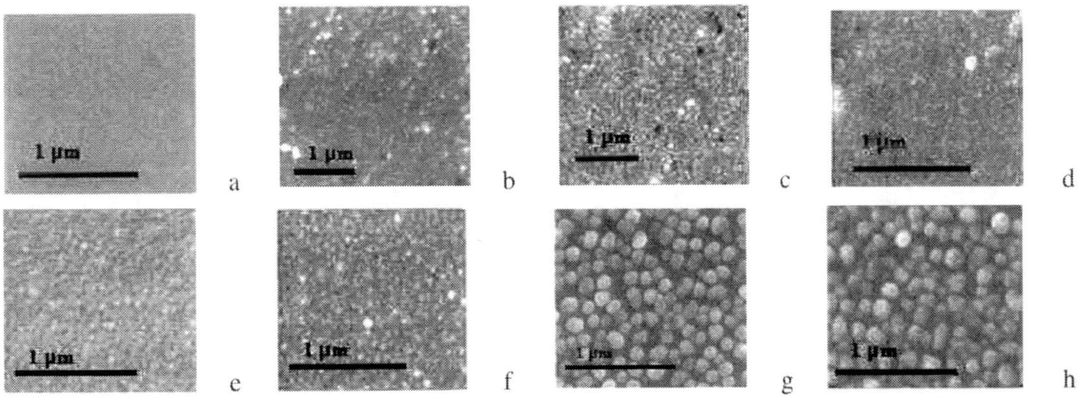

Figure 2. SEM micrographs of as-deposited (a, b, c and d), and annealed (e, f, g and h), multilayer stacks ZnO:Al(20 nm)/Ag(x)/ZnO:Al(20 nm) for x = 6 nm (a and e), 12 nm (b and f), 16 nm (c and g) and 20 nm (d and h). The markers correspond to 1 μm.

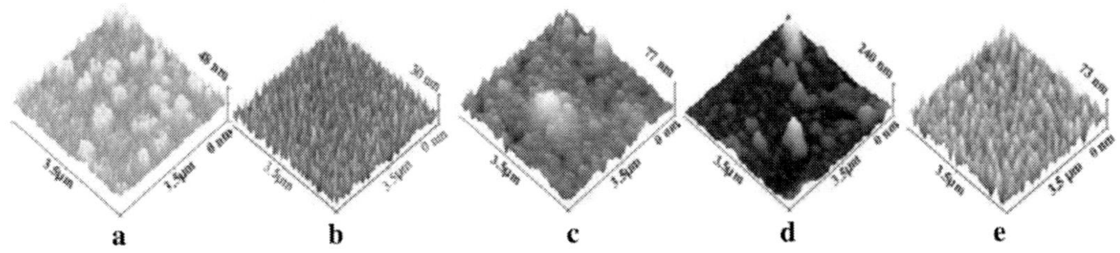

Figure 3. AFM pictures of multilayer structures ZnO:Al(20 nm)/Ag(x)/ZnO:Al(20 nm): as deposited, (a) x = 6 nm, (b) x = 16 nm, (c) x = 20 nm; and after annealing of the two-layer structure ZnO:Al(20 nm)/Ag(x), (d) x =16 nm and (e) x =20 nm.

leads to an increase in the rms roughness to 9 nm due to the formation of NPs of different size and shape partially connected in chains (figure 3c). The annealing results in increasing the surface roughness due to a coalescence of the Ag atoms. After annealing, the rms roughness of the stacks with x = 16 nm is 23 nm; the roughness of the stack with x = 20 nm is 10 nm. The lower value of the rms roughness after annealing of the stack with x = 20 nm can be explained with the denser covering of the surface by Ag grains.

The transmittance spectra of as-deposited stacks ZnO:Al(20 nm)/Ag(x)/ZnO:Al(20 nm) (figure 4a) show a band with a maximum near 365 nm due to the quadrupolar resonance and the near-lying inter-band electron transitions of the electrons of the entirely filled Ag atoms' d-shell [5]. These inter-band transitions were observed in the spectra of all samples. Increasing the Ag layer thickness at $x > 10$ nm leads to this band being red-shifted due to the influence of the quadrupolar resonance as a result of the increased size of the NPs in the middle Ag film. The transmittance has a minimum at about 460 nm due to the dipolar electrons plasma oscillations, deeper and red-shifted as the size of the Ag NPs increases in the samples with Ag layer thickness of 6 nm and 10 nm. A further increase in the Ag layer thickness from 12 nm to 20 nm increases the degree of substrate covering by Ag NPs so that the interaction between them becomes stronger. As a result, an overlapping between the quadrupolar and dipolar resonances occurs [5] and this band is smeared. Above 600 nm in the case of stacks with Ag layer thickness of 6 nm and 10 nm, the transmittance increases as the wavelength is increased; after 700 nm it changes weakly due to the lower degree of substrate covering by Ag NPs. The structures with a Ag layer thickness between 12 nm and 20 nm show a monotonic decrease in the transmittance above 550 nm. This can be explained by the plasmonic absorption of the free electrons in the Ag films, which

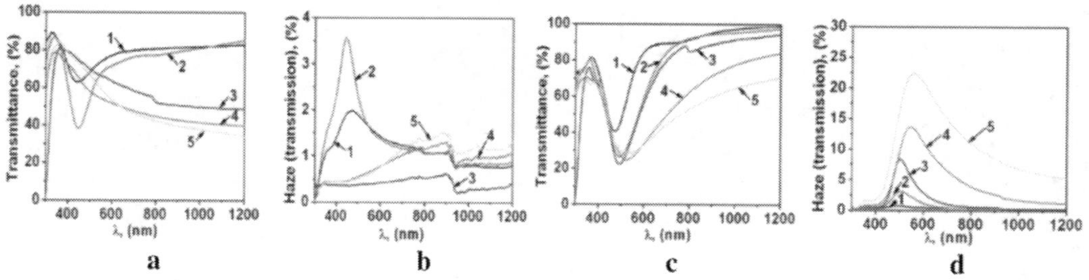

Figure 4. Spectra of transmittance and haze ratio in transmission of multilayer stack structures ZnO:Al(20 nm)/Ag(x)/ZnO:Al(20 nm): as-deposited (a and b) and annealed (c and d), for x =6 nm (curves 1), 10 nm (curves 2), 12 nm curves (3), 16 nm (curves 4) and 20 nm (curves 5).

exhibit a higher degree of substrate covering and conductivity [5]. The same bands of typical inter-band d-shell electrons transitions and electron plasma oscillations are observed in the spectra of the stacks with an annealed ZnO:Al(20 nm)/Ag(x) structure. They are red-shifted compared to the spectra of the as-deposited samples. The changes observed in the transmittance spectra after annealing can be explained by the changes in the size and shape distributions of the Ag NPs during the annealing.

The haze ratio in transmission is higher in the as-deposited stacks with lower Ag layer thicknesses (6 nm and 10 nm), and in the annealed stacks with higher Ag layer thickness (x > 10 nm). It has to be noted that the haze ratio is higher in the whole spectral range in the stacks after annealing. This can be explained by the evolution of the structure of the middle Ag layer and the stronger interaction between the Ag NPs as the thickness is increased during annealing. The transition from a discontinuous to a continuous Ag layer structure in the as-deposited multilayer stacks ZnO:Al(20 nm)/Ag(x)/ZnO:Al(20 nm) when the Ag layer thickness is inncreased leads to a decrease in the diffuse transmittance, and, thus, to a lower haze ratio in comparison with the annealed samples.

The reflectance and the haze ratio in the reflection spectra of the as-deposited (figures 5a and b) and annealed (figures 5c and d) multilayer structures demonstrate typical plasma electron oscillations and inter-band d-electrons transition bands due to the presence of Ag NPs in the film. These bands are more prominent in the spectra of the annealed stacks. The higher reflectance observed in the spectra of the as-deposited stacks with x > 10 nm is due to the higher free carriers density in these stacks with a higher degree of substrate covering by the Ag film and a higher conductivity (figure 6). The haze ratio in reflection is about 15-25% in as-deposited and annealed stacks in the visible region. In the as-deposited stacks with a lower thickness of the middle Ag films (x = 6 and 10 nm), the value of the haze ratio at about 365 nm is higher than in the samples with a higher size of the Ag NPs due to the stronger plasmon resonance. In these stacks, the presence of Ag NPs of a smaller size and the higher surface roughness result in a higher diffuse reflectance and a higher haze ratio. After annealing, the haze ratio

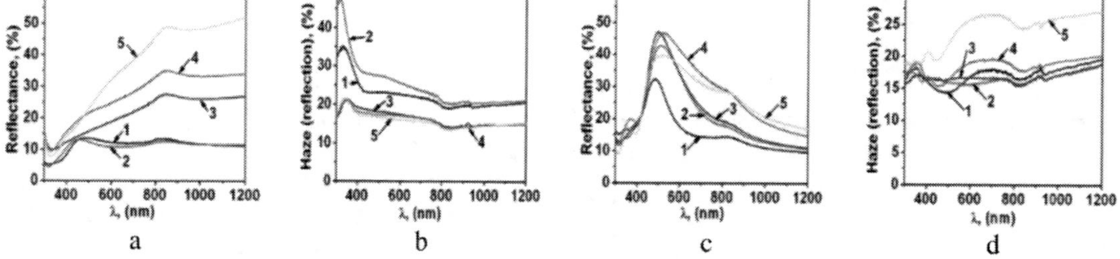

Figure 5. Spectra of reflectance and haze ratio in reflection of multilayer stack structures ZnO:Al(20 nm)/Ag(x)/ZnO:Al(20 nm): as-deposited (a and b) and annealed (c and d), for x =6 nm (curves 1), 10 nm (curves 2), 12 nm (curves 3), 16 nm (curves 4) and 20 nm (curves 5).

of the stacks with thickness of the middle Ag film (x = 6, 10, 12 and 16 nm) does not depend significantly on the thickness. The same spectral bands are seen as in the reflectance spectra. The multilayer structure with x = 20 nm has a higher haze than the haze of the other annealed structures due to the formation of a grain structure in the Ag (20 nm) film and the higher degree of substrate covering leading to higher light scattering.

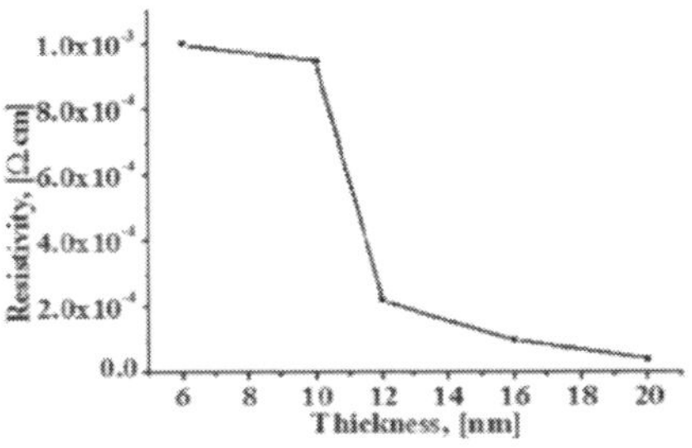

Figure 6. Resistivity of the as-deposited multilayer stack structure ZnO:Al(20 nm)/Ag(x)/ZnO:Al(20 nm) with Ag layer thickness x= 6 nm, 10 nm, 12 nm, 16 nm and 20 nm.

The resistivity of the as-deposited stacks decreases with the thickness of the Ag film from 1×10^{-3} to 4×10^{-5} Ω cm due to the larger size of the Ag NPs and the higher degree of substrate covering (figure 5). The barrier height between the grains decreases and the conductivity of the layer undergoes a transforms from a thermally activated hopping mechanism to a metallic type one [5]. A similar behavior of the resistivity of multilayer ZnO:Al/Ag/ZnO:Al structures has been reported by other authors [6]. After annealing of the ZnO:Al(20 nm)/Ag(x) stacks, the values of the resistivity increase and exceed 10^5 Ω cm, related to the outlined grain structure of the Ag films.

4. Conclusions

The optical, electrical and structural properties were studied of as-deposited and annealed at 180 °C multilayer stack structures ZnO:Al(20 nm)/Ag(x)/ZnO:Al(20 nm) for x = 6 nm, 10 nm, 12 nm, 16 nm and 20 nm. The TEM and SEM analyses demonstrated an increase of the size of the NPs in the middle Ag film with its thickness and after annealing. The structure of the Ag film is transformed from a semi-continuous to a discontinuous islands-like one in the Ag films with a higher thickness after annealing.

The transmittance and reflectance spectra of the as-deposited and annealed films demonstrated bands in the visible spectral region due to Ag electrons plasma oscillations and d-shell electrons inter-band transitions. The position and width of the bands depended on the Ag NPs size and shape distributions in the middle Ag films, which were determined by the thickness of the Ag film and the process of annealing. The same bands associated with the plasmonic effect and d-shell electrons inter-band transitions were observed in the haze ratio spectra in transmission and reflection. The as-deposited stacks with the thickness of the middle Ag films of 20 nm exhibited a high conductivity and high values of the reflectance and plasmonic light scattering, which makes them suitable for application as rear contacts in thin film solar cells thus increasing the light trapping.

Acknowledgements

The work has been supported in part by the 7th European FP – project NanoPV № 246331.

References

[1] Cho J-S, Baek S and Lee J C 2011 *Solar Energy Mater. Solar Cells* **95** 1852
[2] Han H, Theodore N D and Alford T L 2008 *J. Appl. Phys.* **103** 013708
[3] International Centre for Diffraction Data PC PDFW vol. 2.2
[4] Paris A, Vaccari A, Lesina A C, Serra E and Calliari L 2012 *Plasmonics* **7** 525
[5] Kreibig U and Vollmer M 1995 *Optical Properties of Metal Clusters* **25** (Springer)
[6] Sahu D R and Huang J-L 2009 *Solar Energy Mater. Solar Cells* **93** 1923

Dependence of the optical and structural properties of ZnO:Al nanostructured layers deposited electrochemically on seeding multilayer stack substrates

M Petrov[1], K Lovchinov, M Ganchev, O Angelov, H Nichev and D Dimova-Malinovska

Central Laboratory of Solar Energy and New Energy Sources,
Bulgarian Academy of Sciences, 72 Tsarigradsko Chaussee, 1784 Sofia, Bulgaria

E-mail: petrov80@abv.bg

Abstract. We report studies on the optical and structural properties of nanostructured ZnO layers doped with Al (ZnO NW) electrochemically grown on conductive substrates, namely, ZnO:Al/Ag/ZnO:Al multilayer stack structures deposited on glass substrates. Multilayer transparent conductive stack structures with different thickness of the middle Ag film (10 nm, 16 nm and 20 nm) with plasmonic properties were prepared by magnetron sputtering of ZnO:Al and Ag targets. Additionally, a seeding stack annealed at 180 °C for 50 min was prepared in order to study the influence of the seeding stacks properties – resistivity and structure, on the optical and structural characteristics of the electrochemically-grown ZnO layers. A correlation was observed between the surface roughness and the diffuse reflection and the haze ratio in reflection. The results are discussed in terms of the influence of the size of the Ag grains and their distribution in the middle Ag film in the seeding stacks on the nucleation and growth of ZnO NW. The ZnO NW layers exhibited higher values of the diffuse reflection in the 400 – 1100 nm spectral range; they can be applied as rear contacts of thin films solar cells thus increasing the light trapping.

1. Introduction

ZnO is recognized as one of the most promising semiconductor oxide materials due to its good optical, electric and piezoelectric properties. This has prompted a growing interest in its application in solar cells, photonic devices, including laser and light-emitting diodes with low thresholds, optical waveguides and optical switches, sensors, field emission sources and displays [1-4]. Different methods of fabrication of ZnO nanostructured layers have been developed – electrochemical deposition, sol-gel preparation, sputtering deposition, aqueous chemical growth etc. Additionally, noble metal nanoparticles with plasmonic properties are of great interest for application as back reflectors for increasing the light trapping in thin-film solar cells [5, 6]. Nanostructured ZnO films used as electrodes have demonstrated appropriate surface morphology for high light-trapping and optical confinement in thin-film solar cells with increased efficiency [5]. Recently, fabrication was reported of conductive multilayer stack structures with increased diffuse reflection and plasmonic properties in the

[1] To whom any correspondence should be addressed.

Journal of Physics: Conference Series **514** (2014) 012013 doi:10.1088/1742-6596/514/1/012013

visible and IR region on the basis of Al doped ZnO (ZnO:Al) films and Ag nano-particles deposited by magnetron sputtering [5-7].

In this paper we report results from the study of the structural and optical properties of ZnO nano-structured films grown by electrochemical deposition on glass substrates coated by seeding conductive multilayer stack ZnO:Al/Ag/ZnO:Al structures. Studies of the surface structure by SEM and AFM demonstrated changes in the surface morphology and roughness of the electrochemically-deposited ZnO layers as the thickness of the Ag film in the stack structures was varied. A correlation was observed between the diffuse reflection and the haze ratio in reflection.

2. Experimental

ZnO nanostructured films were deposited by means of an electrochemical process from aqueous acid solutions of $ZnCl_2$ (5×10^{-3} M) and KCl (0.1 M) with pH 6.5 at a temperature of 60 °C using a three-electrode electrochemical cell with a saturated calomel electrode (SCE) as a reference electrode. Stoke solution of 4×10^{-5} M $Al_2(SO_4)_3$ was added to the electrolyte to deposit Al-doped ZnO layers. The deposition process of ZnO was carried out at −1.0 V vs. the SCE preventing a metal Zn deposition. ZnO films of good quality were obtained at a redox potential within the range between +0.30 V and +0.40 V vs. the SCE. Zinc peroxide (ZnO_2) was formed on the samples with bad adhesion to the substrate at a redox potential higher than +0.40 V. A spectrally-pure graphite plate electrode was used as an anode. The electrolyte was agitated by a magnetic stirrer. The deposition time was 15 min. The substrates used consisted of glass coated by three-layer stack structures – glass/ZnO:Al(20 nm)/Ag(x nm)/ZnO:Al(20 nm). The thickness of the middle Ag film in the seeding stack structures, x, was varied as follows: 10 nm (substrate S3), 16 nm (substrate S2) and 20 nm (substrates S1 and S4). To compare the properties of the electrochemically deposited ZnO layers on as-deposited (S1) and annealed seeding stacks, a structure (S4) with a Ag thickness of 20 nm was fabricated after annealing the first two layers ZnO(20 nm)/Ag(20 nm) at 180 °C for 50 min followed by deposition of the capping ZnO (20 nm) film.

The surface morphology of the deposited films was studied by a scanning electron microscope (SEM) Philips 515. The AFM analysis was performed by an NT-MDT Solver 47 Pro system in a "semi-contact" (tapping) mode and the root-mean-square (rms) roughness was calculated by the Gwyddion software. The optical spectra were obtained by a Shimadzu UV-3600 spectrophotometer in the 320 – 1200 nm range employing a 60-mm integrating sphere for diffuse transmission and reflection spectra measurements. The resistivity of the seeding stack structures was measured by the four-point probe method using a VEECO apparatus.

3. Results and discussion

Some characteristics of the seeding stack structures and of the ZnO layers deposited electrochemically on them is given in table 1. The stack structures' resistivity increases with the decrease of the thickness

Table 1. List of the substrate structures and of the electrochemically deposited ZnO NW layers, the thickness of the middle Ag films, x, resistivity of the stake structures, ρ, the root-mean-square roughness of the seeding stakes structure, rms, and of the ZnO NW layers, RMS, the haze ratio in reflection of the stake structures, h_R, and of the ZnO NW, H_R, at two wavelengths – 400 nm and 1000 nm.

ZnO NW samples	Stack structures ZnO/Ag/ZnO	x, (nm)	ρ, (Ωcm)	rms (nm)	RMS (nm)	h_R(%) 400 nm	H_R (%) 400 nm	h_R(%) 1000 nm	H_R(%) 1000 nm
1	1S	20	3.3×10^{5}	9	95	20	91	15	35
2	2S	16	1.0×10^{4}	4	106	20	94	14	32
3	3S	10	1.0×10^{3}	6	330	21	85	17	25
4	4S (annealed 1S)	20	2.3×10^{1}	10	85	28	50	38	13

of the middle Ag film in the seeding stacks from 20 nm to 10 nm. Annealing the two-layer structure with a 20-nm Ag film at 180 °C for 50 min before capping with the upper ZnO(20 nm) results in an increase of the resistivity of the three-layer stack to 2.3×10^{-1} Ω cm. Similar dependence has been reported by other authors, e.g. [6].

The higher conductivity can be explained by the increased thickness of the middle Ag layer and the formation of a chain area, as seen from the TEM and SEM analyses published in [7]. Thus, the barrier height between the grains decreases and the mechanism of conductivity of the layer changes from thermally activated hopping to a metallic-type one [8]. After annealing the as-deposited stack with a 20-nm thick Ag film, the island-like areas split and the Ag nano-grains separate, which leads to a rise in the structure's resistivity. The surface roughness value slightly decreases as the thickness of the Ag film is decreased and does not change significantly after annealing.

Figure 1 shows SEM micrographs of electrochemically-deposited ZnO layers. It is seen that the these layers consist of nano-whiskers (ZnO NWs). Their thickness and length increase from about 100 nm to 200 nm and from 1 µm to 3 µm, respectively, while their density decreases as the thickness of the middle Ag films in the seeding stack structures is raised. The electrochemically-deposited ZnO layer on the annealed stack structure exhibits NWs that are smaller in size and distributed more tightly than those on the as-deposited seeding stack.

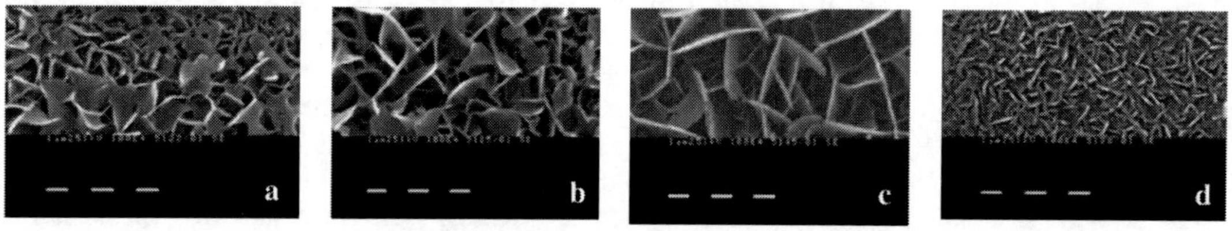

Figure 1. SEM micrographs of the electrochemically-deposited ZnO NW layers on different stack structures: (a) x = 10 nm, sample 3 on seeding stack 3S, (b) x = 16 nm, sample 2 on seeding stack 2S, (c) x = 20 nm, sample 1 on seeding stack 1S and (d) x = 20, sample 4 on seeding stack 4S (annealed). The markers correspond to 1 µm.

One should note that the diverse structures of the seeding layers with different resistance lead to different currents and inhomogeneity in the cathode field. This results in different growing speed and different morphology during the electrochemical deposition of the films. The changes in the RMS values are due to the different surface morphology of the ZnO NWs layers.

Figure 2 displays AFM images of the ZnO NW layers. It is seen that the rms values of the roughness of the seeding stacks are low (in the range 4 – 9 nm) and 10 nm in the annealed stack. The RMS roughness of the ZnO NWs layers is much higher and decrease from 330 nm to 93 nm with the

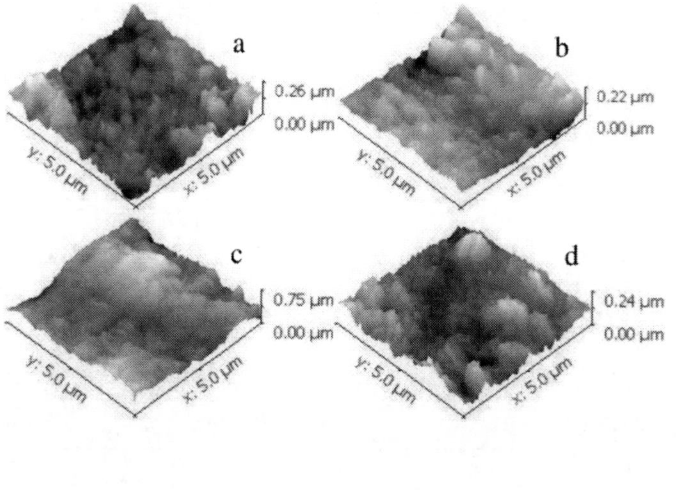

Figure 2. AFM images of the electrochemically-deposited ZnO NW layers on different stake structures: (a) sample 1, x = 20 nm on seeding stack 1S, (b) sample 2, x = 16 nm on seeding stack 2S, (c) sample 3, x = 10 nm on seeding stack 3S and (d) sample 4, x = 20 nm on seeding stack 4S (annealed).

increase of the Ag film thickness in the seeding stacks from 10 nm to 20 nm. The roughness of the ZnO deposited on the annealed stack $x = 20$ nm (sample 1) is the lowest (85 nm), which is probably related to the formation of separated Ag particles in the seeding stack and the resulting smaller size and tighter distribution of the grown NW.

The optical transmittance (a), reflectance (b), diffuse transmission (c) and reflection (d) spectra, and the haze ratio in transmission (e) and in reflection (f) of the ZnO NW layers deposited on different seeding stack structures (solid curves) are shown in figure 3. For comparison, the corresponding spectra of the seeding stack structures are presented as well (dashed curves). The presence is clearly seen of absorption and reflection plasmonic bands in the spectra of the seeding stack structures. The band at about 365 nm corresponds to the quadrupolar resonance and the near lying inter-band electron transitions in the entirely filled electron d-shell of the Ag atoms [8]. The position of the other band at about 460 nm due to the plasmon resonance in the Ag nano-grains is red-shifted as the thickness of the middle Ag film and the Ag grain size in it rise. The stacks deposited with Ag layers of 16 nm and 20 nm, which have a higher conductivity, demonstrate a plasmon absorption due to the free-carrier absorption in the IR region (figure 3a). As the thickness of the middle Ag layer in the stack structures is reduced (and its resistivity rises), the transmittance of the samples in the IR region increases, as is very clearly pronounced in the annealed-stack case (figure 3d). The same bands of absorption and reflection due to the plasmon effect are observed in the electrochemically-deposited ZnO layers, albeit

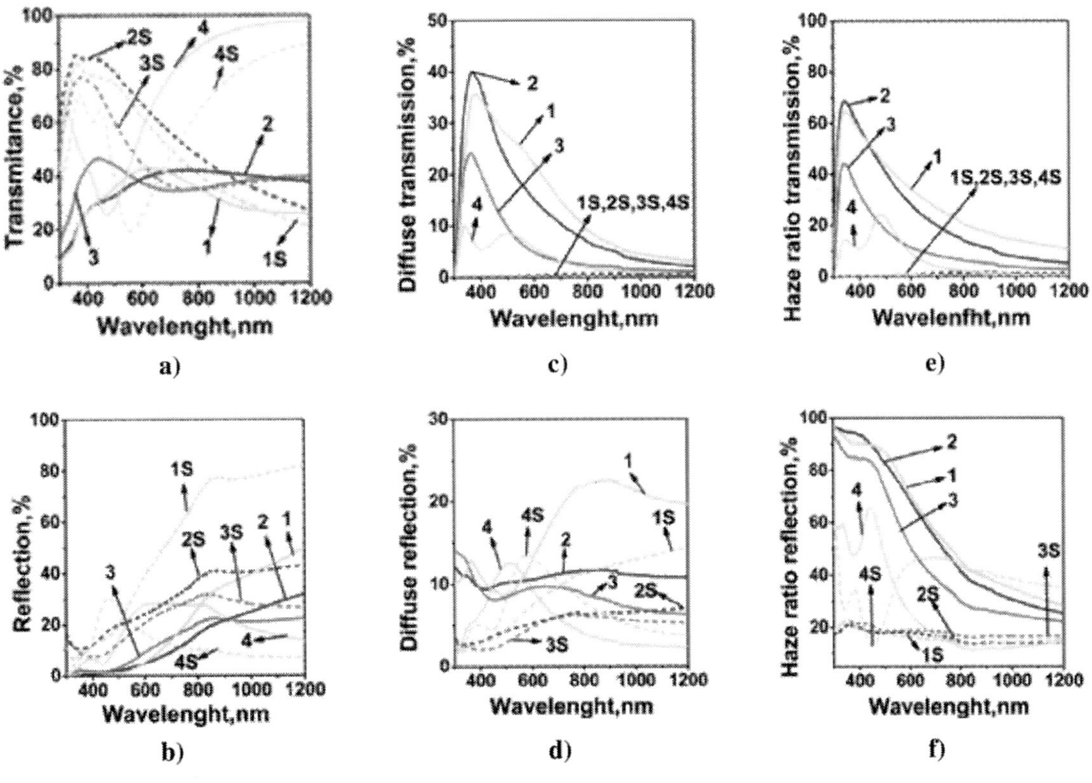

Figure 3. Spectra of optical transmittance (a), reflectance (b), diffuse transmission (c) and reflection (d), haze ratio in transmission (e) and in reflection (f) of the ZnO nanostructured layers (solid curves) deposited on different seeding stack structures: sample 1, $x = 20$ nm on seeding stack 1S, sample 2, $x = 16$nm on seeding stack 2S, sample 3, $x = 10$ nm on seeding stack 3S and sample 4, $x = 20$ nm on seeding stack 4S (annealed); the corresponding spectra of the seeding stack structures (dashed curves) are shown for comparison.

less well pronounced. The ZnO NW layers grown electrochemically on as-deposited stack substrates possess lower transmittance (figure 3a) and reflectance (figure 3b) values. However, they have lower reflectance and higher diffuse transmission and reflection values (figure 3 c and d) than those of the seeding stack. The haze ratios in transmission (figure 3e) and in reflection (figure 3d) are much higher than those of the seeding stack structures, which corresponds with their higher RMS surface roughness (table 1). The haze in transmission and reflection depends weakly on the thickness of the middle Ag film in the seeding stacks for $x = 20$ nm and $x = 16$ nm. The diffuse reflectance values are > 40 % for wavelengths $\lambda < 800$ nm, and 85 – 90 % at $\lambda = 400$ nm, in the case of ZnO NW grown on the as-deposited stack structures. Only in the case of deposition on the annealed seeding stack are the diffuse transmission and reflection and the corresponding haze ratio lower, probably due to the denser distribution of the NW in the grown layers and the lower RMS roughness, (RMS = 85). The ZnO layer deposited electrochemically on ZnO/Ag/ZnO stacks have higher values of the diffuse reflection and the haze ratio in reflectance in the visible range of the spectra, as compared to those of the seeding stacks. This correlates with the values of the surface roughness.

4. Conclusions

The study of nanostructured ZnO layers with nano-wiskers deposited electrochemically on multilayer ZnO:Al/Ag/ZnO:Al stack structures demonstrated that the optical and structural properties of the grown ZnO NW layers depend on the properties of the seeding multilayer stack structures. Changing the thickness of the middle Ag film and annealing the stack structure result in changes of its structural and electrical properties. The size of the nano-grains in the middle Ag film of the seeding stack affects the NW ZnO layers growth – it determines the size of the nano-whiskers and the surface roughness in the nanostructured ZnO layers as well. The stack structures exhibit a plasmonic effect due to the presence of Ag nano-grains. Plasmonic properties of the grown ZnO NW are also observed, although less well pronounced than in the seeding stack structures. The electrochemically-deposited ZnO layers demonstrate very high values of the diffuse reflection and the haze ratio in reflection in the visible and near IR spectral ranges. ZnO layers having a larger size of the nano-whickers and higher values of RMS roughness are grown on the stack structure with a 20-nm thick Ag film, which includes Ag nano-grains connected in chains and has a higher conductivity. They demonstrate higher values of the diffuse reflection and of the haze ratio in reflection in the spectral range of 400 nm – 1100 nm. The metal nanoparticles in the seeding stacks, together with the nano-wiskers in the electrochemically-grown ZnO, could enhance the scattering of the light in the solar cells in the visible and near IR spectral ranges. The nanostructured ZnO layers possessing a higher diffuse reflection, an increased effective area, a higher surface roughness and plasmonic properties can be applied as rear contacts in thin-film solar cells with improved light-harvesting properties.

Acknowledgements

This work supported by the 7 European FP undeer project NanoPV No. 24331 and by the Bulgarian National Science Fund under project No. 859/2012.

References

[1] Granqvist C G 2007 *Solar Energy Mater. & Solar Cells* 2007 **91** 1529
[2] Kopenkamp R, Boedecker K, Lux-Steiner M C, Poschenrieder M, Zenia F, Levy-Clement C, and Wagner S 2000 *Appl. Phys. Lett.* **77** 2575
[3] Chen J Y and Sun K W 2010 *Solar Energy Mater. & Solar Cells* **94** 930
[4] Dimova-Malinovska D, Lovchinov K, Ganchev M, Angelov O, Graff J S and Ulyashin A 2013 *Phys. Stat. Sol.* A **210** 737
[5] Tan H, Santbergen R, Smets A H M and Zeman M 2012 *Nano Lett.* **12** 4070
[6] Sahu D R and Huang J-L 2006 *Thin Solid Films* **515** 876
[7] Lovchinov K, Angelov O and Dimova-Malinovska D 2012 *J. Phys.: Conf. Series* **398** 012017
[8] Kreibig U and Vollmer M 1995 *Optical Properties of Metal Clusters* **25** (Springer)

Journal of Physics: Conference Series **514** (2014) 012014 doi:10.1088/1742-6596/514/1/012014

Asymmetric one-dimensional photonic crystal for optical sensing in the visible spectral range

A Lalova[1] and R Todorov

Acad. J. Malinowski Institute of Optical Materials and Technologies,
Bulgarian Academy of Sciences, Acad. G. Bonchev Str., Bl. 109, 1113 Sofia, Bulgaria

E-mail: alalova@iomt.bas.bg

Abstract. A gas sensor based on an asymmetric one-dimensional (1D) photonic band gap structure with one defect layer was designed and fabricated through layer-by-layer deposition of spin-coated poly methyl methacrylate (PMMA) and vacuum-deposited As_2S_3. Initially, the thickness variations, Δd, were determined of the thin films resulting from the poly methyl methacrylate exposure to chloroform vapor in the concentration range 100 – 9000 ppm. It was found that the value of Δd depends on the gas concentration and the exposure time. A two-layer structure was prepared consisting of PMMA and vacuum-deposited As_2S_3. Further, the permeability of thin As_2S_3 films to chloroform vapors was investigated. The asymmetric photonic structure consisted of 11 alternating layers of As_2S_3 and PMMA. The defect PMMA layer was located before the last high-refractive-index film of chalcogenide glass. The thickness of the defect layer of PMMA was pre-calculated so that the pass band be centered at the wavelength of 550 nm. An offset was observed of the position of the pass band to the larger wavelengths after exposure to chloroform vapor. The multilayered structure proposed is promising for optical sensor applications.

1. Introduction

Photonic crystals can be defined as structures in which the dielectric constant (refractive index) has a periodic variation in one, two or in all three orthogonal directions. 1D photonic crystals, or the so-called multilayer structures, consist of alternating layers of two materials with different refractive indices resulting in a periodically varying refractive index in one direction but homogeneous in the other two directions. Optical elements based on quarter-wave stacks are widely used in different optical devices, such as reflectors, filters, and waveguides [1]. Planar omnidirectional reflectors have been reported with fundamental stop bands centered at the wavelength λ = 1.55 μm made of chalcogenide glass/polymer [2] or chalcogenide glass/chalcogenide glass [3-4].

The thickness of the layers in the Bragg stack is determined by the following equation:

$$nd = \lambda_0/4 \tag{1}$$

Where n is the refractive index of the layer, d is the thickness and λ_0 is the wavelength of the center of the fundamental reflection band of the Bragg stack. Making one of the layers with a different thickness, thee so-called defect layer, results in the appearance of a pass band in the stop band of the

[1] To whom any correspondence should be addressed.

Bragg stack.The defect layer behaves, in principle, as a cavity resonator when the resonant condition is satisfied. To this end, the thickness of this defect layer is usually chosen to be $nd = \lambda_0/2$. The position of the pass band is very sensitive to small changes of the thickness of the defect layer. This phenomenon makes the Bragg stacks with defect films very attractive as sensors for various external stimulus.

The sensitivity of the photonic band gap to small changes of the refractive index of the materials defining a photonic crystal is well known. The photonic crystal systems are responsive to external stimuli, such as solvents, vapors, temperature, pH [5], biomolecules [6], and mechanical force [7]. It is also known that gases possess a refractive index close to that of air, differing in the order of 10^{-4} [2]. It is easily estimated then from formula (1) that different gases permeating into the structure cannot cause a significant change in the position, λ_0, of the fundamental reflection band. Therefore, fabricating a gas sensor based on the changes of the refractive index is not possible and materials must be sought that change their volume under the influence of the gas that would be the object of detection. From the literature it is known that upon contact with chloroform thin PMMA (poly methyl methacrylate) films increase their thickness by $\Delta d = 13.6 - 19.7$ % [8]. In a previous work [9], the possibility was demonstrated of preparing a Bragg stack from As_2S_3/PMMA for gas-sensing application in the infrared spectral range.

The aim of the present work is modelling an asymmetric photonic structure that switches from transparent to Bragg reflector states (characterized by a photonic stop band in the visible range), as well as its subsequent testing as a gas sensor. The chalcogenide glass composition was chosen such that the material would be transparent in the larger part of the visible region while at the same time possessing a sufficiently high refractive index. Our previous studies [10] have shown that thin films of this composition have a band gap of 2.4 eV (516 nm).

2. Experimental details

A multilayered filter was fabricated comprising eleven alternating thermally-evaporated As_2S_3 and spin-coated PMMA layers. The thickness of all films was calculated using equation (1) for $\lambda_0 = 550$ nm with the exception of the last PMMA film before the chalcogenide film which completes the Bragg filter. The thickness of this PMMA film satisfies the relation $nd = \lambda_0/2$. The polymer layers are obtained by pouring a drop of 0.3 ml of a 2.5 wt % polymer solution. The speed and duration of spinning were 2000 rpm and 30 s, respectively. To remove the extra solvent, the samples were annealed for 30 min at a temperature of 60 °C. The chalcogenide films were deposited by conventional thermal evaporation at a base pressure of 2×10^{-3} Pa and a deposition rate of 0.1 nm/s.

Optical transmittance and reflectance measurements at normal incidence of the light beam were carried out in the spectral range from 350 nm to 2000 nm using a Cary 05E UV–VIS–NIR spectrophotometer. The spectrophotometer is equipped with a gas cell which allows *in situ* measurements of spectrophotometric quantities in the presence of a gas with a given concentration.

The optical constants (refractive index, n, and extinction coefficient, k) and thickness, d, of the thin films were calculated using transmittance and reflectance measurements and applying the double methods developed by Abeles and Theye [11] and a procedure developed in [12, 13].

3. Results and discussion

We modelled and prepared a multilayered coating from chalcogenide glass and a polymer operating as a Bragg stack with one defect layer and a stop band in the visible spectral range. As_2S_3 chalcogenide glass and PMMA were used as high- and low-refractive-index materials, respectively. The dispersion of the refractive indices of both materials are given in figures 1a and 1b respectively.

Further, in a separate experiment, the gas sensitivity of a PMMA layer to chloroform was determined. Figure 2a represents a plot of the variation in the reflectance at $\lambda = 450$ nm of a thin PMMA film with a thickness of 92 nm during exposure to chloroform at concentrations ranging from 100 to 2000 ppm.It is seen that notable variations in the reflectance as a result of

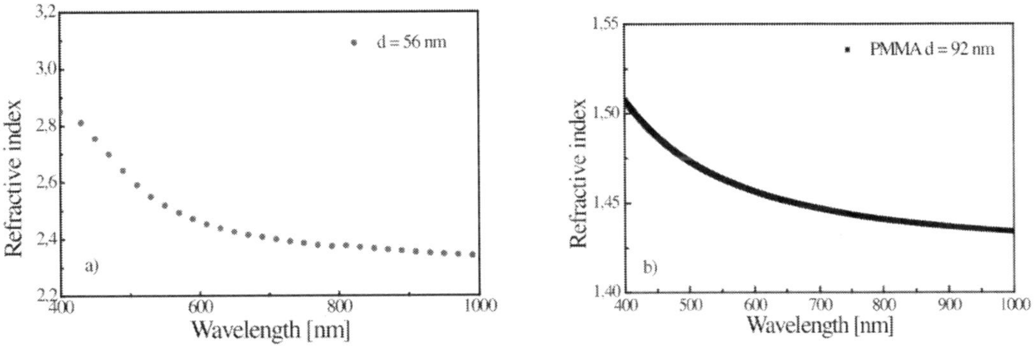

Figure 1. Dispersion of the refractive indices of thin films of As_2S_3 (a) and PMMA (b).

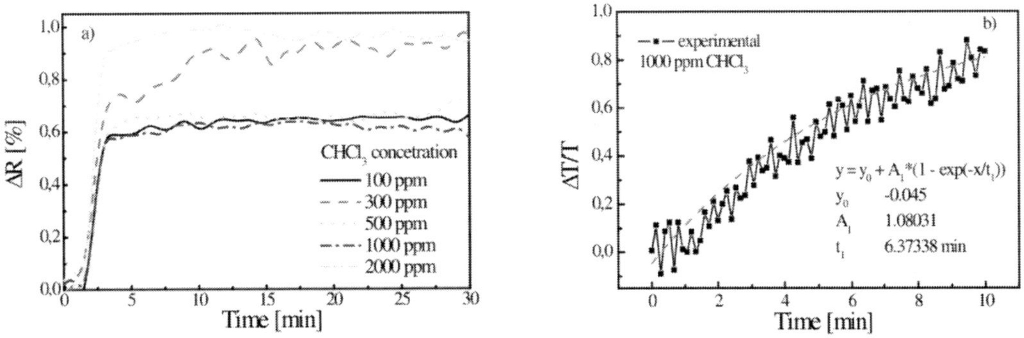

Figure 2. Variations in the reflectance of a single PMMA film exposed to chloroform gas in the concentration range 100 – 2000 ppm (a), and transmittance, ΔT, of a three-layer structure exposed to gaseous chloroform at a concentration of 1000 ppm (b).

the changes in the PMMA thickness start after ~ 30 s of exposure to chloroform and reach saturation after 2.5 – 3 minutes.

In a prior work [9, 10], we demonstrated that chloroform vapor induces negligible variation in the thickness and refractive index of the thin chalcogenide films. This is why our next step was to determine the permeability to chloroform of the thin As_2S_3 film. To this end, three-layer coatings consisting of As_2S_3/PMMA/As_2S_3/substrate were prepared. The relative changes of the transmittance at a $CHCl_3$ concentration of 1000 ppm are shown in figure 2b. It is seen that the time necessary to reach a 1 % variation is approximately 10 min. It is thus concluded that the chalcogenide film delays the reaction of the PMMA film by approximately a factor of 20. It is seen in figure 2b that the time dependence of $\Delta T/T$ follows an exponential law of the type $y = y_0 + A_1(1\text{-}exp(\text{-}t/t_1))$,where y_0 is the initial value of the transmittance, and A_1 = 1.08 and t_1 = 6.37 min are parameters. A similar functional dependence was observed in [14] for the penetration of oxygen in silver films.

Further, the investigations continued with testing the sensitivity of the Bragg stack. The thickness of the layers in the stack is calculated to be a quarter-wavelength at the target wavelength, λ_0, i.e., $n_H d_H = n_L d_L = \lambda_0/4$.The center wavelength in the present work is λ_0= 550nm. The target thickness of the As_2S_3 films was calculated to be 56 nm.The spin coated poly-methyl methacrylate films were with a thickness of ~ 92 nm for the layers in the Bragg stack and $d_D = \lambda_0/2n_L$=183 nm for the defect film. An evaluation of the reflectance as a function of the number of layers in the stack is shown in figure 3. An increase of the reflectance in the spectral range 450 – 700 nm and the formation of a stop band are clearly seen.The maximum reflectance value is R = 98.2 %, which was achieved for a nine-layer coating. The presence of a defect layer creates defect modes located within the photonic band gap, which are analogous to the defect states generated in the forbidden band of a doped semiconductor [15]. The choice of the optical thickness of $\lambda_0/2$ for this layer is the reason for the appearance of a pass band centered at the wavelength $\lambda = 550$ nm in the 11-layer coating. The photonic band is split into two stop bands, the first one having a maximum in the blue spectral range at $\lambda = 490$ nm, with the second one being centered at $\lambda = 602$ nm.

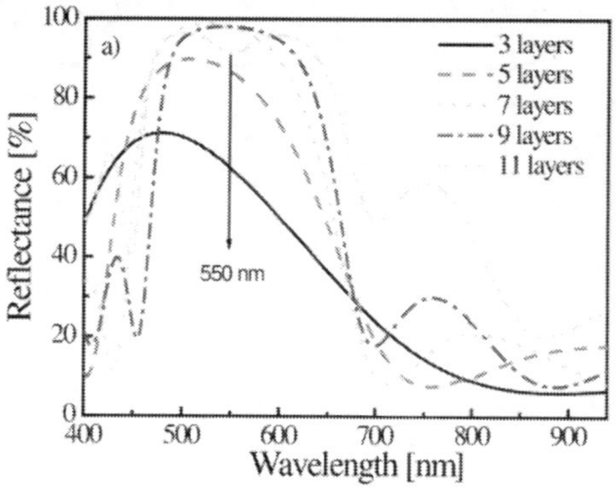

Figure 3. The formation of a stop band in the Brag reflector depends on the number of layers in the stack.

Figure 4 presents the variations in the transmittance spectra in the area of the photonic band gap when the multilayered structure is exposed at different durations to saturated chloroform vapor and to a 1000 ppm concentration of $CHCl_3$ in an argon atmosphere.

It is seen that the changes in the transmission increase with the time. Maximum values for ΔT of ~ 3.5 % and ~ 1.5 % are observed in the spectral range 500 – 650 nm for saturated chloroform vapor of and 1000 ppm gas, respectively.

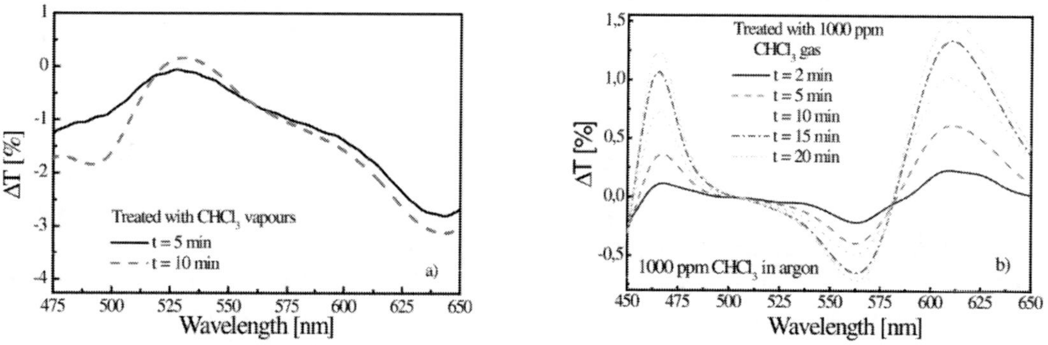

Figure 4. Variations of the transmittance in the area of fundamental reflection of the Bragg stack under exposure to saturated chloroform vapor (a), and to a gas mixture of chloroform at 1000 ppm concentration in an argon atmosphere (b).

4. Conclusions

A Bragg stack was designed with a stop band in the visible spectral range consisting of alternating layers of chalcogenide glass and PMMA with one defect polymer layer. The asymmetric one-dimensional photonic structure was tested as a gas sensor operating on the principle of optical detection. To this end, the penetration of chloroform through a single- and a three-layer coating was initially investigated. An offset of the pass band was observed after exposure to chloroform vapor. The structure fabricated demonstrates variation of the transmission in the blue, green and red spectral ranges after exposure to chloroform. The multilayered structure proposed represents a promising platform for optical sensing applications.

Acknowledgement

The authors would like to acknowledge the financial support under project BG05PO001-3.3-005/0001, Human Resources Development Operational Program, Science & Business Branch, contract D01-4122/17.6.2013.

Journal of Physics: Conference Series **514** (2014) 012014 doi:10.1088/1742-6596/514/1/012014

References

[1] Chattopadhyay S, Huang Y F, Jen Y J, Ganguly A, Chen K H and Chen L C 2010 *Mater. Sci. Engin.* R **69** 1
[2] Kohoutek T, Wagner T, Orava J, Krbal M, Ilavsky J, Vesely D and Frumar M 2007 *J. Phys. Chem. Solids* **68** 1268
[3] Kohoutek T, Orava J, Prikryl J, Wagner T, Vlcek M, Knotek P and Frumar M 2009 *J. Non-Cryst. Solids* **355** 1521
[4] Todorov R, Tasseva J , Babeva Tz and Petkov K 2010 *J. Phys.* D: *Appl. Phys.* **43** 505103
[5] Zakhidov A A, Baughman R H, Iqbal Z, Cui C, Khayrullin I, Dantas S O, Marti J and Ralchenko V G 1998 *Sci.* **282** 897–901
[6] Noda S, Chutinan A and Imada M 2000 *Nature* **407** 608–10
[7] Painter O, Lee R K, Scherer A, Yariv A, O'Brien J D, Dapkus P D and Kim I 1999 *Sci.* **284** 1444
[8] Capan I, Tarimci C, Hassan A K and Tanrisever T 2009 *Mater. Sci. Engin.* C **29** 140-3
[9] Babeva T, Marinov G, Tasseva J, Lalova A and Todorov R 2012 *J. Phys.: Conf. Series* **398** 012025
[10] Lalova A and Todorov R 2012 *J. Phys.: Conf. Series* **398** 012023
[11] Abeles and Theye M L 1966 *Surf. Sci.* **5** 325
[12] Laaziz Y, Bennouna A, Chahboun N, Outzourhit A and Ameziane E L 2000 *Thin Solid Films* **372** 149
[13] Todorov R, Lalova A, Petkov K and Tasseva J 2012 *Semicond. Sci. Technol.* **27**/11 115014
[14] Moore W and Codella P J 1988 *J. Phys. Chem.* **92** 4421
[15] Wu C-J and Wang Z-H 2010 *Prog. Electromagnetics Res.* **103** 169-84

Color tunability in multilayer OLEDs based on DCM and DPVBi as emitting materials

P K Petrova[1], P I Ivanov and R L Tomova

Acad. J. Malinovski Institute of Optical Materials and Technologies, Bulgarian Academy of Sciences, Acad. G. Bonchev Str., Bl. 109, 1113 Sofia, Bulgaria

E-mail: petia@iomt.bas.bg

Abstract. We report studies on the color tunability of a novel type of multilayer organic light-emitting diodes (OLEDs) based on three emitting materials: DCM (4-(Dicyanomethylene)-2-methyl-6-[*p*-(dimethylamino)styryl]-4H-pyran) as a red emitter, DPVBi [4,4′-Bis(2,2-diphenylvinyl)-1,1′-biphenyl] as a blue emitter and zinc bis(2-(2-hydroxyphenyl) benzothiazole) ($Zn(BTz)_2$) as a yellow emitter, and an electron transporting layer. We established that the positions and thicknesses of the different emitting layers determine the efficiencies, luminance and color of the light emitted by the OLEDs.

1. Introduction

During the last two decades, OLEDs have attracted a wide scientific and commercial interest because of their high potential for applications as paper-like light sources, backlight for liquid crystal displays, and full-color OLEDs [1-4]. With the ability to be self-emissive, and being extremely thin and light, OLEDs are particularly suitable to be the future commonly-used light sources. The successful development of OLEDs relies on the capability to obtain emission over the full visible spectrum. OLEDs emit various colors depending on the emitting fluorescent dyes used, which are also the key materials affecting the luminance, efficiency, turn-on voltage and lifetime. White OLEDs (WOLEDs) are of foremost interest for lighting and display applications [5]. Several routes are usually employed to obtain white light, such as, e.g., mixing of three (red, blue, green) or two (complementary) colors in a single host matrix [6,7,8], using several physically separate layers [9] emitting simultaneously different colors [10,11,12], using bimolecular excited species (excimer/exciplex emission [13,14].), etc. However, fine tuning of the color emission and achieving bright white emission are still problematic.

In this work, we report an investigation on the color tunability of a multilayer OLED based on three emitting materials: DCM (4-(Dicyanomethylene)-2-methyl-6-[*p*-(dimethylamino)styryl]-4H-pyran) as a red emitter, DPVBi [4,4′-Bis(2,2-diphenylvinyl)-1,1′-biphenyl] as a blue emitter and zinc bis(2-(2-hydroxyphenyl) benzothiazole) ($Zn(BTz)_2$) as a yellow emitter and an electron transporting layer to achieve white emission. Three types of OLED structures with different sequences and thicknesses of the undoped emitting layers were studied and their electroluminescent characteristics were established.

[1] To whom any correspondence should be addressed.

2. Experimental

We investigated electroluminescent devices with a conventional structure: ITO/HTL/EL/M, where ITO is a transparent anode of In_2O_3:SnO_2; HTL, a hole-transporting layer; EL, emitting layers; and M, a metal Al cathode. As HTL we used an in-house developed composite film of TPD incorporated in a poly(N-vinylcarbazole) (PVK) matrix. Devices with area of 1 cm^2 were prepared on commercial polyethylene terephtalate (PET) substrates coated with ITO (60 Ω/sq). The HTL (31 nm) of PVK:TPD_x (x = 10 w% with respect to PVK) composite films were obtained by spin-coating at 2000 rpm from a 0.75 % solution in dichloroethane. The ELs of DPVBi, DCM and $Zn(BTz)_2$ (with total thickness of 75 nm) and the metal cathode (120 nm) were deposited by thermal evaporation in vacuum better than 10^{-4} Pa at rates 2 – 5 Å/s. An *in-situ* quartz crystal was used to monitor the thickness of the vacuum depositions. The PVK, TPD and DCM were purchased from Sigma-Aldrich, the ITO-covered PET substrate and the DPVBi, from Kintec Compdny, and the $Zn(BTz)_2$, from Yurui (Shanghai) Chemical Co. Ltd.

All measurements were performed with unpackaged devices, at room temperature and under ambient atmosphere. The electroluminescence spectra (El) and the chromaticity coordinates (CIE – Commission Internationale de L'Eclairage) were recorded by an Ocean Optics HR2000+ spectrophotometer. The current-voltage (*I-V*) and luminance–voltage (*L-V*) characteristics were measured by National Instruments Data Aqusition boards controlled by the Labview 7.1 software. The luminance (*L*) was measured in a DC mode, while the light output was detected using a calibrated S2281-01 Hamamatsu silicon photodiode. The electroluminescent efficiency (η_L) was calculated by equation (1) and used for quantifying the properties of the OLEDs:

$$\eta_L = L/I, \qquad (1)$$

where L is the luminance (in cd/m^2) and I is the current density (in A/m^2).

3. Results and discussion

The DCM molecule is a highly fluorescent orange-red laser dye widely used in OLEDs. The red emission can be obtained either through direct excitation of the DCM fluorophores or through Förster energy transfer from the excited host matrix. DPVBi is known as a blue emitting material in OLEDs, usually used as an emissive layer interposed between HTL and EL.

To generate white emission, we mixed the light emitted by separate layers emitting in red (DCM), yellow $Zn(BTz)_2$ and blue (DPVBi). We investigated two types of structures: based on two emitters – blue (DPVBi) and yellow ($Zn(BTz)_2$), and on three emitters (DPVBi, $Zn(BTz)_2$ and DCM). The total thickness of the electroluminescent layers in all the structures was kept constant at 75 nm. Figure 1 shows the schematic energy-band diagrams of these devices. The HOMO and LUMO values are taken from the literature [15-17].

3.1. Structures based on two emitters - DPVBi and $Zn(BTz)_2$

OLEDs with three different thicknesses of the DPVBi layer (5 nm, 15 nm and 60 nm) in the structure HTL/$DPVBi^x$/$Zn(BTz)_2^{75-x}$/Al were investigated. The luminance–voltage curves (*L–V*), the

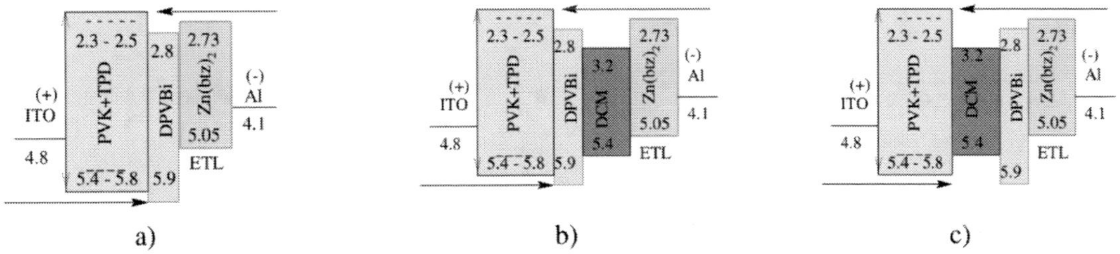

Figure 1. Schematic energy-band diagrams of the two types of OLED structures based on: a) two ITO/ PVK:TPD/ $DPVBi^x$/ $Zn(BTz)_2^{75-x}$/ Al, and three emitting materials in different order b) PVK:TPD/ $DPVBi^x$/ DCM^1/ $Zn(BTz)_2^{74-x}$/ Al and c) PVK:TPD/ DCM^1/ $DPVBi^x$/ $Zn(BTz)_2^{74-x}$/ Al.

electroluminescent efficiency (η_L) and the El spectra of these devices are shown in figure 2 and summarized in table 1. It was found that decreasing the thickness of the DPVBi layer from 60 nm to 5 nm raises the electroluminescent intensity and efficiency of the devices (from 384 cd/m^2 to 1240 cd/m^2 and from 1.61 to 2.47 cd/A, figures 2a and 2b), passing through a maximum at a thickness of the DPVBi layer of 15 nm.

The structure based entirely on DPVBi (60 nm) emits at 496 nm (figure 2c). The device with DPVBi^{15}/Zn(BTz)$_2^{60}$ showed a narrow blue emission band at 497 nm; that with DPVBi^5/Zn(BTz)$_2^{70}$, a broad emission band centered at 513 nm with a full width at half maximum (FWHM) of 27 nm. It was established that the decrease in the DPVBi layer thickness (x) shifts the CIE (x,y) coordinates of the

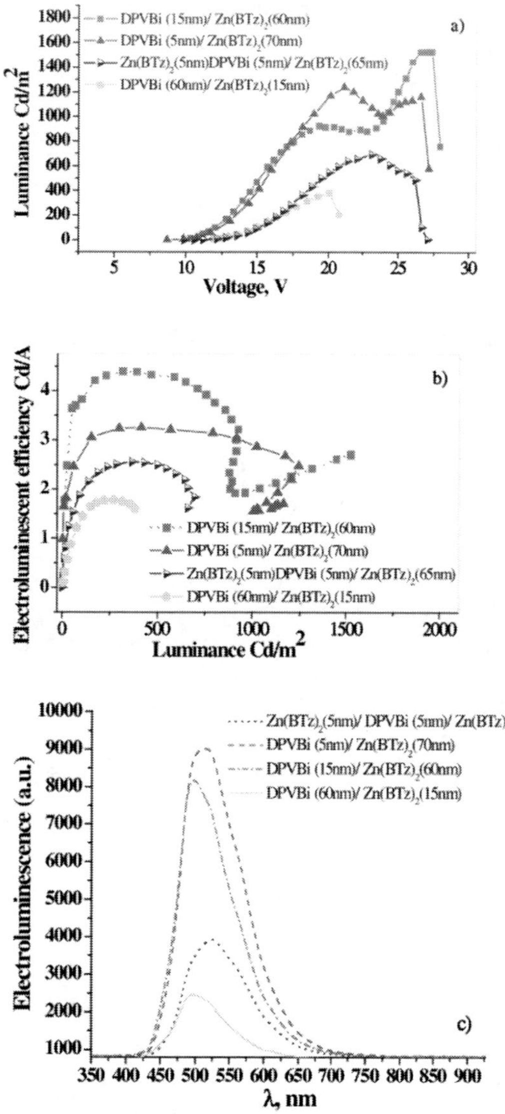

Figure 2. a) Electroluminescent intensity (L), b) Electroluminescent efficiency (η_L) and c) El spectra (at 16 V DC) of devices based on two emitters ITO/PVK:TPD/DPVBi^x/Zn(BTz)$_2^{75-x}$/Al.

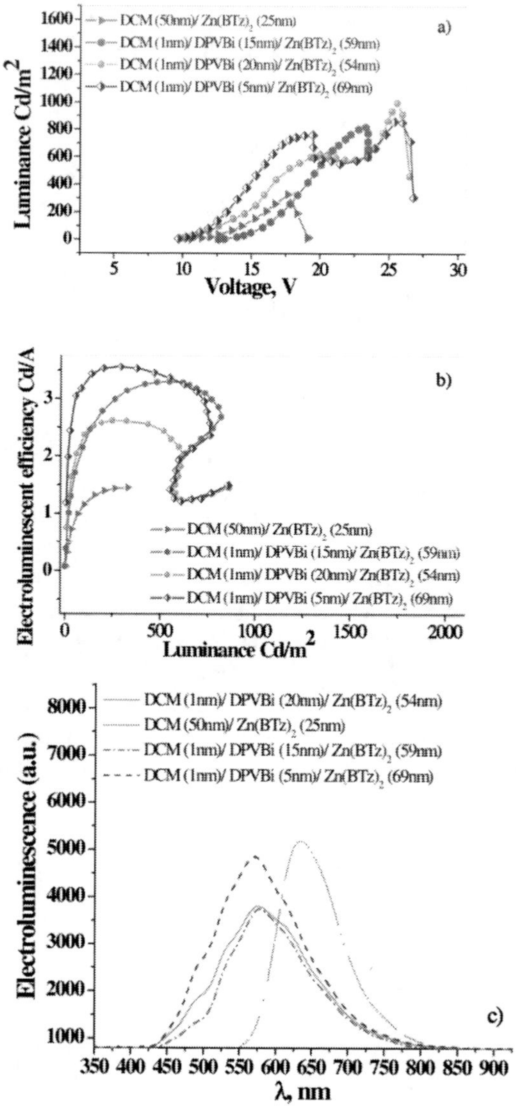

Figure 3. a) Electroluminescent intensity (L), b) Electroluminescent efficiency (η_L) and c) El spectra (at 16V DC) of the devices based on three emitters PVK:TPD/DCM^1/DPVBi^x/Zn(BTz)$_2^{74-x}$/Al.

Table 1. Characteristics of the fabricated OLEDs with the following structure: ITO/ PVK:TPD/ emitting layers/Al where only the sequences of the emitting layers are given.

Sequences of the emitting layers in OLEDs	L_{max} (cd/m^2)	η_L (cd/A)	λ (nm)	λ_{max} at 16 V (nm)	Shoulder (nm)	CIE at 16 V (x,y)
$DPVBi^{60}$/ $ZnBTz^{15}$	384	1.61		496	520	0.2074, 0.3232
$DPVBi^{15}/Zn(BTz)_2^{60}$	929	3.21		497	524 weak	0.2389, 0.3922
$DPVBi^{5}/Zn(BTz)_2^{70}$	1247	2.47		513 broad		0.2698, 0.4435
$Zn(BTz)_2^{5}/DPVBi^{5}/Zn(BTz)_2^{65}$	695	1.85		522 broad	560	0.2863, 0.4551
$DCM^{50}/Zn(BTz)_2^{25}$	325	1.4		634		0.6722, 0.3197
$DCM^{1}/DPVBi^{20}$ $/Zn(BTz)_2^{54}$	616	2.08		575	490 broad, 530, 615	0.4493, 0 4361
$DCM^{1}/DPVBi^{15}/Zn(BTz)_2^{59}$	821	2.69		575	490 broad, 535, 613	0.4851, 0.4288
DCM^{1}/ $DPVBi^{5}/Zn(BTz)_2^{69}$	766	2.58		570 broad	495, 534, 615	0.4223, 0.4311
$DPVBi^{20}$ $/DCM^{1}/Zn(BTz)_2^{54}$	416	1.42	491	620 broad	663	0.5597, 0.3732
$DPVBi^{15}$ $/DCM^{1}/Zn(BTz)_2^{59}$	777	2.22	491	613		0.5389, 0.3783
$DPVBi^{5}/DCM^{1}/Zn(BTz)_2^{69}$	740	2.20	491	612		0.5156, 0.3874
$DPVBi^{5}/DCM^{1}/DPVBi^{10}/Zn(BTz)_2^{59}$	923	2.33		572 broad	493, 613	0.4429, 0.4507

emitted light from the greenish-blue region (0.2074, 0.3232) at $x = 60$ nm to the yellowish-green one (0.2698, 0.4435) at $x = 5$ nm. This shift is due to the displacement of the recombination zone from the HTL/DPVBi interface (for the thick DPVBi layer) to the DPVBi/$Zn(BTz)_2$ interface (for the thinner DPVBi layer).

The insertion of a 5-nm $Zn(BTz)_2$ layer between HTL and a 5-nm DPVBi in the multilayer structure HTL/$Zn(BTz)_2^{5}/DPVBi^{5}/Zn(BTz)_2^{65}$ strongly reduces the electroluminescent efficiency but widens the El spectrum (a broad band at 522 nm and a shoulder at 560 nm). The CIE (x,y) coordinates of the emitted light are 0.2863, 0.4551, which are referred to in the literature as coordinates of "warm white" light.

3.2. Structures based on three emitters – DPVBi, $Zn(BTz)_2$ and DCM – emitting in the blue, yellow and red

3.2.1. Structure with DCM evaporated on the HTL. The idea was to achieve a white emission via mixing red light emitted by a 1-nm thick DCM layer, with blue light emitted by a DPVBi layer with different thicknesses x, and yellow emission from $Zn(BTz)_2$, where the DCM layer lies on the HTL. Devices with structure PVK:TPD/$DCM^{1}/DPVBi^{x}/Zn(BTz)_2^{74-x}$/Al with x = 5 nm, 15 nm and 20 nm (figure 1 b) were fabricated. Their L–V curves, η_L and El spectra are presented in figure 3 a), b) and c), and the results obtained are summarized in table 1.

From the L–V curves and the data in table 1 one can conclude that the device with a 15-nm DPVBi showed the best electroluminescent intensity (821 cd/m^2) and efficiency (2.69 cd/A). As seen in figure 3c, the emission band of the DCM-based OLED is at 650 nm. The addition of DCM to the double emitting DPVBi/$Zn(BTz)_2$ structure leads to a red shift and a widening of the El spectrum of the OLED. For devices in which the DCM lies on the HTL, a broad emission band centered at 575 nm with a shoulder at 650 nm corresponding to DCM, and a weak emission band at 490 nm corresponding to DPVBi, were observed. Increasing the DPVBi layer thickness from 5 nm to 15 nm shifts the CIE (x,y) coordinates of the emitted light from the orange (0.4223, 0.4311) to the orange-red region (0.4851, 0.4288). A further increase of the DPVBi thickness from 15 nm to 20 nm causes a slight reversion to the orange region (0.4493, 0 4361).

3.2.2. Structure with DCM evaporated on a HTL/DPVBi layers stack. Devices with structure HTL/$DPVBi^{x}/DCM^{1}/Zn(BTz)_2^{74-x}$ with $x = 5$ nm, 15 nm and 20 nm were fabricated; their L-V characteristics, η_L and El spectra are presented in figure 4 a), b) and c), respectively. It is seen that the device with 15-nm DPVBi shows again the best performance: an electroluminescent intensity of 777 cd/m^2 and an efficiency of 2.22 cd/A.

As seen in figure 4 c), a red and a weak blue emission band (from DPVBi) at about 491 nm predominate in the El spectra of the devices with this structure. As can be seen, the increase in the DPVBi layer thickness leads again to a shift of the El spectra of the devices to the longer wavelengths and a displacement of the CIE (*x*,*y*) coordinates from the orange-red (0.5156, 0.3874) for the 5-nm DPVBi layer to the red (0.5597, 0.3732) region for the 20-nm DPVBi layer.

Figure 4 presents the performance of another device, for which the thin DCM layer of the structures discussed above is inserted between a 5-nm and a 10-nm DPVBi layers.

The device HTL/DPVBi^5/DCM^1/DPVBi^{10}/Zn(BTz)$_2^{59}$ demonstrates still higher values of the electroluminescent intensity (923 cd/m^2) and efficiency (2.33 cd/A). In the spectrum of this structure, a broad band emission is observed centered at 572 nm with a FWHM of 15 nm. Weaker emissions of DPVBi and DCM are seen as shoulders at 493 and 613 nm, respectively. However, the broad emission range notwithstanding, the device emits most intensely in orange-red.

The variation of the dominating emission color from the OLEDs can be explained by the different position of the recombination zone in the structures ensued from the different mobility of the charges in the emitting layers [18]. Thus, by changing the thickness of the DPVBi, one can adjust the recombination zone and the resulting color of the emitted light.

In the device HTL/DCM/DPVBi/ Zn(BTz)$_2$, where the DCM is inserted between the HTL and DPVBi, there is one barrier for electrons on the HTL/DCM interface (ΔE_{e-} = 9 eV) and one for holes on the DCM/ DPVBi interface (ΔE_{p+} = 0,5 eV). When the DCM is placed between the DPVBi and the Zn(BTz)$_2$ in the device HTL/DPVBi/DCM/ Zn(BTz)$_2$, there are two barriers for electrons: ΔE_{e-} = 0.4 eV on the DCM/DPVBi interface and ΔE_{e-} = 0.5 eV on the DPVBi/HTL interface; and one for holes (ΔE_{p+} = 0.5 eV) on the HTL/DPVBi boundary. As the mobility of electrons in organic compounds is twice as low as that of holes, increasing the thickness of the DPVBi layer impedes the motion of electrons to a larger degree than that of holes, the number of electrons accumulated on the DCM/DPVBi boundary starts decreasing and, as a result, a red shift of the recombination zone toward the DCM is observed for both structures. In this way, by changing the thickness of the DPVBi, fine tuning of the emission spectrum from orange to red can be achieved.

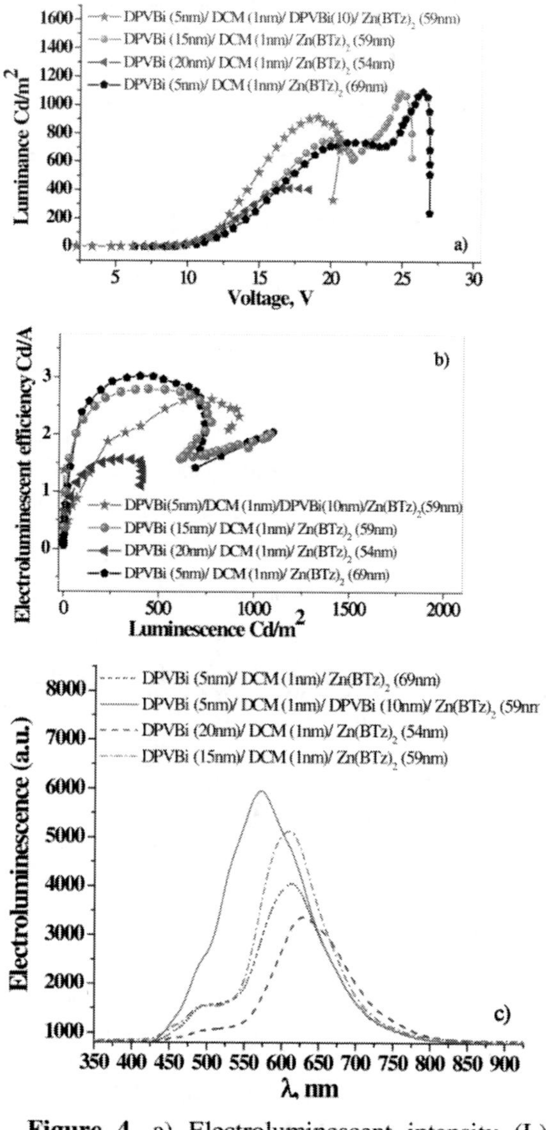

Figure 4. a) Electroluminescent intensity (L), b) Electroluminescent efficiency (η_L) and c) El spectra (at 16V DC) of the devices based on three emitters with DCM evaporated on HTL/DPVBi stack and structure PVK:TPD/DPVBi^x/DCM^1/Zn(BTz)$_2^{74-x}$/Al.

The results from the measurements reported suggest that the thickness of the DPVBi – which can act as a recombination-controlling layer in the multilayer devices – plays an important role in modifying the device's characteristics. It also provides a simple way of fabricating color-tunable OLEDs by simply changing the thickness of this "recombination-controlling" layer.

4. Conclusions

We investigated multilayer OLED structures based on two and three emitters DCM, $Zn(BTz)_2$ and DPVBi, emitting in red, yellow and blue, for different sequence of the emitters and variation of their thicknesses, with the aim to achieve fine tuning of the device's color. We showed that the spectral emission depends on both the location of the emitters and the layers' thicknesses, and also established that an accurate control of the emitted color can be obtained by finely adjusting both of these parameters. We obtained an efficient warm white-light emission (CIE 0.2698, 0.4435) from the device PVK:TPD/ DPVBi (5 nm)/ $Zn(BTz)_2$ (70 nm) with a luminance of 1247 cd/m^2 and a luminous efficiency of 2.47 cd/A.

References

[1] Segal M, Singh M, Rivoire K, Difley S, van Voorhis T and Baldo M A 2007 *Nat. Matters* **6** 374
[2] Reineke S, Lindner F, Schwartz G, Seidler N, Walzer K, Lьssem B and Leo K 2009 *Nature* **459** 234
[3] Nalwa H S and Rohwer L S 2003 *Handbook of Luminescence, Display Materials and Devices* vol 1
[4] *Organic Light-Emitting Devices, Synthesis, Properties, and Applications* 2006 ed K Müllen and U Scherf (WILEY-VCH Verlag GmbH & Co. KGaA Weinheim)
[5] D'Andrade B W and Forrest S R 2004 *Adv. Mater.* **16** 1585
[6] Tang C W, van Slyke S A and Chen C H 1989 *J. Appl. Phys.* **65** 3610
[7] Peng K-Y, Chen S-A and Fann W-S 2001 *J. Am. Chem. Soc.* **123** 11388
[8] Dodabalapur A, Rothberg L J and Miller T M 1994 *Appl. Phys. Lett.* **65** 2308
[9] Sun Y, Giebink N, Kanno H, Wa B, Thompson M and Forrest S 2006 *Nature* **440** 908
[10] Kido J, Kimura M and Nagai K 1995 *Science* **267** 1332
[11] Jordan R H, Dodabalapur A, Strukelj M, Miller T M 1996 *Appl. Phys. Lett.* **68** 1192
[12] Kalinowski J, Marco P D, Cocchi M, Fattori V, Camaioni N and Duff J 1996 *Appl. Phys. Lett.* **68** 2317
[13] Thompson J, Blyth R I R, Anni M, Gigli G and Cingolani R 2001 *Appl. Phys. Lett.* **79** 560
[14] Gupta D, Katiyar M and Deepak M 2006 *Opt. Mater.* **28** 295
[15] Weng-Chang Sh, Su Y, Liang-Wen Ji 2007 *Mater. Sci. Eng.* A **445–446** 509
[16] Shi M-M, Chen H-Z, Sun J-Z, Ye J, and Wang M 2003 *Chem. Phys. Lett.* **381** 666
[17] Morgado J, Alccer L, Esteves M, Pires N and Gigante B 2007 *Thin Solid Films* **515** 7697
[18] Xie W, Hou J and Liu S 2003 *Semicond. Sci. Technol.* **18**/7 L42-L44

Tribological performance of ceramic coatings deposited on metal surfaces for micro-bearing biomedical applications

N Donkov[1], A Zykova[2], V Safonov[2,6], J Smolik[3], R Rogowska[3], V Luk'yanchenko[4] and S Yakovin[5]

[1]Emil Djakov Institute of Electronics, Bulgarian Academy of Sciences, Sofia, Bulgaria
[2]Kharkov Institute of Physics and Technology, National Science Center, Kharkov, Ukraine
[3]Institute for Sustainable Technologies, National Research Institute, Radom, Poland
[4]Inmasters Ltd., Kharkov, Ukraine
[5]Department of Physical Technologies, Kharkov National University, Kharkov, Ukraine

E-mail: v_safonov@kipt.kharkov.ua

Abstract. Modification of metal materials by means of ceramic coating deposition is an effective way of forming alternative bearing surfaces. Ceramic AlN, Al_2O_3 and nanocomposite oxynitride coatings are widely used as protective coatings against wear, diffusion and corrosion. The enhancement of the mechanical properties, such as hardness parameters, effective Young's modulus, toughness, elastic recovery and wear resistance of the coatings, is very important for the tribological performance of the next generation of ceramic-coated ball bearing devices.

1. Introduction

The formation of excessive amounts of metal wear debris, which cause osteolysis followed by loosening of the prosthesis, has brought about a renewed interest in metal bearings and module joint arthroplasty [1-3]. Modification of metal materials by means of deposition of ceramic coating is an effective way of forming alternative bearing surfaces. Ceramic AlN, Al_2O_3 and nanocomposite oxynitride coatings are widely used as protective coatings against wear, diffusion and corrosion [4-6]. The corrosion and wear resistance properties of modern coatings should be accurately tested in view of effective biomedical applications. It is necessary to evaluate the effect of the process variables, such as substrate preparation, surface texture, technological parameters and post-coating treatments, as these influence the tribological performance of the coatings. The successful application of ceramic coatings on metal bearing devices requires a further enhancement of the mechanical properties, such as hardness and effective Young's modulus.

2. Materials and methods

The aim of the study was to analyze the wear mechanism while focusing on the tribological and mechanical properties of the bearing surfaces. We conducted a thorough investigation to evaluate the

[6] To whom any correspondence should be addressed.

tribological parameters of the oxide and composite nitride/oxide coatings. The substrates were load-bearing materials, namely, a titanium-based alloy (Ti_4Al_6V) and stainless steel (1H18N9).

Al_2O_3, AlN and composite nitride/oxide AlN/Al_2O_3 coatings were deposited using a coating apparatus equipped with a high-vacuum pumping system with a base pressure of about 10^{-3} Pa. The current-voltage characteristics (CVC) of the magnetron discharge for different oxygen and nitrogen flow rates are presented in figure 1. In the experiments, the argon pressure was kept constant at 1.8×10^{-1} Pa.

The coatings were deposited at the following process parameters: magnetron discharge power 1 – 4 kW; power of the activated oxygen source up to 1 kW; Ar pressure p_{Ar} = 1,8×10^{-1} Pa; oxygen flux q = 30 sccm; nitrogen flux q = 23,5 sccm; magnetron voltage U_m = 570 V; magnetron current I_m = 8 A; total pressure p = 2,2×10^{-1} Pa. Under these conditions, the coating deposition rate achieved was 8 μm/hour. An ion source was used for etching and cleaning the surface of the samples before the deposition of the ceramic coatings.

Figure 1 presents the current-voltage characteristics of the magnetron with targets of aluminum in a mixture of argon with oxygen or nitrogen at various reactive gas flows.

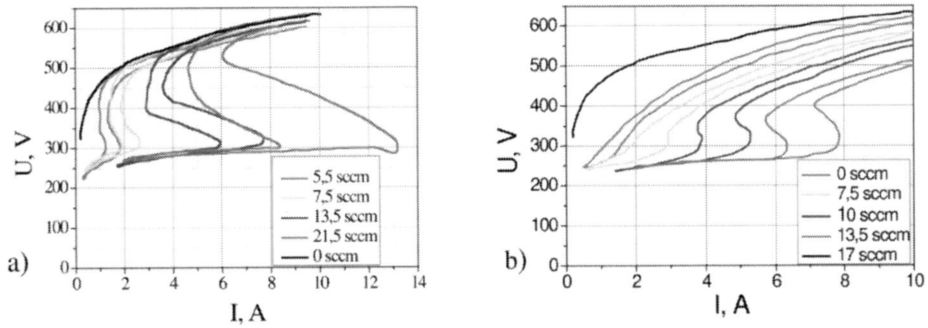

Figure 1. CVC of the magnetron discharge for different flows of oxygen (a) and nitrogen (b), argon pressure $p_{Ar} = 1.8\times10^{-1}$ Pa.

The optimum deposition conditions were realized for the upper part of the CVC curves of the magnetron discharge in argon for both the oxygen and nitrogen gases.

The coatings' adhesion properties, the hardness and the elastic modulus were evaluated by standard methods with the use of a Revetest (CSM Instruments) and a Rockwell indenter with a tip radius of 200 μm, within the load range of 200 N. X-ray diffraction patterns of Al_2O_3, AlN and composite coatings were taken by a DRON-3 diffraction device with filtered Cu-Kα radiation. X-ray photoelectron spectroscopy was carried out using an ESCALAB MkII (VG Scientific) electron spectrometer. The coating surface structure and morphology was estimated by means of a JEM 2100 scanning electron microscope (SEM) and an atomic force microscope (AFM, Quesant Instrument Corporation, USA).

Wear and abrasion resistance tests were performed on a CAT-S-AE (CSM Instruments) micro-scale abrasion tester. A 25-mm diameter micro-blasted ball of hardened steel (SAE 52100, 61 ± 2 HRC, $R_a = 2.5 \pm 0.3$ μm) was used as a counterpart. Suspensions of diamond, corundum and SiC particles (mean size 2 – 4 μm) were used as abrasive slurries for the tests. The rotation speed of the ball was set so as to provide a linear velocity of 0.1 m/s in all tests. The abrasive wear coefficient of the composite ceramic coatings was estimated according to a methodology described elsewhere [7, 8].

3. Results and discussion

The structure of the oxide and the composite magnetron-sputtered thin films was investigated by means of XPS and XRD. The X-ray diffraction patterns of the deposited coatings revealed a low crystallinity. After annealing at 800 °C, a crystalline phase appeared with the major Al_2O_3 peaks corresponding to (311), (222), (400), (440), and the AlN peaks corresponding to (100), (002), (101), (102), (110) reflections. A structural analysis of the coatings by means of XPS was also conducted [9]. The photoelectron spectra of Al2p, O1s, N1s and C1s were observed confirming the stoichiometric composition of the compound. The surface morphology was investigated by means of AFM and SEM. The surface of the Al_2O_3 coatings had a smooth relief with a uniform cross-section structure. The AlN and composite AlN/Al_2O_3 coatings had a crystalline structure with a surface morphology and structure of the fracture cross-section typical for the crystalline phase (figure 2.). The changes observed of the coatings' structure corresponded to the changes in the mechanical properties.

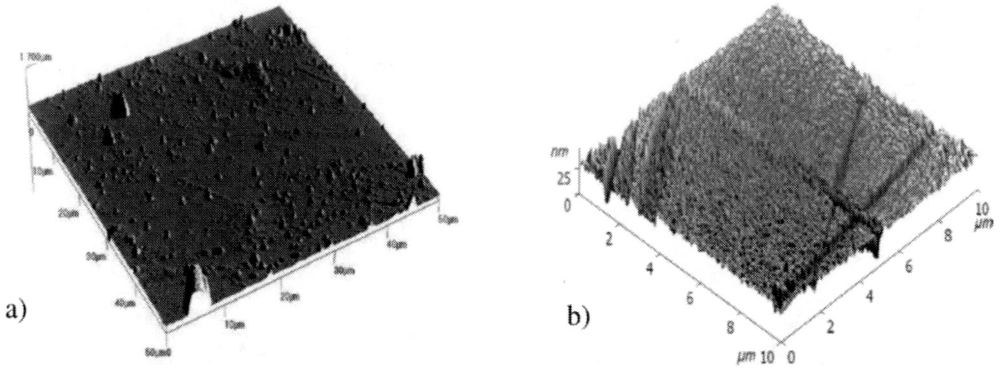

Figure 2. AFM images of the surface structure of the deposited oxide Al_2O_3 (a) and composite AlN/Al_2O_3 (b) coatings.

The mechanical properties of the coatings are the hardness *H* and the effective Young's modulus *E**. The mechanical behavior of the films is characterized by the ratio *H/E**[10, 11]. This ratio is proportional to the fracture toughness of the film and the resistance of the material to plastic deformation. Films with lower values of the effective Young's modulus in general have enhanced resistance to cracking and plastic deformation. It is well known that there is a strong correlation between the mechanical properties and the coating structure. For example, incorporating a nano-crystalline phase transition layer of a composite coating in the amorphous matrix strongly increases the values of *H* and *H/E** and the mechanical properties of the nitride/oxide composite films.

The mechanical parameters of Al_2O_3, AlN and composite nitride/oxide AlN/Al_2O_3 coatings deposited on stainless steel (SS) and Ti alloy substrates are presented in table 1.

Table 1. Mechanical characteristics of composite coatings deposited on SS and Ti alloy substrates.

Material/ Coating type	Mechanical parameters (average results 10 tests)		Mechanical Behavior	
	Hardness *H* [GPa]	Young's Modulus [GPa]	*H/E**	Adhesion [N]
SS/ Al_2O_3	9.7	174.7	0.057	43.1
SS/ AlN	14.3	184.4	0.078	50.3
SS/ AlN/ Al_2O_3	12.9	178.6	0.071	45.9
Ti/ Al_2O_3	9.2	170.3	0.052	37.1
Ti/ AlN	13.8	183.1	0.075	47.2
Ti/ AlN/ Al_2O_3	12.5	177.9	0.069	40.7

The total abrasive wear of composite ceramic coatings subjected to abrasion by various materials (diamond, corundum and SiC particles) is presented in figure 3. The composite coatings deposited on stainless steel and titanium alloy substrates demonstrated an increased wear resistance in comparison with the uncoated substrates and the single oxide and nitride layers.

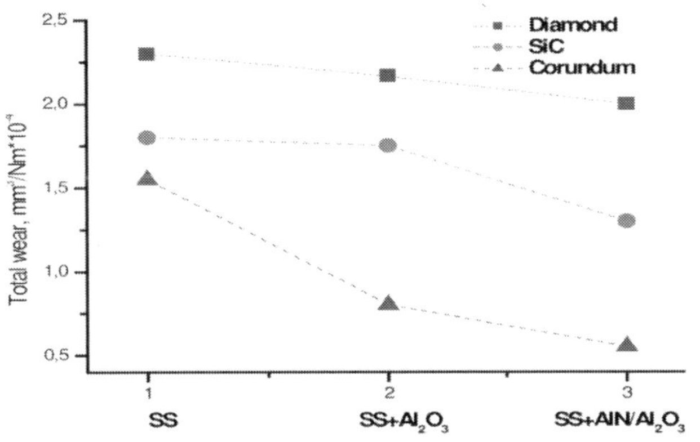

Figure 3. Total surface abrasive wear of SS, Al_2O_3 and AlN/Al_2O_3 composite coatings deposited on stainless steel substrates and abraded by different materials

The addition of a nitride layer to the composite coatings' structure strongly increased the films hardness (from 9 GPa to 14 Gpa) and changed the H/E^* ratio from 0.05 to 0.08. The incorporation of the nano-crystalline phase in the amorphous matrix strongly improved the nitride/oxide composite coatings' mechanical properties. The enhancement of mechanical properties, such as hardness, toughness, elastic recovery and wear resistance of nitride/oxide composite coatings may prove to be of considerable importance for many tribological applications.

4. Conclusions

The results reported in the paper demonstrate an improvement of the tribological characteristics of metal bearing surfaces upon covering by ceramic coatings. The wear resistance, hardness and toughness parameters increase in the case of composite nitride/oxide AlN/Al_2O_3 coatings. The deposition of a ceramic coating on metal substrates (stainless steel 1H18N9, titanium alloy Ti_6Al_4V) allows one to combine the ceramic materials' inertness with the hardness and failure strength of the supporting metals, which is of paramount importance for the next generation of bearing surfaces. The incorporation of a nano crystalline phase in the amorphous matrix strongly increases the mechanical properties of the nitride/oxide composite films. The application of ceramic coatings on metallic materials represents an effective alternative to ball bearing manufacturing.

References

[1] Campbell P *Clinic. Orthop.* 2010 **468**/9 2321
[2] Germain M A 2003 *Biomaterials* **24**/3 469
[3] Hasegawa M, Sudo A, Hirata H and Uchida A 2003 *J. Arthroplasty* **18**/5 658
[4] Klostermann H, Bufcher B, Fietzke F, Modes T and Zywitzki O 2005 *Surf. Coat. Technol.* **200** 760
[5] Trinh D H, Kubart T, Nyberg T, Ottosson M, Hultman L and Hogberg H 2008 *Thin Solid Films* **516** 8352
[6] Zykova A, Safonov V, Smolik J, Rogowska R, Luk'yanchenko V, Virva O and Yakovin S 2012 *Proc. 13th Int. Conf. Plasma Surface Engin.* (Garmisch-Partenkirchen Germany) PO3048
[7] Kusano Y, Van Acker K and Hutchings I M 2004 *Surf. Coat. Technol.* **183**/2–3 312
[8] Camara Cozza R 2013 *Surf. Coat. Technol.* **215** 224–33
[9] Safonov V, Zykova A, Smolik J, Rogovska R, Donkov N, Goltsev A, Dubrava T, Rassokha I and Georgieva V 2012 *J. Phys.:Conf. Series.* **356** 012046
[10] Leyland A and Mathews A 2004 *Surf. Coat. Technol.* **177-178** 317
[11] Musil J, Sklenka J and Cerstvy R 2012 *Surf. Coat. Technol.* **206** 2105

Modification of the structure and composition of $Ca_{10}(PO_4)_6(OH)_2$ ceramic coatings by changing the deposition conditions in O_2 and Ar

N Donkov[1], A Zykova[2,6], V Safonov[2], D Kolesnikov[3], I Goncharov[3], S Yakovin[4] and V Georgieva[5]

[1]Acad. E. Djakov Institute of Electronics, Bulgarian Academy of Sciences, 72 Tsaridradsko Chaussee, 1784 Sofia, Bulgaria
[2] Kharkov Institute of Physics and Technology, National Science Centre, Kharkov, Ukraine
[3] Belgorod State National Research University, Belgorod, Russia
[4]Department of Physical Technologies, Kharkov National University, Kharkov, Ukraine
[5]Acad. G. Nadjakov Institute of Solid State Physics, Bulgarian Academy of Sciences, 72 Tsaridradsko Chaussee, 1784 Sofia, Bulgaria

E-mail: zykova.anya@gmail.com

Abstract. Hydroxyapatite $Ca_{10}(PO_4)_6(OH)_2$ (HAp) is a material considered to be used to form structural matrices in the mineral phase of bone, dentin and enamel. HAp ceramic materials and coatings are widely applied in medicine and dentistry because of their ability to increase the tissue response to the implant surface and promote bone ingrowth and osseoconduction processes. The deposition conditions affect considerably the structure and bio-functionality of the HAp coatings. We focused our research on developing deposition methods allowing a precise control of the structure and stoichiometric composition of HAp thin films. We found that the use of O_2 as a reactive gas improves the quality of the sputtered hydroxyapatite coatings by resulting in the formation of films of better stoichiometry with a fine crystalline structure.

1. Introduction

The preparation of biomaterials for bone substitution, gene delivery and tissue engineering scaffolds applications has been the object of extensive research aimed at fulfilling the clinical requirements, namely, biocompatibility, biodegradability, mechanical and tribological stability [1]. Advances have been recently made in developing nano-crystalline hydroxyapatite ($Ca_{10}(PO_4)_6(OH)_2$ HAp) ceramics for orthopedic and dental applications. Orthopedic implants made of HAp and HAp-coated metal prostheses bond readily to bone and other body tissues without rejection or inflammatory reactions [2-6]. In this respect, the deposition conditions influence strongly the HAp coatings' structure, composition and capability of promoting bone regeneration [7-9]. Parameters, such as initial target

[6] To whom any correspondence should be addressed.

stoichiometry [10], gas pressure, discharge power, annealing and immersion treatment conditions [11], significantly affect the characteristics of the coatings [12]. The present study's aim was to analyze the effect of the different deposition conditions on the properties of hydroxyapatite ceramic coatings.

2. Materials and methods

Hydroxyapatite $Ca_{10}(PO_4)_6(OH)_2$ films were prepared by ion-beam sputtering of a ceramic HAp target in a vacuum chamber equipped with a gridless closed-drift ion-beam source allowing the use of different reactive working gases, such as O_2, and N_2 for etching, pre-cleaning and direct deposition processes [13]. The substrates were made of glass and a titanium alloy (Ti_6Al_4V).

The operating positive anode voltage was 2,0 – 4,0 kV. The working gas pressure (mixture of O_2 and Ar) was in the range $(1 - 6)\times10^{-2}$ Pa. The ion current was in the range 130 – 150 mA, the ion current density on the target being 10 mA/cm^2. The estimated coatings thickness was about 1.5 μm.

Figure 1 presents the current-voltage characteristics of the ion source at various working gas pressures. At pressures $(5 - 6)\times10^{-2}$Pa, a transition takes place from the discharge acceleration mode (curves' linear part) to the magnetron regime with stabilization of the current-voltage characteristics at current values above 100 mA. A pressure of approximately 5×10^{-2} Pa ensured optimal deposition conditions.

The structure and composition of hydroxyapatite $Ca_{10}(PO_4)_6(OH)_2$ films deposited by sputtering in inert Ar and reactive O_2 gases were studied by means of XRD and XPS. X-ray diffraction profiles of HAp were observed on a DRON-3 diffraction device with filtered Cu-Kα radiation. X-ray photoelectron spectroscopy was carried out using an ESCALAB MkII (VG Scientific) electron spectrometer at a base pressure in the chamber 5×10^{-8} Pa (1×10^{-6} P during the measurements) using an Al K-alpha X-ray source (excitation energy hν = 1486,6 eV). The coating surface structure and morphology was observed by AFM (Quesant Instrument Corporation, USA) and SEM (JEM 2100).

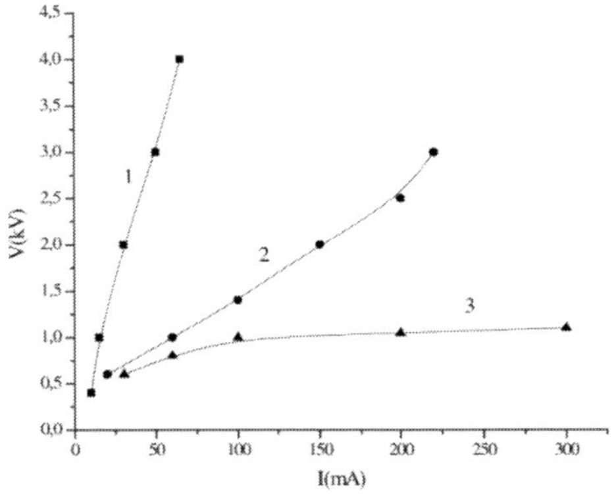

Figure 1. Current-voltage characteristics of the ion source at various working gas pressures. Solenoid current $I_s = 2$ A; pressure: 1 – $p = 4\times10^{-2}$ Pa, 2 – $p = 6{,}5\times10^{-2}$ Pa, 3 – $p = 8\times10^{-2}$ Pa.

3. Results and discussion

The XRD patterns of the target's composition and that of as-deposited and annealed HAp coatings are presented in figure 2. The target's diffraction peaks are close to the standard HAp peak positions. The as-deposited coatings have an amorphous structure. Hydroxyapatite crystalline phases with the major HAp peaks corresponding to (002), (211), (300), (202), (310), (222), (213) reflections were observed after the as-deposited films were annealed at 600 °C and then immersed in deionized water for hydroxyl group formation [11]. We proceeded with a structural analysis of as-deposited hydroxyapatite $Ca_{10}(PO_4)_6(OH)_2$ films produced by sputtering in an inert (Ar) and a reactive (O_2) gas. We also conducted an XPS survey scan of HAp coatings and a narrow scan of the Ca2p, P2p, O1s peaks.

The survey spectrum shows the peaks of core-levels binding energy at about 438,7 eV related to Ca2s, 347,5 eV related to Ca2p, and 133,5 eV corresponding to P2p. The O1s peak was observed at 531,5 eV. The peak observed at about 284,6 eV is assigned to C1s and is due to surface contamination (figure 3). High-resolution XPS spectra of Ca ($2p_{1/2}$), ($2p_{3/2}$) and P ($2p_{3/2}$) peaks are presented in figure 4. Table 1 summarizes the differences in the atomic concentrations in HAp coatings deposited in an inert (Ar) and a reactive (O_2) gas.

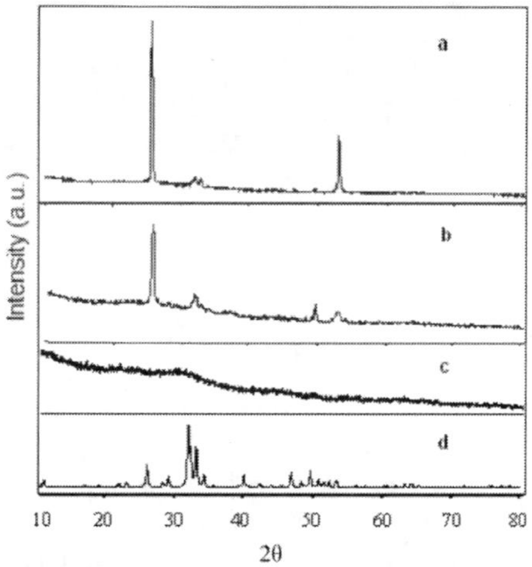

Figure 2. XRD patterns of the target's composition and that of as-deposited and annealed HAp coatings. a – annealed HAp (O_2) coating, b – annealed HAp (Ar) coating, c – as-deposited, d – target composition.

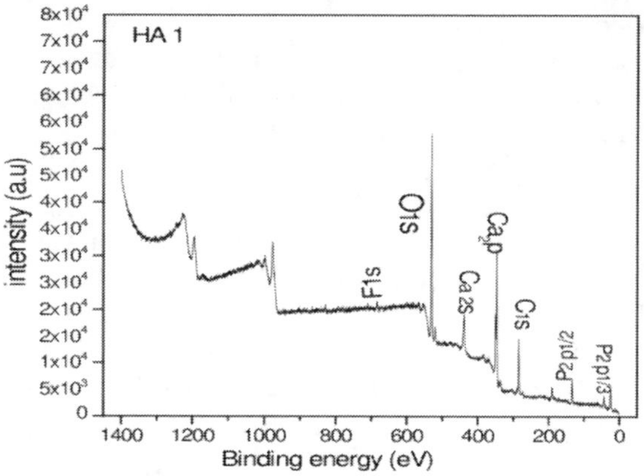

Figure 3. XPS survey scan of HAp coatings deposited in reactive O_2 gas.

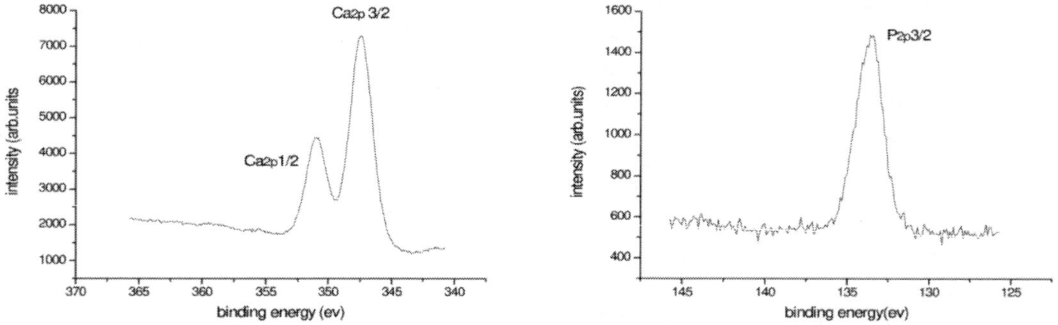

Figure 4. High-resolution XPS spectra of Ca ($2p_{1/2}$), ($2p_{3/2}$) and P ($2p_{3/2}$) peaks of HAp coatings deposited in reactive O_2 gas.

Table 1. Atomic concentrations of HAp coatings deposited in Ar and in O_2.

HAp samples	Peak area results (atomic concentration)				Ca/P ratio
	P2p	O1s	Ca2p	F1s	
HAp deposited in O_2	14.3	59.9	24.7	1.1	1.73
HAp deposited in Ar	13.2	58.5	26.9	1.4	2.04

The atomic ratio percentage Ca/P was 1.73 for deposition in reactive O_2 and 2.04 for deposition in inert Ar. In the standard hydroxyapatite $Ca_{10}(PO_4)_6(OH)_2$ composition, the calcium/phosphate ratio is 1.67. The XPS data showed a higher Ca/P ratio, which is the result of a calcium-enriched surface. The of Ca/P ratio deviation from the stoichiometric value may be due to a surface layer with a composition differing from that of the bulk material. A difference in the surface and bulk stoichiometric ratios has also been reported in [12, 14, 15].

The SEM micrographs of HAp samples showed a well defined platelet morphology. A uniform surface structure was observed in the case of HAp coatings deposited in the reactive O_2 gas. The difference in the surface structure of HAp films deposited by sputtering in inert Ar and in reactive O_2 was revealed by AFM (figure 5). Thus, the estimated average, S_a and mean-square S_q roughness parameters were S_a = 0,46 nm and S_q = 0.73 nm for HAp coatings deposited in O_2; and S_a = 1.61 nm, S_q = 2.18 nm for HAp coatings deposited in Ar.

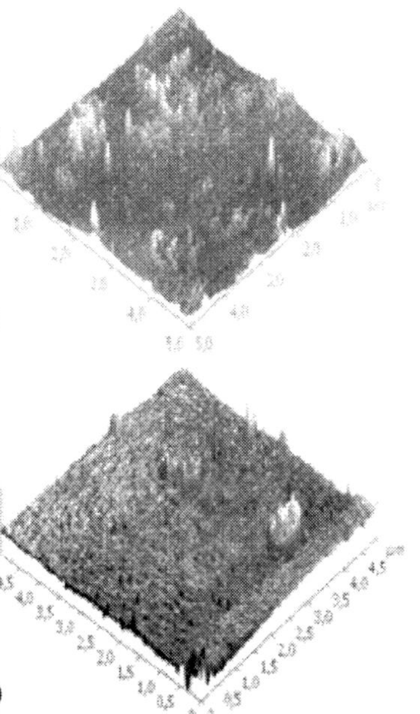
a)
b)

Figure 5. Surface structure of HAp films deposited by sputtering in inert Ar (a) and in reactive O_2 (b) gases.

4. Conclusions

The results demonstrate the strong effect of the sputtering process conditions on the structural and surface properties of the coating produced, so that optimal technological deposition conditions can be recommended. The structure and composition of hydroxyapatite films deposited by sputtering in inert Ar and in reactive O_2 gases were investigated by XPS, XRD, SEM and AFM. The atomic ratio percentage Ca/P was 1.73 for reactive O_2 gas deposition and 2.04 for inert Ar gas deposition. The HAp films formed in a reactive ambient are of better quality as their surface is smoother and their composition is closer to the stoichiometric one. Varying the deposition conditions allows one to improve the quality of the sputtered hydroxyapatite $Ca_{10}(PO_4)_6(OH)_2$ coatings by forming films of better stoichiometric composition with a fine crystalline structure.

Acknowledgement

This research was supported by the bilateral project 2011-2013 between BAS and NASU.

References

[1] Liu D, Savino K and Yates M Z 2011 *Surf. Coat. Technol.* **205** 3975
[2] Leventouri Th 2006 *Biomaterials* **27** 3339
[3] Onder A, Osman E-A and Sabri A 2008 *Surf. Coat. Technol.* **202** 2482
[4] Peng P, Kumar S, Voelcker N H, Szili E, Smart R and Griesser H J 2006 *J. Biomed. Mater. Res.* A **76** 347
[5] Eliaz N and Eliyahu M 2007 *J. Biomed. Mater. Res.* A **80** 621
[6] Suchanek W L, Byrappa K, Shuk R E, Pavel R, Janas V F and Tenhuisen K S 2004 *Biomaterials* **25** 4647
[7] Wang C X, Chen Z Q and Guan L M 2002 *Surf. Coat. Technol.* **130** 39
[8] Kokubo T, Kim H M and Kawashita M 2003 *Biomaterials* **24** 2167-75
[9] Silva C, Rocha H G B, Freire F N A, Santos M R P, Saboia K D A, Goes J C and Sombra A S B 2005 *Marerials Chemistry and Physics* **92** 260
[10] Yamashita K, Matsuda M, Arashi T and Umgeki T 1998 *Biomaterials* **19** 1239
[11] Zykova A V, Dudin S V and Polozhiy K I 2003 *APP Spring Meeting, Book of Papers* (Bad Honnef Germany) 95
[12] Boyd A R, Meenan B J and Leyland N S 2006 *Surf. Coat. Technol.* **200** 6002
[13] Shabalin A, Kishinevsky M and Quinn C 2001 *44th Annual Technical Conf. Proc. Society of Vacuum Coaters* (Philadelphia) 23
[14] Zhang E and Yang K 2005 *Trans. Nonferrous. Met. Soc. China* **5**/5 6326
[15] Zhang Q, Chen J, Feng J, Cao Y, Deng C and Zhang X 2003 *Biomaterials* **24** 4741

Transparent back contacts for P3HT:PCBM bulk heterojunction solar cells

M Sendova-Vassileva[1,3], H Dikov[1], G Popkirov[1], E Lazarova[1], V Gancheva[2], G Grancharov[2], D Tsocheva[2], P Mokreva[2] and P Vitanov[1]

[1]Central Laboratory of Solar Energy and New Energy Sources, Bulgarian Academy of Sciences, 72 Tsarigradsko Chaussee, 1784 Sofia, Bulgaria
[2]Institute of Polymers, Bulgarian Academy of Sciences, Acad. G. Bonchev Str., Bl. 103-A, 1113 Sofia, Bulgaria

E-mail: marushka@phys.bas.bg

Abstract. A new combination of layers functioning as a transparent contact is proposed and tested in real solar cells. The contacts consist of TiO_2 layers and thin metal layers (Ag, Cu) and are deposited by magnetron sputtering. The optical transmission and electrical conductivity of the transparent contact layers (TCL) are measured. The TCLs are applied as back contacts in bulk heterojunction polymer solar cells deposited on ITO covered glass and consisting of the following layers: ITO/PEDOT:PSS/P3HT:PCBM/back contact. The organic layers are deposited by spin-coating. For comparison, the same bulk heterojunction polymer solar cells are prepared with a sputtered Ag back contact. The first results show a dependence of the current-voltage parameters of the studied solar cells on the thickness of the different component layers of the transparent back contacts. There is a balance that has to be observed between the electrical characteristics of the contacts and their optical transparency. Future plans involve their inclusion as intermediate contacts in tandem organic solar cells.

1. Introduction

A large number works have been recently devoted to the development of transparent conductive layers (TCLs) suitable for use as electrodes in organic photovoltaic cells. These include the proposition to use three-layer structures consisting of a thin metal film (Ag, Au, Cu) sandwiched between metal oxide layers (ZnO, ZnO:Al, ITO etc.) [1] The advantages of these TCLs are that they can be thinner and more transparent than ITO for the same conductivity and avoid the use of the expensive element indium. Moreover, they can be deposited by sputtering at a low temperature, which is an advantage when fabricating organic solar cells. A number of papers have been published on the dependence of the conductivity and transparency of such films on the deposition method, composition, thickness of the constituent layers, etc. [2,3]. Such layers have been successfully applied in polymer-based solar cells.

TiO_2 and TiO_x layers have been used in organic solar cells mostly as single hole-blocking layers [4,5]. There are few publications on TiO_2/Ag/TiO_2 multilayers. They include a study of their optical properties and their application in dye-sensitized solar cells [6,7].

[3] To whom any correspondence should be addressed.

Journal of Physics: Conference Series **514** (2014) 012018 doi:10.1088/1742-6596/514/1/012018

Another incentive for the development and application of TCLs has been the attempt to develop inverted organic solar cells in which the solar light falls through the cathode instead of through the anode [4,5]. An advantage of such a structure is the higher stability under ambient conditions.

Transparent contact layers are also necessary for the implementation of tandem organic solar cells as the intermediate contact between two cells connected in series. This contact layer should consist of a hole transporting layer (HTL), an electron transporting layer (ELT) and a recombination layer between them which is often implemented as a thin metal layer or a layer of metal nanoparticles [8]. A number of materials have been experimented in view of these applications, e.g. NiO, MoO_3, V_2O_5, WO_3, ITO as hole transporting layers and ZnO, TiO_2 and Cs_2CO_3 as electron transporting layers. ZnO is not very suitable for some of these applications as it is attacked by the very acidic PEDOT:PSS used as a hole transporting layer in many experimental solar cells.

In this contribution, we present results obtained when transparent conducting layers consisting of multilayer structures of TiO_2 and metal components (Ag or Cu) are used as the cathode back contact of a polymer solar cell with a P3HT:PCBM bulk heterojunction active layer.

2. Experimental

The polymer solar cells studied were prepared in the following manner. Structured ITO covered glass slides from Ossila were used, ITO being the front anode layer. They were cleaned using a 10 % NaOH solution in an ultrasound bath, deionized water and isopropyl alcohol and dried in N_2 gas. Then a layer of PEDOT:PSS (poly(3,4-ethylenedioxythiophene):poly(styrenesulfonate)) Clevious P, purchased from K.C. Stark, was deposited by spin-coating at 1750 rpm for 60 s. After deposition, the film was annealed in Ar for 30 min at 130 °C. Next, the active layer was deposited by spin-coating from a 12 mg/ml 50:50 weight ratio solution in chlorbenzene of rr-P3HT (poly(3-hexylthiophen)) from Rielke Metals and PCBM ([6,6]-phenyl-C61-butyric acid methyl ester) from Solenne. The solution was obtained by stirring for 48 hours at 50 °C under argon and deposited at 400 rpm for 80 s. After deposition, the layer was kept for 10 min under the solvent vapor.

The back cathodes of the solar cells were deposited by magnetron sputtering in a Tokuda CFS-4ES sputtering equipment from two targets. They were either the transparent conductive layers studied in this contribution or a metal Ag contact for comparison. The Ag contact was sputtered for 5' at 200 W from a Kurt J. Lesker 99.99% silver target. The transparent contacts were deposited in one of the following sequences of material and times of deposition: 1) Ag 7"/TiO_2 6'/ Ag 7"/ TiO_2 4', 2) Ag 15"/ TiO_2 90"/ Ag 15"/ TiO_2 90", 3) Ag 7"/ TiO_2 45"/ Ag 7"/ TiO_2 45" and 4) Ag 15"/ TiO_2 60"/ Cu 15"/ TiO_2 60". The TiO_2 and Cu depositions were carried out at a power of 300 W from Kurt J. Lesker 99.99% targets without heating the substrates, while the Ag deposition was conducted as described above.

The solar cells prepared in this way were immediately encapsulated with a cover glass attached using a UV-light curable resin provided by Ossila. No glove box was used in the preparation process. The solar cells were further annealed in air on a hot plate for 10 minutes at 150 °C. This has been found [9] to improve greatly the characteristics of the solar cells. In this way, six solar cells, each with an area of 1.5x3 mm, were produced on the same substrate.

Finally, electric contacts were attached and the *I-V* characteristics were measured using a halogen lamp for illumination.

Separately, the transparent back contact layers were deposited on glass for measuring their conductivity and optical transmission. The sheet resistance was measured by the four-probe method and the optical transmission, using a Shimadzu UV-3600 spectrophotometer.

3. Results

Figure 1 shows the transmission spectra of the transparent back contacts studied. The absorption spectrum of the active layer of the polymer solar cell is also given for comparison with the transparency range of the conducting films. The films are used as back contacts in this instance, but

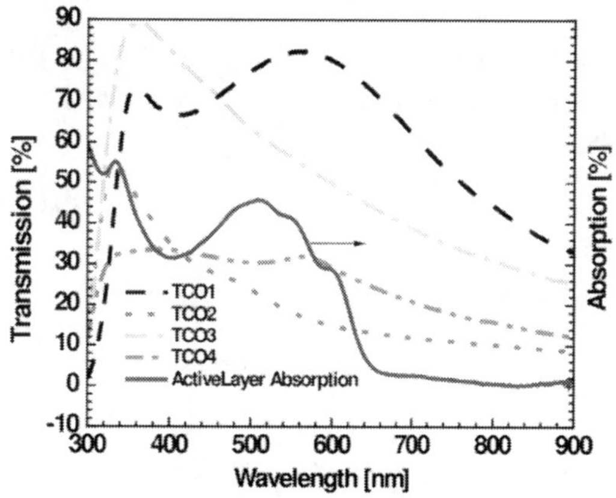

Figure 1. Transmission curves of the TCLs studied. The absorption spectrum of the polymer solar cells' active layer is also shown for comparison.

Table 1. TCL composition and conductivity.

Designation	Type	Sheet Resistance [Ω/sq]
TCO1	$Ag/TiO_2/Ag/TiO_2$ 7"/6'/7"/4'	6
TCO2	$Ag/TiO_2/Ag/TiO_2$ 15"/90"/15"/90"	9
TCO3	$Ag/TiO_2/Ag/TiO_2$ 7"/45"/7"/45"	6
TCO4	$Ag/TiO_2/Cu/TiO_2$ 15"/60"/15"/60"	4

the solar cell could be inverted and illuminated from the back through the contact. The next step of this work is planned to be the construction of tandem solar cells in which this transparent contact would be the recombination layer between the two cells and the second cell would be a P3HT:PCBM bulk heterojunction.

Table 1 shows the conductivity measured by the four-probe method of the different films studied. It should be noted that although we present data for the sheet resistance, measured horizontally, the vertical resistance of these layered films would be low enough. The thickness of the films is between 20 nm and 40 nm.

Figure 2 shows the current-voltage characteristics of typical solar cells made using the films considered as back contacts. The characteristic of an identical cell made with a sputtered Ag back contact is shown for comparison.

4. Discussion

The data presented in the previous section demonstrates that a very careful optimization has to be carried out on the structural, optical and electrical properties of the TCLs proposed for use in polymer solar cells. TCO1 shows a very high transparency but the corresponding solar cell has poor performance. TCO1 and TCO3 have in common the quite high transparency, which is connected with the very thin Ag layers used – sputtered only for 7". However, in both cases, in spite of the good planar conductivity, the current-voltage characteristics demonstrate low V_{oc} and J_{sc} as well as an S-shape deformation. It is interesting to note that the large difference in the thickness of the TiO_2 layers in the two cases has no effect on the planar conductivity and little effect on the *I-V* characteristics.

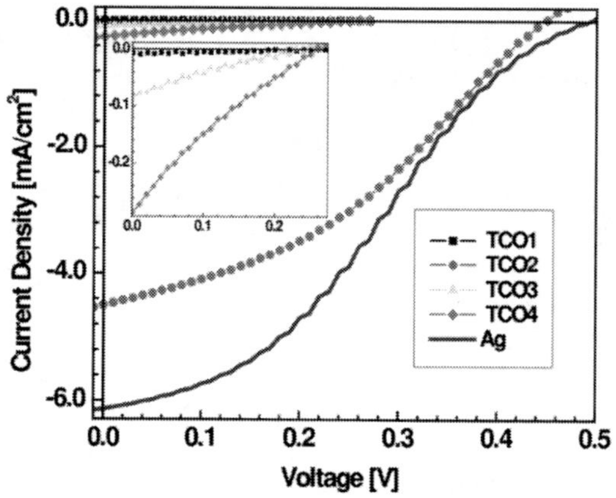

Figure 2. Current-voltage characteristics of polymer solar cells made using the studied TCLs as back contacts. The characteristic of a cell made using sputtered Ag as back contact is given for comparison. The inset shows more details of the characteristics at smaller short current densities.

Journal of Physics: Conference Series **514** (2014) 012018 doi:10.1088/1742-6596/514/1/012018

A thicker metal component sputtered for 15" improves the J_{sc} but lowers the transmission of the layers, as demonstrated by TCO2 and TCO4. The use of Cu in TCO4 leads to a flat transmission spectrum in the region where the P3HT:PCBM active layer has its maximum absorption. Unfortunately, the transmission is less than 40 %. The planar conductivity of that film is the best amongst the TCLs studied. However, the solar cell characteristics in this case are not much better than the ones for TCO1 and TCO3.

The best solar cell parameters, which are comparable to those of the cell with the Ag back contact, are observed for TCO2. In this case, the two metal component layers of the TCL are thicker and both composed of Ag. However this TCL does not exhibit a very good transparency. Its planar conductivity is also not the best one. As a whole, there is no correlation between the planar conductivity of the TCLs and the quality of the *I-V* characteristics of the respective solar cells. On the other hand, no great variation in the planar conductivity of the studied TCLs is observed.

One should bear in mind that the metal Ag back contact of the cell used for comparison serves as a mirror and reflects back to the active layer the light passing through, thus increasing the proportion of light utilized by the solar cell. This would increase the short-circuit current of the device in comparison with the ones with a transparent back contact. Therefore, the *I-V* characteristics of the cells with the TCO2 and the Ag back contacts can be considered comparable.

5. Conclusions

Four types of TCLs consisting of alternating metal and TiO_2 layers deposited by sputtering were studied as a back cathode of P3HT:PCBM bulk heterojunction organic solar cells. They all show a very good planar conductivity of between 4 and 9 Ω/sq. As they could be used as front contacts in inverted solar cells or as intermediate layers in tandem solar cells, their transparency was measured and compared. The best transparency is demonstrated by TCO1 (table 1). On the other hand, the best solar-cell characteristics are observed for the TCO2 containing a thicker Ag component next to the active layer of the solar cell.

References

[1] Guillén C and Herrero J 2011 *Thin Solid Films* **520** 1
[2] Shen L, Ruan S, Guo W, Meng F and Chen W 2012 *Solar Energ. Mater. Solar Cells* **97** 59
[3] Vedraine S, Hajj A E, Torchio P and Lucas B 2013 *Org. Electron.* **14** 1122
[4] Yeo K S, Nakao S, Hirose Y, Hasegawa T and Matsuo Y 2013 *Org. Electron.* **14** 1715
[5] Mohd Yusoff A R B, Kim H P and Jang J 2013 *Solar Energ. Mater. Solar Cells* **109** 63
[6] Dima I, Popescu B, Iova F and Popescu G 1991 *Thin Solid Films* **200** 11
[7] Ito S, Takeuchi T, Katayama T, Sugiyama M, Matsuda M, Kitamura T, Wada Y and Yanagida S 2003 *Chem. Mater.* **15** 2824
[8] Sista S, Park M-H, Hong Z, Wu Y, Hou J, Kwan W L, Li G and Yang Y 2010 *Adv. Mater.* **22** 380
[9] Sendova-Vassileva M, Popkirov G, Vitanov P, Dikov Ch, Gancheva V, Tsocheva D and Mokreva P 2012 *J. Phys.: Conf. Series* **398** 012049

Journal of Physics: Conference Series **514** (2014) 012019 doi:10.1088/1742-6596/514/1/012019

Optical properties of thin merocyanine dye layers for photovoltaic applications

J Dikova[1], S Kitova[1,4], D Stoyanova[1], A Vasilev[2], T Deligeorgiev[2] and S Angelova[3]

[1]Acad. J. Malinowski Institute for Optical Materials and Technologies, Bulgarian Academy of Sciences, Acad. G. Bonchev Str., Bl. 109, 1113 Sofia, Bulgaria
[2]St. Kliment Ohridski University of Sofia, Faculty of Chemistry and Pharmacy, 1 James Bourchier Blvd., 1164 Sofia, Bulgaria
[3]Institute of Organic Chemistry with Center of Phytochemistry, Bulgarian Academy of Sciences, Acad. G. Bonchev str., Bl. 9, 1113 Sofia, Bulgaria

E mail: skitova@iomt.bas.bg

Abstract. The potentiality was studied of our newly synthesized push-pull type merocyanine dye, labeled A1, for use as an electron donating component in solution-processed bulk heterojunction (BHJ) organic solar cells. For the purpose, a soluble n-type fullerene, (6,6)-phenyl C61 butyric acid methyl ester (PCBM), which is currently and in the ear future without an alternative, was chosen as an acceptor. The optical constants (*n* and *k*) of thin films obtained by spin coating from solutions in chlorobenzene of A1 and of an A1/PCBM blend were determined by spectrophotometric measurements. Further, an optical simulation of a standard BHJ cell with an active layer of an A1dye/PCMB blend was performed using a transfer-matrix formalism. Thus, the optimum thickness of the active layer was calculated to be about 80 nm, which provides overlapping of the total absorption with the solar spectrum in the broad range 400 nm – 800 nm. Finally, the maximum current density, J_{sc}, was determined to be 13 mA cm^{-2} assuming that the internal quantum efficiency, IQE, is unity. Comparing the calculated J_{sc} with data on some advanced small-molecule BHJ devices, the prospects for practical applications of the new merocyanine dye are discussed.

1. Introduction

During the past several years, the organic photovoltaics (PV) have been the subject of a continuously growing interest as a promising alternative to the expensive crystalline silicon cells (c-Si), which still dominate the market. In the most successful devices, a soluble n-type fullerene, (6,6)-phenyl C61 butyric acid methyl ester (PCBM), as an acceptor, is blended with a donor of conjugated p-type polymers to form bulk-heterojunction (BHJ) cells [1]. An advantage of these devices, which are processed from a solution, is the ability to deposit active layer precursors over large areas in a single step using techniques ranging from inkjet printing to spin-coating. Small-molecule donors offer the same simple processing, but are easier to synthesize in many variants and to purify, are intrinsically monodisperse, and are often environmentally more stable [2]. The disadvantage of the lower solubility of this class of organic semiconductors can be overcome by functionalizing their molecular structure. This approach has been successfully applied to different dyes and very encouraging results have been

[4] To whom any correspondence should be addressed.

achieved concerning the performance of devices developed on this basis [3]. However, the power conversion efficiency remains considerably lower than that of the polymeric solar cells [4]. This explains the increased scientific efforts devoted recently to developing new small-molecule electron donors for solution-processed BHJ solar cells. In this respect, the so-called "push-pull" chromophores look very promising [5]. The molecular structure of this unique class of dyes comprises electron-donating, π-conjugated bridging and electron withdrawing moieties. In the excited state, a strong intramolecular charge transfer occurs which provides an enlargement of the absorption area of the donor, increases the charge mobility and, therefore, raises the conversion efficiency [6].

The present work represents a first step in the characterization of our newly synthesized push-pull type merocyanine dye, which we labelled as A1. We determined the optical constants (refractive index n and extinctions coefficient k) of spin-coated thin A1 dye and A1/PCMB films using methods developed previously based on spectrophotometric measurements. Further, optical simulation was carried out in order to predict and understand the photovoltaic performance of a bulk heterojunction cell with an active layer consisting of a solid-state blend of A1 and the fullerene C60 derivative PCBM.

2. Experimental

The new push-pull dye A1 was synthesized via the reaction between 3-nitro-7H-benzo[de]benzo[4,5]imidazo[2,1-a]isoquinolin-7-one and 1-benzyl-4-methyl-quinolinium bromide in metoxyethanol in the presence of diizopropyl-ethylamine (DIPEA) and pyridine. The target merocyanine compound (E)-3-((1-benzylquinolin-4(1H)-ylidene)methyl)-7H-benzo[de]benzo[4,5] imidazo-[2,1-a]isoquinolin-7-one was obtained with excellent yield. A detailed description of the reaction procedure and the characterization data on the molecular and electronic structure of the A1 dye will be published in a separate paper.

Single dye films were prepared from a solution of the compound in chlorobenzene by spin coating. In the preliminary experiments, the concentration of the solution, the spin rate and the duration were varied in order to obtain films with thickness suitable for optical measurements. Thus, 25 nm films were deposited from a solution containing 4 mg/ml A1 coated at 1500 rpm for 120 s. Blended films were also prepared by spin coating from chlorobenzene solutions of 1:1 (by weight) mixture of A1 and PCBM (PCBM was commercially available). In both cases, carefully cleaned BK7 optical glass plates and Si wafers were used as substrates.

The optical constants of the A1 dye and the blended A1/PCMB layers were determined on the basis of three spectrophotometric measurements at normal light incidence – transmittance T and reflectance R_f and R_m of the dye and the blended films deposited on transparent (BK7) and opaque (Si wafers) substrates. The spectrophotometric measurements were carried out by a Cary 5E spectrophotometer in the spectral region 350 nm– 800 nm. A value of the dye-film thickness derived through applying the (TR_fR_m) method [7] was used for determining n and k, as well as the (TR_m) or (TR_f) methods using the Newton Raphson iterative techniques [8]. The combination of two methods made it possible to determine the optical constants of the dye and the blended films with a high accuracy [8].

Optical simulation of a BHJ device stack was carried out using the transfer matrix formalism based on the Fresnel formulas for the several interfaces occurring in the cell. This approach has been explained in great detail in the literature [9, 10]. In our work, a Matlab script developed by Burkhard and Hoke was applied, which treats the case of normal light incidence [11]. The absorption distribution for each wavelength over the film thickness was calculated in the 300 nm – 800 nm wavelength range. Multiplying by the light of the AM 1.5 solaar spectrum and integrating over the film thickness yielded the total number of photons absorbed at each wavelength. Finally, summation over all wavelength resulted in the total number of absorbed photons. The maximum current density J_{sc} was calculated assuming that each absorbed photons results in a collected electron, i.e. the internal quantum efficiency, IQE, was set equal to unity [11].

For the optical simulation we chose a common BHJ device structure consisting of ITO (120 nm)/MoO_3 (5 nm)/blend A1/PCBM /LiF (10 nm)/Al (100 nm). The 5-nm MoO_3 and the 10-nm

Journal of Physics: Conference Series **514** (2014) 012019 doi:10.1088/1742-6596/514/1/012019

LiF films are inserted between the active layer and the ITO and Al surfaces to form ohmic contacts for hole and electron collection, correspondingly. In the numerical simulation, the thickness of the blended layer was varied in the 40 nm – 320 nm range. The optical constants of ITO, Al and LiF used in the simulation were taken from the literature [12]. The optical constants of the MoO_3 films were determined by us using the same procedure as for the A1 dye and the blended films. For this purpose, 50-nm thick MoO_3 films were thermally evaporated in vacuum (10^{-4} Pa).

2. Results

Figure 1 shows the molecular structure of the A1 and the absorption coefficient α determined for a 25-nm thick dye film. As it is seen, there are two absorption peaks of $7{,}3\times10^4$ cm^{-1} and $5{,}5\times10^4$ cm^{-1} at λ = 598 nm and 402 nm, correspondingly. Besides, although the absorption of the film is moderate, the absorption area is broad and covers the 400 nm – 750 nm spectral region, where the solar photon flux is maximal. This, as well as the low intrinsic fluorescence of the new push-pull merocyanine dye, is an important prerequisite for its photovoltaic application.

To model the light propagation within the solar cell device, one needs to know the optical constants of all involved layers. The refractive indices (*n*) and the extinction coefficients (*k*) of each layer used in the simulation are plotted as a function of the wavelength in figure 2. The blended film's optical constants were determined for a film with a thickness of about 45 nm.

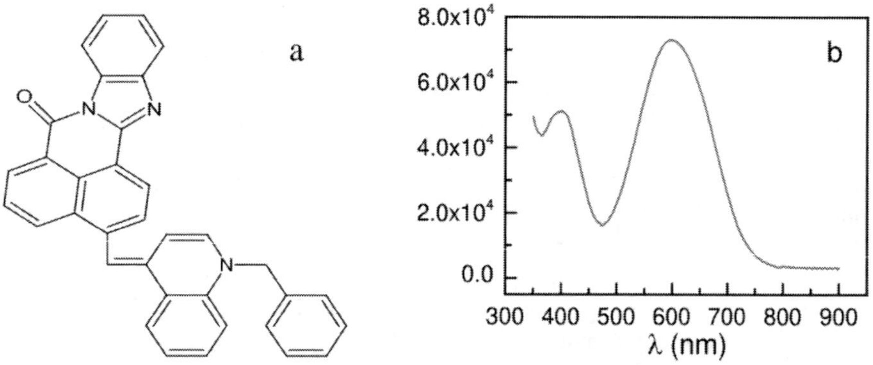

Figure 1. a) Molecular structure of dye A1. b) Absorption coefficient profile ($\alpha = 4\pi k/\lambda$) of a 25-nm thick dye film.

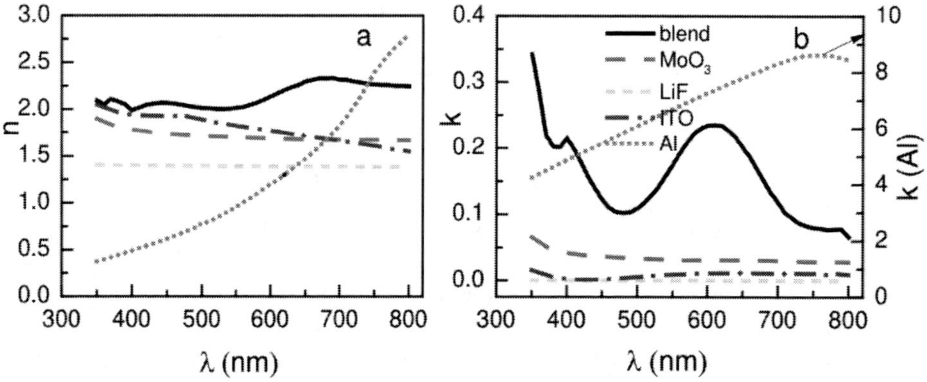

Figure 2. Optical constants of layers used in the simulation: a) refractive indices *n*; b) extinction coefficients *k*.

As already mentioned, the optical simulation was performed for a standard BHJ cell structure, where the light enters through the glass substrate and sequentially passes through the ITO, MoO_3, A1/PCBM blend and LiF layers. The light is then reflected back from the Al electrode and, finally, a part of the light leaves the solar cell through the front.

Figure 3 presents the spectral dependence of the total absorption within the active layer for a set of different layer thicknesses. For comparison, the photon flux of a standard AM 1.5 solar spectrum is shown as a gray solid line. It is clear that the blue light is absorbed mainly by the PCBM, and the red and near infrared, by the A1 dye. It is also seen that the blended film's absorption overlaps with the solar spectrum from 400 nm to 800 nm as required for high-efficiency solar energy conversion.

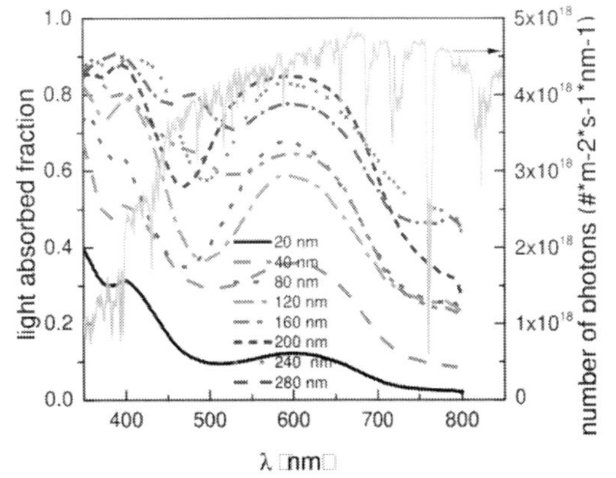

Figure 3. Fraction of incident light absorbed within the dye-PCBM layer of the solar cell for several layer thicknesses. For comparison, the photon flux of a standard AM 1.5 solar spectrum is shown as a gray solid line (right axis).

Assuming an IQE = 1, the maximum short-circuit current density J_{sc} under AM 1.5 illumination for different layer thicknesses is shown in figure 4. Due to the interference effects in the device structure, J_{sc} does not follow a straight line, but shows local maxima at about 80 nm and 230 nm. In general, the photocurrent is higher for a larger active layer thickness, but due to the limited exciton diffusion length in the organic layer, the real photocurrent decreases as the active layer thickness is increased. Therefore, we may expect an optimized overlap between the 80-nm active layer's absorption and the solar spectrum. J_{sc} of about 13 mA cm^{-2} is obtained for this thickness, which is comparable to the maximum current calculated for, e.g., MDMO-PPV (poly(2-methoxy-5-(3,7-dimethyloctyloxy)-1,4-phenylene vinylene) and PCBM bulk heterojunction solar cells [14].

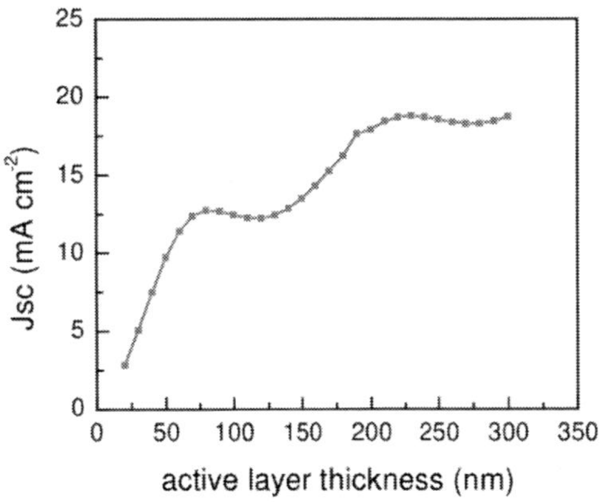

Figure 4. Calculated short-circuit current density J_{sc} under AM 1.5 spectral illumination (100 mW cm^{-2}) versus the active layer thickness (IQE = 1).

Experimental studies of a real device are in progress and are forthcoming in a separate paper.

4. Conclusions

The results of the present study show that thin films of the newly synthesized merocyanine dye A1 is characterized by a moderate absorption in a broad spectral region – between 400 nm and 750 nm. Moreover, the total absorption within a A1dye/PCBM blend layer overlaps with the solar spectrum from 400 nm to 800 nm. Therefore, it could be expected that, as an active layer, the blended film prepared will provide a good sunlight harvesting and, thus, a high-efficiency solar energy conversion. This is confirmed by the estimated value of the maximum density of the short-circuit current J_{sc} under AM 1.5 illumination for the optimum thickness of the blended film. The comparability of this value to literature data for some advanced solar devices demonstrates the potentiality of our new merocyanine dye for application as an electron donor in solution-processed BHJ solar cells.

References

[1] Dennler G, Scharber M and Brabec C 2009 *Adv. Mater.* **21** 1323–38
[2] Walker B, Kim C and Nguyen T 2011 *Chem. Mater.* **23** 470–82
[3] Xin Xo, Wei G, Wang S, Zimmerman G, Renshaw Ch, Thompson M and Forrest S 2012 *Adv. Mater.* **24** 1956-60
[4] Chen H, Hou J, Zhang S, Liang Y, Yang G, Yang Y, Yu L, Wu Y and Li G 2009 *Nature Photonics* **3** 649–53
[5] Ko H, Choi H, Paek S, Kim K, Song K, Lee J and Ko J 2011 *J. Mater. Chem.* **21** 7248–53
[6] So S, Choi H, Ko H, Kim C, Paek S, Cho N, Song K, Lee J and Ko 2012 *Solar Energy Mater. & Solar Cells* **98** 224–32
[7] Panayotov V and Konstantinov I 1994 *SPIE Proc.* **2253** 1070-74
[8] Konstantinov I, Babeva Tz and Kitova S 1998 *Appl. Opt.* **37** 4260-67
[9] Pettersson L, Roman L and Inganä O 1999 *J. Appl. Phys.* **86** 487-96
[10] Peumans P, Yakimov A and Forrest R 2003 *J. Appl. Phys.* **93** 3693-3723
[11] Burkhard G, Hoke E and McGehee M 2010 *Adv. Mater.* **22** 3293-7
[12] Smith D, Shiles E and Inokuti M 1985 *Handbook of Optical Constants of Solids* (New York: Academic)) pp 377-405
[13] Hoppe H, Arnold N, Sariciftci N and Meissner D 2003 *Solar Energy Mater. & Solar Cells* **80** 105-13

Sol-gel TiO_2 films as NO_2 gas sensors

V Georgieva[1], V Gadjanova[1], A Grechnikov[2], N Donkov[3,6], M Sendova-Vassileva[4], P Stefanov[5] and R Kirilov[4]

[1]Acad. G. Nadjakov Institute of Solid State Physics, Bulgarian Academy of Sciences, 72 Tsarigradsko Chaussee, 1784 Sofia, Bulgaria
[2]Vernadsky Institute of Geochemistry and Analytical Chemistry, Russian Academy of Sciences, 19 Kosygin Str., 119991 Moscow, Russia
[3]Acad. E. Djakov Institute of Electronics, Bulgarian Academy of Sciences, 72 Tsarigradsko Chaussee, 1784 Sofia, Bulgaria
[4]Central Laboratory of Solar Energy and New Energy Sources, Bulgarian Academy of Sciences, 72 Tsarigradsko Chaussee, 1784 Sofia, Bulgaria
[5]Institute of General and Inorganic Chemistry, Bulgarian Academy of Sciences, Acad. G. Bonchev Str., Bl. 11, 1113 Sofia, Bulgaria

E-mail: nikolaidd@abv.bg

Abstract. TiO_2 films were prepared by a sol-gel technique with commercial TiO_2 powder as a source material (P25 Degussa AG). After a special treatment, printing paste was prepared. The TiO_2 layers were formed by means of drop-coating on Si-control wafers and on the Au-electrodes of quartz resonators. The surface morphology of the films was examined by scanning electron microscopy (SEM). Their structure was studied by Raman spectroscopy and the surface composition was determined by X-ray photoelectron spectroscopy (XPS). The layers had a grain-like surface morphology and consisted mainly of anatase TiO_2 phase. The sensitivity of the TiO_2 films to NO_2 was assessed by the quartz crystal microbalance (QCM) technique. To this end, the films were deposited on both sides of a 16-MHz QCM. The sensing characteristic of the TiO_2-QCM structure was investigated by measuring the resonant frequency shift (ΔF) of the QCM due to the mass loading caused by NO_2 adsorption. The Sauerbrey equation was applied to establish the correlation between the QCM frequency changes measured after exposure to different NO_2 concentrations and the mass-loading of the QCM. The experiments were carried out in a dynamic mode on a special laboratory setup with complete control of the process parameters. The TiO_2 films were tested in the NO_2 concentration interval from 10 ppm to 5000 ppm. It was found that a TiO_2 loading of the QCM by 5.76 kHz corresponded to a system sensitive to NO_2 concentrations above 250 ppm. On the basis of the frequency-time characteristics (FTCs) measured, ΔF at different NO_2 concentrations was defined, the adsorption/desorption cycles were studied and the response and recovery times were estimated. The results obtained show that the process is reversible in the NO_2 interval investigated. The results further suggested that TiO_2 films prepared by a sol-gel method on a QCM can be used as a sensor element for NO_2 detection.

[6] To whom any correspondence should be addressed.

1. Introduction

Titanium dioxide (TiO_2) is a extensively explored material because of its attractive properties, such as a high refractive index, a high dielectric constant, a high photocatalytic activity, good physical and chemical stability [1]. This is why TiO_2 has found wide applications in various gas sensors, optical coatings and antimicrobial materials [2].Thin TiO_2 films have been prepared by various techniques: reactive sputtering and DC magnetron sputtering [3,4], electron-beam evaporation [5], vacuum arc plasma evaporation [6], chemical vapor deposition [7], chemical spray pyrolysis [8], and sol-gel methods [1-2,9-10]. The latter process has a number of advantages over the others, such as low processing temperature, easy control of the chemical composition, easy way of forming homogeneous thin layers.

In the study presented, by using a QCM we investigated the sorption ability to different NO_2 concentration of TiO_2 thin films prepared on the basis of commercial TiO_2 powder (P25 Degussa AG).

2. Experimental

The source material in our experiments was TiO_2 powder (P25 Degussa AG). Printing paste was prepared by mixing the powder with ethyl cellulose and α-terpineol and crushing in an agate mortar for about 20 minutes, as described in [11]. The paste was diluted by toluene in a weight ratio 1:1. The films were deposited on resonators by drop-coating and dried at 120 °C for 30 min.

The films prepared were analyzed by scanning electron microscopy (SEM), Raman Spectroscopy, X-ray photoelectron spectroscopy (XPS), and a gas-sensing test. The surface morphology of the samples was imaged by SEM after consecutive physical vapor deposition of 10 nm thick conductive films of amorphous carbon (C) and gold (Au). The sample observation was performed by a JEM T 200 device (Jeol) in secondary electron imaging mode. The Raman spectra were recorded by a Horiba Jobin Yvon Lab RAM HR800 spectrometer using a 600 l/mm grating and a HeNe laser for excitation. The samples were placed under the 100× achromatic objective of a BX41 microscope and measured in back scattering configuration. The XPS measurements were performed by an ESCALAB Mk II (VG Scientific) electron spectrometer with aluminum radiation (hν = 1486,6 eV). The X-Ray diffraction spectra were collected on a Bruker D8 Advance diffractometer using CuK radiation and a SolX detector.

The gas sensing properties of the films TiO_2 were studied using 16-MHz QCMs covered with thin TiO_2 films on both sides by drop-coating. The QCMs were fabricated on AT-cut quartz plates with a diameter of 8 mm. Au-electrodes with a diameter of 4 mm and a thickness of 60 nm were evaporated on both sides of the quartz plates thus implementing QCM-TiO_2 systems. The performance of the QCMs and the QCM-TiO_2 systems were evaluated by measuring the motional resistance (R_s) and calculating the quality factor (Q).The sensing properties of QCM-TiO_2 systems were defined by measuring on a laboratory setup; the measurement methodology and the setup were described in detail in [12]. The test gas used in the experiments was 1 % NO_2. The different NO_2 concentrations in the camera were obtained by altering the ratio of the NO_2 gas and diluting the air flow measured by MFCs. The temperature in the camera was maintained at 24 ± 0.2 °C. The frequency change was registered by a counter connected to the QCM and to a computer. The measurement process involved three stages: purging the camera with the QCM-TiO_2 structure with dry air until the frequency reaches a constant value (F_0); switching on the NO_2 gas flow with a certain concentration and reaching the saturation frequency value (F); purging the camera with dry air to reach the initial value of the QCM-TiO_2 system frequency (F_0). The sensitivity was defined by measuring the time-frequency characteristics (FTCs) of the QCM-TiO_2 system at different NO_2 concentrations (10 ppm-5000 ppm).

3. Results and discussion

3.1. Characterization of TiO_2 thin films and of TiO_2-QCM system

Figure 1 shows typical SEM images of as-deposited TiO_2 films at a magnification of 20 000. All films fabricated were found to have a grain-like structure. The average grain size was in the interval from 67 nm to 160 nm. Agglomerates of a size of approximately 250 nm were also observed.

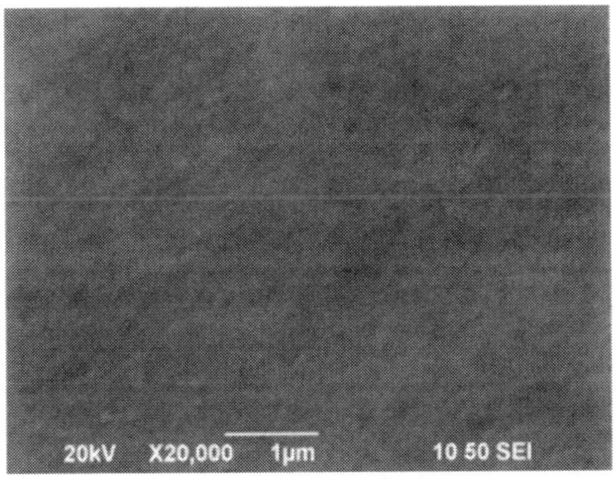

Figure 1. SEM image of TiO_2 films as-deposited on a Au resonator's electrode.

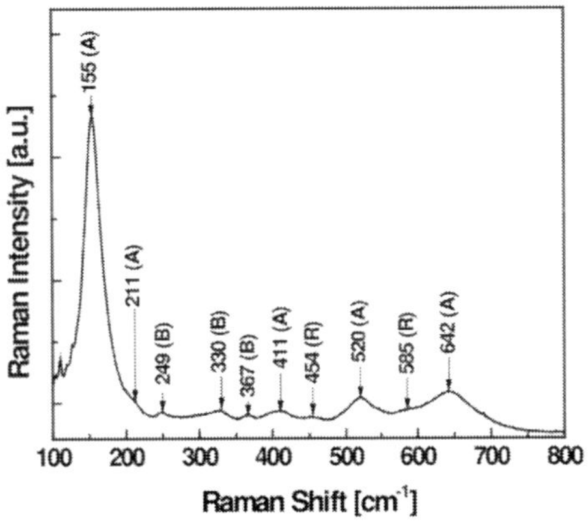

Figure 2. Raman spectrum of a TiO_2 layer; A-anatase, B-brookite, R-rutile.

Figure 2 prsents a typical Raman spectrum of the layers measured directly on the resonator. As cited in [13], the starting material (Degussa P25) for the preparation of the printing paste is a mixture of TiO_2 phases, namely, about 80 % anatase, some rutile and an amorphous phase. In the film studied, the anatase is prevalent again as demonstrated by the characteristic peaks in the Raman spectrum [14]. The fact that the intensive low-energy band is shifted from 144 cm^{-1} to 155 cm^{-1} and its width exceeds 30 cm^{-1} shows that the film contains very small anatase crystallites (with a size of less than 10 nm) [15].The spectrum also exhibits the characteristic rutile bands [15] and the strongest Raman bands of brookite [16]. This demonstrates that the film contains these two phases in smaller amounts.

The Ti2p photoelectron spectrum (figure 3) shows a $Ti2p_{3/2}$ peak at 459,0 eV. The peak position observed and the doublet separation between the $2p_{1/2}$ and $2p_{3/2}$ peaks of ~5.7 eV are characteristic of TiO_2. The O1s spectrum shows a main peak at 530,3 eV, attributed to lattice oxygen, and a shoulder at ~532,2 eV, assigned to hydroxyl groups. The valence band spectrum shows a typical TiO_2 structure with two peaks due to O2p states. The surface concentrations of O and Ti are 71,3 at. % and 28,7 at. %. The inexact stoichiometry is attributed to the presence of adsorbed oxygen species.

The QCM parameters before and after loading were measured to ensure that the sensing response to NO_2 was the actual films' behavior. The QCM basic parameters, R_s and Q, were in the ranges

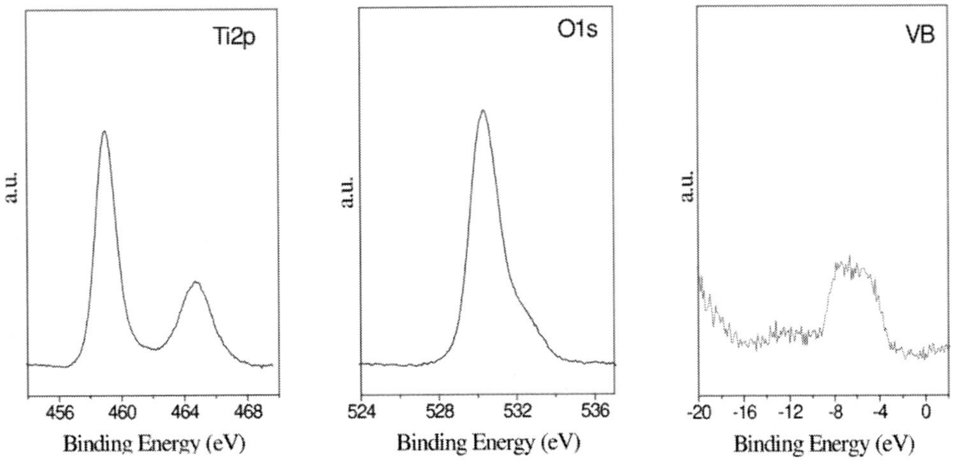

Figure 3. XPS spectra of a sol-gel prepared TiO_2 film.

12 – 20 Ω and 30 000 – 50 000, respectively. The negligible loading (5,76 kHz) of the QCM led to a change in the R_s from 24.95 Ω to 32.24 Ω.

3.2. TiO_2 films' sensing properties to NO_2

All measurement were based on Sauerbrey' formula, who was the first to derive the quantitative relationship between the change in the frequency of an AT-cut piezoelectric crystal (F_0) and the mass change (Δm) caused by mass loading on the piezoelectric crystal surface [17].

$$\Delta F = (2.3\ 10^6\ F_0^2 / A)\,\Delta m\,, \tag{1}$$

where F_0 (106 Hz) is the basic frequency of the unloaded piezoelectric crystal, Δm (g) is the change in the mass on the surface of the crystal, and A (cm^2) is the surface area of the electrode [18]. In our experiments A = 0,1256 cm^2. Thus, using eq. (1) one can calculate the mass (Δm) of the adsorbed gas by measuring the resonant frequency shift ($\Delta F = F - F_0$).

In the NO_2 range measured (10 ppm-5000 ppm), the system's minimal sensitivity was found to be 250 ppm. Figure 4 presents typical FTCs for a TiO_2-QCM structure at two NO_2 concentrations. In a chamber containing a test gas, the TiO_2 film adsorbs NO_2 molecules on its surface, thus mass-loading the QCM and lowering its resonance frequency. The sorption process continues until a dynamic equilibrium is reached. As the test gas flow is switched off, the opposite process begins – NO_2 molecules desorption leading to a diminished mass load and an increased resonance frequency. The two curves in figure 4 have similar behavior in spite of the different concentrations. The FTCs curves obtained proved that the adsorption/desorption process is reversible and typical for physisorption. The frequency shift rises from 18 Hz to 30 Hz as the NO_2 concentration is raised from 1000 to 5000 ppm.

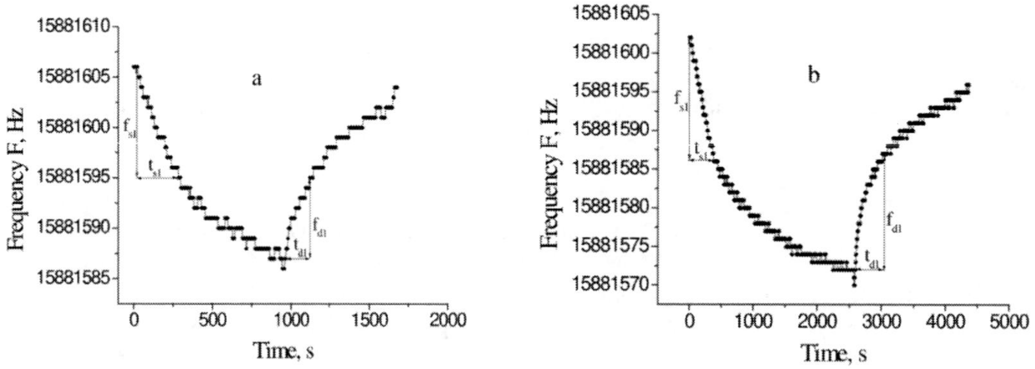

Figure 4. FTC of TiO_2-QCM structure at NO_2 concentration: a)-1000 ppm, b)-5000 ppm.

One can clearly see the existence of a two-step (a fast one and a slower one) sorption-desorption process. In figure 5, the response-time (t_{s1}) variation in the fast sorption step is illustrated as a function of the NO_2 concentration. As the NO_2 concentration is raised, t_{s1} rises from 3 minutes at 250 ppm to 5,6 minutes at 5000 ppm NO_2. The calculated sorption rate during the first step (v_{s1}) of the response time increases from 0,82 Hz/min to 2,85 Hz/min as a function of the NO_2 concentration. We also measured the desorption times and calculated the desorption rates during the first step of the structure's recovery (t_{d1} ,v_{d1}). We established insignificant rises in t_{d1} and v_{d1} – from 4,5 to 5,66 minutes and from 1,77 to 2,47 Hz/min for all NO_2 concentrations.

Figure 6 shows the response of the TiO_2-QCM structure versus the NO_2 concentration. Using the FTCs, and in accordance with Sauerbrey's equation, the total $\Delta F = F - F_0$ was calculated for each concentration, together with the corresponding sorbed mass increases for each ΔF. Thus, figure 6 illustrates that the frequency shift increases from 7 Hz to 30 Hz as the sorbed mass increases (with the NO_2 concentration) from 1.54 ng to 6.53 ng, respectively.

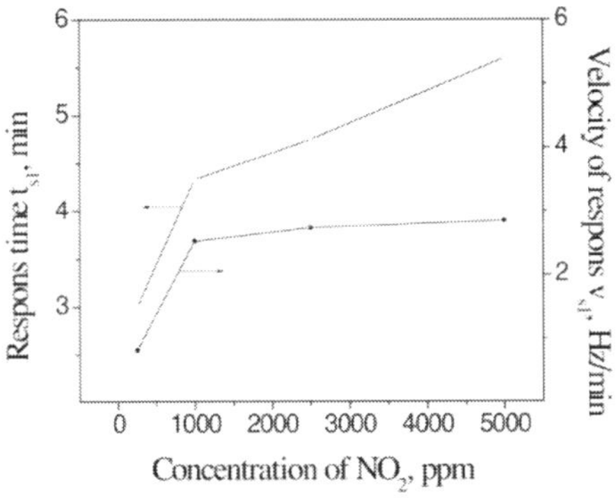

Figure 5. The TiO_2-QCM response and rate of response versus the NO_2 concentration.

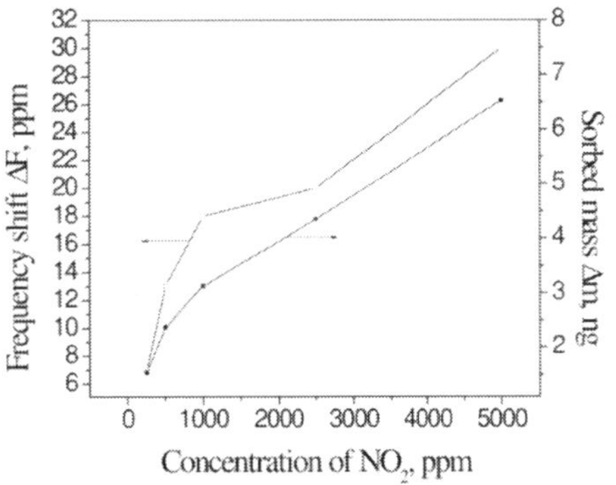

Figure 6. Dependence of the TiO_2-QCM frequency shift and sorbed mass versus the NO_2 concentration.

4. Conclusions

TiO_2 films prepared by a sol-gel technique on QCMs were investigated in view of application as NO_2 sensors. The films had a grain-like structure and contained predominantly an anatase phase. The adsorption/desorption process of NO_2 was studied and was found to be reversible and typical for physisorption. The lower sensitivity limit of the TiO_2 films was was determined to 250 ppm. A TiO_2 loading (5,76 kHz) of a QCM is sufficient to implement a NO_2 sensitive system. The results obtained are very promising, but it is necessary to optimize the preparation conditions, the structure of the films, and to define the higher loading of the QCM in order to improve the TiO_2 sensitivity.

Acknowledgement

The work is supported partly by the bilateral project (2011-2013) ISSP, BAS and GEOKHI and Project ISSP-BAS - № 05-13/ 13.04.2013.

References

[1] Ahn Y, Kim E, Kim H and Hahn S 2003 *Mater. Lett.* **57** 4660
[2] Sumana K, Rao K, Murthy C, Rao Y and Phani A 2012 *Int. J. Eng .Res .Appl.* **2** 681
[3] Boyadzhiev S, Georgieva V and Rassovska M 2010 *J. Phys.: Conf. Series* **253** 012040
[4] Horprathum M, Eiamchai P, Chindaudom P and Limsuwan P 2012 *Proc. Eng.* **32** 676
[5] Georgieva V, Donkov N, Spassov L, Petkov P and Gadjanov V 2008 *J. Phys.: Conf. Series* **113** 012053
[6] Miyata T, Tsukada S and Minami T 2006 *Thin Solid Films* **496** 136
[7] Babelon P, Dequiedt A, Mostafa H, Sibillot P and Sacilotti M 1998 *Thin Soli Films* **322** 63
[8] About-Helal M and Seeber W 2002 *Appl. Surf. Sci.* **195** 53
[9] Kozlowska K, Lukowiak A, Szczurek A, Dudek K and Maruszewski K 2005 *Opt. Appl.* **35** 4
[10] Phani A and Santucci S 2006 *J. Phys.: Condensed Matter.* **18** 6965
[11] Kussovski V, Stefchev P and Kirilov R 2011 *Comp. Ren. de l'Acad. Bulg. Sci.* **64** 693
[12] Georgieva V, Stefanov P, Spassov L, Raicheva Z, Atanassov M and Tincheva T 2009 *J. Optoelectr. Adv. Mater.* **11** 1363
[13] Bakardjieva S, Šubrt J, Štengl V and Sayagues M 2005 *Appl. Catal. B-Environ.* **58** 193
[14] Swamy V, Kuznetsov A, Dubrovinsky L, Shchukin D and Muddle B 2005 *Phys. Rev.* B **71** 184302
[15] Swamy V and Muddle B 2006 *Appl. Phys. Lett.* **89** 163118
[16] Moreta M, Zallena R, Vijayb D and Desu S 2000 *Thin Solid Films* **366** 8
[17] Sauerbrey G 1959 *J. Physik* **155** 206
[18] Zhu Y, Yuan H, Xu J, Hu P and Pan Q 2010 *Sensors and Actuators* B **144** 164

Influence of the agglomeration in the initial suspension (ferrofluid) on the oriented magnetic structure

S Kolev[1] and T Koutzarova

Acad. E. Djakov Institute of Electronics, Bulgarian Academy of Sciences, 72 Tsarigradsko Chaussee, 1784 Sofia, Bulgaria

E-mail: svet_kolev@yahoo.com

Abstract. The process was studied of preparation of oriented $BaFe_{12}O_{19}$ films via deposition of $BaFe_{12}O_{19}$ nanoparticles from a suspension in a magnetic field. The films' structural, microstructural and magnetic properties were investigated by X-ray diffraction, scanning electron microscopy and magnetic measurements. The influence was explored of the particles's agglomeration in the ferrofluid on the films' properties and microstructure. Using time delay deposition, we obtained thick films with a high degree of orientation and good density.

1. Introduction

Barium ferrite ($BaFe_{12}O_{19}$) is a high-performance material widely used as a permanent magnet because of its large magnetocrystalline anisotropy along the *c*-axis of its hexagonal structure, high Curie temperature and relatively large magnetization, as well as excellent chemical stability and corrosion resistance [1, 2].

Magnetically oriented barium ferrite films have found various uses, such as high-density magnetic recording [3], magneto-optical applications [4] and mm-wave applications [5, 6]. In what concerns the latter of these, barium ferrite-based compounds are without alternative. Due to their high magnetocrystalline anisotropy, they exhibit a ferromagnetic resonance at mm-wavelengths, while, at the same time, being magnetically self-biased. The most suitable form of barium ferrite for mm-wave applications is magnetically oriented thick films. Several techniques have been used to prepare such films, including thermal spraying [7], electrophoretic deposition [8] and screen printing [9].

We report a study on the influence of the aggregation of barium hexaferrite magnetic particles in a ferrofluid on the properties of the thick film formed in a magnetic field. This method consists in dropping a suspension of barium ferrite particles (ferrofluid) on an Al_2O_3 substrate, followed by drying the film in an external magnetic field between two permanent magnets. The technique allows one to grow magnetically oriented thick films while controlling the magnetic properties depending on the agglomeration process in the ferrofluid.

2. Experimental

The $BaFe_{12}O_{19}$ powder was synthesized using the hydrothermal method [10] in collaboration with the K8 Department, Jozef Stefan Institute, Ljubljana, Slovenia. To prevent the attraction between the particles and their subsequent agglomeration, they were stabilized through adsorption of a surfactant

[1] To whom any correspondence should be addressed.

on the particles's surface, as described in [11]. The particles were dispersed in an aqueous acid solution with 10 mass % of the surfactant dodecilybenzenesulphornic acid per mass of powder. The ferrofluid consisted of particles with an average size of 10 nm and 80 nm. Such a mixture proved to be the best for obtainig the highest degree of films' density when this stable suspension was subsequently used for deposition of oriented films.

The deposition process was carried out using the experimental setup described below. The barium ferrite suspension prepared was dropped on an Al_2O_3 substrate placed between two permanent magnets. The as-deposited sample was slowly dried between these two permanent magnets for 6 hours. The magnetic attraction force between the particles is too strong even in the presence of a surfactant in the ferrofluid. In order to prevent the particles agglomeration, the suspension was subjected to ultrasound homogenization. Two approaches were employed to investigate the influence of the agglomeration process on the physical properties of the thick films obtained. In the first one (type I), the deposition process was performed immediately after the ultrasound homogenization (without delay). In the second (type II), the deposition process was conducted at a certain time delay after the ultrasound treatment, namely, after 3, 6, 12, 15 and 18 hours. The greatest thickness and homogeneity and the lowest roughness of the films were observed after the 15-hour delay. The film deposited under these optimal conditions was used to study the influence of the particles agglomeration process in a suspension on the thick films' properties. After drying, the samples (from type I and type II depositions) were annealed at 1050 °C for 10 hours. This annealing temperature was so chosen since below it the degree of crystallization of the ferrite sample is not sufficient, while exceeding it leads to diffusion of Al from the substrate [12].

The barium ferrite thick film samples were characterized by X-ray diffraction (XRD) using an X-ray diffractometer. The microstructures of the samples were observed by a scanning electron microscope and the magnetic properties were measured by a vibrating-sample magnetometer. The magnetic measurements were performed at a maximum magnetic field of 10 kOe applied perpendicular to the sample plane.

3. Results and discussion

Figure 1 shows a typical XRD pattern of barium hexaferrite films annealed at 1050 °C for 10 hours. Excluding the XRD lines corresponding to Al_2O_3 (the substrate used), all other XRD features observed are assigned to $BaFe_{12}O_{19}$.

After the thermal treatment at 1050 °C, the films (both types) were well-crystallized with a high degree of orientation corresponding to the *c*-axis being perpendicular to the film plane, as suggested by the high relative intensities of the 00l peak.

Figure 2a shows a SEM image of the film deposited from the suspension immediately after ultrasound homogenization (type I) and then annealed at 1050 °C. The film has a good density with a thickness around 1 μm. The SEM image in figure 2b is of the film deposited from the suspension with a 15-hour delay after ultrasound homogenization and annealed at 1050 °C (Type II deposition). As can be seen, a type of a sintering process has occurred. The film surface is porous. This film is considerably thicker, with a thickness of 10 μm. The barium hexaferrite is a hard magnetic material, so that the magnetic attraction between the $BaFe_{12}O_{19}$ particles is very strong even in the presence of the surfactant in the suspension. The ultrasound treatment prevents the agglomeration of the magnetic particles in the suspension.

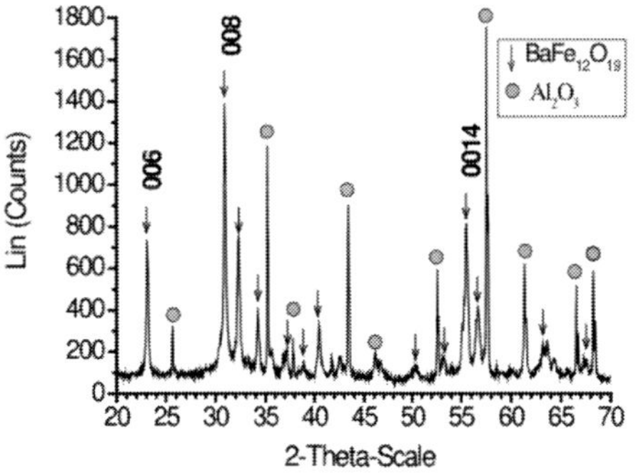

Figure 1. XRD pattern of a type I film annealed at 1050 °C for 10 hours.

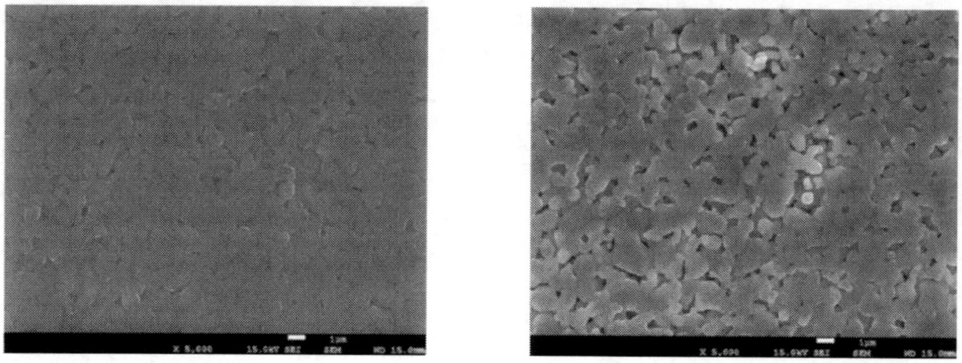

a) b)

Figure 2. SEM images of the surface of films annealed at 1050°C (a) type I – after ultrasound homogenization without time delay and (b) type II – with a 15-hour time delay after ultrasound homogenization.

In type I deposition, the particles are homogeneously dispersed in the suspension and no agglomeration takes place during the deposition process. Due to the orientation of the deposited particles induced by the external magnetic field applied during the drying process, with the easy magnetic axis (i.e. the crystallographic *c*-axis) perpendicular to the film plane, the grains grow in the film plane, thus improving its density. In the second kind of deposition – type II, the time delay after the ultrasound treatment is sufficient for the particles to agglomerate, which results in preparing thicker films of lower densty. This affects the magnetic properties of the films.

The hysteresis loops for a magnetic field applied perpendicularly to the as-deposited $BaFe_{12}O_{19}$ films prepared by the two deposition types are presented in figure 3. Figure 4 shows the hysteresis loops for a magnetic field applied perpendicularly to the samples annealed at 1050 °C. For the as-deposited type-I sample, the values of the saturation magnetization (M_s = 4,98 emu/g) and coercivity (H_c = 183 Oe) are considerably lower than those of the corresponding type-II sample, namely, 29,21 emu/g and 2000 Oe, respectively. This can be related to the agglomeration and the greater film thickness. In the first case, the type-I sample consists mainly of small particles (10 nm) with a size on the superparamagnetic limit. The type-II film contains both fractions, 10 nm and 80 nm (stable monodomain particles), so that it exhibits higher coercivity and saturation magnetization.

Figure 4 illustrates the magnetic properties of both type of films, with the external magnetic field applied perpendicularly and in parallel to the film. The annealing dramatically improves the magnetic properties of the type-I $BaFe_{12}O_{19}$ films, while affecting only slightly the type-II deposited films. The

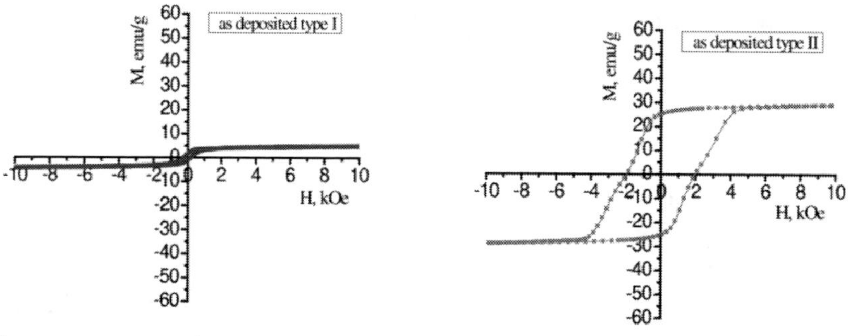

a) b)

Figure 3. Magnetic properties of as-deposited films prepared by (a) type I – after ultrasound homogenization without time delay and (b) type II – with a 15-hour time delay after ultrasound homogenization.

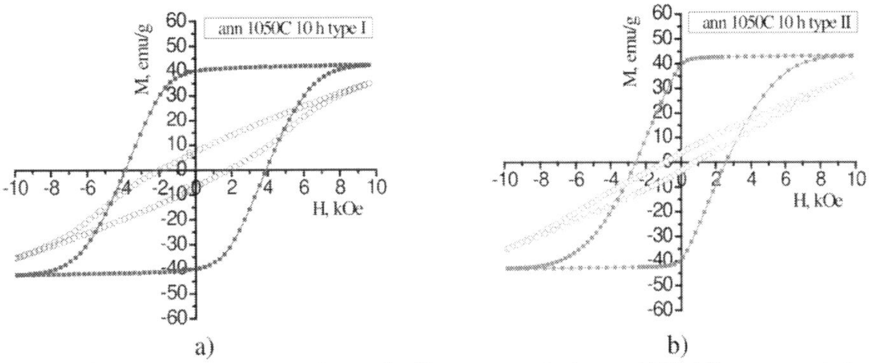

Figure 4. Magnetic properties of films annealed at 1050 °C prepared by (a) type I – after ultrasound homogenization without time delay and (b) type II – with a 15-hour time delay after ultrasound homogenization.

M_s and H_c of the film obtained by the type-I deposition and annealed at 1050°C are 42,18 emu/g and 3917 Oe, respectively. The M_s and H_c of the film obtained by type-II deposition and annealed at 1050°C are 43.50 emu/g and 2593 Oe, respectively. i.e., they are higher than those of the as-deposited film. The coercivity does not change significantly after the high-temperature annealing of these films. The possible reason can be the higher porosity in the films obtained by type-II deposition. Both samples show a high degree of orientation, as the squareness ratio SQR = 0,96 (the ratio between the remanent and saturation magnetizations).

4. Conclusions

Polycrystalline $BaFe_{12}O_{19}$ films prepared by deposition of $BaFe_{12}O_{19}$ particles in a magnetic field followed by a high-temperature treatment at 1050 °C. The preferential orientation of the crystallites in the film plane resulted in anisotropic magnetic properties. Using a ferrofluid with a different degree of particles' agglomeration (type I and type II depositions), $BaFe_{12}O_{19}$ films of different thickness can be deposited with a very high degree of orientation. The oriented barium ferrite films are promising candidates for perpendicular magnetic recording and for mm-wave application.

Acknowledgments

The authors wish to thank the staff of K8 department of Jozef Stefan Institute, Ljubljana, Slovenia for the support in this common work. The work was supported in part by the Bulgarian National Science Fund under grant DNTC/SLOVENIA01/4.

References

[1] Coey J M D 1999 Whither Magnetic Materials? *J. Magn. Mater.* **196-197** 1
[2] Rossignol M and Yonnet J P 2005 Permanent magnets *Magnetism: Materials and Applications* part 2 ed É. du Trémolet de Lacheisserie, D. Gignoux and M. Schlenker (Springer) pp 3-88
[3] Sharrock M P and Carlson L W 1995 *IEEE Trans. Magn.* **31** 2871
[]4 Masterson H J, Lunney I C and Coey J M D 1998 *J. Phys.: Condensed Matter.* **9** 4761
[5] Fal T J and Camley R E 2008 *J. Appl. Phys.* **104** 023910
[6] Chen Z, Yang A, Mahalingam K, Averett K L, Gao J, Brown G J, Vittoria C and Harris V G 2010 *Appl. Phys. Lett.* **96** 242502
[7] Altenburg H, Plewa J, Plesch G and Shpotyuk O 2002 *Pure Appl. Chem.* **74** 2085
[8] Ovtar S, Lisjak D and Drofenikz M 2011 *J. Am. Ceram. Soc.* **94** 3373
[9] Chen Y, Smith I, Geiler A L, Vittoria C, Zagorodnii V, Celinski Z and Harris V G 2008 *J. Phys.* D: *Appl. Phys.* **41** 095006

[10] Primac D, Makovec D, Lisjak D and Drofenik M 2009 *Nanotechnol.* **20** 315605
[11] Ovtar D, Lisjak D and Drofenik M 2010 *Surf. Interface Anal.* **42** 1217
[12] Kolev S, Lisjak D and Drofenik M 2011 *J. Experim. Nanosci.* **6** 4

Structural and adhesional properties of thin MoO_3 films prepared by laser coating

Mihail Mihalev[1,5], Chavdar Hardalov[2], Christo Christov[2], Michail Michailov[3], Bogdan Ranguelov[3]and Harald Leiste[4]

[1]Faculty of Mechanical Engineering, Technical University of Sofia, 8 Kliment Ohridski Blvd., 1000 Sofia, Bulgaria
[2]Department of Applied Physics, Technical University of Sofia, 8 Kliment Ohridski Blvd., 1000 Sofia, Bulgaria
[3]Institute of Physical Chemistry, Bulgarian Academy of Sciences, Akad. G. Bonchev Str., Building 11, 1113 Sofia, Bulgaria
[4]Insitute of Applied Materials – Applied Materials Physics, Karlsruhe Institute of Technology, Karlsruhe, Germany

E-mail: mmihalev@tu-sofia.bg

Abstract. Laser marking plays an important role in numerous technological applications because of its flexibility, fastness and versatility. The present study deals with the structural and adhesional properties of thin MO_3 layers on stainless steel substrates prepared by a specific modification of the laser coating technology known as "laser bonding". This approach consists in the local laser sintering of an initially deposited proper powder material, which forms under laser irradiation a layer with a definite graphical and topological design. The coatings, prepared of only MoO_3 powder irradiated by a CO_2 laser beam, are well bonded to the substrate and exhibit diffusive light reflection. Through applying a variety of methods for surface structural analysis, as micro-indentation, XRD, micro-Raman and SEM, this study also provides detailed information about the coatings' chemical bonding and composition. Our results reveal a good adhesion to the steel due to the formation of an amorphous interface between the MoO_3 and the substrate. This amorphous interface arises from a "quenching" process of the molten MoO_3 acting both as an oxidant and flux. Depending on the laser beam's intensity, energy and scanning velocity, we also observed recrystallization in specific areas of the coatings. The present study contributes to the better understanding of the adhesion, wear-resistance and hardness of MoO_3 coatings obtained by laser bonding.

1. Introduction

A new versatile technology [1], which is similar to the laser cladding – the so called "laser bonding", has been known since 1997. A complex powder mixture, consisting of pigments, oxides, glazes and silicates is irradiated by a laser source, thus heating the powder. As a result, a stable coating with different optical, mechanical and structural properties is formed on the steel surface. According to the patent cited, the chemical bond is realized through iron silicate.

[5] To whom any correspondence should be addressed.

We have implemented a modification of the technology of laser bonding, based on a laser treatment of MoO_3 powder on stainless steel. This material was chosen as a typical transition metal oxide. It is well known that MoO_3 is a very aggressive oxidant in its liquid phase [2, 3], acting as a flux ([3] and references therein) as well. It produces an iron molybdate interface on the substrate [4, 5, 6]. Moreover, the colour of the powder is changed to dark violet/black, so that no additional pigments are necessary, which is aim of the direct laser marking.

2. Experimental setup and procedure for laser coating

The equipment used for laser coating was a 50-W cw sealed CO_2 laser. The laser beam was steered by a standard XY fly optics moving system (with a high scanning velocity of up to 30 m/min) to a 50-mm focusing ZnSe lens, thus forming a focal spot of approximately 200 μm diameter on the sample.

For the substrate, grade 304 stainless steel was chosen. It was pre-coated with 99.9 % α-orthorhombic MoO_3 (CAS Number 1313 – 27 – 5) using a spray gun [7].

Two types of samples were prepared: 1) a static laser beam was used (forming a nearly circular structure, see figure 1 left and middle), when the laser beam does not move over the sample, called by us "point samples" and 2) the laser beam scanned the sample at a given sweep rate (see figure 2 right), called by us as "raster-scanned samples". The sweep rate and the raster step were controlled by a CNC unit. More details were presented in [8, 9].

In the case of point samples, CO_2 laser pulses of different durations were applied: 10 ms, 100 ms, 1000 ms. The pulses of 10 ms and 100 ms produced similar SEM images, thus the image with 10-ms pulse duration was omitted.

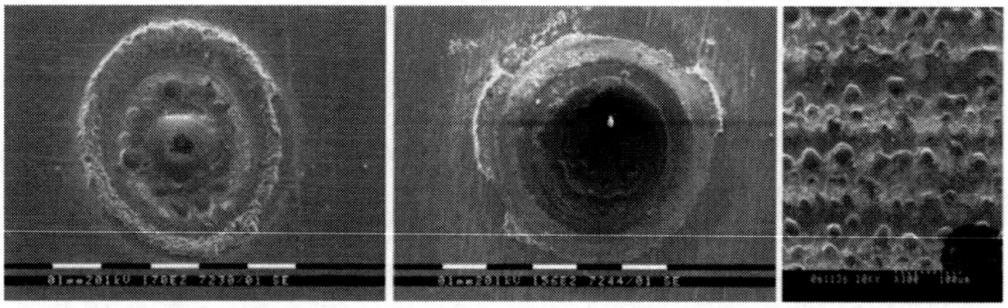

Figure 1. SEM of typical spots at 100 ms (left) and 1000 ms (middle) exposure time, and (right) SEM image of a typical raster-scanned laser bonded coating.

3. Experimental results and discussion

3.1. Mechanical properties

3.1.1. Vickers hardness. The Vickers hardness test was conducted using a CSEM Revetest scratch tester with a Vickers shaped indenter.

The hardness test of the substrate and coating, averaged over 3 measurements at 3 different points, yielded values of 302 HV0.05 and 251 HV0.06, respectively. However, it must be noted that these data should be considered as indicative only, because 1) the substrate is not undeformable, and 2) in a hardness measurement, the coating thickness must be at least one order of magnitude larger then the indentation depth, which was not the case in our samples. Nevertheless, it can be assumed that the coating possessed hardness comparable with that of the steel.

Figure 2. Micrograph of scratch trace: top: 50× magnification; bottom: 500× magnification – at the beginning, middle and end of test (from left to right).

3.1.2. Adhesion. To the best of our knowledge, no methods exist that provide an unambiguous quantitative evaluation of the adhesion. All methods describe more or less phenomenologically the process of coating failure.

One standard method is the ramp scratch test [10]. A CSEM Revetest scratch tester was used again.

In the ramp test, the load was varied from 0 to 100 N at a rate of 10 N/min, while the sample was moved at 5.7 mm/min.

The micrographs of the scratch trace at 50× and 500× microscope magnification show no material failures, even at 100 N (figure 2). The material was dragged by the indenter, building a groove of increasing depth and width. This fact is indicative of strong plastic deformations in the material due to the well-known layered structure of MoO_3 [11], which implies usages as surfaces that are locally tribologically changed. No critical load was found at which coating failure could be observed. Moreover, at the end of the test, when the maximal load of 100 N was reached, the indenter scraped the substrate, but traces of the coating on the substrate could still be noticed. These facts support unambiguously the assumption for an excellent coating-to-substrate adhesion.

3.2. Raman spectroscopy

3.2.1. Point sample. Micro-Raman spectroscopy was performed by a HORIBA LabRAM HR spectrometer with an incorporated He-Ne laser (wavelength 633 nm).

Figure 3a) presents micro-Raman spectra acquired from three point samples irradiated by the CO_2 laser at different exposure times – 10 ms, 100 ms and 1000 ms.

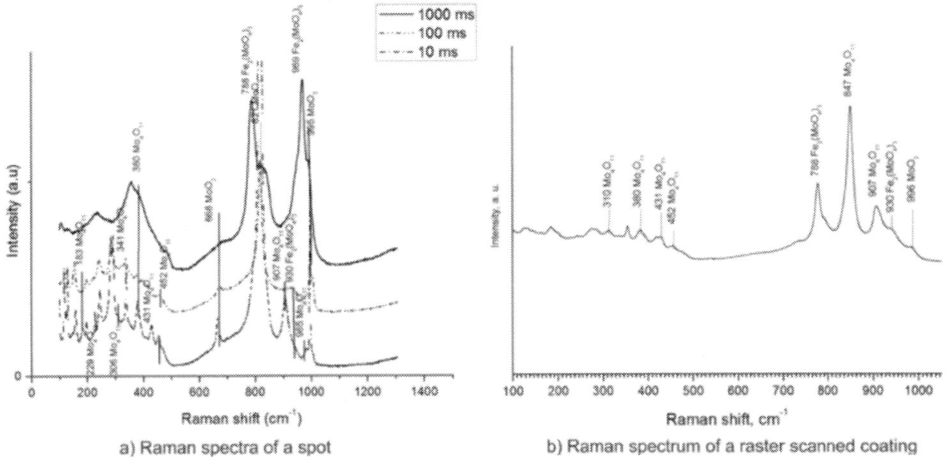

a) Raman spectra of a spot b) Raman spectrum of a raster scanned coating

Figure 3. Raman scattering spectra of a point sample (left) and a raster-scanned sample (right).

In all spectra (10 ms, 100 ms and 1000 ms) the most typical Raman lines of orthorhombic MoO_3 at 668 cm^{-1}, 822 cm^{-1} and 995 cm^{-1} [12] are seen.

In the spectrum of the sample with the shortest exposure time of 10 ms, the following lines are seen in addition to the MoO_3 lines: 183 cm^{-1}, 229 cm^{-1}, 306 cm^{-1}, 341 cm^{-1}, 380 cm^{-1}, 431 cm^{-1}, 452 cm^{-1}, 907 cm^{-1} and 985 cm^{-1}, which can be associated with a more complex orthorhombic molybdenum oxide, namely, Mo_4O_{11} [13].

When the exposure time was raised to 100 ms, a new line, that of iron molybdate $Fe_2(MoO_4)_3$, appeared at 930 cm^{-1} [14] in addition to the lines of Mo_4O_{11} and MoO_3.

The longest exposure time resulted in broadening and in lowering the intensity of Mo_4O_{11} and MoO_3 lines, while the lines of $Fe_2(MoO_4)_3$ 788 cm^{-1}, 930 cm^{-1} and 969 cm^{-1} [14] became narrower and with higher intensities. No MoO_2 Raman lines were detected.

In the case of the very short CO_2 laser exposure time of 10 ms, the presence of MoO_3 lines can be explained by the existence of initial material that has not been melted. No $Fe_2(MoO_4)_3$ molybdate lines are observed. This exposure time is too short for formation of substantial amount of iron molybdate for a detectable Raman scattering. We believe that molybdate serves as an interface between the steel and the coating. The presence of the most typical lines of Mo_4O_{11} can be explained with the exposure time leading to a rapid non-equilibrium re-crystallization process.

Under the conditions of a longer exposure time (100 ms), the amount of $Fe_2(MoO_4)_3$ increases, resulting in the appearance of the new line at 930 cm^{-1}.

At 1000 ms, the intensities of the $Fe_2(MoO_4)_3$ lines are much higher the those of the lines of all other oxides. We believe that substantial amounts of MoO_3 and Mo_4O_{11} are vaporized at the high temperature in the zone affected by the laser light (as evidenced by the concave meniscus in the point sample), so that the He-Ne laser beam can reach the $Fe_2(MoO_4)_3$ interface. At the shorter exposure times, the oxide scale absorbs the He-Ne beam and the Raman signal is much lower.

It must be noted that, at the centre of all point samples (10, 100, 1000 ms), an amorphous phase is seen (wide lines in their bases, ending with sharp peaks, which is an indication of Raman scattering on different non-collective and collective bonds, respectively), due to the extremely high temperature and time gradients and the lack of re-crystallization as a result of the Gaussian intensity distribution of the CO_2 laser beam.

3.2.2. Raster-scanned samples. Figure 3b) presents a typical spectrum of a coating produced by raster-scanning the CO_2 laser beam. Depending on the raster-scanning step (distance between individual scanning lines), the CO_2 laser beam passes through a given point several times, leading to multiple re-crystallization of the molten bead. Generally, lines of all above-mentioned compounds, MoO_3, Mo_4O_{11}, $Fe_2(MoO_4)_3$, are observed in the spectrum with various intensities, due to the convolution of concurrent Raman peaks from both crystallites and amorphous phases during the laser processing.

3.3. Analysis of the XRD spectra

Unlike Raman spectroscopy, XRD is less sensitive to absorption by the coating due the high photon energy; thus, the signal originates from all depths of the coating, including the interface and the substrate.

The measurements were performed on a Seifert PTS300 spectrometer with a Cu-Kα line (λ = 1,5405 nm). A typical XRD pattern of a raster-scanned sample is presented in figure 4.

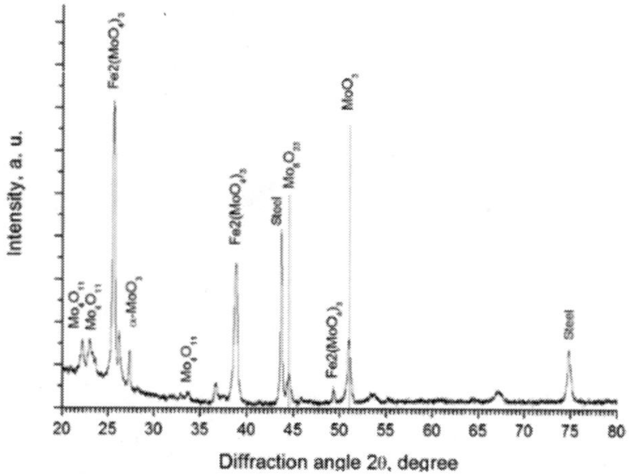

Figure 4. XRD pattern of a raster-scanned coating.

In accordance with the JPCPDS (card numbers 5-508, 5-503, 5-452, 5-337, 89-2367, 89-2368) data bases, the presence of the following compounds was established: MoO_3, Mo_4O_{11}, $Fe_2(MoO_4)_3$, as well as steel [15] (figure 4), which is in agreement with the Raman scattering data for the same sample. Reference to the JPCPDS data bases for MoO_2 and $Fe(MoO_4)$ did not reveal the presence of these compounds.

The experiments on Vickers hardness showed that the coatings possessed hardness against normal penetration comparable to the metal hardness. Moreover, no material failure was observed during the scratch test, which can be explained by the typical layered structure of MoO_3 [11].

The Raman spectroscopy and XRD measurements showed the formation of additional compounds, namely, Mo_4O_{11} and $Fe_2(MoO_4)_3$ from the initial powder, as well as the presence of some initial unmelted MoO_3. This can be explained by the occurrence of complex non-equilibrium physical and thermo-chemical processes, induced by i) enormous temperature gradients, ii) fast temperature rise time and iii) fast quenching processes in the melt.

The heating of the initial powder leads to a very high temperature in the laser-affected zone (in the range of 1500 – 2000 °C, as the FEM model in [9] predicts). In its molten phase, MoO_3 acts as an aggressive oxidant, as well as effective flux ([16] and references therein). As a result, an interface consisting of $Fe_2(MoO_4)_3$ is formed. $Fe_2(MoO_4)_3$ is chemically bonded to the substrate, as all our tests confirm. The presence of MoO_3 and Mo_4O_{11} and the absence of MoO_2 is, in our opinion, in accordance with the catalytic and autocatalytic properties of MoO_3 [17]. At the longest CO_2 laser exposure time, a well expressed erosion of the substrate surface was observed (see figure 1, middle), which also supports the assumption of an intensive oxidation.

4. Conclusions

The formation of the coating is a result of complex phenomena with a number of processes acting simultaneously. The very high temperature and the fast temperature rise time initiate the chemical transformation of MoO_3 to the intermediate oxide Mo_4O_{11}, which is a stable phase. The formation of the amorphous phase is a result of the fast quenching of the molten powder. Due to the properties of MoO_3, acting both as an oxidant and a flux, an interface of $Fe_2(MoO_4)_3$ appears. The inter-diffusion between $Fe_2(MoO_4)_3$, MoO_3 and intermediate oxides seems to be responsible for the good adhesion of the coating to the substrate. Taking into account: i) the good adhesion, ii) the absence of material failures at scratch and iii) surface roughness leading to diffusive reflection, this technology, in our opinion, can be used for direct laser marking meeting the respective requirements [18], as well as for improving the wear-resistance of stainless steels.

To the best of our knowledge, all existing technologies for laser marking that use laser bonding are based on multi-component liquid or dry mixtures, known in the industry as "laser marking inks". This paper contributes to the better understanding of the physical and thermo-chemical processes in such type of technologies.

The study of high-temperature physical and chemical transient processes in the millisecond range requires sophisticated "in-situ" equipment and will be the subject of future works.

Acknowledgments

The support of KNMF/ANKA, KIT, Germany and of Prof. Miroslav Abrashev, University of Sofia, are gratefully acknowledged.

References

[1] Harrison and Wollcott P 1997 *US Patent 6075223*

[2] Brenner S S 1955 Catastrophic oxidation of some molybdenum alloys *J. Electrochem. Soc.* **102**/1 16-21

[3] de Van J H 1961 Catastrophic oxidation of high-temperature alloys *Oak Ridge National Laboratory* **10**

[4] Berdzenishvili I G 2012 Functional corrosion-resistant enamel coatings and their adherence strength *Proc. Int. Congress Adv. Appl. Phys. Mater. Sci.* (Antalya Turkey 2011) pp 178-80

[5] *Fundamentals of Adhesion* 1991 ed L. H. Lee (Springer Verlag)

[6] Fick D, Layne C, Gnizak D and Evele H 2001 Iron oxide interfacial reactions as related to enamel bonding *Proc. 63rd Porcelain Enamel Institute Technical Forum: Ceramic Engin. Sci.* vol 22 issue 5

[7] Mihalev M, Hardalov Ch, Chirstov C G and Leiste H 2013 Microstructural characterization of thin laser bonded coatings based on transitional metal oxides powder *Proc. 23d Nat. Symp. Metrology and Metrology Assurance* (Sept. 2013 Sozopol Bulgaria) pp 219-25

[8] https://ecad.tu-sofia.bg/e-publ/files/1102_69_1.pdf

[9] Christov C G, Mihalev M S and Hardalov Ch M 2010 Characterization of thin MoO_3 layers

obtained by laser bonding *J. Technol.* (Assumption University Bangkok Thailand) **14**/1 1-10

[10] Taube K 1994 Mechanische charakterisierung an duennen schichten 1994 *J. für Oberflächentechnik* **11** 83-7

[11] Rao M C, Ravindranadh K, Kasturi A and Shekhawat M S. 2013 Structural stoihometry and phase transitions of MoO_3 thin films for solid state microbatteries *Res. J. Recent Sci.* **2**/4 67-73

[12] Dieterle M and Mestl G 2002 Raman spectroscopy of molybdenum oxides: Structural characterization of oxygen defects in MoO_{3-x} by DR UV/VIS Raman spectroscopy and X-ray diffraction *Phys. Chem. Chem. Phys.* **4** 812-21

[13] Dieterle M and Mestl G 2002 Raman spectroscopy of molybdenum oxides: Resonance Raman spectroscopic characterization of the molybdenum oxides Mo_4O_{11} and MoO_2 *Phys. Chem. Chem. Phys.* **4** 822-6

[14] Xu Q, Jia G, Zhang J, Feng Z and Li C 2008 Surface phase composition of iron molybdate catalysts studied by UV Raman spectroscopy *J. Phys. Chem.* C **112** 9387–93

[15] *pcpdfwin software for XRD diffraction patterns*

[16] Young D J 2008 *High Temperature Oxidation and Corrosion of Metals* (Elsevier Linacre House UK)

[17] Thijs F, Piet M and Paul G J 1979 Monolayer- and crystal-type MoO_3 catalysts: Their catalytic properties in relation to their surface structures. *J. Colloid and Interface Sci.* **70**/1 97-104

[18] NASA technical standard *NASA-STD-6002A* Sept. 2002 NASA USA

Room-temperature sensitivity to NO_2 exposure of electrochemically-deposited nanostructured ZnO layers

K Lovchinov[1,3], M Petrov[1], M Ganchev[1], V Georgieva[2], H Nichev[1], B Georgieva[1] and D Dimova-Malinovska[1]

[1]Central Laboratory of Solar Energy and New Energy Sources, Bulgarian Academy of Sciences, 72 Tsarigradsko Chaussee. 1784 Sofia, Bulgaria
[2]Acad. G. Nadjakov Institute of Solid State Physics, Bulgarian Academy of Sciences, 72 Tsarigradsko Chaussee. 1784 Sofia, Bulgaria

E-mail: lov4@abv.bg

Abstract. This paper reports studies on the sensitivity of ZnO layers to NO_2 exposure. ZnO layers were fabricated by electrochemical deposition on the surface of a quartz crystal microbalance (QCM) with Au electrodes. The sensitivity was estimated using the frequency-time characteristics of the QCM. For this purpose, the resonance frequency shift was measured. The sorption process was investigated in a NO_2 gas flow. The change in the resonance frequency, *f*, of the QCM as a function of the loaded mass of NO_2 was followed for a NO_2 concentration of 500 ppm. Under gas exposure, the frequency decreased and reached saturation in five min. A frequency shift of 38 Hz was measured and a mass loading of 8.39 ng was calculated. The resonance frequency showed a very good recovery within two minutes after the NO_2 flow was switched off. The results demonstrate that the electrodeposited nanostructured ZnO layers have a potential for application as NO_2 gas sensors.

1. Introduction

Recently, TCO materials, and especially ZnO, have shown a growing potential for applications in the development of new optoelectronic and electronic devices. Having a high chemical stability, a low dielectric constant and a high optical transmittance, ZnO has found numerous uses as dielectric ceramics, pigment, catalyst, transparent conductive electrode and sensing material in solar cells, gas sensors, etc. [1- 4]. In particular, an increase in the gas sensitivity and selectivity for different gases has been reported by doping ZnO films with different elements [3-5].

In this work, the sensitivity of undoped thin films of ZnO to exposure to NO_2 was estimated by means of the quartz crystal microbalance (QCM) method.

2. Experimental

Thin ZnO nanostructured films were deposited by an electrochemical process from a slightly acidic aqueous solution of $ZnCl_2$ (5×10^{-3} M) and KCl (0.5 M) with pH 4.0 at 80 °C and −1000 mV (vs. SCE) using a three-electrode electrochemical cell [6]. The electrolyte was agitated by magnetic stirrer. The quartz crystal resonator with a Au electrode was used as a cathode. A spectrally-pure graphite plate

[3] To whom any correspondence should be addressed.

electrode was used as an anode. The deposition was carried out for between 30 and 60 minutes while controlling the redox potential of the system. In addition, the total oxygen content in the solution was controlled by a DO&T meter (Hanna Instruments 9146). The thickness of the ZnO films thus prepared was in the range 0.2 μm – 0.5 μm.

Quartz crystal resonators were used as microbalances to monitor the speed of electrodeposition of the ZnO nano-layers. The oscillators consisted of thermo-stable polished AT-cut quartz plates with a diameter of 8 mm with gold electrodes with a thickness of 600 nm and a surface area of 0,1256 cm^2 in an HC-49 type package. A 16-MHz base resonance frequency was thus obtained.

The gas sensitivity of the electrochemically-deposited ZnO on QCM was studied by exposure to NO_2. The resulting structure was kept in air until saturation in the frequency value was reached. The temperature in the test chamber was kept constant (25 °C ± 0.5 °C) by a thermostat. The concentration of the NO_2 gas flowing into the thermally-stabilized measurement chamber was 500 ppm. The resonance frequency was measured by a frequency counter and recorded on a PC. The experimental setup for measuring the sensing properties of the ZnO-QCM system was described in detail in [7]. The frequency accuracy was ~ 0.1 Hz.

3. Results and discussion

To determine the films' growth rate, several experiments were performed involving deposition of ZnO on a quartz resonator surface, the ZnO layers being deposited for four different time durations – 30, 35, 45 and 60 min. Figure 1 shows the dependence of the ZnO films' thickness on the deposition time. The layers' thickness was calculated from the change in the resonance frequency caused by the ZnO film deposited using the Sauerbrey equation [8]. The dependence could be fit by a straight line:

$$d = 0.01123t - 0.17837, \tag{1}$$

where d is the Sauerbrey thickness of the layer (in nm) and t is a the plating time (in minutes).

The slope of the line yields a value of the film growth rate of 11 nm/min. The negative term of equation (1) indicates a delay in the film growth onset (nucleation time), whose value (at d = 0) is about 15 min [9].

Figure 2 shows a scanning electron microscopy (SEM) micrograph of a nanostructured ZnO thin film deposited at constant charge density for 60 min. The structure of this thin film consists of planar hexagonal ZnO nano-whiskers (NW) with a thickness of about 100 – 300 nm and different lengths (0.5 – 1μm), stacked with the narrow side to the substrate surface. The thickness of the ZnO film is 0.5 μm. The SEM micrograph reveals the porous surface morphology of the deposited layer.

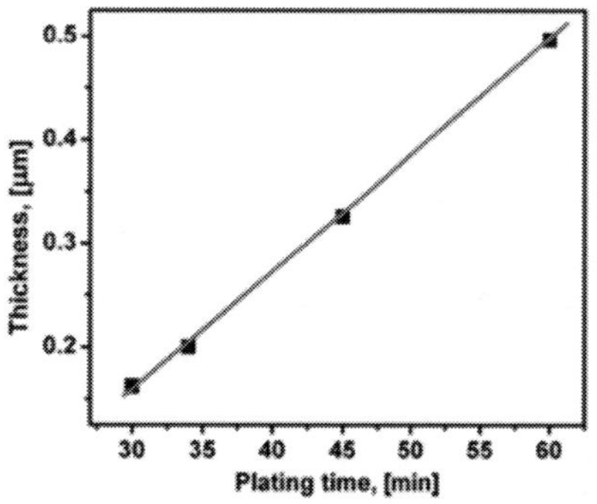

Figure 1. Dependence of a ZnO film thickness deposited on a quartz resonator on the plating time.

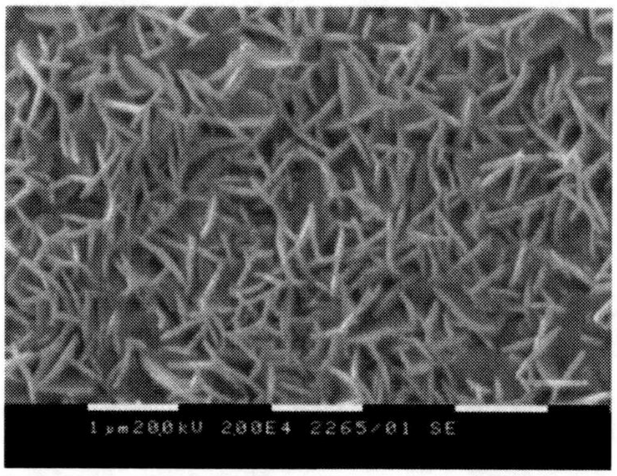

Figure 2. SEM image of a ZnO layer deposited electrochemically for 60 minutes on a quartz resonator.

The ability of the ZnO thin films to sorb NO_2 was evaluated by the QCM method. For this purpose, the resonance frequency shift was measured. The sorption process was investigated in a gas flow of NO_2 with a concentration of 500 ppm by following the change in the frequency, f, of the QCM as a function of the mass loading. Figure 3 shows this change as a function of the exposure time. The highest rate of frequency change, S_p = 11.5 Hz/min, was registered in the first two minutes. After that, the rate decreased more slowly, with the frequency reaching a constant value within about five minutes after exposure to NO_2. When the gas flow was switched off, the frequency returned to its initial value within two minutes due to the desorption of the gas, the fastest change occurring in the first minute. The rate of change was calculated for the first five minutes after exposure to the NO_2 in order to estimate the sensitivity of the electrodeposited ZnO layer. The average value of the frequency-change rate, S_p, and of the total sorbed mass, Δm, were evaluated on the basis of the time-frequency characteristics measured using the relationship between Δf and Δm for AT-cut quartz plates as given by the Sauerbrey equation [6, 8]:

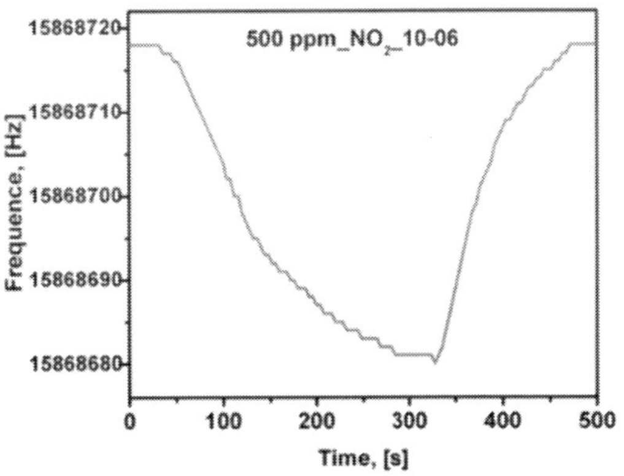

Figure 3. Frequency-time characteristic of a quartz crystal microbalance with an electrodeposited nanostructured ZnO layer under and after exposure to a 500-ppm NO_2 flow.

$$\Delta f = -(2{,}26\times10^{-6} f^2\, \Delta m)/A, \qquad (2)$$

where f (MHz) is the frequency of the QCM before exposure to NO_2, Δm (g) is the mass of the gas sorbed on the surface of the ZnO films, and A (cm^2) is the surface area of the Au electrodes. In our experiments, the frequency shift measured, Δf, was 38 Hz, and the sorbed mass, Δm, was calculated to be 8.39 ng. The S_p value of 7.6 Hz/min was calculated as the ratio of the frequency shift to the time of saturation (five minutes).

The electrochemically-deposited ZnO layer with grown nano-wiskers had a highly developed surface interacting with the NO_2 molecules. When the concentration of the gas in the chamber is raised, the NO_2 molecules are adsorbed in the ZnO layer; as a result, the resonator's mass increases and its frequency decreases. When the gas flow in the chamber is stopped, the NO_2 molecules start desorbing and the frequency of the quartz resonator returnes to its initial value.

4. Conclusions

The sensing characteristic to NO_2 of nanostructured ZnO layer electrochemically deposited on the surface of a QCM was studied. The structure of the ZnO layer, as revealed by SEM, consisted of nano-whiskers and had a developed surface morphology. The NO_2 mass sorbed in the ZnO layer at room temperature and 500 ppm concentration of the gas flow was calculated from the change in the frequency of the QCM after exposure to NO_2. The change in the frequency reached saturation within five minutes after exposure to the NO_2 flow; the absorbed mass was calculated to be 8.39 ng. The time of frequency recovering once the gas flow is switched off was high – about 2 minutes. The results obtained demonstrated a potential application of the electrochemically-deposited nanostructured ZnO layers as room-temperature gas sensors of NO_2.

Acknowledgement

The work was funded under the 7 European FP – project NanoPV No. 24331

References

[1] Dimova-Malinovska D, Tzenov N, Tzolov M and Vassilev L 1998 *J. Mater. Sci. Technol.* B **52** 59
[2] Cao Y, Miao L, Tanemura S, Tanemura M, Kuno Y, Hayashi Y and Mori Y 2006 *Jap. J. Appl. Phys.* **45** 1623
[3] Rao G S T and Rao D T 1999 *Sensors Actuators* B **55** 166
[4] Aslam M, Chaudhary V, Mulla I, Sainkar S, Mandale B, Belhekar A and Vijayamohanan V 1999 *Sensors Actuator* **75** 162
[5] Koshizaki N and Oyama T 2000 *Sensors Actuators* B **66** 119
[6] Dimova-Malinovska D, Nichev H, Georgieva V, Angelov O, Pivin J C and Mikly V 2008 *Phys. Status Solidi* A **205** 1993-7
[7] Georgieva V, Stefanov P, Spassov L, Raicheva Z, Atanassov M and Tincheva T 2009 *J. Optoelectr. Adv. Mater.* **11** 1363
[8] Sauerbrey G 1959 *Zeitschrift für Physik* **155** 206
[9] Lovchinov K, Ganchev M, Petrov M, Nichev H, Rachkova A, Angelov O, Mikli V and Dimova-Malinovska D 2013 *Phys. Status Solidi* A **210** 743-7

Journal of Physics: Conference Series **514** (2014) 012024 doi:10.1088/1742-6596/514/1/012024

Noble metallic nanostructures: preparation, properties, applications

P A Atanasov[1,5], N N Nedyalkov[1], A Og Dikovska[1], Ru Nikov[1], S Amoruso[2,3], X Wang[3], R Bruzzese[2,3], K Hirano[4], H Shimizu[4], M Terakawa[4] and M Obara[4]

[1]Institute of Electronics, Bulgarian Academy of Sciences, 72 TsarigradskoChaussee, 1784 Sofia, Bulgaria
[2]Dipartimento di Scienze Fisiche, Università degli Studi di Napoli Federico II, Complesso Universitario di Monte S. Angelo, Via Cintia, I-80126 Napoli, Italy
[3]CNR-SPIN, Complesso Universitario di Monte S. Angelo, Via Cintia, I-80126, Napoli, Italy
[4]School of Integrated Design Engineering, Keio University, 3-14-1, Hiyoshi, Kohoku-ku, Yokohama-shi, 223-8522, Japan

E-mail: paatanas@ie.bas.bg

Abstract. The process of formation and the characteristics are studied of noble metal nanostructures created by pulsed laser ablation in vacuum. Femtosecond (fs) and nanosecond (ns) laser systems lasing at different wavelengths are used. Several different modifications of the pulsed lased deposition (PLD) technique, as off-axis deposition and glancing angle deposition configurations are used to create nanostructures. Laser annealing of single or bimetal thin films is used to fabricate alloyed nanostructures. The possibility is demonstrated of tuning the optical properties of gold nanostructures on flexible substrates. Different experimental techniques, as fast photography, optical emission spectroscopy, FE-SEM, AFM, TEM, and Raman spectroscopy are applied to characterize the noble metallic nanostructures produced. The optical spectra of the Au and Ag nanostructures are also studied experimentally and theoretically. The theoretical simulation methods used are: molecular dynamic (MD), finite difference time domain (FDTD) and a method based on the generalized multi-particle Mie (GMM) theory. Applications of noble metal nanostructures to surface enhanced Raman spectroscopy (SERS) and biophotonics are briefly considered.

1. Introduction

The properties of the noble metal nanostructures have been the object of considerable fundamental and technological interest. The excitation spectrum of noble metallic sub-wavelength structures is determined by its surface plasmon resonances. The energies of plasmon resonances depend strongly on the shape and composition of the nanostructure. The tunability of the plasmon resonances of noble metallic nanoparticles can be exploited to position the optical resonances at specific wavelength regions of interest and has led to a wide range of applications. The strong local electro-magnetic field enhancement accompanied by the surface plasmon resonances has also been used to manipulate light-

[5] To whom any correspondence should be addressed.

matter interactions; noble metallic sub-wavelength structures have been widely applied to surface enhanced Raman spectroscopy (SERS). In addition to their fundamental importance, plasmonic nanostructures have received a great deal of attention for their potential applications in areas such as sub-wavelength waveguides, optical nanoantennas, photovoltaic technology for efficient light coupling into solar cells, metamaterials, chemical and biological sensing, and in biomedical applications.

The interaction of intense nanosecond (ns) or femtosecond (fs) laser pulses with solid targets has generated a lot of interest in recent years due to the variety of possibilities offered both to basic science and technology. The use of fs laser pulses opens up the possibility to decouple the initial photo-excitation of the target from the subsequent dynamics of the removed material, in contrast with the case of longer pulses (ns). In the former case, the pulse width is less than the typical electron-lattice thermalization time, which is ~ few ps. And finally, the irradiated material is driven into an extremely excited state followed by a rapid quenching, which results in a material blow-off composed of fast ions and atoms as well as clusters and nano-aggregates of the target material, depending on the processing parameters. Fs-laser ablation of metals and semiconductors in a vacuum environment has been investigated extensively. Recent research has in particular been focused on the formation mechanism of nanoparticles (NPs) created by the irradiation of a solid target by intense fs laser pulses of different duration in vacuum [1–4].

As a result of the progress in nanotechnologies during the last two decades, nanosystems find nowadays application in many areas, such as chemistry, optics, biology, medicine, microelectronics, etc. This multilateral usage of the NPs in their quality of fundamental structures in nanotechnology, is based on their unique properties, differing considerably from those of bulk materials. The appearance of characteristic bands in the absorption spectra of metallic NPs is related to the electron gas excited by light oscillations. For noble metals (Ag, Au), these bands are in the visible and near UV region. The collective oscillations of electrons are sensitive to the NPs size, shape and surrounding medium. Both Ag and Au NPs have attracted intensive interest because of peculiarities in their absorptrion spectra in the region of the surface plasmon resonance and the specific and very important applications in nanobiotechnology.

The purpose of this paper is to describe the development of several methods for generating noble metal nanostructures. Ns and fs laser ablation of solid metal targets or thin films in vacuum is studied and used, respectively, to manipulate the production of nanostructures. Moreover, results on laser annealing of single or bimetal thin films for fabrication of alloyed nanostructures are also presented. Tuning of the optical properties of gold nanostructures fabricated on flexible substrates is also demonstrated. The properties of the structures produced are experimentally characterized by a variety of methods, as fast photography, optical emission spectroscopy, FE SEM, AFM, TEM, Raman spectroscopy. Moreover, molecular dynamic (MD) simulation, finite difference time domain (FDTD) and a method based on the generalized Multiparticle Mie theory (GMMT) are applied and the results are compared with the experimental ones. Three possible applications of the Au NPs are briefly described, namely, SERS, photothermal cancer cell therapy and plasmonic nanometric optical tweezers.

2. Experimental

Two laser systems are used in the experiments. The fs laser system consists of a Nd: Glass laser lasing at $\lambda = 1055$ nm ($\tau_p \approx 0.9$ ps and rep rate 33 Hz), SHG - $\lambda = 527$ nm ($\tau_p \approx 300$ fs) and FHG - $\lambda = 263$nm ($\tau_p \approx 300$ fs), respectively. The second experimental set up used is based on a ns Nd:YAG laser working at $\lambda = 1064$, 532, 355 or 266 nm ($\tau_p = 15$ ns and rep rate 10 Hz).

The experimental study of the plume and its evolution in time in vacuum are described in detail in [3, 5-7]. Briefly, the optical emission from the ablation plume is collected along an axis perpendicular to the target surface normal and imaged onto the entrance slit of a spectrograph coupled to an intensified charge-coupled device camera. The evolution of the emitting species inside the expanding plume is followed by measuring their emission spectra at different time delays with respect to the arrival time of the laser pulse on the target, and at different distances from the target surface. Fast photography is applied to follow the vacuum expansion dynamics of the ablated species. The morphology of the deposited NPs is analyzed by AFM or SEM and the shape of the NPs is viewed by TEM. XRD and EDS analyses are performed to check the elemental properties of the NPs.

3. Theoretical

The numerical model is based on the classical MD simulation technique together with two-temperature (2T) modeling. Details are given elsewhere [3, 5-8]. A Morse potential is used to describe the interaction between the particles, and its specific parameters are taken from Ref. [9]. In order to simulate an infinite material in lateral directions, periodic boundary conditions are imposed in the x–y directions. Specific boundary conditions are applied to the bottom of the computational cell in order to prevent the reflection of shock waves [10]. The laser energy absorption and the heat redistribution into the metal is governed initially by the free electrons. Since the electrons are not included explicitly in the MD model, their role, and the heating of the material, is taken into account by using the one dimensional 2T diffusion model. The temperature into the material predicted by the two temperature model is implemented in the MD by way of scaling the atoms' velocities.

The interaction of light with the metal nanoobjects can be simulated by the FDTD approach [11,12]. It has recently become a state-of-the-art method for solving Maxwell's equations in complex geometries, variety of materials and environments. Being a direct time and space solution, it offers the user a unique insight into all types of problems in electromagnetism and photonics. Using the FDTD algorithm is necessary when the characteristic sizes are in the order of the wavelength.

The near-field properties around a metallic particle can be described by the Mie scattering theory [13]. Mie theory, also called Lorenz-Mie theory or Lorenz-Mie-Debye theory, is an analytic solution of Maxwell's equations for the scattering of electromagnetic radiation by spherical particles (also called Mie scattering). Mie theory provides rigorous analytic solutions for light scattered by an isotropic sphere embedded in a homogeneous medium.

4. Control of the nanostructure's properties

We have developed several methods for producing noble metal nanostructures and manipulating their properties – size distribution, shape and composition. One of the important parameters for controlling the size or size distribution of the NPs produced in vacuum is the photon energy (laser wavelength) of the laser radiation used for ablation. This was proven in the case of laser ablation of a solid target by fs pulses [6]. In particular, the size distribution of the NPs is quite broad in the case of visible light (527 nm), and becomes significantly narrower and slightly shifted towards smaller sizes for ultraviolet light (263 nm). This concept being only verified theoretically and experimentally in the case of Ni notwithstanding, we assumed that the laser wavelength is an effective parameter in controlling the NPs size distribution in fs laser ablation of other metals.

4.1. Ultrafast-laser ablation of thin gold film targets

A way to control the size of the nanoparticles is proposed in [14]. Fs-laser ablation of a Au thin film is studied in order to create NPs and compared with that of the bulk target. The experimental results are confirmed by MD simulation of the process and by optical spectroscopy of the plumes. A reduction of the NPs size and a narrowing by a factor two of their size distribution in the case of ablation of a thin film target is observed – figure 1. This fact is attributed to the more uniform heating of the target material – thin film compared to the bulk material – figure 2. In addition, a significant difference is observed between the two cases in terms of the plume's expansion dynamics. Thus, ultrashort-laser ablation of thin films provides a promising way of controlling the plume's characteristics and the NPs size.

4.2. Nanosecond pulsed laser deposition of thin films and laser heating

A method is developed for production of Au and Ag nanostructures which consists of two steps: 1st one – PLD of a thin (50÷100 nm) metal film and 2nd – laser annealing with ns pulses [15]. At certain conditions, the annealing leads to a homogeneous decomposition of the film into NPs with diameters

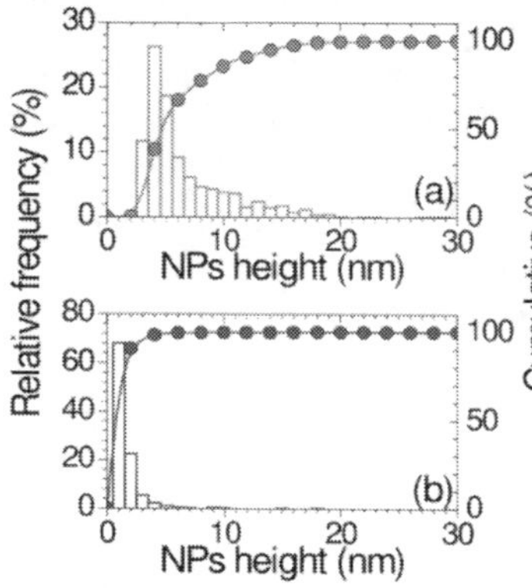

Figure 1. Size histogram and cumulative distribution of the NPs height produced by ultrashort laser ablations of (a) a Au bulk target, and (b) a Au thin film. The thickness of the thin film is 100 nm.

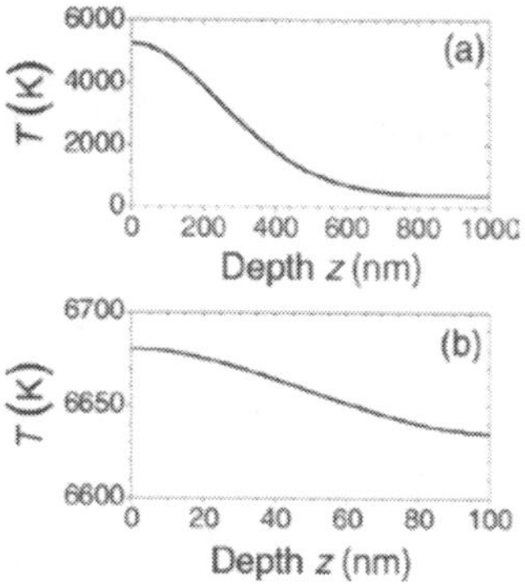

Figure 2. Temperature profiles for a Au target obtained by one-dimensional 2T diffusion model calculated at 40 ps after the laser pulse: (a) bulk material; (b) 100 nm thick film. The scales of z are different in plots (a) and (b).

in the range of few tens of nanometers – figure 3. The fragmentation into droplets occurs if the liquid phase wets poorly the substrate. This is a relatively simple way to produce metal NPs on different substrates. The nanostructured surfaces fabricated can be used successfully as substrates for SERS analyses.

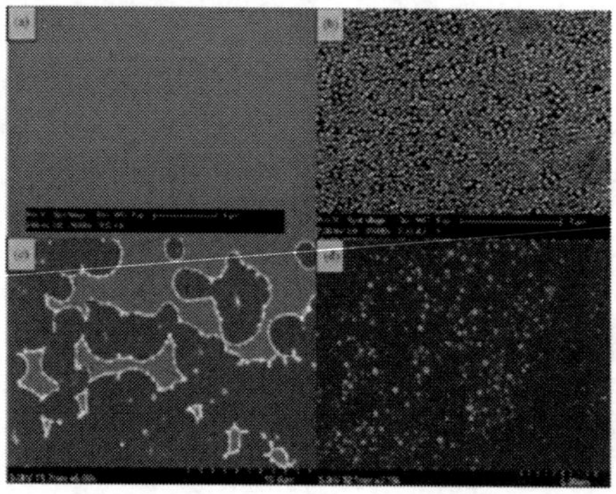

Figure 3. SEM images of a Au thin film with a thickness of $d \sim 60$nm annealed at a fixed number of pulses (10): (a) - as deposited film; (b) - annealed at $F = 130$ mJ/cm^2, (c) – $F = 200$ mJ/cm^2; and (d) - $F = 300$ mJ/cm^2.

4.2.1. Different modifications of ns-pulsed laser deposition of nanostructures. Two different experimental modifications of ns-pulsed laser deposition are applied in order to create Au, Ag or ZnO nanorods – off-axis PLD and deposition at glancing angle of the plume with respect to the substrate (GLAD) configurations.

Using off-axis PLD, where the target center is at a distance of 1÷2 cm from the plume's axis, ZnO nanorods grow when Au NPs are initially deposited on the substrate [16]. The experimental conditions are: HG or FHG Nd:YAG laser with energy density of ~3.5 J/cm^2, substrate temperature $T_s = 0$°C or 150°C and oxygen pressure in the chamber $P_{O2} = 5$ Pa. These Au NPs act as growth nuclei – figure 4.

Figure 5 depicts Au nanocolumns grown in a GLAD configuration. The glancing angle between the substrate surface and the plume axis is $\alpha \sim 5^o$. As is seen, the nanocolumns are parallel are inclined at a small angle with respect to the substrate's surface normal. Some of these structures are successfully used in the SERS experiments after covering with a thin Au film (~ 5 nm).

4.2.2. Fabrication of alloyed nanostructures. A technique is developed for fabricating alloyed nanostructures [17] allowing one to control and manipulate the optical properties of the nanostructures produced. The films are deposited by the classical on-axis PLD technology using targets consisting of

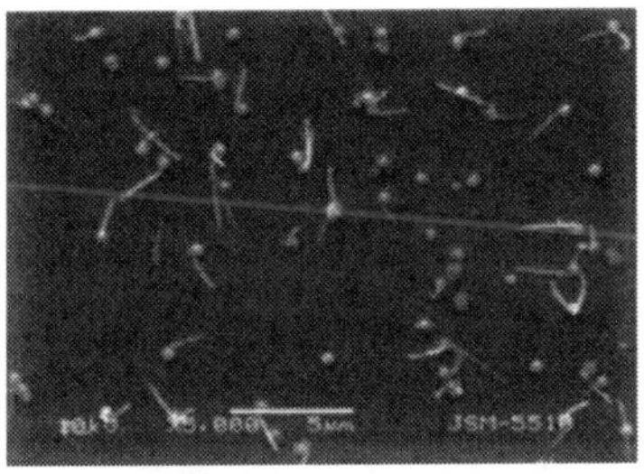

Figure 4. SEM image of ZnO nanorods grown using 200 nm Au NPs (as growth nuclei) at 150 C substrate temperature during deposition.

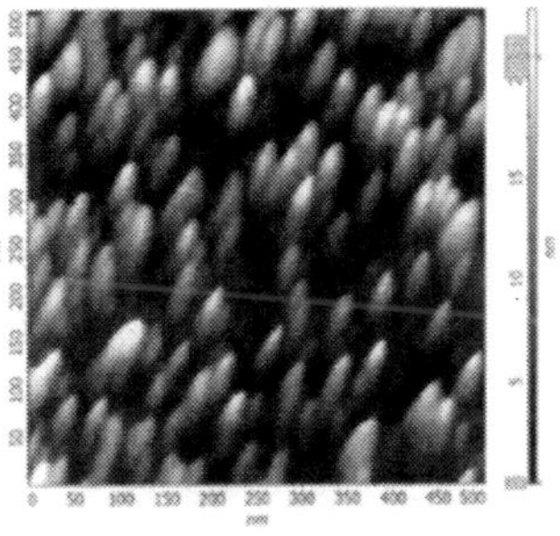

Figure 5. AFM images of Au nanocolumns grown at a GLAD configuretion and Nd: YAG laser THG at 1 J/cm^2 and room temperature.

two sections composed by different metals. Au/Ni or Au/Ag thin films are thus deposited on quartz substrates. By changing the area of the different sections of the target, thin alloyed films with different concentrations of the two metals are obtained. The concentration in the films affects the optical behavior of the films (experimentally obtained extinction spectra) – figure 6. As it is seen, the optical spectra (extinction) depend strongly on the ratio of the elements' concentration.

The average particle diameter in all samples (film thickness of ~80 nm) is in the range between 50 and 30 nm; the size decreases with the increase of the laser fluence. The structures fabricated exhibit a pronounced plasmon resonance behavior in the optical spectra. In combination with a magnetic material, as in the case of Au/Ni structures, a possibility will be open for the development of new technologies in the magneto-optics and spintronics.

4.2.3. Ns-pulsed laser deposition and Ns-annealing of noble metal nanostructures on polymers. An interesting approach is to deposit noble-metal thin films on polymers, then anneal them thus creating 2D or 3D nanostructures [18]. The problem encountered here is the low melting or burning points of the polymers. Bending the flexible substrate will change the distance between the NPs and, thus, tune the optical properties of the structure. Figure 7 presents SEM images of gold nanostructures produced on PMMA and PVC substrates.

Bending the substrate in the range 0÷1,5 mm results in a continuous change of the position of the transmission minimum by ~ 30 nm. The influence of the structure's geometry change on its optical spectra has to do with the change of the inter-grain distance, which affects significantly the optical properties of system via plasmon coupling and multimode plasmon excitation – figure 8.

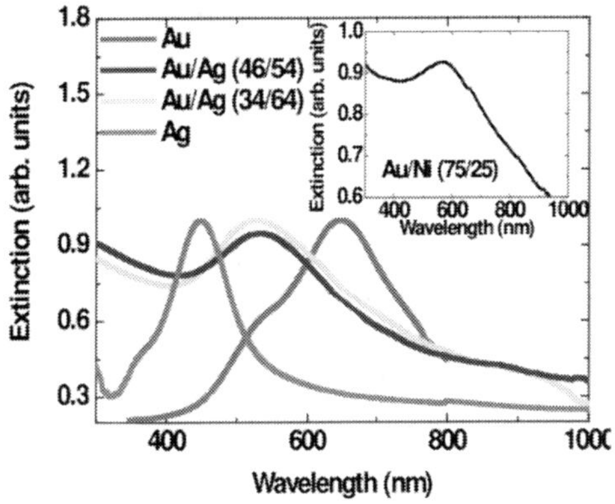

Figure 6. Extinction spectra of Au, Ag and Au/Ag alloy structures. The inset depicts the extinction spectra of a Au/Ni structure. The films deposited are annealed by 5 ns pulses at a fluence of 470 mJ/cm^2. The numbers in brackets indicate the composition in atomic percents.

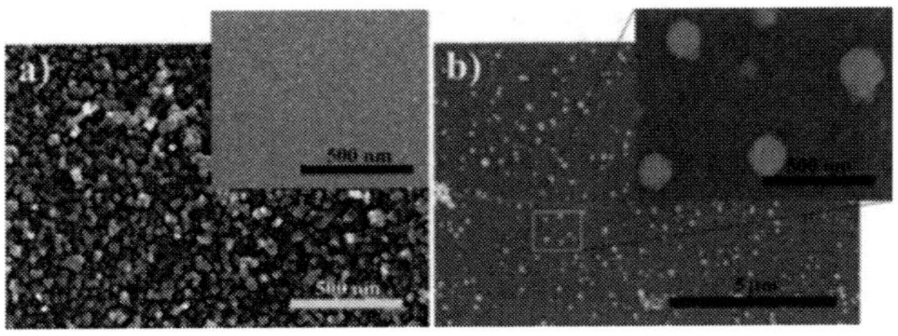

Figure 7. SEM images of Au nanostructures on: a) PMMA and b) PVC substrates. Annealing by a single pulse at F = 90 mJ/cm^2 is used. Insets a) as-deposited Au film on PMMA; b) a detailed view of the structure.

Journal of Physics: Conference Series **514** (2014) 012024 doi:10.1088/1742-6596/514/1/012024

5. Application of gold nanostructures

We present three possible advanced applications of Au nanostructures in SERS spectroscopy, photothermal HeLa cancer cell therapy and development of plasmonic nanometric optical tweezers in an asymmetric gap on Au nanostructured substrates.

5.1. Application of gold nanostructures to SERS spectroscopy

Gold nanostructures produced by the different modification methods described above are successfully applied to SERS spectroscopy. As an example, the Raman spectrum of Rhodamine 6G (R6G) on Au nanocolumns produced by a modified GLAD PLD technology (figure 5) is presented in figure 9 a). Figure 9 b) shows Raman spectra of R6G-covered gold nanostructures and a Au thin film produced on PVC. Obviously, no enhancement is present in the case of the gold film (blue line in figure 9 b). Generally, the Raman spectrum of R6G at a concentration as low as 1 nM is measured [19]. The maximal field-enhancement effect induced by surface plasmon polaritons of Au-nanostructured surfaces exceeds 10^5.

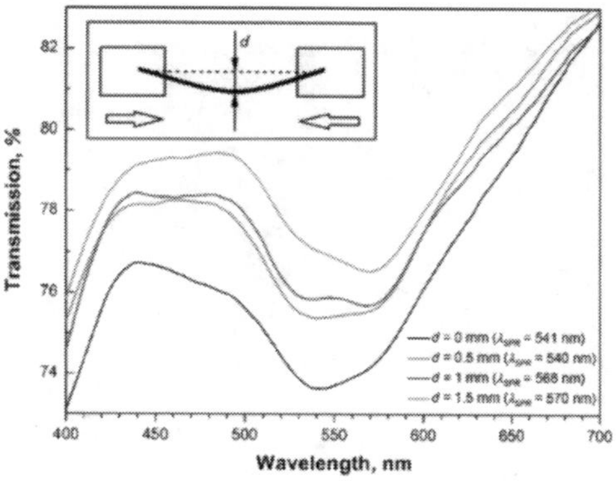

Figure 8. Dependence of the optical transmission spectra of a Au nanostructure on a PMMA substrate on the degree of bending. The as-deposited Au film is annealed by 5 pulses at F = 90 mJ/cm^2. The inset shows the experimental setup for the bending experiments. λ_{SPR} is the wavelength of the transmission minimum.

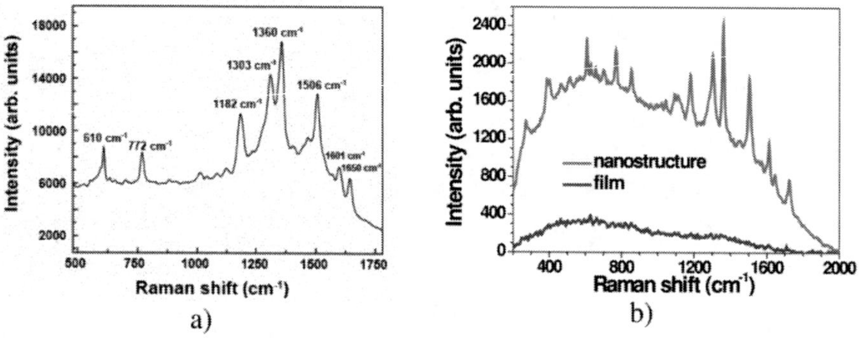

Figure 9. Enhanced Raman spectra of R6G covering: a) nanocolumns produced by a modified GLAD PLD technology; b) red line – spectrum of Au NPs produced by PLD on a polymer and subsequently annealed (the structure is the same as in figure 7 b); blue line – spectrum of an as-deposited Au thin film. The R6G concentrations are equal.

5.2. Nanosecond laser heating of Au nanoparticles: application in photothermal cancer cell therapy

The effect is studied of gold nanoparticles with diameters between 40 and 200 nm on several types of cancer cells [20]. Based on the Mie scattering theory, the optimal conditions of particle absorption at different wavelengths in the visible spectrum are estimated. The absorption cross-sections calculated are used in th 2D heat-diffusion equation to describe the heating process of the particle and the surrounding medium. Using this model, the dependence is obtained of the maximal particles' temperature reached on the laser fluence applied. Figure 10 shows a TEM image of a HeLa tumor cell with Au NPs and optical spectra of this NPs configuration.

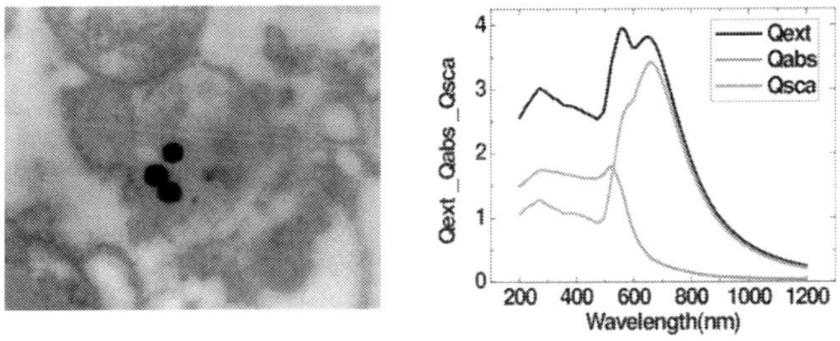

Figure 10. TEM image of a slice of a HeLa tumor cell with deposited gold NPs with diameter of 100 nm and the calculated optical spectra of the system.

Figure 11 a) and b) present the HeLa cell viability after nanoparticle-assisted photothermal treatment. The influence is seen of the Au NPs diameter on the cytotoxic effect – the maximum effect is achieved for 100 nm (fgure 10). Moreover, HeLa cell viability decreases with the increase of the laser fluence – figure 11 b). Further experiments are needed to clarify these effects.

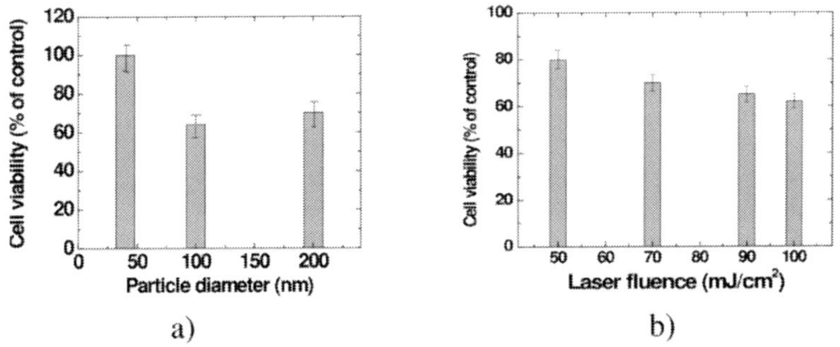

Figure 11. a) HeLa cell viability after NP-assisted photothermal treatment at a laser fluence of 90 mJ/cm^2 and irradiation time of 5 s. b) HeLa cell viability as a function of laser fluence at irradiation time of 5 s. The nanoparticles' diameters are 100 nm. The viability is estimated 6 hours after the laser treatment.

5.3. Plasmonic nanometric optical tweezers in an asymmetric space of Au nanostructured substrates

A possible application of Au nanoparticles is proposed [21], namely, plasmonic near-field tweezers in water with gold nanosphere pairs on various substrates. The enhanced near field (figure 12) localized in the nanometric gap pumped by fs laser light (800 nm) linearly-polarized along the axis between thee nanosphere pair can trap and kill small viruses.

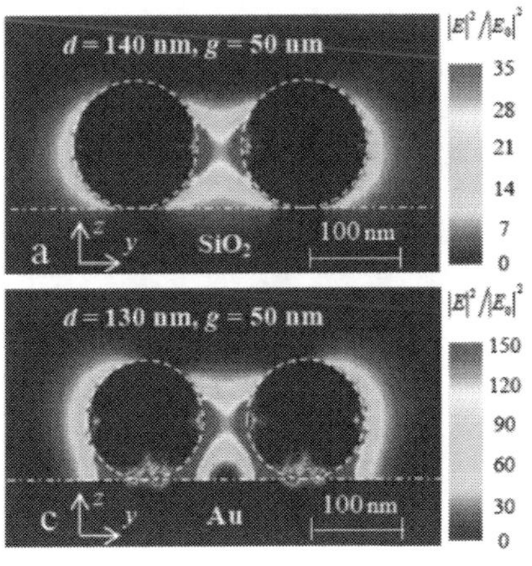

Figure 12. Enhanced field intensity distributions at the gap center for three different substrate materials - SiO_2, Si and Au. The gap distance is 50 nm.

The maximal optical trapping force obtained is larger than 20 pN at an incident optical peak intensity of 1 mW/μm^2. In the system, the enhanced near field stems mainly from the image charge effect, exhibiting an optical trapping in an asymmetric space. The trapped viruses may easily be inactivated using an unfocused 800 nm fs laser. The system proposed may be used for plasmonic dialysis of viruses *ex-vivo* in the blood flow of patients suffering from sepsis and/or an infectious disease, e.g. of the kidney. Further improvements of the method are the subject of a forthcoming work.

6. Conclusions

Results on formation and characterization of noble metal nanostructures created by pulsed laser ablation in vacuum are presented. Fs and ns laser systems are used in the experiments. Several methods for nanostructuring are employed and presented. Tuning of the composition and optical properties of the structures is achieved. The characterization includes several experimental techniques, while the theoretical study includes advanced simulation methods, as MD, FDTD and the GMM theory. Applications of the noble nanostructures obtained are described to SERS and biophotonics, namely, photo-thermal therapy of HeLa cancer and plasmonic optical tweezers.

References

[1] Glover T E, Ackerman G D, Lee R W and Young D A 2004 *Appl. Phys.* B **78**/7-8 995
[2] Glover T E 2003 *J. Opt. Soc. Am.* B **20**/10 125
[3] Amoruso S, Bruzzese R, Vitiello M, Nedialkov N N and Atanasov P A 2005 *J. Appl. Phys.* **98** 044907
[4] Lorazo P, Lewis L J and Meunier M 2006 *Phys. Rev.* B **73**/13 134108
[5] Nedialkov N N, Atanasov P A, Amoruso S, Bruzzese R and Wang X 2007 *Appl. Surf. Sci.* **253** 7761
[6] Amoruso S, Bruzzese R, Wang X, Nedialkov N N and Atanasov P A 2007 *J. Phys.* D: *Appl. Phys.* **40** 331
[7] Amoruso S, Bruzzese R, Wang X, Nedialkov N N and Atanasov P A 2007 *Nanotechnology* **18** 145612
[8] Allen M P and Tildesley D J 1987 *Computer Simulation of Liquids* (Clarendon Oxford)
[9] Girifalco L A and Weizer V G 1959 *Phys. Rev.* **114**/3 687
[10] Zhigilei L V and Garrison B J 1999 *Appl. Phys. Lett.* **74**/9 1341
[11] Taflove A 2005 *Computational Electromagnetics: The Finite-Difference Time-Domain Method.* (Boston, Artech House)
[12] Sullivan D M 2000 *Electromagnetic Simulation Using The FDTD Method* (New York, IEEE Press Series)
[13] Liu Z, Wei Q and Zhang X 2005 *Nano Letters* **5** 957
[14] Amoruso S, Nedyalkov N N, Wang X, Ausanio G, Bruzzese R and Atanasov P A 2011 *J. Appl. Phys.* **110** 124303
[15] Imamova S, Nedyalkov N, Dikovska A, Atanasov P, Sawczak M, Jendrzejewski R, Sliwinski G and Obara M 2010 *Appl. Surf. Sci.* **257** 1075
[16] Dikovska A Og, Nedyalkov N N and Atanasov P A 2011 *Materials Sci. Engin.* B **176** 1548
[17] Nedyalkov N N, Nikov Ru, Dikovska A Og, Atanasov P A, Obara G and Obara M 2012 *Appl. Surf. Sci.* **258** 9162
[18] Nikov Ru, Nedyalkov N, Atanasov P A, Terakawa M, Shimizu H and Obara M 2013 *Appl. Surf. Sci.* **264** 779
[19] Sakano T, Tanaka Y, Nishimura R, Nedyalkov N N, Atanasov P A, Saiki T and Obara M 2008 *J. Phys.* D: *Appl. Phys.* **41** 235304
[20] Nedyalkov N N, Imamova S E, Atanasov P A, Toshkova R A, Gardeva E G, Yossifova L S and Alexandrov M T 2010 *Comptes rendue de l'Acad. Bulg. des Sci.* **63**/5 767
[21] Hirano K, Shimizu H, Enami T, Terakawa M, Obara M, Nedyalkov N N and Atanasov P A 2013 *J. of Nanotech. in Diagnosis and Treatment* **1** 2

Fabrication of 2D arrays of multi-component nanoparticles

Ru G Nikov[1], N N Nedyalkov, N E Stankova and P A Atanasov

Acad. E. Djakov Institute of Electronics, Bulgarian Academy of Sciences, 72 Tsarigradsko Chaussee, 1784 Sofia, Bulgaria

E-mail: rumen_nikov24@abv.bg

Abstract. The paper presents a study of a physical method for fabrication of two-dimensional (2D) arrays composed of multi-component nanoparticles on a dielectric substrate. The method consists of two steps. In the first one, thin films composed of different metals are deposited by a classical PLD technique by using targets consisting of sections of different materials. Thin films composed of mixtures of different metals, as gold, silver, nickel, cobalt, iron, platinum, are thus deposited on a quartz substrate. By changing the area of the different sections in the target, thin films with different concentration of the metals are produced. The films fabricated are then annealed by nanosecond laser pulses delivered by a Nd:YAG laser system operating at its third harmonic. The modification of the films is studied as a function of the parameters of the incident radiation, as number of pulses and laser pulse fluence. It is found that the laser annealing can lead to a decomposition of the film into a monolayer of nanoparticles with a narrow size distribution. The optical properties of the structures produced are analyzed on the basis of their transmission spectra. The structures can be used in surface enhanced Raman spectroscopy (SERS) and magneto-optics.

1. Introduction

Following the development of reliable methods of production of nanoparticles, the research efforts have lately been directed to exploring the possibilities of fabricating multicomponent nanostructures containing two or more nanometer-scale components, such as composites, alloys and multilayered particles [1–5]. Such structures have attracted much attention recently due to the synergistic properties induced by the interactions between the different components that may enhance some specific characteristics compared to single-material particles and even give rise to new properties [6–9]. Progress in nanomaterial synthesis has made it possible to mix uniformly nanoscale components in hybrid materials [10, 11]. A special interest has been devoted to the fabrication of multicomponent nanoparticles composed of noble and ferromagnetic metals with potential applications in catalysis, biophotonics and magneto-optics [12–15]. Ag/Co, Ag/Ni, and Au/Ni particles show well-expressed plasmon effects combined with magnetic properties [16]. The position of the plasmon absorption maximum of nanoparticles of an Au/Ag alloy can be tuned precisely in a wide spectral range by changing the ratio between the two metals. A linear dependence has been shown of the plasmon resonance wavelength on the composition of such nanoparticles [17]. This effect can be used to increase the efficiency of the surface-enhanced Raman spectroscopy, in photothermal cell therapy, and in bio-imaging based on such structures.

[1] To whom any correspondence should be addressed.

The fabrication of multi-component particles can make use of various techniques, the most popular among them being wet chemical synthesis, sonochemical synthesis, photochemical synthesis, sputter deposition, electroless plating, and electrochemical synthesis [8]. Depending on the specific conditions, particles of core–shell, random or separate structures can be produced [8, 18]. Multi-component materials can also be prepared via self-assembly of NPs from their binary colloidal mixtures [19–21]. Although highly productive, these synthesis techniques usually consist of several steps and require toxic reagents. In this respect, in addition of being flexible and inexpensive, the laser-treatment-based methods exhibit important advantages in the fabrication of nanoparticle arrays. E.g., the de-wetting instabilities produced by laser annealing may result in a homogeneous decomposition of thin metal films directly into 2D nanoparticle arrays [22–24]. The method can also be applied successfully for fabrication of alloyed and bimetallic structures [24–26]. Moreover, the characteristics of the nanostructures formed can be controlled by varying the laser fluence applied and the films' thickness and composition. Furthermore, by controlling appropriately the laser beam parameters, one can produce complex alloy structures with desirable 2D configurations.

In this article we report a study on a method of forming 2D arrays of multi-component nanoparticles using laser structuring of thin Au/Ag/Ni and Au/Fe_3O_4 films on a SiO_2 substrate. We found that laser annealing of the films leads to fabrication of 2D arrays composed by particles whose size, shape and density on the substrate surface depend on the processing parameters and the initial films' composition. The morphology, composition and optical spectra of the structures prepared are studied and discussed.

2. Experimental

The multi-component films were prepared by on-axis pulsed laser deposition. The targets consisted of sections of different metal foils (see table 1), namely, Au, Ag and Ni of 99.99 % purity. By changing the area of the different sections in the rotating target, thin films with different compositions were obtained. An Nd:YAG laser system operating at the wavelength of 355 nm at a pulse repetition rate of 10 Hz and a pulse duration of 15 ns was used to ablate material from the rotating target. The laser fluence on the target was 1,2 J/cm^2. The films were deposited in a vacuum chamber at an ambient pressure of about 10^{-3} Pa for a deposition time of three minutes. The thickness of the bimetallic films thus produced was about 80 nm as measured by AFM.

Table 1. Composition of the thin films studied. The target structure is shown schematically.

	Ni, %	Au, %	Ag, %	Fe_3O_4, %	
Sample 1	35	40	25	-	
Sample 2	20	30	50	-	
Sample 3	26	37	37	-	
Sample 4	-	50	-	50	

The laser treatment of the as-deposited films was performed using the laser system dscribed above. The films were annealed by five laser pulses in the fluence range from 60 mJ/cm^2 to 130 mJ/cm^2 in order to investigate the influence of the processing conditions on the thin film modifications observed. The composition of the films and structures produced after the laser treatments were analyzed by EDX (Quantax 200 Bruker). The morphology of the nanostructures fabricated was studied by SEM (LYRA Tescan). The optical properties of the samples were analyzed on the basis of the transmission spectra taken by an optical spectrometer (Ocean Optics HR 4000). The films' composition was measured at ten points within the 5 mm × 5 mm central area. The EDX analysis showed a change of the atomic percentage of the materials of up to 5 %.

3. Results and discussion

Multi-component samples with different metal ratio were obtained by varying the areas containing the different metals in the target. Table 1 describes the samples with different concentrations used in this study. The table also shows the structure of the target corresponding to the different film compositions. The composition of each sample was confirmed by EDX analysis of the as-deposited thin films. The morphology of the annealed films was observed by SEM. Figure 1 shows SEM images of Au/Ag/Ni (a – c) and Au/Fe_3O_4 (d) films annealed by five laser pulses at different laser fluences.

The structural changes of the films were followed in the fluence range from 60 mJ/cm^2 to 130 mJ/cm^2. At the lowest fluence, only slight changes of the surface relief were observed without decomposition of the film being evident (figure 1a). Annealing of the same alloy at a higher laser fluence (F = 90 mJ/cm^2) results in a nanoparticle array containing particles with a nearly uniform size and density on the substrate surface, as can be seen in figure 1b. Laser treatment at F = 130 mJ/cm^2 leads to the formation of nanoparticles with a larger size and a lower film's density (figure 1c) compared to the structure fabricated at F = 90 mJ/cm^2 due to agglomeration. This effect is expressed even more strongly in the case of sample 4, as evidenced by the absence of material in the close vicinity

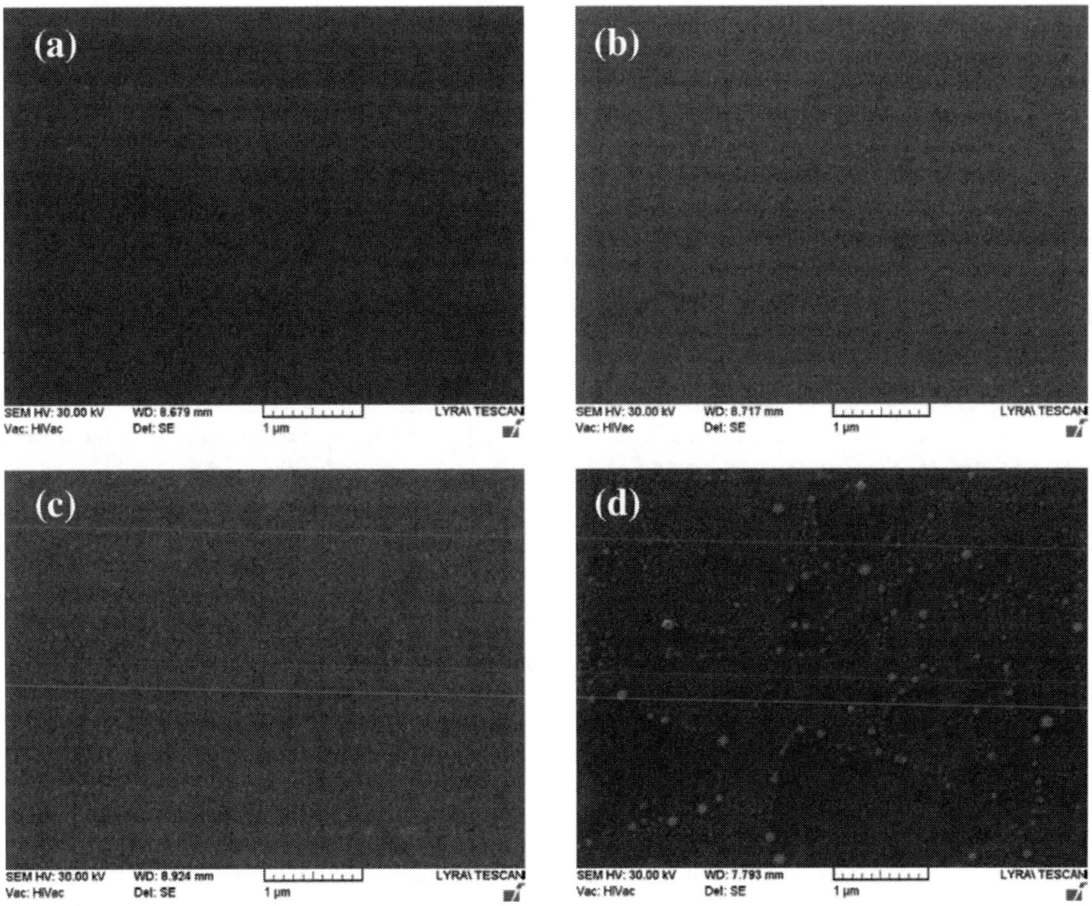

Figure 1. SEM images of multi-component films after treatment by five laser pulses. Sample 3 is annealed at (a) F = 60 mJ/cm^2, (b) F = 90 mJ/cm^2, (c) F = 130 mJ/cm^2 and sample 4 at (d) F = 90 mJ/cm^2.

Journal of Physics: Conference Series **514** (2014) 012025 doi:10.1088/1742-6596/514/1/012025

of the big particles (figure 1d). At laser fluences over 130 mJ/cm^2, a process of film removal off the substrate begins. Since the goal of the study was to define the conditions for formation of a nanoparticle monolayer, we focused our attention on the above-mentioned processing window. It should be mentioned that in the cases of all samples presented in table 1, the laser annealing led to a decomposition of the thin film into a nanoparticle array in the given fluence window. The change of the content of metals in the Au/Ag/Ni films did not influence significantly the characteristics of the structures produced under the same processing conditions.

The fabrication of multi-component nanoparticles containing noble metals presupposes the presence of a surface plasmon resonance in their optical spectrum. In order to determine the specific optical properties of the fabricated structures, we studied their dependence on the processing parameters. Figure 2a presents the optical transmission spectra of a Au/Fe_3O_4 film (sample 4) annealed by five laser pulses at different laser fluences. The well-expressed minima located at 566 nm, 554 nm and 551 nm for the nanostructures fabricated at F = 60 mJ/cm^2, F = 90 mJ/cm^2, F = 130 mJ/cm^2, respectively, are related to plasmon excitation in the nanoparticle array. It is seen that increasing the laser fluence results in a blue-shift of the resonance wavelength. Figure 2a shows also that no plasmon resonance behavior is observed in the spectrum of the unannealed Au/Fe_3O_4 film.

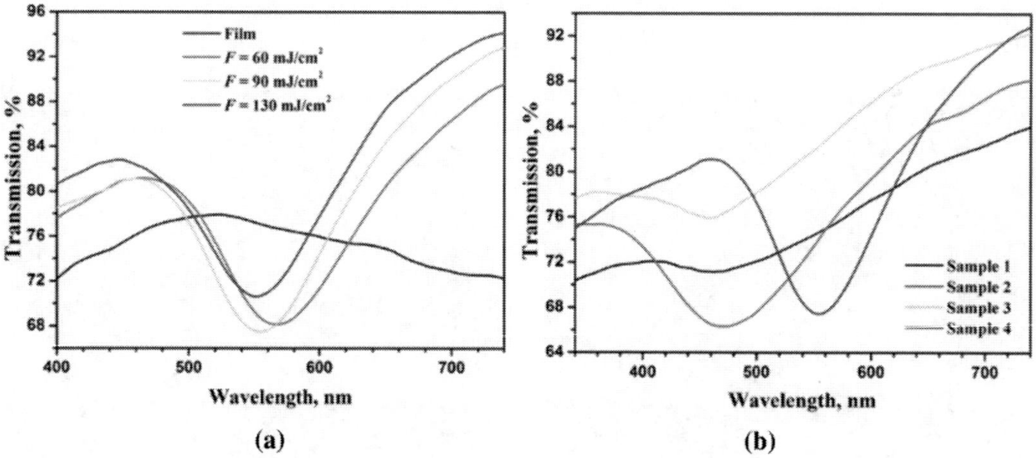

Figure 2. Optical transmission spectra of (a) Au/Fe_3O_4 film (sample 4) annealed by five laser pulses at different laser fluences and (b) multi-component nanostructures fabricated at F = 90 mJ/cm^2, N = 5 at different compositions.

The optical properties of the multi-component nanostructures depend not only on the laser treatment but also on the material's composition. Figure 2b shows the optical transmission spectra of the structures of the samples used in this study (see table 1). The nanostructures with different structural composition were fabricated by five laser pulses at a laser fluence of 90 mJ/cm^2. Plasmon excitation effect is observed in the optical spectrum of each alloy described in table 1. For the Au/Fe_3O_4 nanostructure, a well-expressed SPR minimum is located at 554 nm. In the case of the Au/Ag/Ni alloy, the resonance wavelength varies in the range 459 – 472 nm as a function of the metal content ratio. The SPR depths (with respect to the out-of-resonance level) here are lower compared to the case of the Au/Fe_3O_4 mixture. The influence of the material's composition on the plasmon band characteristics is realized through the change of the dielectric function of the system. Inclusion of Ni leads to an increase of the imaginary part of the dielectric function [24], which is related to a rise in the absorption of the incident radiation. This leads to a dampening of the plasmon minimum. It should be mentioned that the absence of two minima in the transmission spectra of the alloyed samples that would correspond to two single-metal particles also confirms the fact that the fabricated structures are composed by particles of a multi-component material [26].

4. Conclusions

We proposed and demonstrated a technique for fabrication of 2D arrays of Au/Ag/Ni and Au/Fe_3O_4 nanoparticles. The method is based on laser annealing of multi-component thin films deposited by PLD. Under certain conditions, the annealing leads to a decomposition of the films into a monolayer of nanoparticles. The EDX and the optical properties analyses confirm that the particles consist of a mixture of the initial metals used in the deposition process. Treatment of the Au/Ag/Ni film by five laser pulses at a laser fluence of 90 mJ/cm^2 results in the formation of a 2D array composed of nanoparticles with a nearly uniform size and density on the substrate surface. Processing at $F = 60$ mJ/cm^2 does not result in a clear film decomposition and formation of a monolayer of nanoparticles. Surface plasmon resonance is observed in the optical transmission spectra of the fabricated nanostructures depending on the processing conditions. In the case of laser annealing of the Au/Fe_3O_4 film, changing the laser fluence from 60 mJ/cm^2 to 130 mJ/cm^2 results in a blue-shift of the plasmon band position from 566 nm to 551 nm. The material's composition is also found to affect the SPR. The resonance wavelength of the Au/Ag/Ni alloy varies in the range 459 nm – 472 nm as the ratio of the metals' content is varied.

References

[1] Pacholski C, Kornowski A and Weller H 2004 *Angew. Chem. Int.* **43** 4774
[2] Gu H W, Zheng R K, Zhang X X and Xu B 2004 *J. Am. Chem. Soc.* **126** 5664
[3] Pellegrino T, Fiore A, Carlino E, Giannini C, Cozzoli P D, Ciccarella G, Respaud M, Palmirotta L, Cingolani R and Manna L 2006 *J. Am. Chem. Soc.* **128** 6690
[4] Yang J, Elim H I, Zhang Q, Lee J Y and Ji W 2006 *J. Am. Chem. Soc.* **128** 11921
[5] Zeng H and Sun S 2008 *Adv. Funct. Mater.* **18** 391
[6] Sun S H, Murray C B, Weller D, Folks L and Moser A 2000 *Science* **287** 1989
[7] Salem A K, Searson P C and Leong K W 2003 *Nat. Mater.* **2** 668
[8] Major K J, De C and Obare S O 2009 *Plasmonics* **4** 61
[9] Bochicchio D and Ferrando R 2010 *Nano Lett.* **10** 4211
[10] Zeng H, Li J, Liu J P, Wang Z L and Sun S H 2002 *Nature* **420** 395
[11] Zheng H, Wang J, Lofland S E, Ma Z, Mohaddes-Ardabili L, Zhao T, Salamanca-Riba L, Shinde S R, Ogale S B, Bai F, Viehland D, Jia Y, Schlom D G, Wuttig M, Roytburd A and Ramesh R 2004 *Science.* **303** 661
[12] Baletto F, Mottet C and Ferrando R 2003 *Phys. Rev. Lett.* **90** 135504
[13] Ortigoza M A and Rahman T S 2008 *Phys. Rev.* B **77** 195404
[14] Wang L L and Johnson D D 2009 *J. Am. Chem. Soc.* **39** 14023
[15] Kilimis D A and Papageorgiou D G 2010 *Eur. Phys. J.* D **58** 189
[16] Gaudry M, Cottancin E, Pellarin M, Lermé J, Arnaud L, Huntzinger J R, Vialle J L, Broyer M, Rousset J L, Treilleux M and Mélinon P 2003 *Phys. Rev.* B **67** 155409
[17] Link S, Wang Z L and El-Sayed M A 1999 *J. Phys. Chem.* B **103** 3529
[18] Toshima N and Yonezawa T 1998 *New J. Chem.* **22** 1179
[19] Kiely C J, Fink J, Brust M, Bethel D and Schiffrin D J 1998 *Nature* **396** 444
[20] Shevchenko E V, Talapin D V, O'Brien S and Murray C B 2005 *J. Am. Chem. Soc.* **127** 8741
[21] Kalsin M, Fialkowski M, Paszewski M, Smoukov S K, Bishop K J M and Grzybowski B A 2006 *Science* **312** 420
[22] Henley S J, Carey J D and Silva S R P 2005 *Phys. Rev.* B **72** 195408
[23] Trice J, Favazza C, Thomas D, Garcia H, Kalyanaraman R and Sureshkumar R 2008 *Phys. Rev. Lett.* **101** 017802
[24] Wu K, Fowlkes J D and Rack P D 2011 *J. Mater. Res.* **26** 277
[25] Alloyeau D, Ricolleau C, Langlois C, Le Bouar Y, Loiseau A, Beilstein J 2010 *Nanotechnol.* **1** 55
[26] Beliatis M J, Henley S J and Silva S R P 2011 *Opt. Lett.* **36** 1362

Laser measurements of the refractive index of polymer solutions, thin films and bulk materials

N G Sultanova[1,3], **S N Kasarova**[1] **and I D Nikolov**[2]

[1]Department of Mathematics and Physics, Assen Zlatarov University of Burgas, 1 Prof. Yakimov Str., 8010 Burgas, Bulgaria
[2]Department of Optics and Spectroscopy, Faculty of Physics, St. Kliment Ohridski University of Sofia, 5 J. Bourchier Blvd., 1164 Sofia, Bulgaria

E-mail: sultanova@btu.bg

Abstract. We present two laser-based measurement techniques for investigating the refraction of polymer solutions, thin polymer films and bulk samples. Our results confirm the fact that the refractive indices of polymer films differ from those of bulk polymer samples and polymer solutions. However, the analysis of the dispersion yields similar results for all states of the polymer.

1. Introduction

Polymer materials are presently used in the design and fabrication of both consumer products and precise optical instruments. They offer advantages over optical glasses, including low cost and weight, high impact resistance, and ability to integrate proper mechanical and optical characteristics [1]. Optical polymers (OPs) are manufactured in view of a wide spectrum of photonic applications in which bulk polymer bodies with various dimensions as well as thin polymer films (TPFs) are used [2].

The possible usages of OPs are defined mainly by their refraction index (RI) value, which should be determined with a sufficiently high accuracy. The proper choice of a measuring technique depends on the material and the sample's size and thickness. Our earlier results showed that TPFs refraction differs from that of bulk specimens and should be studied separately [3]. We propose here a set-up of a laser microrefractometer (LMR), based on the critical angle determination [4], to obtain refractive data of polymer solutions and films. The results are compared with measured indices of bulk samples obtained by a He-Ne laser and a goniometric set-up. One can thus analyze the refractive and dispersive properties of different states of polymers.

2. Experimental

2.1. Measuring the refractive indices of bulk polymer samples

We have studied the optical properties of various types of plastics, including principal and some newly developed polymers [5, 6]. The principal OPs are polymethyl methacrylate (PMMA), polystyrene (PS), polycarbonate (PC), methyl methacrylate styrene copolymer (NAS), and styrene acrylonitrile (SAN). Different trade-marks of OPs as NAS-21 Novacor, CTE-Richardson, Zeonex, Optorez, Bayer,

[3] To whom any correspondence should be addressed.

etc. and polymers produced by the Eastman Chemical Company, USA (ECC) were also examined.

To obtain more extensive refractometric data, we have used different measuring techniques to determine RIs of OPs in the entire visible (VIS) and near-infrared (NIR) spectrum. The classical Zeiss Pulfrich refractometer with its V-type SF3 glass prism (VoF3) was used to measure the RIs of bulk OPs at five emission wavelengths λ of the spectral lamps of the device in the VIS region. An additional goniometric set-up with the VoF3 prism and a white-light module (a 250-W halogen lamp and a condenser system) with interference filters with spectral bandwidths of about 7 nm – 9 nm was assembled for measurements in the entire VIS and NIR regions up to 1060 nm. In both cases, the deviation angle method was used. Our metrological tests revealed that the goniometric measurements agree with the refractometric results with an accuracy of ± 0.001 in the VIS spectrum [7].

The measurements of the RIs were carried out by a conventional 1-mW He-Ne laser and a goniometric set-up (figure 1). A G5-LOMO goniometer with an accuracy of one arc second was used to measure the angle α. The angle of deviation γ is formed by the sample located into the V-shaped prism. Polymer specimens were produced as injection moulded plates or cubes having two well-polished, mutually perpendicular surfaces. An immersion emulsion with a proper RI is required to ensure the optical contact during the measurements [7]. The thermostatic housing of the VoF3 prism allowed us to maintain a temperature of 20 °C with a stability of 0.2 °C. A photo detector was assembled with the aid of a plane silicon diode, an operational amplifier and an indicator for registering the output beam. The RI n_λ of the polymer examined is calculated as follows:

$$n_\lambda^2 = N_\lambda^2 - \cos\gamma\left(N_\lambda^2 - \cos^2\gamma\right)^{1/2}, \quad \gamma = 90^\circ - \alpha , \tag{1}$$

where $N_\lambda = 1.7347$ is the RI of the VoF3 prism at λ = 632.8 nm. This value is determined by means of a modified Caushy's dispersion formula (4) and the data published in [8]. A metrological analysis was carried out in [9] and a maximal combined uncertainty of $u_C = 3.65\times10^{-4}$ was obtained for the case of laser goniometric measurements of polymer bulk samples.

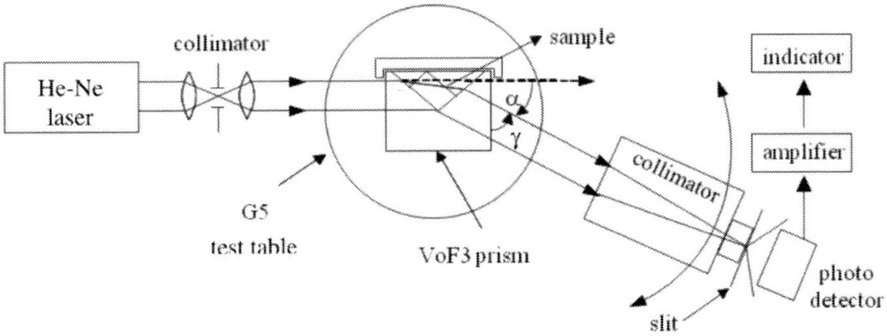

Figure 1. Experimental set-up for refractive index measurements of bulk optical plastics.

2.2. *Refractive index measurement of polymer solutions and thin films*

Among the OPs studied, solutions and TPFs could only be obtained for the ECC materials, which were delivered in the form of pellets. Polyester, polyarylate, polyacrylate, and cellulose polymers were dissolved in chloroform and the two types of copolyesters A and B, in 1,1,2,2-tetrachloroethane. Solutions of the first four materials were prepared with concentration of 10 wt% while the concentration of the copolyesters' solutions was 1 wt%. TPFs were obtained by dropping a certain amount of the polymer solutions on glass substrates. A TK21 glass plate was used for the low refractive cellulose material and heavy flint TF4 glass, for the other OPs. The samples were dried at a temperature of 20 °C for 48 hours and then heated up to 60 °C during 6 hours to evaporate the remaining solvent. The TPFs were preserved in a desiccator. The film thickness *d* was measured by a

Journal of Physics: Conference Series **514** (2014) 012026 doi:10.1088/1742-6596/514/1/012026

digital micrometer (Mitutoyo Corporation) with an accuracy of ± 1 μm.

A three-wavelength LMR was assembled to obtain the RIs of polymer solutions and films. The experimental scheme is presented in figure 2. A 0.5-mW He-Ne laser 1 and two diode lasers 2 (7 mW and 25 mW) with emission wavelengths 632.8, 532 and 790 nm are used. The three beams are directed by splitters 3 to illuminate the internal surface of the prism 5 which is positioned on the goniometric table 4. The sample measured 6 is placed between the prism and a chromium diffraction grating 7 with a 40-μm period and 0.8-μm depth of the grooves. Unit 5 is made of TF4 optical glass with a vertex angle A = 64.76 °. In case of solutions, a drop of the examined liquid is poured on the grating and then pressed tightly to the prism. For TPFs, the polymer layer with the glass substrate is fixed between the prism and the grating. The optical contact is realized using methylene iodide. At small angles of incidence, the diffraction orders are observed in reflection on the screen 8. At the critical angle of the material, total internal reflection occurs and the diffraction pattern disappears. The incident angle φ_c is then measured on the goniometric scale. The RI of the examined material is calculated by:

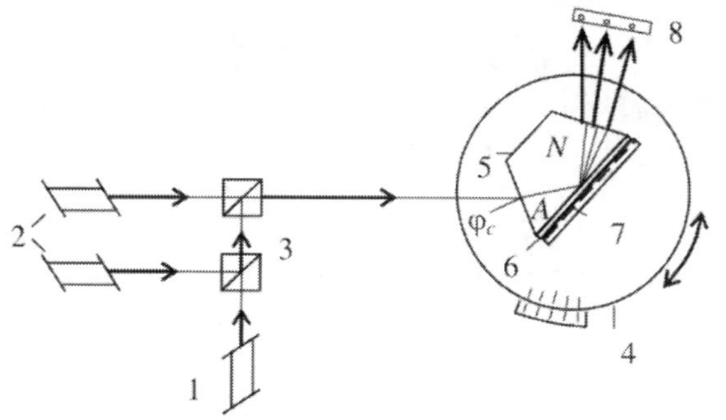

Figure 2. Experimental set-up for refractive index measurements of polymer solutions and TPFs.

$$n_\lambda = N_\lambda \sin\left[A \pm \arcsin\left(\sin\varphi_c / N_\lambda\right)\right], \tag{2}$$

where N_λ is the RI of the prism at the illumination wavelength. The corresponding values are found by fitting with the data published in [10]. The signs "+" and "–" in equation (2) correspond to counterclockwise or clockwise rotation of the incident beam with respect to the normal to the entrance surface of the prism.

A metrological analysis of the results for the polymer solutions is given in [9]. The standard uncertainty, connected with the determination of N_λ, A and φ_c in equation (2) and the temperature effects, was found to be $\pm 2\times10^{-4}$ for standard liquids, as distilled water, ethanol and methanol.

In the case of microrefractometric measurements of TPFs, the accuracy is less because of the extra noise introduced by the multiple surface reflections in the film and the glass substrate. Some light scattering in the polymer medium also occurs. Several measurements of each polymer film were completed and the maximal value of the standard deviation was estimated to be $\pm2\times10^{-3}$.

3. Results and discussion

The RIs measured are presented in table 1 and illustrated in figure 3. The refractometric data of solutions, TPFs and bulk specimens differ substantially. A correlation between the RIs of polymer solutions and of volume samples is possible to be established based on the Lorentz–Lorenz formula, since the specific refraction of solutions r_s is an additive quantity:

$$r_s = \frac{n_s^2 - 1}{n_s^2 + 2}\frac{1}{\rho_s} = \left(\frac{n^2 - 1}{n^2 + 2}\frac{1}{\rho}\right)w + \left(\frac{n_0^2 - 1}{n_0^2 + 2}\frac{1}{\rho_0}\right)w_0, \tag{3}$$

where n_s, n, n_0 are RIs of solution, material and solvent, respectively; ρ_s, ρ, ρ_0 represent their density; w and w_0 are the mass fractions of the polymer and solvent. For example, in the case of the PC material, the calculated value of n is 1.5690 (ρ = 1.195 g/cm^3, ρ_s = 1.452 g/cm^3, ρ_0 = 1.492 g/cm^3;

measured RIs at 632.8 nm – table 1; n_0 = 1.4417). Thus, the result obtained is very close to the value of 1.5684 for the bulk sample measured by means of the goniometric set-up. In the case of the polyarylate material, the RIs of the bulk specimen and of the thickest film are very close. The results for the cellulose and polyester are similar. Some differences are observed for the other polymers. It should be underlined that the deviation angle method gives the average RI value for the volume of the specimen while the critical angle determination is a way of obtaining the local RI near the surface of the TPFs. The depth of light penetration during total internal reflection was evaluated in [4] to be about several microns.

The transmittance of TPFs measured in [11], as well as the RIs obtained, reveal normal dispersion

Table 1. Measured refractive indices of polymer solutions, TPFs and bulk samples.

Optical polymer	Solutions			TPFs				Bulk
	λ (nm)			d (μm)	λ (nm)			λ (nm)
	532	632.8	790		532	632.8	790	632.8
Polyarylate	1.4761	1.4675	1.4592	15	1.656	1.644	1.639	1.6041
				80	1.638	1.626	1.619	
				140	1.616	1.603	1.595	
Polyester	1.4741	1.4694	1.4652	40	1.506	1.500	1.496	1.5064
				44	1.514	1.509	1.505	
Polycarbonate	1.4621	1.4552	1.4489	35	1.599	1.595	–	1.5684
Copolyester A	1.5016	1.4882	1.4786	3	1.538	1.525	1.510	1.5521
				6	1.541	1.526	1.515	
Copolyester B	1.5006	1.4882	1.4793	33	1.649	1.635	1.621	1.5675
				42	1.595	1.562	1.550	

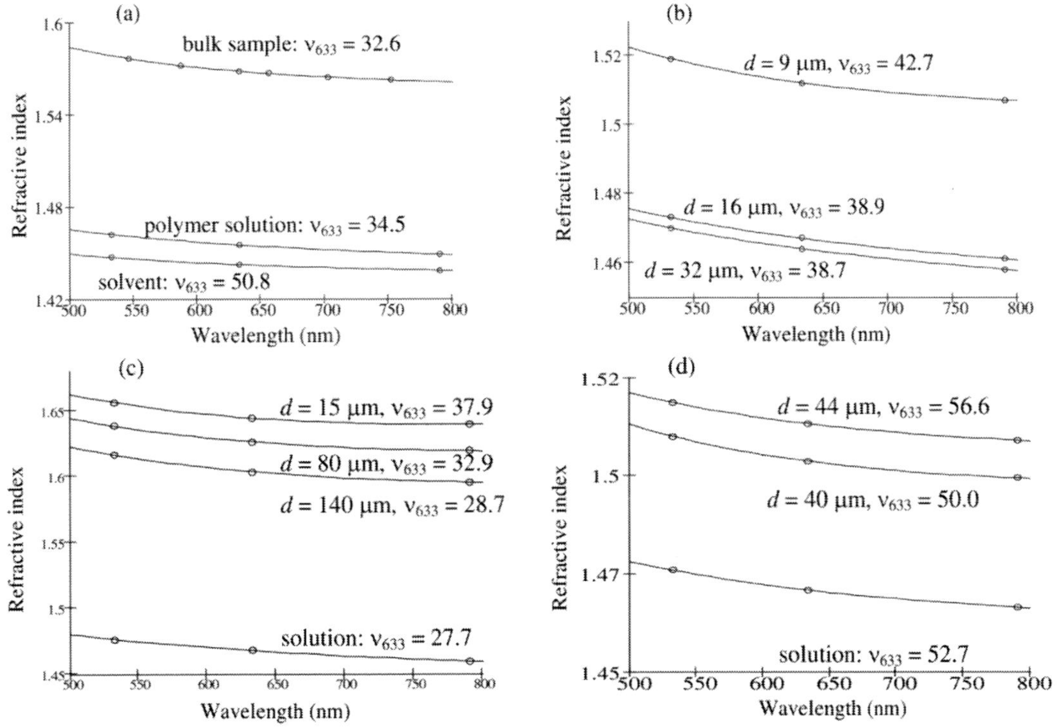

Figure 3. Dispersion curves of: a) PC bulk and solution with chloroform; b) cellulose TPFs, c) polyarylate films and solution; d) polyester films and solution.

of OPs in the spectral range considered. We applied a modified Caushy's formula involving six dispersion coefficients, which ensures a calculation accuracy better than ± 0.0001 [6]:

$$n_\lambda^2 = A_1 + A_2\lambda^2 + A_3/\lambda^2 + A_4/\lambda^4 + A_5/\lambda^6 + A_6/\lambda^8 . \quad (4)$$

In the case of bulk polymer samples, the dispersion coefficients A_1, ... , A_6 are determined with the aid of a system consisting of six linear equations [7]. The dispersion of polymer solutions and films is fitted by the first three terms in the series (4), the calculation accuracy then being ± 0.001.

The dispersion curves of the OPs studied are presented in figure 3. A comparison between the results for the PC bulk sample and solution (figure 3a), cellulose TPFs (figure 3b), polyarylate (figure 3c) and polyester (figure 3d) TPFs, as well as their solutions, yield an equal slope of the dispersion curves of the materials, although the RIs differ depending on the polymer state and the TPFs thickness. The Abbe numbers, fitted to the measuring spectral range of the LMR, are calculated by the expression:

$$\nu_{633} = (n_{633} - 1)/(n_{532} - n_{790}) . \quad (5)$$

The results are included in figure 3. Their values are very close for one and the same material. The only substantial difference is seen for the dispersion of the solvent (figure 3a) and its Abbe number.

The results presented in table 1 and figure 3 b, c, d show a dependence of the TPFs refraction on the film thickness *d*. In the case of polyarylate, copolyester B, and cellulose films, the RIs values decrease as the film thickness is increased. A similar dependence for TPFs was established by other authors [4, 12]. The tendency for the RI to increase as the film *d* diminishes could be explained by an increase in the polymer packing and a reduction in the free fractional volume near the layer surface, due to the surface tension forces. This dependence was not established for the styrenic films in [12]. We also observed some differences in the fluctuations of the RI values with the thickness for the copolyester A and polyester TPFs. As was explained in [13], the film thickness determines the over-molecular structure of TPFs and their structural non-uniformity at the surface, while inside the layer the structure is isotropic.

4. Conclusions

The experimental refractometric data of the OPs examined confirm the existence a dependence on the state of the polymer. The RIs of bulk samples, polymer solutions and TPSs differ substantially. A correlation is only found for the polymer solutions and the bulk samples by means of the Lorentz–Lorenz formula. A variation of the refraction of the polymer layers with their thickness is established. For some of the materials (cellulose and polyarylate), the RIs increase as the film thickness is decreased. This could be explained by the higher packing density of the macromolecules on the TPFs surface and alteration of the over-molecular structure of the material in the layer. The differences between the results for bulk and film RIs depend on the measuring method as well. The deviation angle method of the goniometric set-up measures an average volumetric RI, while in the case of LMR, the RI values near the surface are obtained.

Our study on dispersion of polymers in the measuring spectral range 500 nm ÷ 800 nm confirms a dependence only on the type of the material. Regardless of the state of the polymer, the dispersion curves and Abbe numbers of bulk samples, solutions and films are very close for one and the same material.

References

[1] Tolley P 2003 *Phot. Spectra* **10** 76-9
[2] Doushkina V 2011 *BioPhotonics* **18** 32-5
[3] Sultanova N, Kasarova S and Nikolov I 2013 *J. Optoelectron. Adv. Mater.* **15**/1-2 58-61
[4] Sainov S, Sarov Y and Petrova Tz 2000 *Proc. ISCMP* 416-9

[5] Sultanova N, Kasarova S and Nikolov I 2013 *Opt. Quant. Electron.* **45**/3 221-32
[6] Kasarova S, Sultanova N, Ivanov C and Nikolov I 2007 *Opt. Mater.* **29** 1481-90
[7] Sultanova N, Ivanov C and Nikolov I 2003 *Opt. Quant. Electron.* **35** 21-34
[8] Carl Zeiss JENA 1976 *The manual of Pulfrich-Refraktometer PR2* (Jena D-77830, Germany)
[9] Sultanova N and Kasarova S 2013 *J. Int. Sci. Publ.* **7** Part 1 432-43
[10] Russian GOST Standard 1980 *Russian standard: 13659-78* (Moscow)
[11] Sultanova N, Kasarova S and Nikolov I 2012 *J. Phys. Conf. Series* **356** 0120409
[12] Petrova K, Nikolova K, Leonkieva V, Kitova S and Sainov S 2007 *J. Optoelectron. Adv. Mater.* **9** /2 464-7
[13] Gancheva T 1982 *Structure and properties of polymer materials* (Sofia Tehnika) p 220

Journal of Physics: Conference Series **514** (2014) 012027 doi:10.1088/1742-6596/514/1/012027

Development of topologically structured membranes of aluminum oxide

A Bankova[1], **V Videkov and B Tzaneva**

Technical University of Sofia, 8 Kliment Ohridski Blvd., 1000 Sofia, Bulgaria

E-mail: a_bankova@tu-sofia.bg

Abstract. In recent years, nanomembranes have become one of the most widely used construction material for ultrasensitive and ultrathin applications in micro-electromechanical systems (MEMS) and other sensor structures due to their remarkable mechanical properties. Among these, the mechanical stability is of particular importance. We present an approach to the analysis of the stability of nanostructured anodic aluminum oxide free membranes subjected to mechanical bending. The membranes tested were with a thickness of 500 nm to 15 μm in various topological shapes; we describe the technological schemes of their preparation. Bends were applied to membranes prepared by using a selective process of etching and anodizing. The results of the preparation of the membranes are discussed, together with the influence of the angle of deflection, and the number of bendings. The results obtained can be used in designing MEMS structures and sensors which use nanostructured anodic aluminum oxide.

1. Introduction

The progress in micro-and nanotechnology stimulates the development of products with unique properties applicable in engineering, medicine, scientific research and everyday life. Some of them are employed to create a variety of sensors and micro- and nano-electromechanical devices. The predominant part of these devices use components made of membranes or beams [1]. Membranes are used in some gas sensors [2] and LED modules [3], and beams, in, e.g., acceleration sensors [4] or resonant sensors [5]. These structures are implemented as building blocks because they are easy to model, design and manufacture. Figure 1 presents schematically the different types of mechanical constructions used in the MEMS technology.

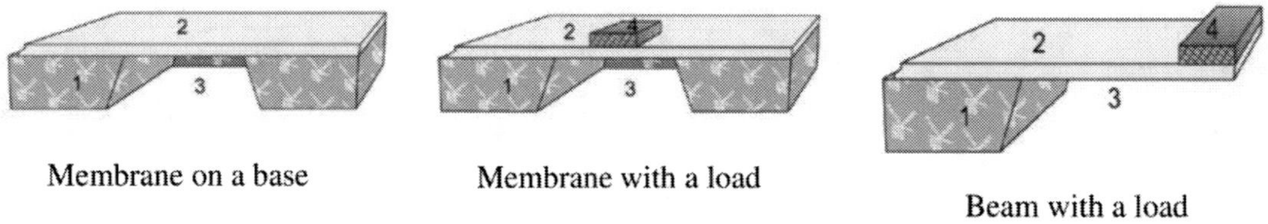

Figure 1. Mechanical constructions.

[1] To whom any correspondence should be addressed.

The aluminum base (1) is anodized and thus covered with an anodic oxide (2); this is followed by a selective etching of the aluminum oxide surface (3). Depending on the application, the mechanical constructions can be with (4), or without a load. In sensors of acceleration (accelerometers), for example, the load can be the mass of the oxide. The load can also be an area of metal or an additional element, such as an absorber of molecules and gases, a piezo element if constructing a pump, etc. Typically, the anodic aluminum oxide consists of a close-packed hexagonal array of nanopores [6]. During the anodization, the main parameters of the electrolytic process can be varied, which allows a precise control of the layer thickness, the pore size and the associated electrical and optical properties [7].

A significant advantage of the process of aluminum anodizing is that it can be easily combined with the planar technology used in microelectronics. Some of the sensors, and some MEMS devices as well, use membranes based on aluminum and its oxide, which is heated or subjected to mechanical stress. A good example is a gas sensor implemented on a membrane, with the membrane being heated to obtain the sensitivity needed [8]. This requires knowledge of the mechanical properties of the oxide and a long-term stability analysis of the membrane's deformation (fatigue). Of particular interest are membranes or beams of small thickness. Examination of these elements requires the construction of appropriate test samples and the application of appropriate methods of examination.

2. Technological schemes for preparation of membranes

Two-sided photolithographic processing is used to implement topologically-structured membranes. Membranes with thicknesses of 400 nm – 500nm to tens of microns can thus be prepared. Figure 2 illustrates the main processes involved in the preparation of a monolithic membrane with a certain topology.

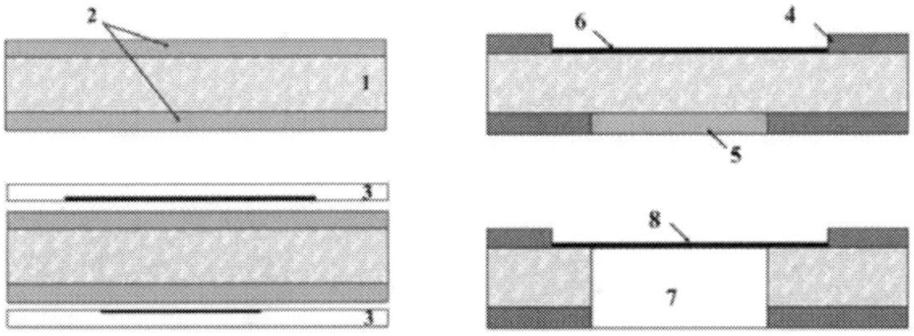

Figure 2. Technological scheme of membrane preparation.

The preparation of a complete continuous membrane begins by laminating an aluminum foil (1) with a dry negative photoresist (2), which provides the necessary resolution and adhesion to the aluminum substrate. Using a photoresist, the precise dimensions in the implementation of the various configurations are defined. Before the exposure, the template (3) is aligned with the aluminum base using additional fiducial marks. The exposure of both sides is conducted through the protective foil for two minutes. The photoresist is developed only on one of the membrane's sides (4), the other side of the sample being exposed (5) without developing in order to preserve the protective sheet. Then, anodic oxide (6) of different thicknesses is grown in the hole developed in the photoresist. Subsequently, the second side of the sample is developed using a spray developing technique, which is followed by etching off the aluminum in the resulting opening in the second side (7). The etching is performed by a selective etcher reacting with the aluminum, but not dissolving the oxide. The resulting anodic oxide layer (8) forms a beam or a membrane depending on the topology.

3. Experimental results

The technological experiments were carried out using a free strip of aluminum with a 99 % purity. To reach the exact dimensions of the various configurations of the samples, we used a dry negative photoresist manufactured by Shipley Laminar GA (General Applications), with a thickness of 37 μm [9] and intended for for alkaline etching and acid electrochemical systems. This photoresist is characterized by high performance, good resolution (50 μm) and high repeatability. The dry negative photoresist was applied by hot rubberized rollers using an Ozatek laminator. The selected mode of laminating was as follows: temperature of the rollers of 100 ± 10 °C, speed of motion of the rollers 0,4 – 1,5 m/min, air pressure 2 – 4 bar. After lamination, as recommended by the manufacturer, the layer was left to homogenize for a minimum of 15 minutes before any further technological process was initiated. The exposure was carried out by UV light through a mask (Figure 3) at intensities 55 mJ/cm^2 to 95 mJ/cm^2. The exposure time on both sides of the template was two minutes. The photoresist on the first side of the test specimen was developed in a 1-% solution of sodium carbonate (Na_2CO_3) containing 1 ml to 1,5 ml defoamers at 30 ± 2 °C for three minutes. The process was controlled visually by illuminating by a selective light. The second side of the sample was not developed in order to preserve the protective sheet. In the opening developed in the photoresist, anodic oxide was grown with thicknesses ranging from 0.5 μm to 15 μm. The oxide layers were produced in electrolyte of 5 % oxalic acid at a temperature of 15 °C and a potential of 4,0 V. The oxide layer formed in 10 minutes was dissolved selectively in a solution of phosphoric and chromic acid to obtain regular nanosized grooves. When re-anodizing an aluminum surface prepared in the way described, a cell structure is grown with a high degree of order. Under the specified conditions of the electrolysis process, the growth rate of the oxide was 0.1 μm/min and the average pore diameter, about 30 nm.

To create different configurations of membranes and beams, structures were used produced in one production cycle. Figure 3a presents the test photomask used in the experiment, where 1 is the topology of the oxide, 2 is the topology of the opposite etching, and 3 is a fiducial mark. Figure 3b shows samples of slugs with different oxide thickness for test beams.

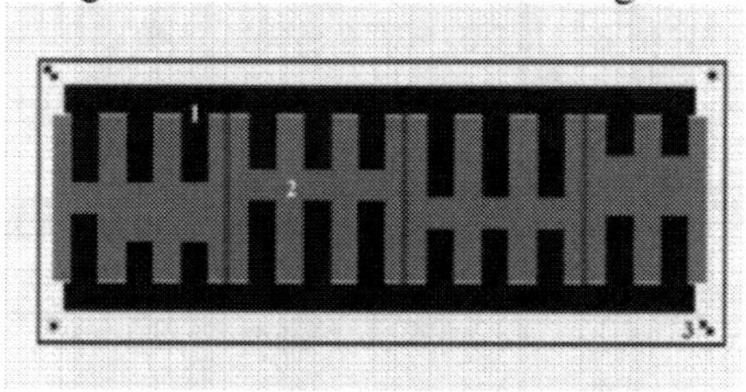

Figure 3a. Test photomask.

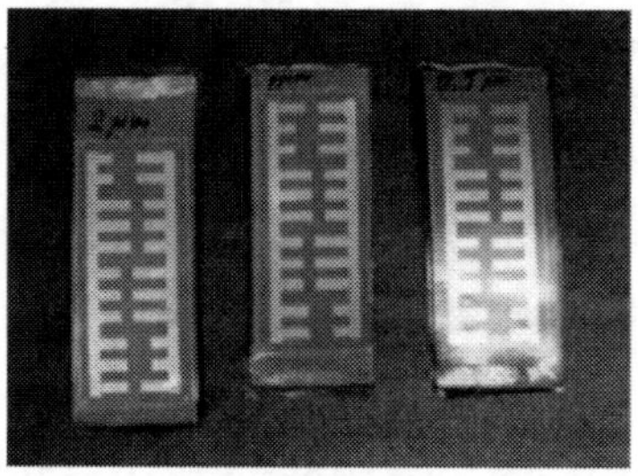
Figure 3b. Samples of slugs with different oxide thickness.

Once the anodic aluminum oxide was grown, the second side of the test specimen was developed using a spray development technique and the aluminum in the resulting opening was etched off. The etching was performed by a selective etcher that reacts with aluminum but does not dissolve the oxide. Figure 4 illustrates the second stage of etching the back side of the test specimen.

Experiments showed that the process of photoresist removal from the front side and the simultaneous etching of both sides of the free aluminum surface is uncontrollable due to the exothermic reaction and the considerable deformation of the oxide. This is why we chose the one-side etching (from the bottom side). Figure 5 shows a stage in the samples preparation, where: 1 – aluminum is etched and a (transparent) oxide beam is formed; 2– meniscus of solution not wetting

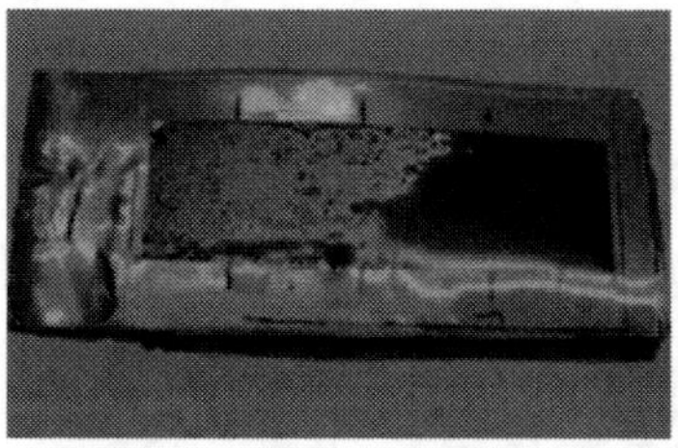
Figure 4. Etching.

Figure 5. Process stage.

Figure 6. Fragment of the process.

the photoresist, 3 – photoresist (free) under which the aluminum is etched; 4 – oxide layer under which the metal has not been removed yet; 5 – photoresist under which the metal has been removed. A detailed fragment of the process, observed under a microscope, is shown in figure 6. A MIK 4 microscope was used at a magnification of 150 and side lighting.

The resulting detailed fragment of the process is formed by: 1 – oxide with etched metal; 2 – residue from the process of aluminum etching; 3 – photoresist without metal underneath; 4 – photoresist with aluminum underneath; 5 – oxide with aluminum underneath.

One of the major problems encountered in implementing this scheme was to prepare free thin films with thicknesses of less than 4 μm. As a result of the wetting of the oxide and of its thermal response under its own weight, a substantial number of the beams were braking in their base. To avoid this problem, we developed a new series of test items containing different configurations of beams and membranes. Figure 7 shows the resulting elements, where: 1 – a channel for separation of groups of samples; 2 – a comb of membranes; 3 – a free membrane held by the photoresist; 4 – residue of the etcher; 5 – a broken beam.

Of particular interest was to study the membranes retention by the resulting support frame of dry photoresist. We found that the anodizing and the concurring processes result in lifting the photoresist (1) while below it another anodizing process takes place where a gradual thinning is performed (3) (figure 8). This tucked oxide (2) with thickness over 4μm is sufficient to support the membrane, depending on its mass.

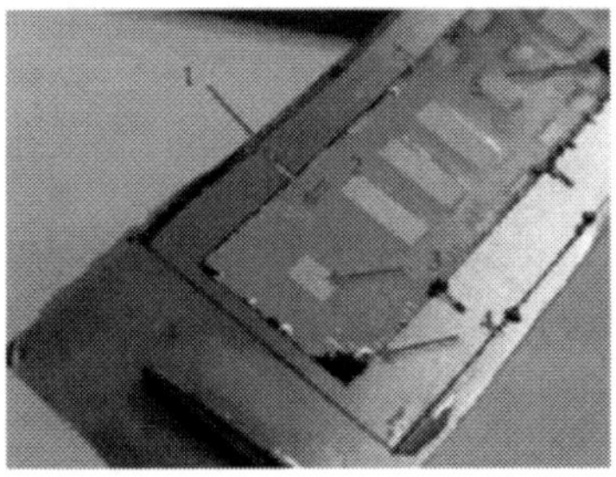

Figure 7. Resulting elements.

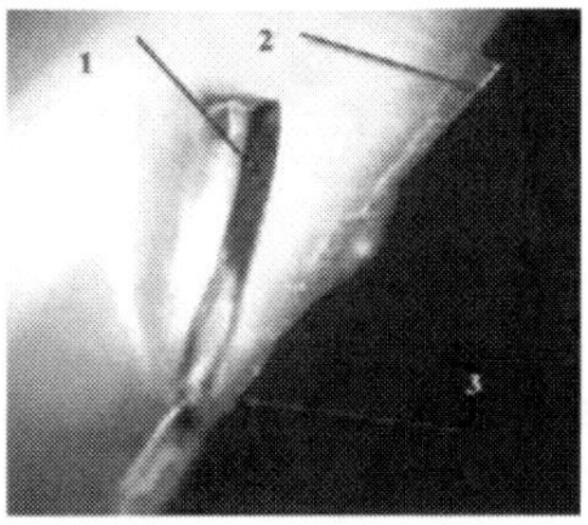

Figure 8. Membrane holding.

Figure 9 shows various kinds of beams glued to holders, which were developed specifically for the purposes of the experiment. Each beam is checked prior to testing for the presence of cracks (figure 10): 1 – photoresist; 2 – oxide on aluminum; 3 – a beam of oxide; 4 – a crack. Evaluation of the mechanical stability of free membranes of nanostructured anodic aluminum oxide can be carried out by measuring the deflection of the test sample at different loads.

Figure 9. Beams.

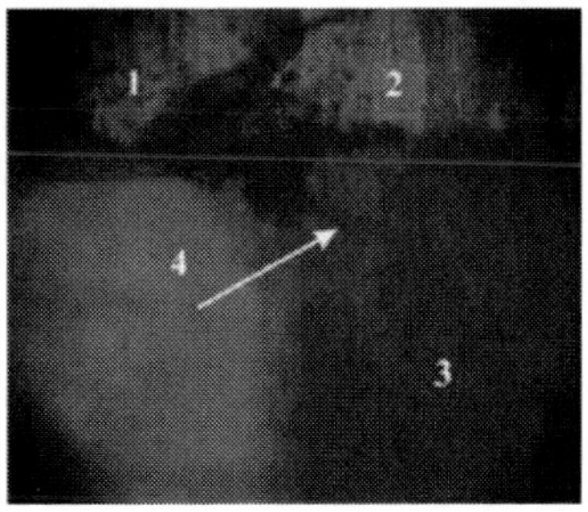

Figure 10. Cracks.

Figure 11 illustrates the experimental setup for mechanical impact studies. Initial experiments of measuring the load were conducted by lowering the needle attached to the analytical scale which measures a change of the weight of 0.1 milligram (1 – aluminum base level; 2 – direction of the oxide beam – 8 μm). In this case, the load caused by lowering the needle leads to bending of the beam at an angle of $\alpha = 15.7°$ (for a total load of 701.3 mg) and $\beta = 19{,}2°$ until impact at the base. The overall weight of the sample and the holder was 683.1 mg.

Detailed studies with various loads, thicknesses and directions of bending will be the subject subject of a future work.

Figure 11. Experimental setup.

4. Conclusions

In recent years, nanomembranes have become one of the most widely used major structural components in the production of various types of nanoscale sensors, optical and micromechanical devices. Based on literature data, we developed a methodology for preparation of topologically-structured membranes of aluminum oxide. The main techniques used were electrochemical anodizing, photolithography and selective etching of aluminum. Using the technological scheme developed, we were able to fabricate membranes with a wide range of thicknesses and topological structures. The resulting oxide elements can be used in various applications in the micro and nanoelectronics.

Acknowledgement

This report presents results obtained under contract No. DFNI – T01/6 "Innovative approach to obtaining structured catalysts for the disposal of methane emissions" and contract 122PD0080-03 for support of graduate students funded by an internal TU competition in 2012.

References

[1] Videkov V, Tsaneva B and Yachovski A 2009 Tecnological schemes of membrane of anodic oxide *Proc. 18th Nat. Conf. Automation of Discrete Production* (Sozopol Bulgaria May 2009) pp 350-5

[2] Vasiliev A A, Lipilin A S, Mozalev A M, Lagutin A S, Pisliakov A V, Zaretskiy N P, Samotaev N N, Sokolov A V and Soloviev S A 2012 Gas sensor MEMS platform for harsh conditions *Proc. 14th Int. Meeting Chem. Sensors* (May 2012) pp 164–7

[3] Aleksandrova M P 2013 Effect of the luminescent layer thickness on the conduction mechanism in light-emitting devices *Materials and Manufacturing Methods* **28** 91-5

[4] Gao Y L, Jin P, Liu N and Tan J B 2006 The optimized design of accelerometer for the UV-LIGA process *J. Phys.: Conf. Series* **48** 1465–70

[5] Agustín L, May H, Luz A, Aguilera-Cortés, Pedro J, García-Ramírez, Manjarrez E 2009 Resonant magnetic field sensors based on MEMS technology *Sensors* **9** 7785-7813

[6] Park J, Fattaccioli J, Fujita H and Kim B 2012 Fabrication of aluminum/alumina patterns using localized anodization of aluminum *Int. J. Prec. Engin. Manufacturing* **13** 765-70

[7] Bai A, Hu C, Yang Y and Lin C 2007 Pore diameter control of anodic aluminum oxide with ordered array of nanopores 2008 *J. Electrochem. Acta* **53** 2258-64

[8] Huang J-W, Lu K C-C, Huang Y-S and Wu S-P 2009 Novel fabrication process using nanoporous anodic aluminum oxidation and MEMS technologies for gas detection *Proc. 23^{d} Eurosensors Conf.* (September 2009) pp 56-9

[9] http://www.interflux-polska.pl/wwwpages/products/shipley/laminar-ga.pdf

Light-induced atomic desorption under different types of illumination

M Taslakov, S Tsvetkov and S Gateva[1]

Acad. E. Djakov Institute of Electronics,
Bulgarian Academy of Sciences, 72 Tsarigradsko Chaussee, 1784 Sofia, Bulgaria

E-mail: sgateva@ie.bas.bg

Abstract We studied the light-induced atomic desorption (LIAD) at different geometries and homogeneities of cell illumination. An original system for homogeneous illumination of the cells was developed and the Rb D2 line absorption spectra in several uncoated and coated cells were compared without illumination, with single-side illumination and with homogeneous illumination. The transmission as a function of the illumination intensity and the tuning direction was measured. The absorption coefficients with and without illumination of the stem were compared. The system developed for homogeneous illumination increases the efficiency of the LIAD, enables the investigation of LIAD without lateral diffusion, prevents cluster formation and reduces the influence of the stem. A good knowledge of the factors influencing the alkali atom spectra will be useful not only for the development of new LIAD-loaded atomic devices and their miniaturization, but also for the study of atom-dielectric surface interactions and the development of new all-optical methods for surface and coating diagnostics.

1. Introduction

Light-induced atomic desorption (LIAD) is a non-thermal process whereby atoms adsorbed at a surface are released under illumination. LIAD has been investigated in a large number of works and has found various applications – vapor density stabilization, magneto-optical trap (MOT) loading, surface nanostructuring etc. Among the parameters affecting the LIAD are the cell dimensions and geometry, the wavelength of the illuminating light, the atom species, the cell's history, the stem-opening area etc. [1,2 and references therein]. LIAD was reported for the first time by Gozzini et al. [3] and has been subsequently investigated extensively both experimentally and theoretically. There exist several models describing LIAD, which take into account different processes concerning its dynamics [2,4-6]. Reduction of the factors influencing LIAD will help the better understanding of the dynamics of this process. A special model describing the LIAD dynamics in a completely illuminated cell was developed in [7].

The present paper presents a study on LIAD at different geometries and homogeneities of illumination. To this end, a special system for homogeneous illumination of the cell was developed. The Rb D2 line absorption spectra in different uncoated and coated cells were compared without illumination, with single-side illumination and with homogeneous illumination. The transmission as a function of the illumination intensity and the tuning direction was also measured. Further, the

[1] To whom any correspondence should be addressed.

Journal of Physics: Conference Series **514** (2014) 012028 doi:10.1088/1742-6596/514/1/012028

absorption coefficients with and without illumination of the stem were compared. The dynamics of the LIAD in an uncoated cell was measured and analysed. An application is proposed for the results obtained to the analysis of vacuum cells' quality.

2. Experimental setup

The experimental setup is given in figure 1; in principle, it is the same as the one described in [8]. The frequency of a 780-nm diode laser is tuned by the laser current and the transmitted spectra are registered by a power meter. A special sphere with a diameter of 150 mm provides homogeneous illumination for LIAD activation. The sphere consists of two hemispheres allowing easy mounting of the cells. The high reflection coating on the inner surface ensures a near Lambertian scattering. A 60-W light emitting diode (LED) (Luminus PT120 with a central wavelength of 460 nm and FWHM of 20 nm) is mounted on the sphere. When the sphere is closed, the efficiency of the blue light illumination improves dramatically, with the inhomogeneity of the illumination of the sphere's surface being less than 10 %. The blue LED current is controlled by means of a laboratory-designed electronic circuit. Two photodetectors are placed on the inner surface of the sphere; one of them providing a negative feedback, the other monitoring the blue light intensity. All measurements are performed at room temperature (25 °C). In order to minimize the influence of the optical pumping, a relatively low 780 nm laser power (3 – 4 μW) is used. Table 1 presents some parameters of the cells used, namely, an uncoated one, a silane-based dry film (SC-77 [9]) coated one, and a porous clay heterostructure (PCHS [10]) coated one.

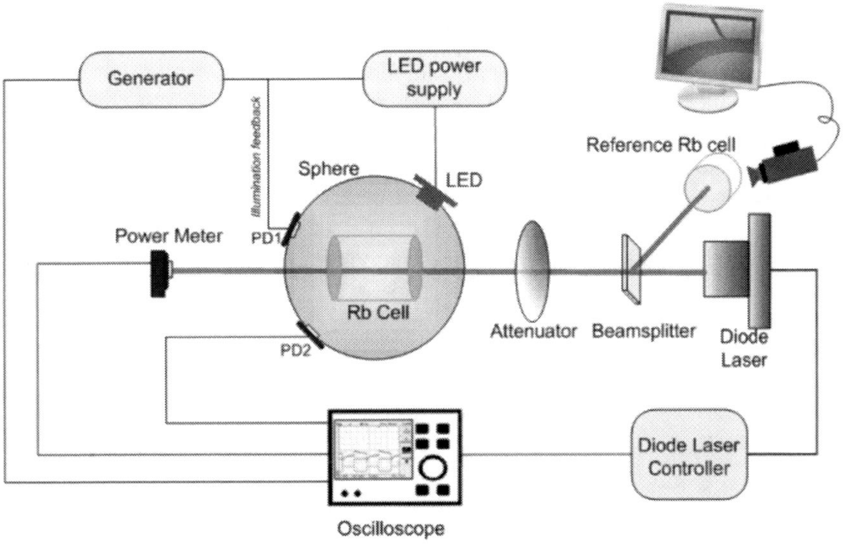

Figure 1. Experimental set up.

Table 1. Parameters of the Rb cells: length, diameter, volume, surface.

Cell coating	Length [cm]	Diameter [cm]	Volume [cm^3]	Surface [cm^2]	Surf/Vol [cm^{-1}]
uncoated 1	4.8	3.2	38.34	64.31	1.68
uncoated 2	2.0	1.8	5.05	16.39	3.24
uncoated 3	4.6	2.4	20.67	43.71	2.11
SC-77	6.0	2.6	31.63	59.60	1.88
PCHS	6.0	2.6	31.64	59.60	1.88

3. LIAD at different illumination

In figure 2, the transmission spectra at different illumination in an uncoated cell and an SC-77 coated one are shown at 4 μW 780 nm laser power and 1 A blue LED current. The spectrum of the 780 nm (Rb D_2) line contains four components. The optical spectrum of each Rb isotope consists of two sets of lines: (i) ^{87}Rb, with two groups of hyperfine transitions starting from $F_g = 1$ and $F_g = 2$ ground levels and (ii) ^{85}Rb, with two sets originating from $F_g = 2$ and $F_g = 3$. The spectrum is the sum of the Doppler profiles (with FWHM ~500 MHz) of all F_g-F_e hyperfine transitions (see for example [8]).

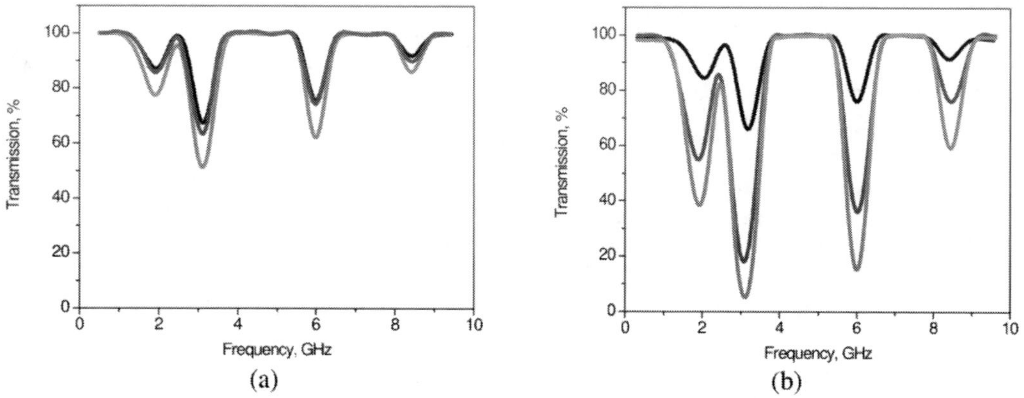

Figure 2. Influence of the illumination geometry in an uncoated cell (a) and an SC-77 coated cell (b) without (black), with one side illumination (blue) and at homogeneous illumination (pink).

The absorption coefficients in uncoated cell 1 and the SC-77 coated cell were compared at different illuminations. Without illumination, the absorption coefficients are comparable (slightly smaller in the coated cell). The absorption coefficient is the largest in the case of homogeneous illumination in both cells. As a result of the different ratio of adsorption due to lateral diffusion and the stem influence, the single-side illumination of the uncoated cell 1 is not efficient, while it is efficient for LIAD in the coated cell.

The response of the three different uncoated cells from table 1 to blue-light rectangular pulses is shown in figure 3. The peak-to-peak values of the detected signal are in a ratio of 250/40/1. This difference is the result of the stem geometry and the technology of preparing the vacuum cells [2]. The third cell is the cleanest one, as it was prepared using an oil-free vacuum system and baked for a long time.

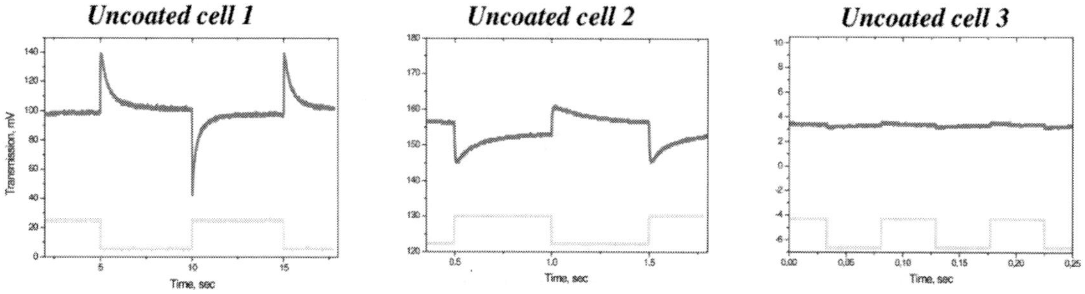

Figure 3. LIAD dynamics in different uncoated cells (the amplitudes ratio is 250/40/1).

In Figure 4 the dynamic characteristics of the LIAD in uncoated cell 1 are presented. The time constants are smaller than in paraffine-coated cells [1]: τ_1=2ms, τ_2=260ms, τ_3=4ms. Another difference with the results obtained before is that when switching on and off the blue light, the changes in the Rb densities are comparable and we can define and measure a fourth time constant τ_4 connected to the processes after the blue light is turned off. In our case τ_4=370ms and it can be attributed to the recovering of the alkali metal vapor density and the initial Rb atom density on the cell surface defined by the temperature of the cell.

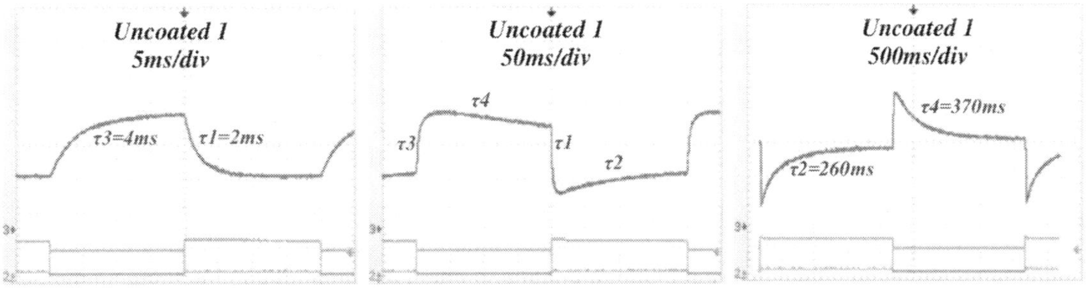

Figure 4. Time constants of LIAD process dynamics.

In the SC-77 coated cells and the PCHS coated one, no modulation of the transmission spectrum (alkali metal density) was detected at these frequencies of modulation of the blue light intensity.

Figure 5 illustrates the tuning of the second and third components of the transmission amplitude when the blue light intensity is increased and decreased in uncoated cell 1. (The component number increases with the tuning frequency [8].) As the light intensity is increased, the typical saturation of the absorption coefficient is registered [6,7]. When the current decreases, the transmission behavior is nonlinear, too, but closer to a linear one.

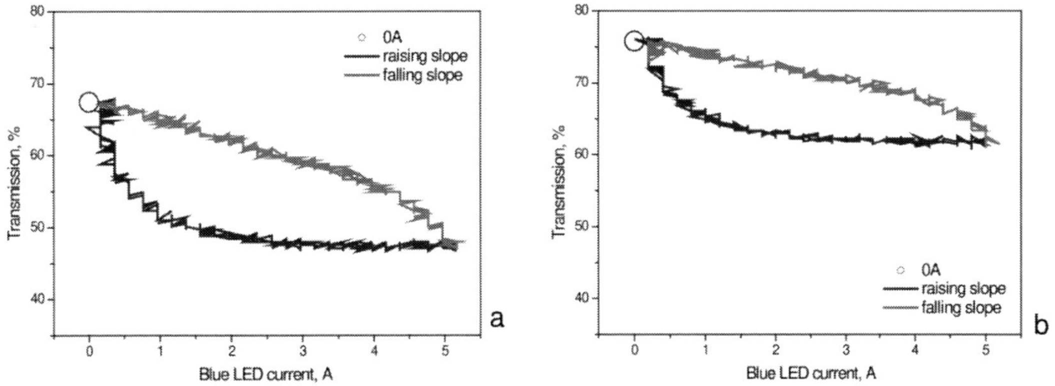

Figure 5. Transmission as a function of the blue LED illumination intensity and the direction of its tuning at 4 μW 780 nm laser light and 0.2 Hz in uncoated cell 1.

4. Stem illumination

The measurement of the transmission spectra when the metal film in the stem of a PCHS coated cell was illuminated and not illuminated by a blue light show that the absorption coefficient is higher when the metal was illuminated; for one of the components, the transmission changed by more than 30 %. Additional investigations of this influence and its possible applications for compensation of the losses due to the open stem are planned.

5. Conclusions

A new original system for homogeneous illumination of the alkali metal cells is developed, where the illuminated cell is mounted in a sphere with a high-reflection coating on its inner surface functioning as a Lambertian screen.

The investigation of the LIAD effect at different illumination in uncoated and coated cells shows that in the case homogeneous illumination, which enables investigation of LIAD excluding the lateral diffusion, the alkali metal atom concentration is the highest.

A much faster modulation of the density of alkali metal vapors is obtained in uncoated cells compared with coated cells.

The different amplitude of modulation of LIAD in the uncoated cells investigated can be explained as being the result of the technology of their preparation; this can be used as an indicator of the cell surface cleanness.

The illumination of the stem increases the alkali metal atoms density. Additional investigations are planned in order to understand the origin of this fact.

In conclusion, the proposed system for homogeneous illumination of the cells investigated increases the LIAD efficiency, enables one to study LIAD excluding the lateral diffusion, prevents cluster formation and reduces the influence of the stem. A better knowledge of the factors influencing the LIAD in coated cells will be useful not only for the development of new LIAD-loaded atomic devices, but also for the study of atom-dielectric surface interactions and the development of new all-optical methods for surface and coating diagnostics.

Acknowledgements

This work was supported by a Marie Curie International Research Staff Exchange Scheme Fellowship within the 7th European Community Framework Program.

References

[1] Karaulanov T, Graf M T, English D, Rochester S M, Rosen Y J, Tsigutkin K, Budker D, Alexandrov E B, Balabas M V, Kimball D F J, Narducci F A, Pustelny S and Yashchuk V V 2009 Controlling atomic vapor density in paraffin-coated cells using light-induced atomic desorption *Phys. Rev.* A **79**/1 012902

[2] Villalba S, Failache H and Lezama A 2010 Light-induced atomic desorption and diffusion of Rb from porous alumina *Phys. Rev.* A **81**/3 032901

[3] Gozzini A, Mango F, Xu J, Alzetta G, Maccarrone F and Bernheim R 1993 Light-induced ejection of alkali atoms in polysiloxane coated cells *Il Nuovo Cimento* D **15**/5 709-22

[4] Atutov S N, Biancalana V, Bicchi P, Marinelli C, Mariotti E, Meucci M, Nagel A, Nasyrov K A, Rachini S and Moi L 1999 Light-induced diffusion and desorption of alkali metals in a siloxane film: Theory and experiment *Phys. Rev.* A **60**/6 4693-700

[5] Alexandrov E B, Balabas M V, Budker D, English D, Kimball D F, Li C H and Yashchuk V V 2002 Light-induced desorption of alkali-metal atoms from paraffin coating *Phys. Rev.* A **66**/4 042903

[6] Rebilas K and Kasprowicz M J 2009 Reexamination of the theory of light-induced atomic desorption *Phys. Rev.* A **79**/4 042903

[7] Rebilas K 2009 Light-induced atomic desorption dynamics: theory for a completely illuminated cell *Phys. Rev.* A **80**/1 014901

[8] Gateva S, Taslakov M, Sarova V, Mariotti E and Cartaleva S 2013 Light-induced atomic desorption for miniaturization of magneto-optical sensors *Proc. SPIE* **8770** 877000

[9] Stephens M, Rhodes R and Wieman C 1994 Study of wall coatings for vapor-cell laser traps *J. Appl. Phys.* **76**/6 3479-88

[10] Arellano-Cárdenas S, Gallardo-Velázquez T, Osorio-Revilla G and López-Cortez M S 2010 Study of the surface charge of a porous clay heterostructure (PCH) and its adsorption capacity of alkaline metals *J. Mex. Chem. Soc.* **54**/2 92-7

Journal of Physics: Conference Series **514** (2014) 012029 doi:10.1088/1742-6596/514/1/012029

Spin randomization of light-induced desorbed Rb atoms

Y Dancheva[1], C Marinelli[1,2], E Mariotti[1], S. Gozzini[2], L Marmugi[1,2], M R Zampelli[1], L Moi[1], P N Ghosh[3], S Gateva[4], A Krasteva[4] and S Cartaleva[4,5]

[1]CNISM and Department of Physical, Earth and Environmental Sciences, University of Siena, Via Roma 56, 53100 Siena, Italy
[2]Istituto Nazionale di Ottica CNR, UOS "A.Gozzini", Via Moruzzi 1, 56124 Pisa, Italy
[3]Department of Physics, University of Calcutta, 92APC Road, 700 009 Kolkata, India
[4]Acad. E. Djakov Institute of Electronics, Bulgarian Academy of Sciences, 72 Tsarigradsko Chaussee, 1784 Sofia, Bulgaria

E mail: stefka-c@ie.bas.bg

Abstract. We present the first experimental observation of atomic spin randomization of Rb atoms released by light-induced atomic desorption (LIAD). A natural mixture of Rb atoms contained in paraffin and PDMS coated glass cells is irradiated by a free-running diode laser light tuned to the Rb D_2 resonance line. The transmission spectrum of the Rb vapor is thus modified and shows a strong enhancement of the hyperfine optical pumping as the light intensity is increased and the laser-frequency scanning rate is decreased. The D_2 line spectra are compared for two cases: without and with illumination of the walls of the cell by a UV lamp centered around the wavelength of 404 nm. A simple theoretical model based on the solution of the rate balance equations is introduced in order to analyze the experimental results.

1. Introduction

In order to suppress the spin randomization of vapor-phase atoms during their collisions with the cell walls, a coating procedure is frequently used [1,2]. Among the most common coatings, polysiloxane or paraffin films show particularly low atomic adsorption energies [1-3]. Furthermore, it has been found that the antirelaxation coating adsorbs a significant number of alkali atoms, which can be easily released by any low intensity, non-resonant light beam, as a consequence of a process called light induced atomic desorption (LIAD). Most of the papers devoted to this phenomenon mainly concern the vapor density dynamics in coated cells. Even if some works have indirectly addressed the effects of external illumination on the coatings' behavior [4,5], the question remains open concerning the details of the influence of LIAD effect on the key property of the coatings, namely their ability to significantly reduce the spin relaxation of atoms desorbed and introduced in the cell volume [1].

In this communication we present the results of studying the hyperfine optical pumping depending on the light intensity and the laser-frequency scanning rate, which made it possible to observe experimentally for the first the spin randomization of Rb atoms released by LIAD. A theoretical model

[5] To whom any correspondence should be addressed.

Journal of Physics: Conference Series **514** (2014) 012029 doi:10.1088/1742-6596/514/1/012029

based on the solution of the rate balance equations was developed in order to analyze the experimental results.

2. Experimental results

A natural mixture of Rb atoms contained in a paraffin or a PDMS coated pyrex cell is irradiated by a free-running diode laser light tuned to the Rb D_2 line ($\lambda = 780{,}24$ nm) and scanned around it by applying a triangular waveform to the input current. The transmission spectrum of Rb is measured for different laser intensities as well as for various laser-frequency scanning rates. The D_2 line spectra are compared for: (i) paraffin and PDMS coated cells and (ii) without or with illumination of the walls of the cell by a UV lamp at a wavelength around 404 nm and a bandwidth of 20 nm (figure 1). All experiments are performed at room temperature. The laser beam is split in two parts: the first beam is expanded and directed to the coated cell, while the second beam passes through an uncoated vacuum cell used as a reference. Linearly or circularly polarized light is used.

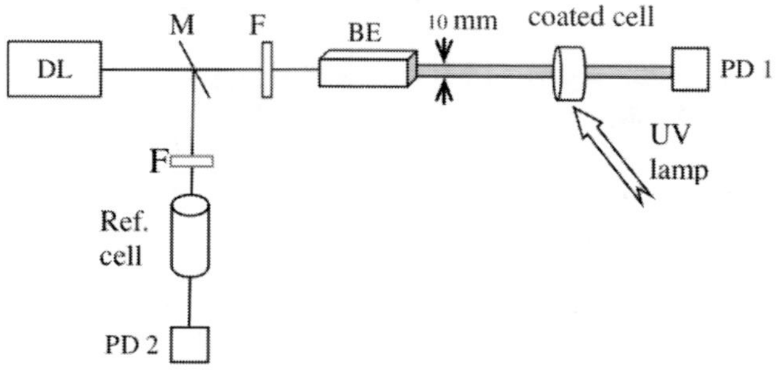

Figure 1. Experimental set up: DL - diode laser, M – mirror, F – neutral density filter, BE – beam expander, PD 1 and PD 2 – optical detectors.

In figure 2, the transmission spectra of both Rb isotopes are shown for a paraffin coated cell at a scanning frequency of 10 Hz and in both scanning directions. All measurements presented are performed by recording simultaneously the transmission signals from the reference and the coated cell. It is seen that the spectrum of the reference cell does not depend on the direction of the laser frequency scan.

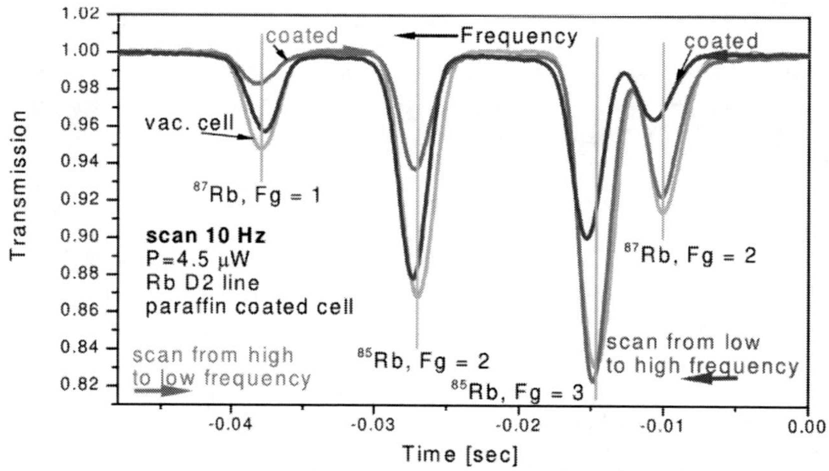

Figure 2. Transmission spectra of the D_2 line of Rb, for linear polarization, low laser power and laser-frequency scanning rate of 10 Hz (grey trace) – the reference spectrum of Rb atoms contained in an evacuated glass cell; (**blue** trace) – the spectrum of Rb in a paraffin-coated cell at increasing laser frequency; (red trace) – the spectrum of Rb at decreasing laser frequency; the frequency difference between the $F_g = 2$ and $F_g = 3$ levels (^{85}Rb) is 3036 MHz. The difference in the transmission spectra for frequency scanning in the two opposite directions is clearly seen in each line profile.

In the case of the coated cell, however, the experimental study shows that the scanning direction strongly influences the transmission of each absorption line. For example, in the case of ^{85}Rb, the frequency scan over the $F_g = 2$ line produces atomic accumulation on the $F_g = 3$ level, which is measured while scanning the frequency over the $F_g = 3$ set of transitions (figure 2, red trace). In the opposite direction (figure 2, blue trace), however, the absorption of the $F_g = 2$ line is larger than that of the $F_g = 3$ line. Reducing the laser frequency scanning rate results in a stronger optical pumping of the ground state hyperfine levels not interacting with the laser light. Due to this process, in addition to the difference in the transmission amplitudes for the blue and red spectra, a well pronounced shift of the transmission profile maximum occurs for each line (figure 3). The shift of the profile maximum depends on the laser frequency scanning direction, as recently reported in ref. [6]. It is attributed to the competition between the hyperfine optical pumping that depletes the atomic population of one ground level through the laser light excitation, followed by decay to the other ground level and the absorption enhancement during the scan when the laser frequency approaches the center of the absorption line. Thus, the maximum of the absorption profile is formed before the center of the hyperfine transition, i.e. the absorption maximum shifts to the starting point of the frequency scan.

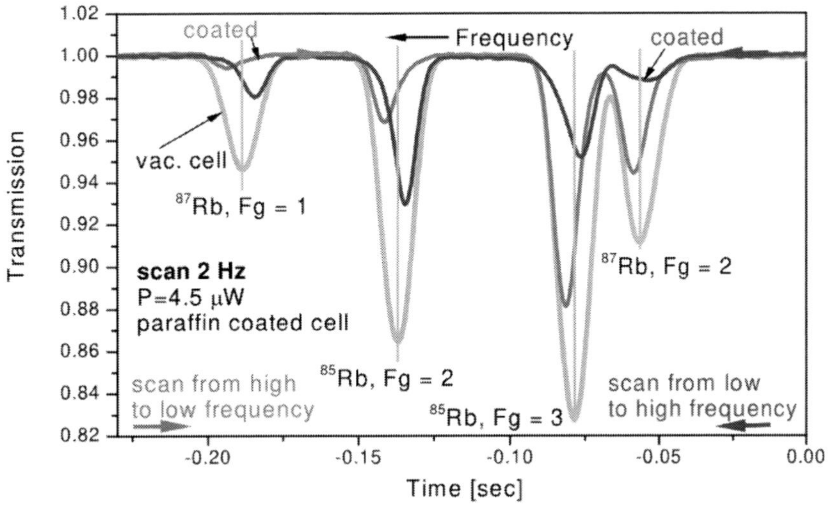

Figure 3. Transmission spectra for linearly polarized light and lower laser-frequency scanning of 2 Hz. (grey trace) – the reference spectrum of Rb atoms contained in an evacuated cell; (**blue** trace) – the spectrum of Rb in a paraffin coated cell at increasing laser frequency; (red trace) – the spectrum of Rb in a paraffin coated cell for frequency scanning in opposite direction.

Our experimental investigation shows that both types of coated cells (paraffin and PDMS) exhibit a significant hyperfine optical pumping (figure 4). Thus, the first irradiated ground state level of each Rb isotope is strongly depleted and its absorption is almost cancelled (figure 4, red line). The fact that absorption from the level interacting with the light vanishes represents a proof of the strong orientation of the atomic spins caused by the laser light. Under such experimental condition, it is important to know if the atoms released by the LIAD effect will also be oriented or their spins will suffer some randomization of the orientation due to the time spent within the cell wall coating.

Our experiment shows that the flow of LIAD atoms results in a reduction of the hyperfine optical pumping. Such a reduction of the Rb atoms orientation in the cell volume is attributed to the random spins distribution of the atoms desorbed from the cell wall. In the case of LIAD involvement, the measured shift of the absorptions profile maximum is smaller, and both the absorption ratio of the ^{85}Rb ($F_g = 2$) set of transitions to the ^{85}Rb ($F_g = 3$) set and that of the ^{87}Rb ($F_g = 1$) set of transitions to the

^{87}Rb ($F_g = 2$) set change. The spin randomizing of atoms desorbed by LIAD is illustrated for the PDMS coated cell that shows a stronger LIAD effect than the paraffin coated one.

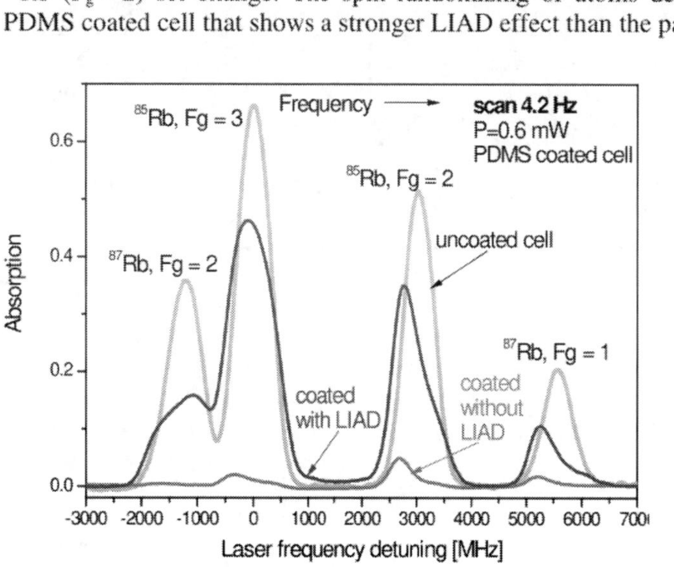

Figure 4. Absorption spectra for circular polarization and without/with modification by the LIAD effect; (grey trace) – reference spectrum of Rb atoms; (red trace) – the spectrum of Rb in the PDMS coated cell without LIAD effect, exhibiting strong hyperfine optical pumping: (i) the population of the $F_g = 2$ level of ^{87}Rb is almost depleted and the absorption of the $F_g = 3$ level of ^{85}Rb is strongly reduced; (**blue** trace) – the spectrum of Rb in the PDMS coated cell, in the case of a strong LIAD effect; the enhancement of the absorption at each line due to LIAD is clearly seen.

3. Theoretical model and discussion

The theoretical model presented is based on a system of rate balance equations describing the accumulation and loss of atomic population from energy levels and is used to analyze the optical pumping processes, considering for simplicity only the D_2 line of one isotope (^{85}Rb). A slightly modified energy levels scheme is shown in figure 5. The two sets of hyperfine transitions, separated by 3036 MHz, are depicted in such a way to emphasize the scanning frequency direction as realized in the experimental approach. The two systems are connected through spontaneous emission from excited levels to a ground level not interacting with the laser light, namely the open transition $F_g = 3 \rightarrow F_e = 2', 3'$ and $F_g = 2 \rightarrow F_e = 2, 3$. Note that the $F_e = 2', 3'$ and $F_e = 2, 3$ are the same excited levels.

The system of rate balance equations describes the evolution of the energy levels' atomic population as a result of the spontaneous and stimulated transitions [7,8]. The system of rate equations is formulated taking the laser frequency as a variable and is solved for the steady-state case as follows:

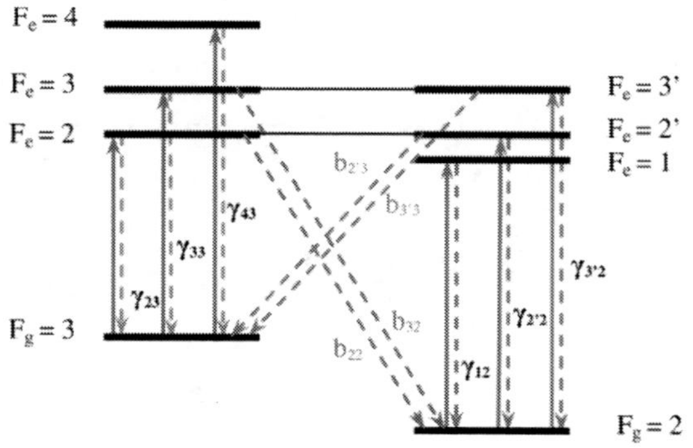

Figure 5. Energy level diagram of ^{85}Rb on D_2 line, the hyperfine ground state splitting is 3036 MHz.

$$\frac{\partial N\{e_2\}}{\partial t} = \left(N\{g_3\} - N\{e_2\}\right) * \sigma_{32} - \gamma_{23} * N\{e_2\} - b_{22} * N\{e_2\}, \tag{1}$$

$$\frac{\partial N\{e_3\}}{\partial t} = \left(N\{g_3\} - N\{e_3\}\right) * \sigma_{33} - \gamma_{33} * N\{e_3\} - b_{32} * N\{e_3\}, \tag{2}$$

$$\frac{\partial N\{e_4\}}{\partial t}=\left(N\{g_3\}-N\{e_4\}\right)*\sigma_{34}-\gamma_{43}*N\{e_4\}, \tag{3}$$

$$\begin{aligned}\frac{\partial N\{g_3\}}{\partial t}=&\left(N\{e_2\}-N\{g_3\}\right)*\sigma_{32}+\gamma_{23}*N\{e_2\}+\left(N\{e_3\}-N\{g_3\}\right)*\sigma_{33}+\gamma_{33}*N\{e_3\}+\\&+\left(N\{e_4\}-N\{g_3\}\right)*\sigma_{34}+\gamma_{43}*N\{e_4\}+b_{2'3}*N\{e_{2'}\}+b_{3'3}*N\{e_{3'}\}+\\&+\left[n_0G(\upsilon)-N\{g_3\}\right]*\Gamma\end{aligned}. \tag{4}$$

$$\frac{\partial N\{e_1\}}{\partial t}=\left(N\{g_2\}-N\{e_1\}\right)*\sigma_{21}-\gamma_{12}*N\{e_1\}, \tag{5}$$

$$\frac{\partial N\{e_{2'}\}}{\partial t}=\left(N\{g_2\}-N\{e_{2'}\}\right)*\sigma_{22'}-\gamma_{2'2}*N\{e_{2'}\}-b_{2'3}*N\{e_{2'}\}, \tag{6}$$

$$\frac{\partial N\{e_{3'}\}}{\partial t}=\left(N\{g_2\}-N\{e_{3'}\}\right)*\sigma_{23'}-\gamma_{3'2}*N\{e_{3'}\}-b_{3'3}*N\{e_{3'}\}, \tag{7}$$

$$\begin{aligned}\frac{\partial N\{g_2\}}{\partial t}=&\left(N\{e_1\}-N\{g_2\}\right)*\sigma_{21}+\gamma_{12}*N\{e_1\}+\left(N\{e_{2'}\}-N\{g_2\}\right)*\sigma_{22'}+\gamma_{2'2}*N\{e_{2'}\}+\\&+\left(N\{e_{3'}\}-N\{g_2\}\right)*\sigma_{23'}+\gamma_{3'2}*N\{e_{3'}\}+b_{22}*N\{e_2\}+b_{32}*N\{e_3\}+\\&+\left[n_0G(\upsilon')-N\{g_2\}\right]*\Gamma\end{aligned}. \tag{8}$$

In the rate equations system, the terms σ_{ij} describe the cross-sections of optical transition and light-induced emission and absorption, while γ_{ij} and b_{ij} represent the spontaneous decay rates from the excited state to the ground state interacting and non-interacting with the laser field, respectively [7-9]. The last terms in the equations for the ground levels introduce the non-radiative population relaxation Γ; in a coated cell such a process is caused by collisions of atoms with the coating. Γ assumes different values depending on the type of coating and the magnitude of spin randomization after atoms collide with the coated wall of the cell. In our case, we set Γ = 10 MHz, including velocity-changing collisions of Rb atoms with the cell wall without spin randomization [6].

We assume that initially, before irradiation (I = 0), the population densities of both ground states are equal and the excited states are not populated: $N\{g_3\}=n_0G(\upsilon)$, $N\{g_2\}=n_0G(\upsilon')$, where $G(\upsilon)$ and $G(\upsilon')$ are the Doppler profiles with $\Delta\upsilon_D$ = 400 MHz. The spontaneous decay rate is taken as γ = 60 MHz, involving also the spectral width of the free-running diode. The saturation parameter is assumed to be s = 50, as it includes also the hyperfine transition saturation due to the optical pumping.

In the steady state, the analytical solution of the system of rate-balance equations in this regime yields the following expressions:

$$N\{g_3\}=n_0\frac{B\left[G(\upsilon)+G(\upsilon')\right]+\Gamma*G(\upsilon)}{A+B+\Gamma} \quad \text{and} \quad N\{g_2\}=n_0\frac{A\left[G(\upsilon)+G(\upsilon')\right]+\Gamma*G(\upsilon')}{A+B+\Gamma}, \tag{9}$$

where A and B are $A=\frac{0.42\gamma*\sigma_{23}}{\sigma_{23}+0.54\gamma}+\frac{0.34\gamma*\sigma_{33}}{\sigma_{33}+0.77\gamma}$ and $B=\frac{0.12\gamma*\sigma_{22'}}{\sigma_{22'}+0.55\gamma}+\frac{0.43\gamma*\sigma_{23'}}{\sigma_{23'}+0.78\gamma}$. (10)

The model is illustrated by its first result describing the ground levels populations, at a very low laser-light intensity and in the strongly saturated regime. For low saturation, the theoretical result is in agreement with the experimental spectrum obtained in an uncoated vacuum cell where the optical pumping effect is very small (figure 6). At high saturation, the theoretical modelling shows a difference between the populations of the ground levels due to the hyperfine optical pumping.

In order to simulate the atomic spectra's dependence on the scanning frequency direction, two types of calculations are performed: (i) starting from the $F_g = 3$ set of transition and going to the $F_g = 2$ set (figure 6, **blue** curve) and (ii) vise versa, i.e. starting from the $F_g = 2$ set of transitions (figure 6, red curve). The calculations are in a qualitative agreement with the experimental results, showing that the maximum population of each ground state level depends on the initial laser frequency. As in the experiment, different shifts of the atomic population profile maximum are demonstrated depending on the beginning of the calculation. Again, the results of the model are in a qualitative agreement with the experimentally observed spectra. Note that the dip in the atomic population of the $F_g = 3$ level (figure 6, **blue** curve) is not observed experimentally, which can be attributed to an additional broadening of the hyperfine transitions in the coated cell due to velocity-changing collisions of atoms with the cell walls taking place without disorientation of the atomic spin. This effect is not included in our theoretical description.

As a next step, the model proposed will be developed further in order to explain other peculiarities of the experimental spectra, particularly the spin randomization of the atoms desorbed by LIAD.

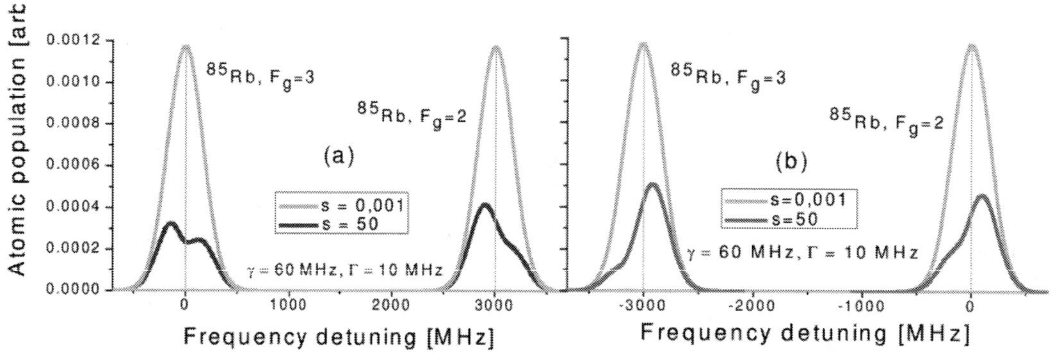

Figure 6. Theoretical atomic population of the $F_g = 3$ and $F_g = 2$ levels, for low and high saturation of the hyperfine transitions. (grey trace) – atomic population at very low saturation intensity; (**blue** trace) – atomic population at high saturation intensity and starting calculation from the Fg = 3 level; (red -trace) – atomic population at high saturation intensity and starting calculation from the $F_g = 2$ level. Now the $F_g = 2/F_g = 3$ population ratio is opposite to this of the previous case. Note also the shift of the atomic population profile maximum of the respective set of transitions at high saturation compared to the low saturation.

4. Conclusions

Our study shows that, in both types of coated cells, the Rb atoms undergo a significant hyperfine optical pumping in the saturated regime. The ^{87}Rb ($F_g = 1$) and ^{85}Rb ($F_g = 2$) ground state levels can be almost completely depleted. The maxima of the spectral profiles of the four absorption lines are frequency-shifted from the respective reference lines that do not suffer optical pumping. No significant difference is obtained between irradiation with linear and circular polarizations The absorption maximum shift observed for Rb atoms confirms the one obtained in the very recent work [6] for the case of Cs atoms.

Based on the effect of optical pumping, we demonstrate experimentally that the flow of fresh LIAD atoms results in a reduction in the atomic spin orientation due to the random spin distribution of the desorbed atoms. In the case of LIAD involvement, the measured shift of absorptions profile is smaller, and the absorption ratio of the ^{85}Rb ($F_g = 2$) set of transitions to the ^{85}Rb ($F_g = 3$) set changes approaching the result obtained in an evacuated cell.

Acknowledgements
This work was supported by a Marie Curie International Research Staff Exchange Scheme Fellowship within the 7th European Community Framework Program and by the Italian National Research Council and the Bulgarian Academy of Sciences (Bilateral Cooperation Program CNR/BAS 2013-2015).

References

[1] Aleksandrov E B and Balabas M V 2005 *Optics and Spectroscopy* **98**/6 800–4
[2] Graf M T, Kimball D F, Rochester S M, Kerner K, Wong C and Budker D 2005 *Phy. Rev.* A **72** 023401
[3] Marmugi L, Gozzini S, Lucchesini A, Bogi A, Burchianti A and Marinelli C 2012 *J. Opt. Soc. Am.* B **29**/10 2729-33
[4] Marinelli C, Nasyrov K A, Bocci S, Pieragnoli B, Burchianti A, Biancalana V, Mariotti E, Atutov S N and Moi L 2001 *Eur. Phys. J.* D **13** 231-5
[5] Marinelli C, Burchianti A, Bogi A, Della Valle F, Bevilacqua G, Mariotti E, Veronesi S and Moi L 2006 *Eur. Phys. J.* D **37** 319–25
[6] Sevost'yanov D I, Yakovlev V P, Kozlov A N, Vasil'ev V V, Zibrov S A and Velichansky V L 2013 *Quantum Electronics* **43**/7 638–45
[7] Siddons P, Adams C S, Ge C and Hughes L G 2008 *J. Phys.* B*: At. Mol. Opt. Phys.* **41** 155004
[8] Gozzini S, Slavov D, Cartaleva S, Marmugi L and Lucchesini A 2009 *Acta Physica Polonica* A **116**/4 489-91
[9] Alzetta G, Cartaleva S, Dancheva Y, Andreeva Ch, Gozzini S, Botti L and Rossi A 2001 *J. Opt.* B*: Quantum Semiclass. Opt.* **3** 181-8

Journal of Physics: Conference Series **514** (2014) 012030 doi:10.1088/1742-6596/514/1/012030

Light-induced atomic desorption in cells with different PDMS coatings

S Tsvetkov[1], S Gateva[1,3], M Taslakov[1], E Mariotti[2] and S Cartaleva[1]

[1]Acad. E. Djakov Institute of Electronics,
Bulgarian Academy of Sciences, 72 Tsarigradsko Chaussee, 1784 Sofia, Bulgaria
[2]Physics Department, University of Siena and CNISM,
via Roma 56, 53100 Siena, Italy

E mail: sgateva@ie.bas.bg

Abstract. Light-induced atomic desorption (LIAD) is a non-thermal process in which atoms adsorbed at a surface are released under illumination. It is applied mostly to implementing optical dispensers in the cases when high atomic density at low temperature is needed – for example, for loading atomic devices as atomic magnetometers, atomic clocks, magneto-optical traps and their miniaturization. However, as the desorption depends on the atom-surface interaction, it can also be used for optical characterization and manipulation of alkali metal nanoparticles. The paper describes an experimental investigation of the shape of the transmission spectra and their dependence on the illuminating blue-light power in PDMS coated cells prepared with two different concentrations of PDMS in ether. A comparison is conducted with the LIAD effect in SC-77 coated and in uncoated cells. All measurements are performed on the Rb D2 line in vacuum. The potential is discussed for application of these dependences to analyzing the quality of the coating surface and its optimization.

1. Introduction

Light-induced atomic desorption (LIAD) is a non-thermal process whereby atoms adsorbed on a surface are released under illumination. It is applied mostly to implementing optical atomic dispensers in the cases when high atomic density at low temperature is needed – for example, for loading atomic devices as atomic magnetometers, atomic clocks, magneto-optical traps and their miniaturization [1,2 and references therein]. However, as the desorption depends on the atom-surface interaction, it can be applied to the optical characterization and manipulation of alkali metal nanoparticles [3]. LIAD was reported for the first time in a polydimethylsiloxane (PDMS) coated cell with Na vapor [4]. Since then, LIAD in PDMS has been observed with Rb, Cs and K alkali atoms as well.

Polydimethylsiloxane ($CH_3[Si(CH_3)_2O]_nSi(CH_3)_3$) is a polymer with a chemical behavior intermediate between organic and inorganic materials and a good stability after dehydration. It has been used as an antirelaxation coating (to reduce the spin relaxation resulting from atom-wall collisions). The PDMS molecules are highly flexible and PDMS is amorphous. The glass transition temperature is 144 K, with the melting temperature being 232 K. In our experiments, the temperature was kept above these values, so that PDMS was in a rubbery state. On the one hand, the structure of

[3] To whom any correspondence should be addressed.

PDMS is suitable for alkali metal atoms adsorption. On the other hand, it causes problems, such as bubble formation and sample evaporation. The AFM measurement of the roughness values for various concentrations of PDMS in ether reported in [5] showed that the surface roughness increases as the PDMS concentration is raised, namely, from 20 nm to 50 nm for an increase from 2 % to 5 %. The thickness of the PDMS film also increases with the concentration.

This paper presents an experimental investigation of the shape of the transmission spectra and their dependence on the illuminating blue-light power in PDMS coated cells prepared with two different concentrations of PDMS in ether. A comparison is carried out with the LIAD effect in SC-77 coated and in uncoated cells. All measurements were conducted on the Rb D2 line in vacuum cells with similar dimensions and geometry.

2. Experimental setup

The experimental setup for measuring the influence of the LIAD on the transmission spectra in a Rb vacuum cell is described in detail in [6,7]. The frequency of a 780-nm diode laser was tuned by varying the current through it (using a waveform generator) and the transmitted spectra were registered by a photodiode. To activate LIAD, a special sphere with a diffusely-reflecting inner surface was made and an 18-W light emitting diode (LED) (Luminus PT120 with a central frequency of 460 nm and FWHM of 20 nm) was mounted on it. The optical power was regulated and measured, after a proper calibration, by varying the current supplied to the LED. All measurements were performed at a temperature of 27 °C. A continuous wave (cw) illumination was performed, which provided the possibility to examine the absorption spectrum profile. Moreover, the cw approach is suitable for various practical applications. The measurements were performed in four different Pyrex cells. Table 1 summarizes some parameters of the cells (length, diameter, volume, surface, surface-to-volume ratio). The LIAD effect in the PDMS cells was compared with that in SC-77 cells, because the SC-77 coating (a silane-based dry film coating) ensures a very low equilibrium desorption rate coefficient, which maintains the best vacuum level in the cell and it has been widely used for magneto-optical trap loading [8].

Table 1. Parameters of the Rb cells: length, diameter, volume, surface.

Cell coating	Length [cm]	Diameter [cm]	Volume [cm^3]	Surface [cm^2]	Surf/Vol [cm^{-1}]
PDMS 5	4.7	2.6	25	49	1.98
PDMS 2	6.0	2.6	32	60	1.88
SC-77	6.0	2.6	32	60	1.88
uncoated	4.8	3.2	38	64	1.68

The transmission spectra were measured at a relatively low 780-nm light intensity (3 – 4 μW) in order to minimize the optical pumping so that the absorption characteristics could be deduced.

For an optically-thin medium, the change of the Rb density due to the LIAD is measured by the absorption coefficient κ_ω of Rb vapor. According to the Beer's law, the transmission $T(\omega) = T_\omega$ is:

$$T_\omega = I/I_0 = exp(-\kappa_\omega L) = exp(-\sigma_\omega NL), \quad (1)$$

where I_0 is the 780-nm laser power entering the vapor sample; I, the transmitted power; κ_ω, the frequency-dependent absorption coefficient; σ_ω, the non-saturated, frequency-dependent absorption cross-section; N, the atomic density and L, the cell length.

In the case of an optically-thick medium, the 780-nm light intensity is reduced along the optical cell. Thus, one measures the average absorption coefficient $\kappa_\omega = (1 - T_\omega)/L$.

The absorption coefficients were measured in order to compare the LIAD effect in cells with different length.

Journal of Physics: Conference Series **514** (2014) 012030 doi:10.1088/1742-6596/514/1/012030

3. Transmission spectra

Figure 1 provides a comparison of the transmission spectra of the Rb D2 line in two PDMS coated cells prepared with 2 % and 5 % concentrations in ether, an SC-77 coated cell and an uncoated cell without illumination at 3 μW 780-nm laser power. The spectrum of the 780 nm (Rb D_2) line contains four components. The optical spectrum of each Rb isotope consists of two sets of lines: (i) ^{87}Rb involves two groups of hyperfine transitions starting from $F_g = 1$ and $F_g = 2$ ground levels and (ii) two sets of ^{85}Rb transitions originating from $F_g = 2$ and $F_g = 3$. The spectrum is the sum of the Doppler profiles (with FWHM ~560 MHz) of all $F_g \rightarrow F_e$ hyperfine transitions (see for example [6]). The spectra of all cells are similar to that of the uncoated cell, but the absorption coefficient is smaller as a result of the optical pumping and depends on the coating. The absorption coefficients of the two PDMS cells are equal within the experimental error (5 %). The absorption coefficient of the SC-77 coated cell is higher by a factor of about two.

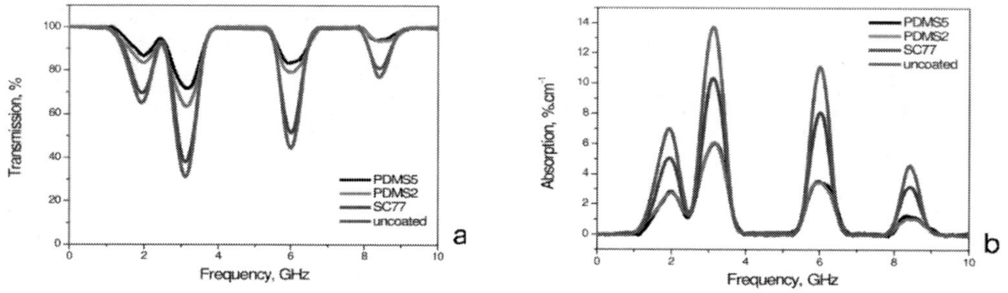

Figure 1. Transmission and absorption spectra of the Rb D2 line in different coated cells without illumination at 3-μW 780-nm laser power.

Figure 2 displays the transmission spectra with blue-light illumination at 1-A and 5-A LED current. The comparison of the shape and amplitude of the spectra from the three coated cells (PDMS2, PDMS5 and SC-77) shows that the Rb density depends on the coating and that the spectra have different behavior depnding on the blue-light intensity. The maximum density was observed in the SC-77 coated cell. The Rb D2 line components widths and amplitudes for the case of LIAD in SC-77 coated cells were investigated in [6].

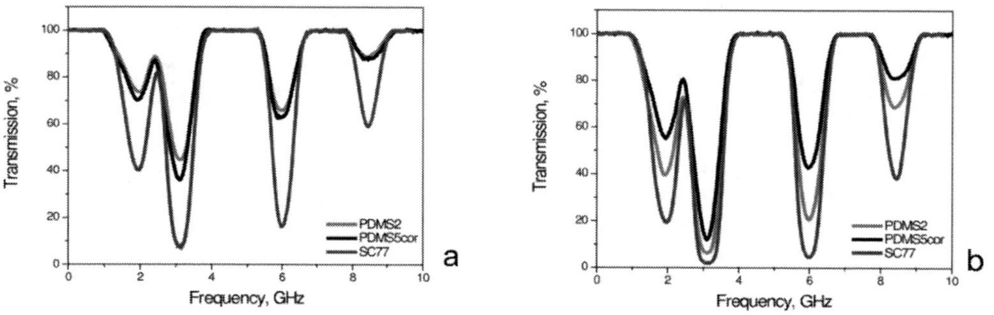

Figure 2. Transmission spectra of the Rb D2 line with 1 A (a) and 5 A (b) blue LED illumination.

Figure 3 presents the transmission spectra of the PDMS coated cells prepared with PDMS concentrations in ether of 2 % (figure 3a) and 5 % (figure 3b). The measurements were registered as the LED current was increased after illumination time of three minutes. In figure 3b, as the cell is shorter, the transmission is corrected in such a way as to have a quantity proportional to the absorption coefficient. The component number increases with the tuning frequency. The width of the second and third component does not change within the experimental error limit. Only the second component of the SC-77 coated cell at 5 A is broader, which can be explained by atomic emission re-absorption [6].

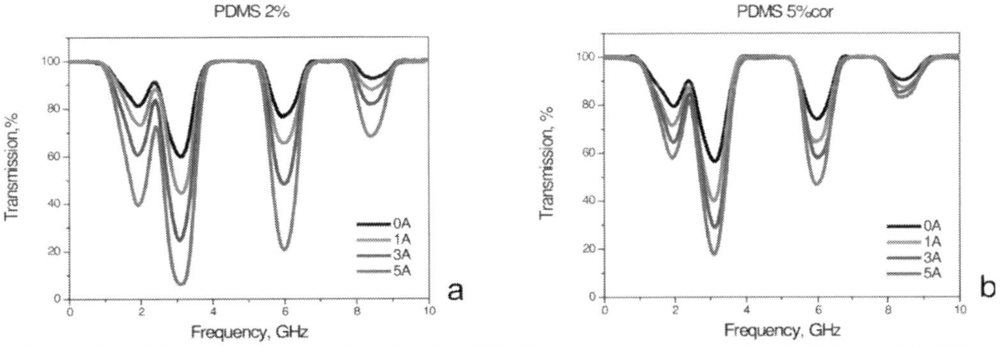

Figure 3. LIAD spectra in coated cells with PDMS concentrations in ether 2 % (a) and 5 %(b).

Figure 4 compares the absorption coefficients of the four components in PDMS coated cells with PDMS concentrations in ether 5 % (solid line) and 2 % (dashed line). At a LED current of less than 2 A, the absorption coefficient is slightly higher in the 5-% PDMS coated cell, while the dependence of the amplitude on the illumination power has the typical saturation shape. At powers exceeding that at a LED current of 2 A, the absorption coefficient is higher in the 2-% PDMS coated cell, while the dependence of the amplitude on the illumination power does not saturate in the range of blue-light powers investigated.

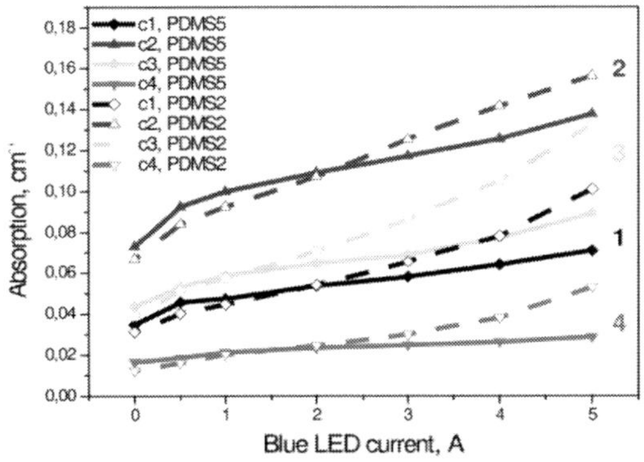

Figure 4. Absorption coefficients of the four components in PDMS coated cells with PDMS concentrations in ether 5% (solid line) and 2% (dashed line).

4. Summary and conclusions

The experimental study performed of the shape of the Rb D2 line transmission spectra under homogeneous blue-light illumination and their dependence on the illuminating power in PDMS coated cells prepared with two different concentrations of PDMS in ether (5 % and 2 %) showed that without illumination and at low blue-light powers (at LED currents of less than 2 A), the absorption coefficient is higher in the 5-% PDMS coated cell within the experimental error limit; the dependence of the amplitude on the illumination power has the typical saturation shape [9]. At powers higher than that achieved at a LED current of 3 A, the absorption coefficient is higher in the 2-% PDMS coated cell, while the dependence of the amplitude on the illumination power does not show a tendency to saturation up to 5-A LED current. A possible explanation of this increase is the different probability for alkali atoms to find adsorption sites on substrates with a different volume density. On the other hand, the widths of the components do not change with the illumination power, except for the second component at 5 A, which can be explained with atomic emission re-absorption [6]. As is known, increasing the PDMS concentration causes an increase in the surface roughness and the film thickness [5]. Additional studies are necessary to find the best roughness needed for preparation of an anti-relaxation coating and to understand the influence of the film thickness. Knowing the influence of the coating roughness on the absorption spectra will enable one to use the absorption spectra for analysis of the quality of the coating surface and its optimization.

Acknowledgements
This work was supported by a Marie Curie International Research Staff Exchange Scheme Fellowship within the 7th European Community Framework Program and by the Italian National Research Council and the Bulgarian Academy of Sciences (Bilateral Cooperation Program CNR/BAS 2013-2015).

References

[1] Karaulanov T, Graf M T, English D, Rochester S M, Rosen Y J, Tsigutkin K, Budker D, Alexandrov E B, Balabas M V, Kimball D F J, Narducci F A, Pustelny S and Yashchuk V V 2009 Controlling atomic vapor density in paraffin-coated cells using light-induced atomic desorption *Phys. Rev.* A **79**/1 012902
[2] Villalba S, Failache H and Lezama A 2010 Light-induced atomic desorption and diffusion of Rb from porous alumina *Phys. Rev.* A **81**/3 032901
[3] Burchianti A, Bogi A, Marinelli C, Maibohm C, Mariotti E, Sanguinetti S and Moi L 2008 Optical characterization and manipulation of alkali metal nanoparticles in porous silica *Eur. Phys. J.* D **49** 201-10
[4] Gozzini A, Mango F, Xu J H, Alzetta G, Maccarone F and Bernheim RA 1993 Light-induced ejection of alkali atoms in polisiloxane coated cells *Il Nuovo Cimento* **15D**/5 709-22
[5] Brewer J, Burchianti A, Marinelli C, Mariotti E, Moi L, Rubahn K and Rubahn H-G 2004 Pulsed laser desorption of alkali atoms from PDMS thin films *Appl. Surf. Sci.* **228** 40-7
[6] Gateva S, Taslakov M, Sarova V, Mariotti E and Cartaleva S 2013 Light-induced atomic desorption for miniaturization of magneto-optical sensors *Proc. SPIE* **8770** 877000
[7] Taslakov M, Tsvetkov S and Gateva S 2014 Ligh-induced atomic desorption at different illumination (*available in this volume of J. Phys.: Conf. Series*)
[8] Stephens M, Rhodes R and Wieman C 1994 Study of wall coatings for vapor-cell laser traps *J. Appl. Phys.* **76**/6 3479-88
[9] Rebilas K and Kasprowicz 2009 Reexamination of the theory of light-induced atomic desorption *Phys. Rev.* A **79** 042903

Effect of the plasmon-exciton coupling on the optical response of a ZnO/Ag/ZnO nanocomposite

M E Koleva[1], N N Nedyalkov and P A Atanasov

Acad. E. Djakov Institute of Electronics, Bulgarian Academy of Sciences, 72 Tsarigradsko Chaussee, Sofia 1784, Bulgaria

E-mail: mihaela_ek@yahoo.com

Abstract. The study prsented is focused on the optical properties of ZnO/AgNPs/ZnO multilayer nanocomposites. The optical properties of the nanostructures depend on the silver layer thickness. The optical spectral characteristics of Ag nanoparticles (AgNPs) nanostructured by laser irradiation in a ZnO medium were investigated theoretically and experimentally. The extinction spectrum was determined as a function of the nanoparticles size distribution and of the surrounding medium. The extinction efficiency calculated is red-shifted and broadened when the AgNPs are introduced in a ZnO environment. This result is in agreement with the experimental measurements, where the plasmon resonance peak position of the ZnO/AgNPs/ZnO composite is red-shifted, compared to the AgNPs/ZnO sample. Raisng the thickness of the Ag layer increases the resonance extinction. We also propose a method for controlling and tuning the SPR-based sensing of Ag nanoparticles in a ZnO medium.

1. Introduction

Metal-semiconductor nanocomposites are of great interest because of their potential applications in various nanodevices that cannot be designed based on single individual components [1, 2]. Hybrid nanomaterials composed of plasmonic metal nanostructures and semiconductor thin films provide the unique combination of a highly-localized and enhanced electromagnetic field in, e.g., Ag nanoparticles (NPs) with the wide bandgap and large exciton binding energy of a semiconductor, such as ZnO [3, 4]. Zinc oxide is an important material used in solar sells as transparent conductive contacts, in thin film transistors and in optoelectronic components. One of the approaches to developing novel materials and improving their functions is to combine two or more functional phases. The research efforts have been focused on determining the influence of the coupling mechanism attributed to the combined effects of surface modification, band alignment and charge transfer on the optical response in composite heterostructures. The theoretical and experimental studies of the properties of noble metal nanoparticles have shown that they could be modified by varying the plasmon resonant frequency through varying the size, the shape, the interparticle distance and the dielectric properties of the surrounding environment [5-7]. The AgNPs behavior in a ZnO ambient has in recent years attracted a great interest [8-10]. By analyzing composites containing metal nanoparticles, one could explore the possibilities of fabricating sensors based on surface plasmon resonance (SPR) that operate in various spectral ranges. The interpretation of the results could be further extended to other semiconductors and

[1] To whom any correspondence should be addressed.

be useful for creating a variety of nanocomposite materials with novel optical properties. In this paper, we describe the formation of a ZnO/AgNPs/ZnO nanocomposite by laser deposition and laser annealing and a study of the morphology and plasmon resonance properties.

2. Theoretical model

The simulation model considered an ideal array, where all Ag nanoparticles have the same diameter of 60 nm and the medium is ZnO. Nanoparticles array with the mean diameter of 60 nm and a size distribution in the range of 30 nm – 90 nm was also the object of the theoretical investigations. For comparison, calculations for the case of an air ambient were also performed. The theoretically estimated optical properties were compared with the experimentally measured ones. The optical efficiencies, (Q_{ext}, Q_{sca} and Q_{abs}) were calculated by applying the generalized multiparticle Mie (GMM) method [11]. Each simulated system consisted of 27 nanoparticles.

3. Experimental

A ZnO sintered target was ablated for ten minutes in a vacuum chamber at an oxygen pressure of 2.4×10^{-1} Torr. The chamber was evacuated down to 3.9×10^{-4} Torr before introducing the oxygen. The ZnO thin film was deposited on a SiO_2 (100) substrate. A third-harmonic Nd:YAG laser (Lotis LS-2147, $\lambda = 355$ nm, pulse duration $\tau = 18$ ns) was used at a repetition rate of 10 Hz and a fluence of 1.7 J/cm^2. The ZnO layers thus formed on two substrates were covered by a Ag thin film for deposition times of two and three minutes. The thickness of the silver films was 120 nm and 180 nm as determined in previous experiments [12]. The silver layer was decomposed into nanoparticles by laser annealing in air by ten laser pulses of high fluence (1 J/cm^2) incident normally to the surface. The NPs morphology was observed by scanning electron microscopy (SEM) (dual beam system SEM/FIB, Lyra\Tescan). The AgNPs mean diameter and size distribution were calculated. The optical transmission spectra of the fabricated samples in the range from 200 nm to 800 nm were obtained by an Ocean Optics HR 4000 spectrophotometer.

4. Results

Laser annealing leads to a decomposition of the thin silver film into a layer of nanoparticles. Figure 1a presents a SEM image of AgNPs formed as described above on a ZnO layer. The laser annealing results in well-formed NPs in the case of the thicker layer (figure 1b), while the thinner film shown in figure 1a does not exhibit a well-defined nanoparticles structure. Most probably, the laser annealing leads to embedding of silver in the ZnO matrix. Figure 1c shows the histogram of the size distribution of the AgNPs, which is in the 30 nm – 90 nm range. The silver nanoparticles's mean diameter for the thicker layer is 60 nm.

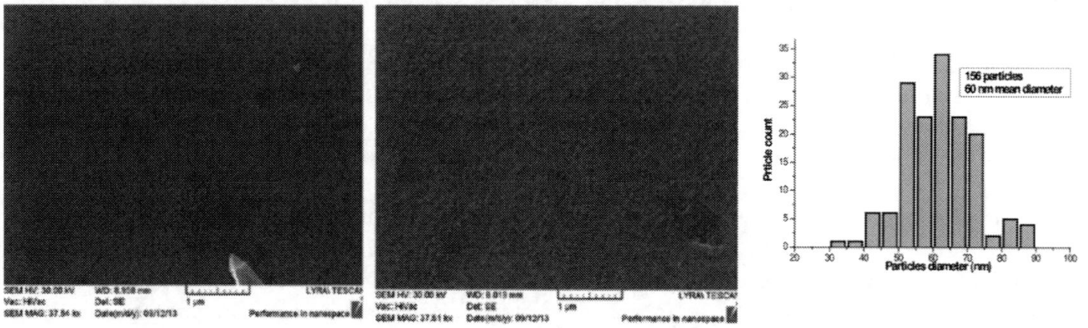

(a) (b) (c)

Figure 1. SEM images of AgNPs on a ZnO film: (a) deposited for 2 min; (b) deposited for 3 min; (c) size histogram of the AgNPs layer deposited for 3 min.

The strong optical extinction band associated with the SPR of AgNPs was registered for all samples after laser annealing (figure 2). Resonance extinction was registered also for the sample with the thinner AgNPs layer (figure 2a, curve 2), where the SEM image does not show a good nanoparticles' morphology. Obviously, interaction with silver nanoclusters embedded in ZnO provokes a resonance response with a lower intensity compared to the signal arising from the other sample shown in figure 2b (curves 2 and 3). Figure 2 presents the transmission spectra of Ag layers of different thicknesses and a fixed thickness of the upper and lower ZnO layers. The spectrum shows a resonance extinction in the Ag/ZnO sample before laser annealing, when the silver layer is thinner (figure 2a, curve 1). The sample having a thicker silver

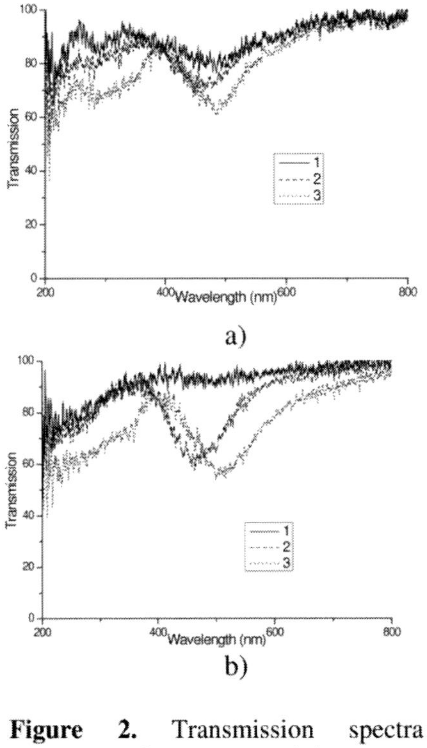

Figure 2. Transmission spectra of nanocomposites containing AgNPs produced by laser annealing of a silver layer (a) deposited for 2 min.; (b) deposited for 3 min.; curve 1 – Ag film/ZnO on SiO_2 (100); 2 – AgNPs/ZnO on SiO_2 (100); 3 – ZnO/AgNPs/ZnO on SiO_2 (100).

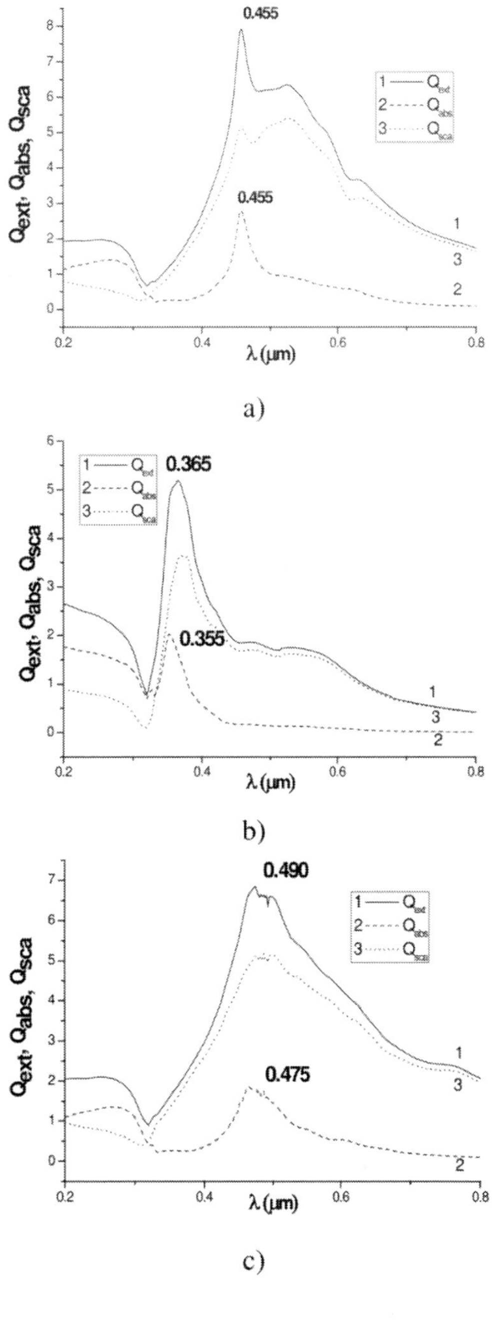

Figure 3. Extinction, absorption and scattering efficiency spectra of AgNPs arrays composed of 27 nanoparticles, where (a) all AgNPs in ZnO environment have the same diameter of 60 nm (an ideal system); (b) AgNPs with mean diameter of 60 nm in air environment; (c) AgNPs with mean diameter of 60 nm in ZnO environment.

layer does not exhibit any resonance band before annealing (figure 2b, curve 1). This result is associated with a rough surface formation in the case of two-minute Ag deposition and a discontinuous (island-like) structure of the silver layer. The surface plasmon resonance response in the ZnO/AgNPs/ZnO nanostructure is obtained at about 490 nm – 500 nm. A higher value was measured for the sample with the thicker silver layer. The excitonic band of ZnO at about 300 nm in figure 2a is detected for the sample where the silver layer is thinner. This band disappears when the silver film thickness is increased in the multilayer structure, as demonstrated in figure 2b.

The nanocomposite heterostructure under consideration consists of Ag nanoparticles (figure 1b) with a mean NPs diameter of 60 nm arranged between two ZnO layers. In all spectra calculated and presented in figure 3, the scattering maximum dominates over the absorption one by a factor of more than two; also, the maximum of the scattering spectra is red-shifted with respect to the absorption one. The scattering dominates the optical spectra and provides the main contribution to the overall extinction for the nanoparticles array with a diameter of about 60 nm. The extinction calculated corresponds to the plasmon resonance extinction measured of the fabricated nanostructures. The NPs size distribution has a strong impact on the overall extinction. This multiparticle system consists of peaks related to excitation of different plasmon modes in a single particle within the array, as well as to collective excitation. Figure 3a presents the extinction efficiencies for the ideal system of NPs with a diameter of 60 nm, to be compared to the real NPs array system (size distribution in the 30 nm – 90 nm range) in figure 3b. The extinction calculated of the real system in figure 3c exhibits one broad peak due to the overlapping of the individual peaks.

The change from an air environment (figure 3b) to ZnO leads to a broadening of the extinction peaks (figure 3c). The calculation results show a red-shift of the extinction efficiency in a ZnO environment. This shift is accompanied by an increased intensity, as was also demonstrated by the SPR results in figure 2. The maximal extinction efficiency shown in figure 3c and the maximal resonance extinction of AgNPs in a ZnO medium in figure 2b were located at about 490 nm.

5. Conclusions

Multilayer composite structures of ZnO/AgNPs/ZnO were fabricated by pulsed laser deposition and laser annealing. The nanocomposites produced with AgNPs of 60-nm mean diameter and size distribution in the 30 nm – 90 nm showed strong plasmon resonance properties. The experimental results were validated by the simulation conducted. The optical properties of the nanostructures explored depend on the AgNPs mean diameter, their size distribution and the surrounding environment. The ZnO medium causes a SPR shift to the higher wavelengths in the VIS spectrum to about 490 nm and broadens the bandwidth. Resonance wavelength tuning could be performed by varying the AgNPs size distribution in an ZnO environment. The results may contribute to the development for SPR-based sensors covering various spectral regions.

References

[1] Alammar T and Mudring A V 2009 *J. Mater. Sci.* **44** 3218
[2] Lam S W, Chiang K, Lim T M, Amal R and Low G K C 2007 *Appl. Catal.* B **72** 363
[3] Schaadt D M, Feng B and Yu E T 2005 *Appl. Phys. Lett.* **86** 063106
[4] Achermann M 2010 *J. Phys. Chem. Lett.* **1** 2837
[5] *Optical Properties of Metal Clusters* 1995 ed U. Kreibig and M. Vollmer (Berlin: Springer)
[6] Noguez C 2007 *J. Phys. Chem.* C **111** 3806
[7] Quinten M 2011 *Optical Properties of Nanoparticle Systems: Mie and Beyond* (Weinheim: Wiley-VCH)
[8] You J B, Zhang X W, Fan Y M, Yin Z G, Cai P F and Chen N F 2008 *J. Phys.* D*: Appl. Phys.* **41** 205101
[9] Sahu D R, Lin S-Y and Huang J-L 2006 *Appl. Surf. Sci.* **252** 7509
[10] Gao M, Xing G, Yang J, Yang L, Zhang Y, Liu H, Fan H, Sui Y, Feng B, Sun Y, Zhang Zh, Liu Sh, Li S and Song H 2012 *Microchimica Acta* **179** 315-21
[11] www.astro.ufl.edu/~xu
[12] Koleva M E, Dikovska A Og, Nedyalkov N N, Atanasov P A and Bliznakova I 2012 *Appl. Surf. Sci.* **258** 9181–85

PLD fabrication of ZnO nanostructures on metal-coated substrates

A Og Dikovska[1,3], M E Koleva[1], G B Atanasova[2], T R Stoyanchov[1], N N Nedyalkov[1] and P A Atanasov[1]

[1]Acad. E. Djakov Institute of Electronics, Bulgarian Academy of Sciences, 72 Tsarigradsko Chaussee, 1784 Sofia, Bulgaria
[2]Institute of General and Inorganic Chemistry, Bulgarian Academy of Sciences, Acad. G. Bonchev Str., Bl. 11, 1113 Sofia, Bulgaria

E-mail: dikovska@ie.bas.bg

Abstract. In this work, ZnO nanostructures were fabricated on metal (a metal alloy containing Fe, Cr, Mn and Ni) coated silicon substrates by applying pulsed laser deposition. The samples were prepared at substrate temperatures in the range of 550 – 650 °C, oxygen pressure of 5 Pa, and laser fluence ≤ 1 J cm^{-2}, i.e., process parameters usually used for thin-film deposition. We found that the metal layer's role is substantial in the preparation of nanostructures, the morphology of the catalyst layer determines the growth of the ZnO nanowalls and the increase of the process temperature leads to nanorods formation on the nanowalls.

1. Introduction

The synthesis of semiconducting nanostructures has attracted considerable interest due to their potential importance in the fabrication of nanodevices. ZnO is one of the semiconducting materials of great current interest because of its wide direct bang gap (3.37 eV) and large exciton binding energy (60 meV). The potential uses of such materials include gas sensors, UV emitting diodes and UV sensitive photocells, whose efficiency benefits from a large surface area [1,2].

A variety of fabrication techniques have been employed to synthesize nanostructures; these include thermal evaporation, laser-assisted ablation, template-based methods, solution-phase syntheses, etc. [3,4]. Pulsed laser deposition (PLD) has also proved to be a suitable method for the synthesis of nano-scale materials [5-7]. The method involves direct laser ablation of the target material into a gas phase, which is then transferred to a substrate held at a specific temperature. Because of the high deposition rate as a result of the energetic plasma and the high absorption rate of the target materials, especially suitable for those with high melting point, the PLD method has been extensively applied for deposition of oxide materials. By modifying the deposition conditions (including the process gas and its pressure, the substrate temperature, the target-to-substrate distance, the laser wavelength and fluence, the pulse repetition rate, etc.), it is possible to produce a variety of nanostructures, such as nanorods, nanowires, nanoparticles, etc. [8,9,6,10].

Basically, nanostructures can be formed with or without catalysts. A catalyst-free approach allows the fabrication of nanostructures directly or by a seed layer on the substrate [11,6,12]. A catalyst

[3] To whom any correspondence should be addressed.

approach requires a metal catalyst, namely, a stable and effective nucleation agent deposited on the substrate [13,9,14,15]. The Au catalyst has been widely used for growing 1-D ZnO [13,14,10,9]. Investigating the growth of different nanostructures using catalysts other than gold is important from the viewpoint of both fundamental research and applications. Other catalysts that have been used include Ag, Cu, Ni, and Fe [15-17]. Zhu et al. made a comparative study of the growth of ZnO nanowires with Au, Ag, Fe, and Ni as catalysts [18]. Iron has been extensively used in the catalytic growth of carbon nanotubes, but its application to the fabrication of other nanostructured materials has rarely been studied. Randomly oriented Fe-related ZnO nanowires were successfully grown on Si and sapphire substrates [18]. Kuo et al. demonstrated the growth of uniform and vertically aligned ZnO nanorods with a diameter of 80 – 100 nm on a sapphire substrate covered by a Fe/ZnO bilayer [17]. The authors discussed also the mechanism of ZnO nanorods and nanowalls growth in the presence of a Fe-based catalyst. However, to the best of our knowledge, pure metals have only been used for the deposition of a catalyst. This is why we focused our investigations on the formation of nanostructures by means of deposition of a metal-alloy catalyst.

The aim of this work was to fabricate ZnO nanostructures by PLD using a Fe-based catalyst layer deposited on the substrate before the nanostructure growth. The morphology of the ZnO nanostructures depends on the substrate temperature and on the morphology of the pre-deposited catalyst layer.

2. Experimental

The experiments were performed in a standard on-axis PLD configuration. The third harmonic of a Nd:YAG laser (λ = 355 nm, τ = 18 ns full-width at half-maximum (FWHM), and a repetition rate of 10 Hz) was used for ablating a ZnO ceramic target. The laser fluence applied was 0.5 J cm^{-2}. (100)-oriented silicon (Si) covered by a metal-alloy layer was used as a substrate. Catalyst layers with an average thickness of 40 nm were produced by thermal evaporation of a metal target while the substrate was kept at room temperature. The distance between the target and the substrate was maintained constant at 60 mm. Two different substrate temperatures were applied in the experiments, namely, 550 °C and 650 °C. All experiments were performed in an oxygen atmosphere at a pressure of 5 Pa for and deposition time of 90 min.

The surface morphology of the nanostructured ZnO samples was analyzed using scanning electron microscopy (SEM). The elemental composition and chemical state of the sample surfaces were studied using X-ray photoelectron spectroscopy (XPS). Electron dispersive X-ray spectroscopy (EDX) was employed in order to determine the elemental content of the metal-alloy target.

3. Results and discussions

As a first step in the nanostructure growth, all silicon substrates were covered by thermal evaporation of the metal-alloy target. The elemental analysis of the metal target material is presented in figure 1. As it is seen, the metal target is a Fe-Cr alloy with small additions of Mn and Ni. After deposition of a thin metal film with a thickness of about 40 nm, XPS analysis on the surface was performed in order to determine the elemental content of the as-prepared catalyst layer. The results obtained show that the metal film has a native surface-oxide layer. Furthermore, the thickness of this layer undoubtedly increases in the process chamber while the substrate is heated in the oxygen atmosphere.

The surface morphology of the covering layer after heating the substrate for 10 min up to 650 °C is presented in figure 2. The nanostructure growth temperature used in this work (550 or 650 °C) is much lower than the melting point of Fe or Fe-oxide (1538 °C or 1377 °C). This makes it unlikely that melting of the thin film or formation of metallic nanocluster occurred during the heating procedure before the growth of ZnO nanostructures. As it is seen in the figure, no nanoparticles or nanoclusters exist. However, one can observe the initial stage of the formation of holes in the film. The holes formation is probably due to the poor wetting of the Si surface. It should be mentioned that no Si compounds were detected by the XPS analyses performed after heating the Fe-Cr alloy film.

In summary, the increase of the substrate temperature during the heating (before ZnO deposition) changes the morphology but preserves the elements' content in the metal layer (no Si-compounds).

The second step of the nanostructure growth was PLD of ZnO material on the metal-covered substrate. The surface morphology of the ZnO nanostructures fabricated at different substrate temperatures is presented in figure 3. The growth of ZnO follows the nanowalls formation. The top view of the structures for both temperatures used resembles a net (figure 3a and 3b). The higher substrate temperature applied during the experiments forms nanorods on top of the nanowalls (figure 3b). The diameter of the nanorods is approximately 50 nm, their length being defined by the deposition time.

When the ZnO target ablation starts, the laser-generated plume ensures the presence of charged (Zn^{2+}, Zn^{+}, O^{+} ions) and neutral (Zn and O atoms) species, which react with the metal layer. The use of a metal layer and high process temperatures suggests a vapor-liquid-solid (VLS) mechanism for the nanostructures growth. Typical for the VLS mechanism is the presence of metal particles on the tips of the nanostructures. In our case, no elements originating from

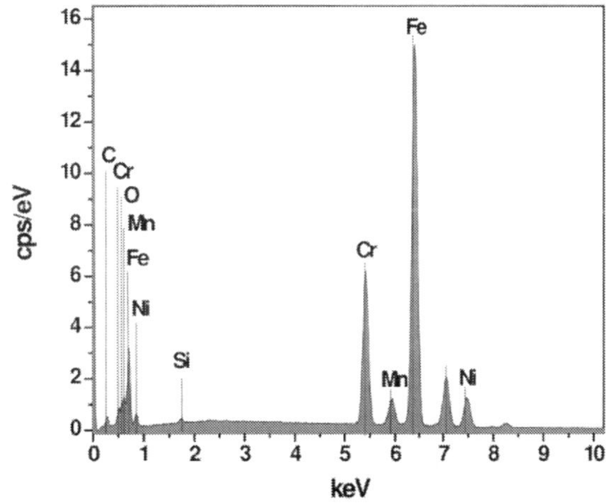

Figure 1. EDX analysis of the metal target material.

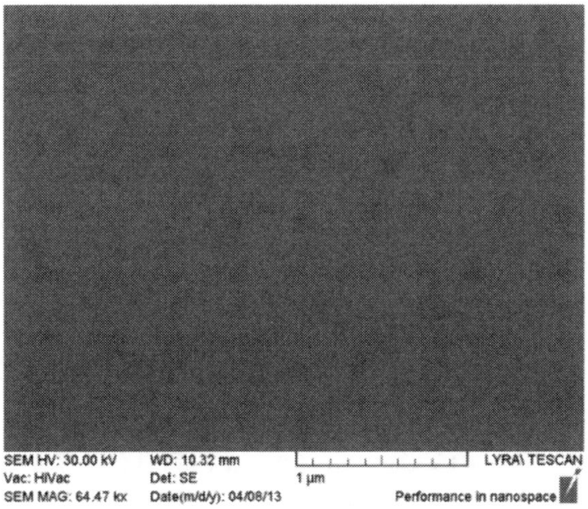

Figure 2. SEM image of the metal-alloy layer after heating up to 650 °C.

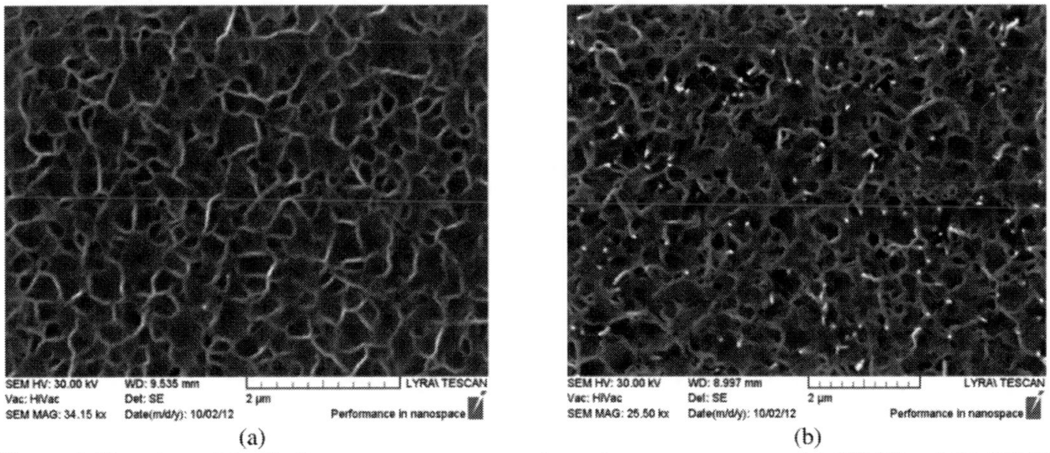

(a) (b)

Figure 3. Top view of the ZnO nanostructure prepared at substrate temperature (a) 550 °C and (b) 650 °C.

the metal-alloy layer were detected by the XPS analyses performed on the samples' surfaces. These results give us a reason to suggest that the growth of the ZnO nanostructures follows a vapor-solid (VS) growth mechanism.

The metal layer's role is substantial for the preparation of nanostructures. A thin ZnO film was produced on a pure Si substrate (without a metal-covered layer) under the same experimental conditions. The morphology of the covering layer defines the nanostructure's growth and development. The use of different metals or alloys with differing physical and thermodynamic properties for forming a cover layer allows one to produce a variety of surface morphologies of the covering layer and, subsequently, of the nanostructures.

4. Conclusions

ZnO nanostructures were grown with a Fe-based catalyst deposited on silicon by PLD. The morphology of the catalyst layer defines the further evolution of the nanostructure. Heating the Fe-alloy layer leads to the formation of holes in the surface. Subsequently, the growth of ZnO on the catalyst layer follows the nanowalls formation. A higher substrate temperature applied results in the formation of nanorods on top of the nanowalls. The ZnO nanowalls growth follows a vapor-solid growth mechanism.

Acknowledgements

This work was supported in part by the Bulgarian National Science Fund under Contract No. FFNNI001/43.

References

[1] Chen T Y, Chen H I, Hsu C S, Huang C C, Wu J S, Chou P C and Liu W C 2012 *IEEE Electron Dev. Lett.* **33** 1486
[2] Park S H, Kim S H and Han S W 2007 *Nanotechnol.* **18** 055608
[3] Dai Z R, Pan Z W andWang Z L 2003 *Adv. Func. Mater.* **13** 9
[4] Morales A M and Lieber C M 1998 *Science* **279** 208
[5] Bazargan S and Leung K T 2012 *J. Phys. Chem.* C **116** 5427
[6] Sun Y, Doherty R P, Warren J L and Ashfold M N R 2007 *Chem. Phys. Lett.* **447** 257
[7] Fuge G M, Holmes T M S and Ashfold M N R 2009 *Chem. Phys. Lett.* **479** 125
[8] Ajimsha R S, Manoj R, Aneesh P M and Jayaraj M K 2010 *Current Appl. Phys.* **10** 693
[9] Shen Y, Xu N, Lai J, Sun J, Wu J, Ying Z and Okada T 2009 *J. Vac. Sci. Technol.* B **27** 1856
[10] Zhang Y, Russo R E and Mao S S 2005 *Appl. Phys. Lett.* **87** 133115
[11] Hartanto A B, Ning X, Nakata Y and Okada T 2004 *Appl. Phys.* A **78** 299
[12] Conley J F J, Stecker L and Ono Y 2005 *Nanotechnol.* **16** 292
[13] Huang M H, Wu Y, Feick H, Tran N, Weber E and Yang P 2001 *Adv. Mater.* **13** 113
[14] Chen Y X, Lewis M and Zhou W L 2005 *J. Crystal Growth* **282** 85
[15] Yang S H, Chen P C and Hong S Y 2009 *Current Appl. Phys.* **9** E 180
[16] Li S Y, Lee C Y and Tseng T Y 2003 *J. Crystal Growth* **247** 357
[17] Kuo D H, Fang J F, Chen R S, Chen C A and Huang Y S 2011 *J. Phys. Chem.* C **115** 12260
[18] Zhu Z, Chen T L, Gu Y, Warren J and Osgood R M J 2005 *Chem. Mater.* **17** 4227

Influence of the thickness and thermal treatment on the humidity and ethanol sensing properties of Sn-O-Te layers

B Georgieva[1], **J Pirov and I Podolesheva**

Acad. J. Malinowski Institute of Optical Materials and Technologies,
Bulgarian Academy of Sciences, Acad. G. Bonchev Str., Bl.109, 1113 Sofia, Bulgaria

E-mail: biliana@iomt.bas.bg

Abstract. Thin Sn-O-Te layers were prepared by thermal vacuum co-evaporation of Sn and TeO_2. Depending on the atomic ratio of Sn to Te ($R_{Sn/Te}$) and on the subsequent treatments, layers sensitive towards humidity or ethanol vapor were obtained. The as-deposited amorphous layers with $R_{Sn/Te} \approx 0.8$ and a thickness of 60 nm showed very good characteristics as humidity sensors operating at room temperature. Layers with $R_{Sn/Te} \approx 2.3$, additionally doped with Pt and crystallized by thermal treatment up to 360 °C, were found to be sensitive to ethanol vapor at 120 °C and 240 °C.

The $R_{Sn/Te}$ ratio in the films was estimated by EDS. The composition and structure of the thermally treated layers were studied by X-ray diffraction (XRD). The sensitivity to humidity and ethanol was determined on the basis of the electrical resistance changes in the presence of humidity and ethanol vapor. The influence was studied of the layer thickness and of the sequence of thermal treatment and Pt doping on the layer composition and gas sensing properties.

1. Introduction

The use of Sn-O-Te semiconductor thin films prepared by co-evaporation of Sn and TeO_2 as humidity and ethanol sensors is of substantial technological interest. It is known that the as-deposited amorphous layers with a thickness of 60 nm and $R_{Sn/Te} \approx 0.8$ exhibit high humidity sensitivities at room temperature [1]. Layers with the same thickness, but with $R_{Sn/Te} \approx 2.3$, thermally treated up to 360 °C and doped with Pt, are in turn sensitive to ethanol at 120 °C [2, 3]. Moreover, the processing sequence of the thermal treatment and Pt-doping affects the composition of the layers [3, 4].

In this paper, we present studies on the influence of the thickness of as-deposited amorphous Sn-O-Te layers with $R_{Sn/Te} \approx 0.8$ on their humidity sensing properties, as well as of the thickness and the sequence of the thermal treatment and Pt doping on the ethanol sensing properties of crystalline Sn-O-Te layers with $R_{Sn/Te} \approx 2.3$. The results obtained could lead to a better understanding of the relationship between structure, composition and sensing properties of Sn-O-Te layers.

[1] To whom any correspondence should be addressed.

Journal of Physics: Conference Series **514** (2014) 012033 doi:10.1088/1742-6596/514/1/012033

2. Experimental

Sn-O-Te layers with thicknesses of 60 nm and 200 nm and of $R_{Sn/Te} \approx 0.8$ and 2.3 were prepared by a method previously reported by us in [5]. The crystallization of the layers with $R_{Sn/Te} \approx 2.3$ was achieved by thermal treatment – non-isothermal up to 360 °C (with a step of 40 °C and a 15 minutes delay at each temperature) on a heated plate in the test chamber equipped with a temperature controller. Doping with Pt (2 – 3 at. %) was performed by thermal evaporation in a separate vacuum deposition process either before or after crystallization of the layers. Electron dispersive spectroscopy (EDS) in SEM was used for determining the atomic ratio ($R_{Sn/Te}$) and the Pt content in the samples. The layer thickness was established using a Talystep profilometer (Rank Taylor Hobson). The crystallographic structure and composition of the thermally treated layers with a thickness of 200 nm were investigated by X-ray diffraction (XRD). For these studies, the layers were evaporated on Si substrates. The results obtained by XRD were compared with those achieved for thin layers (60 nm) by selected area electron diffraction (SAED) [3, 4].

To test the layers as humidity and ethanol sensors, Cr comb-like electrodes were patterned using the photolithographic technique described in [6]. The layers were exposed to humidity and ethanol vapor in a test chamber equipped with temperature- and *RH*-controllers. The electrical resistance R [Ω] of undoped and Pt-doped samples was measured at room temperature as a function of the relative humidity (*RH* [%]) or at elevated temperatures (up to 280 °C) in the presence of ethanol vapor. For electrical measurements, a multichannel ohmmeter (National Instruments) was used. The data acquisition and processing, as well as the control of the ohmmeter, were computerized using Lab View. The sensitivity (S) to ethanol was calculated as the ratio of the difference between the electrical resistance of the layers in dry air (R_{air}) and in ethanol vapor (R_{gas}) to the electrical resistance of the layers in ethanol vapor (R_{gas}):

$$S[\%] = \frac{(R_{air} - R_{gas})}{R_{gas}} * 100. \quad (1)$$

3. Results and discussion

3.1. Structure and composition

All as-deposited layers were amorphous, regardless of the value of $R_{Sn/Te}$. The selected area electron diffraction did not indicate the presence of any crystalline phase [7]. The thermal treatment of the layers led to crystallization processes and resulted in the formation of different crystalline Sn- and Te-oxide phases, depending on the temperature and the layer thickness.

Figure 1 illustrates the XRD pattern of a Sn-O-Te layer with a thickness of 200 nm thermally treated after doping with Pt. As is clearly seen, the layer is crystalline. The diffraction peaks can be indexed to SnO_2 and TeO_2 (since the layers were deposited on Si substrates and Pt-doped, there are peaks of Si and Pt as well). Differences in the composition are observed in the layers thermally treated in the same manner, but with a thickness of 60 nm [3, 4]. Moreover, the composition depends on the sequence of heating and Pt-doping. The phases are presented in table 1. Besides TeO_2 and SnO_2, the thinner layers contain non-stoichiometric Sn oxides, namely Sn_2O_3 and Sn_3O_4, if they have been doped with Pt before the thermal treatment on the one hand, and Sn_3O_4, if the doping has been conducted prior to the heating, on the other. As shown below, these differences in the composition correlate with differences in the sensitivity to ethanol.

3.2. Humidity and ethanol sensing properties

Figure 2 presents the resistance R [Ω] of an amorphous Sn-O-Te layer with $R_{Sn/Te} \approx 0.8$ and a thickness of 200 nm as a function of the relative humidity *RH* [%]. As seen, the layer is sensitive to humidity at *RH* [%] > 70 %. The inset shows the same dependence for samples with a thickness of 60 nm. The resistance change covers five decades when the relative humidity varies in the range of 30 – 100 %.

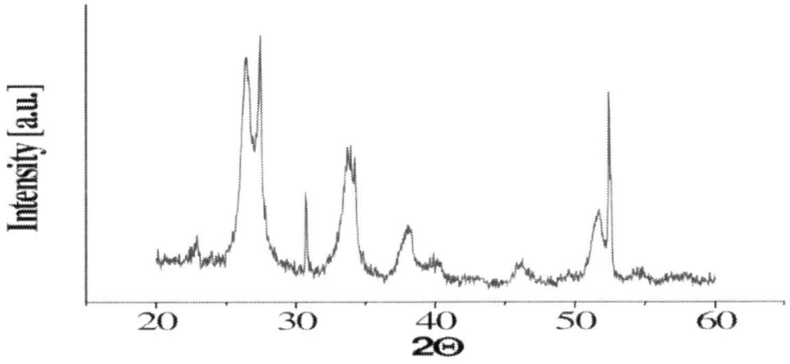

Figure 1. XRD spectra of a Sn-O-Te layer with a thickness of 200 nm.

Phases	PDF
SnO_2	77-0451
TeO_2	76-0680

Obviously, increasing the layer thickness leads to a decrease in the humidity sensitivity. This dependence has also been observed by other researchers [8].

Fgure 3 presents the sensitivity *S* (calculated by equation (1) to ethanol vapor as a function of the substrate temperature for layers with $R_{Sn/Te} \approx 2.3$ and a thickness of 60 nm doped with Pt before the thermal treatment. For comparison, in the same figure are also shown results for layers of the same composition and thickness, both doped and undoped with Pt after the thermally-induced crystallization. As it can be seen, the undoped layers are practically insensitive to ethanol. The Pt-doping results in a substantial sensitivity increase at 120 °C and 240 °C. Moreover, the sequence of doping and thermal heating plays an important role in improving the sensitivity; layers crystallized in the presence of Pt exhibit a higher sensitivity to ethanol than those doped after crystallization. As already shown, they differ in composition as well (see table 1).

Figure 4 presents the sensitivity of layers with a thickness of 200 nm that were non-isothermally treated and doped with Pt before the thermal treatment. They exhibit a significantly lower sensitivity to ethanol than similar layers with a thickness of only 60 nm. It should be noted that the thicker layers do not contain non-stoichiometric Sn-oxides, and there exists evidence that the latter could improve the sensitivity and selectivity of SnO_2 based sensors [9].

Table 1: Crystalline phases in non-isothermally treated layers with a thickness of 60 nm doped with Pt.

Doping with Pt	Phases	PDF
Before thermal treatment	TeO_2	Paratellurite 42-1365
	SnO_2	Cassiterite 41-1445
	Sn_2O_3	25-1259
	Sn_3O_4	20-1293
After thermal treatment	SnO_2	Cassiterite 41-1445
	TeO_2	Paratellurite 42-1365
	Sn_3O_4	20-1293

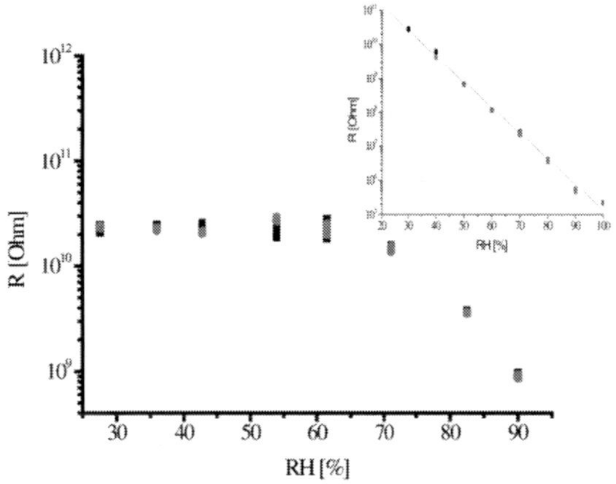

Figure 2. *R* [ohm] as a function of *RH* [%] of an as-deposited Sn-O-Te layer with $R_{Sn/Te} \approx 0.8$ and a thickness of 200 nm.
Inset: *R* [ohm] as a function of *RH* [%] of a sample with the same composition, but with a thickness of only 60 nm [1].

Investigations regarding the selectivity of the Sn-O-Te ethanol sensors showed no cross-sensitivity to humidity and ammonia at a substrate temperature of 120 °C.

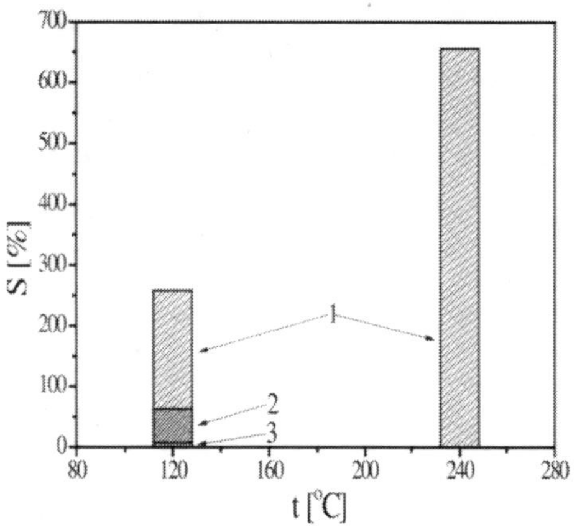

Figure 3. Ethanol sensitivity (S) of Sn-O-Te layers with $R_{Sn/Te} \approx 2.3$ and $d = 60$ nm as a function of the substrate temperature (t): (1) Pt-doped before and (2) after thermal treatment; (3) undoped layers.

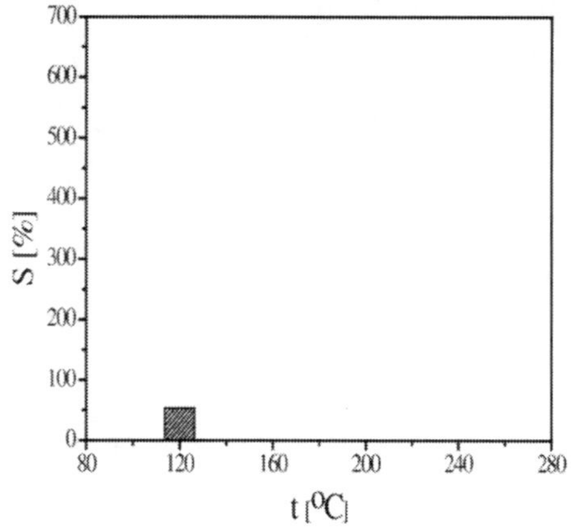

Figure 4. Ethanol sensitivity (S) of Sn-O-Te layers with $R_{Sn/Te} \approx 2.3$ doped with Pt before thermal treatment and $d = 200$ nm at 120 °C.

4. Conclusions

Sn-O-Te layers prepared by co-evaporation of Sn and TeO_2 were investigated in view of applications as humidity and ethanol sensors. It is shown that the sensitivity depends on the layer thickness and (for ethanol sensors) on the sequence of Pt-doping and thermal treatment.

The as-deposited amorphous layers with $R_{Sn/Te} \approx 0.8$ show very good characteristics as humidity sensors operating at room temperature. Increasing the thickness leads to a decrease in sensitivity – layers with $d = 200$ nm are sensitive to humidity at $RH > 70$ %.

Layers with $R_{Sn/Te} \approx 2.3$ that were non-isothermally treated up to 360 °C and doped with Pt are sensitive to ethanol at 120 °C and 240 °C. The layers' thickness affects the gas sensitivity – thinner layers of a thickness of about 60 nm have a gas sensitivity higher than that of thicker layers with $d = 200$ nm. Moreover, the sensitivity is significantly higher if doping precedes the thermal treatment.

Analysis of the crystallized layers revealed that their composition differs not only with the thickness but also with the sequence of Pt-doping and thermal treatment. Thinner layers doped after annealing contain non-stoichiometric Sn-oxide (Sn_3O_4), besides the stoichiometric SnO_2 and TeO_2. If the Pt-doping precedes annealing, a second non-stoichiometric Sn-oxide (Sn_2O_3) is detected in the layer. These results could explain the higher sensitivity of the thin layers and of the layers doped with Pt prior to the thermal treatment.

Acknowledgments

This work was supported by the Bulgarian Ministry of Education and Science under Contract DO 02-148/16.12.2008.

References

[1] Georgieva B, Podolesheva I, Pirov J and Platikanova V 2005 *J. Optoelectr. Adv. Mater.* **7** 2595
[2] Maekawa T, Tamaki J, Miura N, Yamazoe N and Matsushima S 1992 *Sensors and Actuators* B: *Chemical* **9** 63
[3] Georgieva B, Podolesheva I and Spasov G 2010 *Solid State Phenomena* **159** 133
[4] Georgieva B, Nihtianova D, Pirov J and Podolesheva I 2010 *J Phys: Conf. Series* **223** 012018
[5] Podolesheva I, Gushterova P, Platikanova V and Konstantinov I 1998 *J. Vac. Sci. Technol.* A **16** 674

[6] Mednikarov B and Sahatchieva M 2005 *J. Optoelectr. Adv. Mater.* **7** 1415
[7] Podolesheva I, Dimov V, Tarassov M, Platikanova V and Konstantinov I 1998 *J. Vac. Sci. Technol.* A **16** 2608
[8] Yamazoe N, Sakai G and Kengo Shimanoe K 2003 *Catalysis Surveys from Asia* **7** 63
[9] Zhao J, Tan R, Shen W, Yang Y, Guo Y, Li J, Zhou Z, Jian J, Song W, 2012 *Materials Lett.* **84** 94

Journal of Physics: Conference Series **514** (2014) 012034 doi:10.1088/1742-6596/514/1/012034

Interaction of slow highly charged ions with hard dental tissue: studies of fluoride uptake and reminalization efficacy

A Daskalova[1,4], **G Kasperski**[2], **P Rousseau**[3], **A Domaracka**[3] **and A Lawicki**[3]

[1]Acad. E. Djakov Institute of Electronics, Bulgarian Academy of Sciences, 72 Tsarigradsko Chaussee, 1784 Sofia, Bulgaria
[2]Gdańsk University of Technology, Narutowicza 11/12, 80-233 Gdańsk Wrzeszcz, Poland
[3]CEA-CIMAP, GANIL, BP 5133, F-14070 Caen Cedex 5, France

E-mail: a_daskalova@code.bg

Abstract. TOF-SIMS mass spectroscopy data are presented on ion irradiation of hard dental tissue using a beam of $^{129}Xe^{20+}$ (15 kV) ions delivered in the ARIBE facility by an ECR source. The investigation was focused on the mass distribution of the fragment ions. A comparison is made between the mass spectra from hard dental tissue treated by olaflur-($C_{27}H_{60}F_2N_2O_3$) and untreated hard dental tissue obtained under irradiation by low-energy highly-charged ions (HCIs). We found significant differences between the mass spectra of enamel after introducing amine fluoride (olaflur) and the mass spectra of pure untreated enamel. Further, we separated out the effects caused by radiation induced in the tooth enamel from those induced in dentin, which has not been performed before. In order to conduct a further detailed analysis, it is necessary to extend the research scope to include the influence of fluorine compounds on enamel and dentin.

1. Introduction

Collision induced fragmentation using highly charged ions from an ECR source has been demonstrated to be a powerful method providing information complementary to that of other experimental methods. The specifity of experiments with an ECR source is related to the large choice of the type and charge of the ion, the easy scan and the high selectivity of the kinetic energy [1]. Slow highly-charged ions, while approaching a surface, can capture electrons and (in addition to projectile kinetic energy deposition by elastic and inelastic collisions) a strong localized electronic excitation is induced close to the surface. Various studies have shown that due to their low kinetic energy, slow HCIs do not penetrate deeper layers, thus opening new possibilities for surface analysis. Secondary ion mass spectrometry (SIMS) has been proved to be a powerful tool for microanalytical characterization of biomineralized tissues. The technique permits quantitative analysis of more than 40 elements in tooth tissues [2]. In some earlier studies, secondary ion mass spectrometry was used to determine the distribution of Na, K, Sr, F and Cl in the buccal enamel of human deciduous incisors [3]. The work of Lundgren et al. [4] allowed the estimation of the incorporation of some critical elements into the enamel in correlation with the differentiation stages of the ameloblasts throughout the whole tooth.

[4] To whom any correspondence should be addressed.

The elemental composition pattern of dental enamel was examined and a picture was presented describing a limited region. Later on Jaelevik et al. [5] studied the hypomineralization in children's first molars with respect to the chemical composition of the disturbed enamel. The concentration gradients for F, Cl, Na, Mg, K and Sr in hypomineralized enamel were examined by means of secondary ion mass spectrometry and were compared to normal enamel. Gotliv and colleagues have utilized the technique of time-of-flight secondary ion mass spectrometry (TOF-SIMS) to map the distribution of positive and negative inorganic ions, as well as the organic components in the fully mineralized intact peritubular dentin (PTD) structure in bovine tooth cross-sections [6].

The present study aims at obtaining data on the fundamental interaction processes of multiply charged ions with a hard biological tissue (teeth). Slow multiply charged ions can be a powerful tool for biological damage studies due to the localized energy deposition.

We have previously acquired good experimental results on ultra-short laser ablation of hard tissue (bone and tooth). These preliminary studies have provided initial information on how to alter the chemical composition of molecules of biological origin [7-9]. However, for a full understanding of the processes of the molecular fragmentation pathways, it is especially important to study the interaction process by slow multiply charged ions. Most approaches to studying the ion-induced damage to biomolecules involve mass-spectrometric techniques. In this study, a comparison is performed between the mass spectra from treated with olaflur and from untreated tooth samples obtained under irradiation with low-energy highly-charged ions (HCIs). Previous experiments have shown the possibility of avoiding the effect of charging the insulator samples by wrapping them with a metallic wire, as this is an important problem in studying insulator targets. Moreover, the HCI-induced mass spectrum will be a reference of the blank sample before fluorine covering.

The tooth enamel is a hard substance covering the crown of the tooth (figure 1); it consists of 98 % of non-organic compounds, mainly hydroxyapatite [$Ca_{10}(PO_4)_6(OH)_2$], magnesium, carbonate, sodium, calcium and fluorine. The remaining 2 % are organic compounds and water.

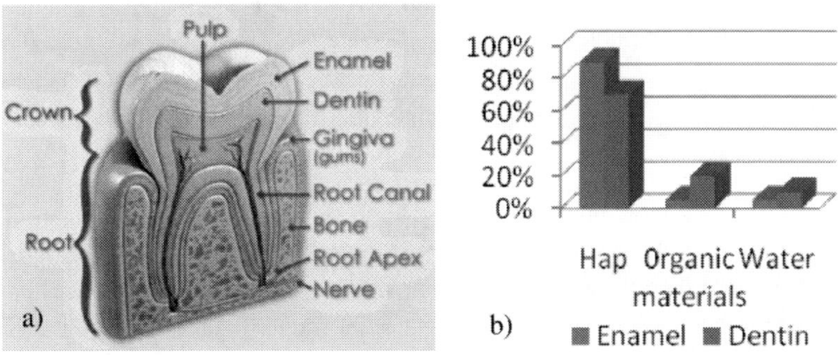

Figure 1. a) Tooth structure, b) Enamel and dentin major components.

Fluorine, which is introduced from an outside source, combines with the enamel thus increasing the enamel's resistance to acids; fluorine also facilitates the enamel's remineralization and accelerates the growth of new, greater hydroxyapatite crystals. However, too high an amount of fluorine may lead to osteoporosis. Hydroxyapatite is mainly used as a biomaterial for covering orthopedic or dental prostheses and producing bones substitutes. Up to the present, studies concerning the enamel have been focused on the influence of fluorine compounds (e.g. MgF_2, CaF_2 and ions Mg^{2+}, Sr^{2+}, Ba^{2+}) on changes in its physico-chemical properties [10]. However, research is lacking on the influence of the interaction between carbonate–fluorine compounds (with relatively low dissociation energy) and the hydroxyapatite contained in the enamel.

In view of the above, we studied the sputtering process of hard biological tissue under various ion beam parameters (velocity and charge state, i.e. potential energy) as the results could form the basis of optimizing the desorption process. The sputtered species were characterized using time-of-flight SIMS, yielding information on the physical processes involved during exposure to ionizing radiation.

Journal of Physics: Conference Series **514** (2014) 012034 doi:10.1088/1742-6596/514/1/012034

The interaction of keV ions with biomolecules or clusters can depend on a variety of properties of both the projectile and the target.

The results of the measurements performed may contribute to the better understanding of enamel remineralization mechanisms and to determining the mechanisms of increasing the immunity acquired (to caries decay) by the fluorization process (enamel assimilation of fluorine); interest in such studies have been declared by many pharmaceutical companies. The results may also be beneficial to optimizing the processes of producing new composite mixtures used in hydroxyapatite-containing implants. Further, it is our believe that this study's results will contribute to a better recognition of the origin of osteoporosis, thus leading to its prevention.

2. Materials and methods

Freshly extracted human third molars with clinically sound enamel were provided by dentists and were stored in an olaflur solution. From each tooth, three longitudinal enamel or dentine samples were prepared by using a water-cooled diamond saw. The separation planes of the halves were ground with a successively decreasing grain size. This procedure ensured optimal surface flatness. Then they were brushed with non-fluoridated pumice powder and rinsed with distilled water. Surfaces without any visible defects were only selected for the treatment procedure. The fluoridation process was performed by treating the samples for three days with an amine fluoride solution – olaflur ($C_{27}H_{60}F_2N_2O_3$) for three minutes. Subsequently, the samples were rinsed in distilled water for 30 s, air dried at room temperature and stored in closed plastic bags.

The principle of the experimental procedure is given in figure 2. The experiments were performed in an ultra-high vacuum chamber at a base pressure of 1.3×10^{-6} Pa. The samples were irradiated using an electron gun.

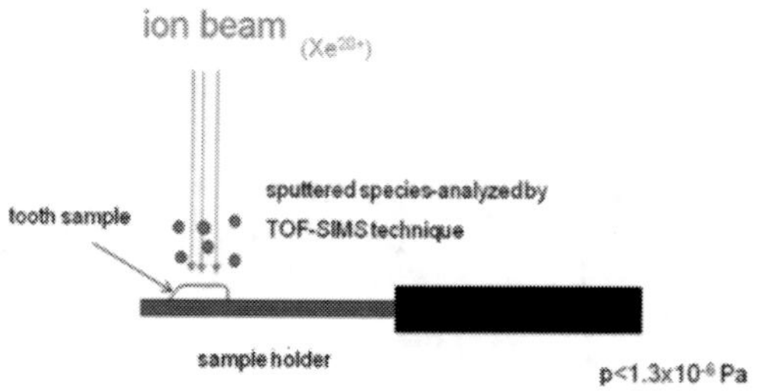

Figure 2. Schematic diagram of the geometrical arrangement of the experiment.

The biological sample under investigation was bombarded by ions and the interaction products were analyzed by the time-of-flight method. The ions generated in the collision region were extracted through an electric field lens system into a Wiley-McLaren time-of- flight (TOF) spectrometer and detected by a microchanel plate (MCP) detector. We used an electron cyclotron resonance ion source to generate ions with charge states q ranging from 1 to 30 at kinetic energies from 1 keV to 15 keV. A crucial feature of the experiment was cooling the UHV chamber by liquid nitrogen. The base pressure during experiments was thus kept around 1.3×10^{-6} Pa and the contributions of the residual gas to the experimental data were negligible. The charging effect being the main issue when insulator targets are studied, we avoided it by wrapping the insulator samples with a metallic wire.

3. Results and discussion

Amine fluoride is a surface-active fluoride beneficial to dental plaque control and contained in a number of complex formulations of oral hygiene products. Amine fluorides such as olaflur and dectaflur are used in tooth pastes. Fluoride ions can combine rapidly with calcium in dental enamel to form calcium fluoride [11, 12]. The uptake of fluoride into the almost insoluble crystal structure of the tooth is a very slow process, but it binds the ions very tightly. The enamel is, therefore, also called the "stable" fluoride reservoir. The surfactant characteristics of amine fluorides promote also a homogeneous distribution of the calcium fluoride globules. Thus, Nelson et al. [13] suggested that the particle size of CaF_2 crystallites may be an important factor in determining the effectiveness of a topical fluoride agent. Therefore, it could be assumed that the larger globules observed after treatment with low fluoride concentrations will dissolve more slowly and serve as a reservoir for more extended periods of time contributing to a caries preventive effect. This process is called dental remineralization with fluorides.

The choice of elements for analysis in this study was based on the assumption that they might have a role during the remineralization of the hard dental tissues [14]. The secondary ion mass spectrometric analyses only showed differences in the elemental composition between normal and olaflur treated enamel for Ca, C, F, but no notable differences in the elemental composition for Mg, Na.

The SIMS mass spectrum of a tooth dentin sample (figure 3) exhibits dominating peaks of Na^+ and Ca^+ ion at *m/z* = 23, 40. The peaks at *m/z* = 52, 56 do not originate from the biological sample, but rather from Cr and Fe, these compounds being present in the metallic wires used to prevent charging of the organic sample.

Table 1. Recorded mass numbers and corresponding ion species.

Mass number	Ion species
12	C
16	O
19	F
23	Na
24	Mg
31	P
39	K
40	Ca
56, 57	CaO, CaOH
59	CaF, CaO, CaOH
60	CO_3

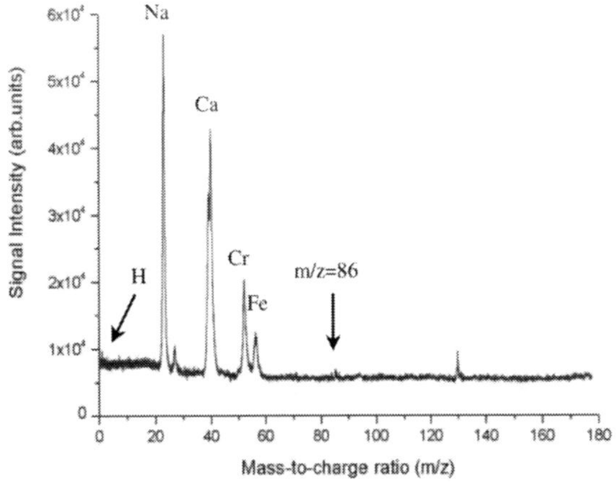

Figure 3. ToF SIMS mass spectrum of tooth dentin sputtered by $^{129}Xe^{20}$ ions bombardment.

Figure 4 shows a typical mass spectrum obtained with $^{129}Xe^{20+}$ projectiles colliding with tooth enamel (table1).

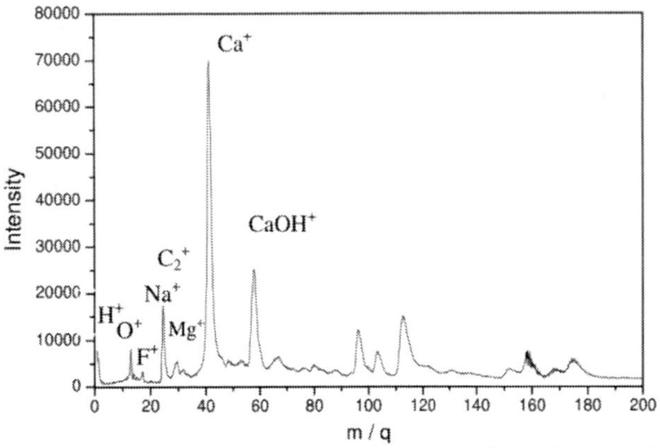

Figure 4. TOF-SIMS mass spectrum of tooth enamel sputtered by $^{129}Xe^{20+}$ ions at normal incidence.

The principal inorganic constituent of enamel is hydroxyapatite $Ca_{10}(PO_4)_6(OH)_2$, which tends to incorporate a number of trace elements. Some of the elements present in the mass spectrum here might reflect events that occur during hard tissue formation where huge a number of ions are transported through the odontoblast or ameloblast layers. One possible role for some of the different inorganic elements, other than Ca^+, might be as co-transport ions in the plasma-membrane transporters of the mineralizing cells. Sodium, for example, is transported in parallel to phosphorus in many cell types. Other elements are probably exogenous in origin and accumulate principally by ionic exchange at the tooth surface.

Differences between normal and treated enamel are expressed in the elemental composition and in the Ca^+, F^+, CO_3^+, $CaOH^+$ peaks ratios. In treated enamel there was a considerable increase in fluorine content (figure 5a). In the present study, the higher concentration of fluorine in the treated enamel areas might point to the possibility that deeper penetration of exogenous elements from the surface had been facilitated by the porosity of the enamel.

The content of C^+ increased in irradiated dentin in contrast to enamel (figure 5b). Carbon is a component of olaflur - ($C_{27}H_{60}F_2N_2O_3$). A possible explanation could be that these ions are incorporated into the dentin as a consequence from the treatment with olaflur.

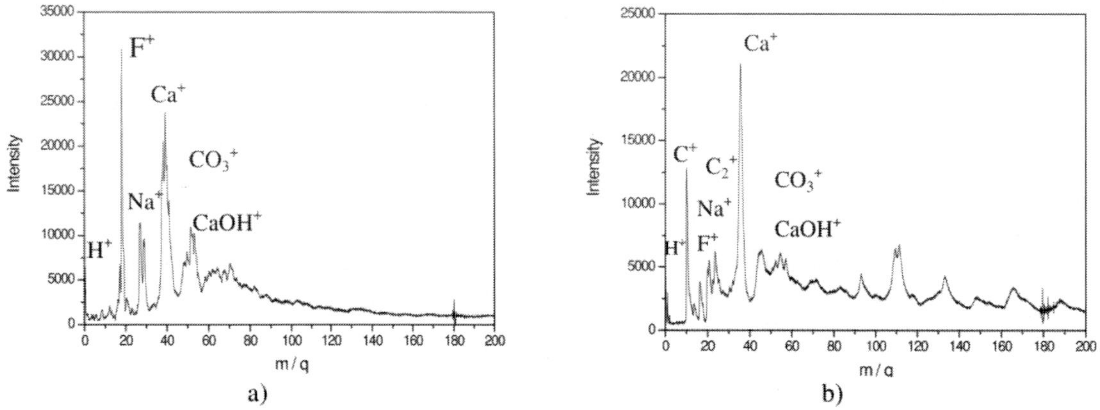

Figure 5. a) TOF-SIMS mass spectrum of tooth enamel + olaflur; b) TOF-SIMS mass spectrum of tooth dentin + olaflur.

In olaflur treated enamel and dentin, the general tendency was a slight reduction in calcium content in relation to untreated enamel. However, despite the considerable variation, the Ca ratio was rather constant throughout. The presence of magnesium is closely related to calcium. It is also possible that some magnesium ions had been excluded during crystallite formation and had finally become located on crystallite surfaces, in hydration layers or as parts of a separate phase. There might also be a correlation between magnesium and the enamel protein content.

However, there is a need for further studies of normal and treated enamel and dentin and for further evaluation of HCI SIMS method in hard tissue research.

4. Conclusions

We conclude that the treated enamel analyzed here by secondary ion mass spectrometry has a higher amount of fluorine. The concentration of Ca level stays proportional in both treated enamel and dentin. We demonstrated the sensitivity of the HCI SIMS method and its capability in examining hard dental tissue. We report obtaining secondary ion mass spectra of untreated (pure) enamel, pure enamel with olaflur and pure dentin with olaflur under the impact of Xe^{20+} ions. Conducting further detailed analysis would necessitate extending the research scope to include the influence of fluorine compounds on enamel and dentin (i.e. amine fluoride - fluorinol and pure fluoride protector).

Acknowledgements

This experiment was performed at the distributed LEIF-Infrastructure at ARIBE, GANIL, France, supported by Transnational Access, under a grant by the European Project P09073A.

References

[1] Briand J-P, Giardino G, Borsoni G, Le Roux V, Béchu N, Dreuil S, Tüske O and Machicoane G 2000 The interaction of slow highly charged ions on surfaces *Rev. Sci. Instrum.* **71** 627

[2] Lodding A, Fischer P, Odelius H, Norén J, Sennerby L, Johansson C, Chabala J and Levi-Setti R 1990 Secondary ion mass spectrometry in the study of biomineralizations and biomaterials *Analytica Chimica Acta* **241** 314

[3] Norén J, Lodding A, Odelius H and Linde A 1983 Secondary ion mass spectrometry of human deciduous enamel distribution of Na, K, Mg, Sr, F and Cl *Caries Res.* **17** 502

[4] Lundgren T, Persson L G, Engström E, Chabala J, Levi-Setti R and Norén J 1998 A secondary ion mass spectroscopic study of the elemental composition pattern in rat incisor dental enamel during different stages of ameloblast differentiation *Archives of Oral Biology* **43** 841–8

[5] Jälevik B, Odelius H, Dietz W and Noren J 2001 Secondary ion mass spectrometry and X-ray microanalysis of hypomineralized enamel in human permanent first molars *Archives of Oral Biology* **46** 247

[6] Gotliv B and Veis A 2008 The composition of bovine peritubular dentin: matching TOF-SIMS, scanning electron microscopy and biochemical component distributions *Cells Tissues Organs* **189** 19

[7] Daskalova A and Husinsky W 2004 Ultra-short laser ablation of biological tissue *Proc. SPIE* **5830** 473-8

[8] Daskalova A and Husinsky W 2006 Detection of biological molecules from hard biological tissue using ion-beam-induced desorption and femtosecond laser ablation *Plasma Process. Polym.* **3** 257-61

[9] Daskalova A and Husinsky W 2007 Ionization of biomolecules in high-intensity laser fields *Proc. SPIE* 6604 66042G1-5

[10] Weir M D, Chow L C and Xu H K 2012 Remineralization of demineralized enamel via calcium phosphate nanocomposite *J. Dental Res.* **10** 979-84

[11] Holler B E, Friedl K H, Jung Hiller K A and Schmalz G 2002 Fluoride uptake and distribution in enamel and dentin after application of different fluoride solutions *Clin. Oral Investig.* **6** 144

[12] Schemehorn B R, Wood G D, McHale W and Winston A E 2011 Comparison of fluoride uptake into tooth enamel from two fluoride varnishes containing different calcium phosphate sources *J. Clin. Dent.* **22** 51-4

[13] Nelson D, Jongebloed W and Arends J 1983 Morphology of enamel surfaces treated with topical fluoride agents: SEM considerations *J. Dent. Res.* **62** 1201-24

[14] Roveri N, Battistella E, Bianchi C, Foltran I, Foresti E, Iafisco M, Lelli M, Naldoni A, Palazzo B and Rimondini L 2009 Surface enamel remineralization: biomimetic apatite nanocrystals and fluoride ions different effects *J. Nanomater.* http://dx.doi.org/10.1155/2009/746383

Journal of Physics: Conference Series **514** (2014) 012035 doi:10.1088/1742-6596/514/1/012035

XPS study of nanoscale SiO_xN_y layers synthezised by plasma immersion implantation of nitrogen

S Alexandrova[1], E Halova[1,4], S Bakalova[2], A Szekeres[2], A Marin[3], P Osiceanu[3], M Gartner[3] and N Koujuharova[1]

[1]Technical University of Sofia, 8 Kl. Ohridski Blvd., 1797 Sofia, Bulgaria
[2]Acad. G. Nadjakov Institute of Solid State Physics, Bulgarian Academy of Sciences, 72 Tsarigradsko Chaussee, 1784 Sofia, Bulgaria
[3]Ilie Murgulescu Institute of Physical Chemistry, 202 Splaiul Independentei, 060021 Bucharest, Romania

E-mail: ehalova@tu-sofia.bg

Abstract. Plasma immersion implantation (PII) of nitrogen into Si was used to prepare nanoscale SiO_xN_y layers. The SiO_xN_y layers were synthesized in a cycle of PII followed by a high-temperature treatment in O_2 ambient. Nitrogen was implanted into the Si substrate with energy of 2 keV and a fluence in the range of 10^{16} – 10^{18} cm^{-2}. The implanted structures were subjected to a high-temperature annealing at 1050 °C for 10 or 20 minutes in dry oxygen at atmospheric pressure. X-ray photoelectron spectroscopy (XPS) revealed the chemical bonds and surface layer composition. The atomic concentrations of the SiO_xN_y constituents were obtained. By a combination with a sputter experiment the N concentration was found to be about 2,6 at. %. The low N content is due to the nitrogen escaping from Si during the oxidizing annealling. The thickness of the SiO_xN_y layer for the highest implantation fluence did not exceed 10 nm as revealed from ellipsometric and XPS data analyses.

1. Introduction

Silicon oxynitride SiO_xN_y thin layers have found different applications in the semiconductor industry. They have been used as a substitute for SiO_2 gate oxide layers to reduce the gate leakage and boron penetration through the oxide and to adjust the threshold voltage. Nanoscaled SiO_xN_y layers have also found applications as a low-defect interface layer between a Si substrate and high-*k* layers deposited over it. Resistance to oxidation and a lower mechanical stress are other important properties of the SiO_xN_y layers. The large flexibility in tuning the refractive index between 1.46 of SiO_2 and 2.0 of Si_3N_4 is important for use in optical communication devices. An attractive opportunity of major significance is the possibility to integrate electronic and optical components on the same Si chip.

The growth of ultrathin oxynitride layers by plasma ion implantation (PII) of nitrogen into Si, followed by thermal processing in an oxidizing ambient, has been reported for different electronic and optoelectronic applications [1-3]. Recently, we suggested that the presence of N in the Si promotes the formation of Si-Si bonds during the synthesis of SiO_xN_y in an O_2 ambient [4]. Clustering of the Si-Si bonds can result in the formation of nanoparticles. Different N concentrations and distributions

[4] To whom any correspondence should be addressed.

through the oxide can be achieved using a variety of processing conditions, such as implantation energy and fluence, annealing temperature and ambient. A better knowledge of the structural properties of the grown SiO_xN_y layers is required in view of process development and engineering and for process control.

The optical and structural properties of SiO_xN_y have been thoroughly studied mainly in the form of CVD deposited layers [5,6]. Information has been obtained on the chemical bonds and their concentration in dependence of the deposition parameters, such as gas flow ratio and temperature [7]. The formation of Si nanocrystals has also been discussed [8, 9]. The results reported have mostly been related to films containing hydrogen and the latter's effect on the film's properties [10, 9]. In our study, the deposition in a PII module allowed us to produce hydrogen-free oxides.

The problem of finding the chemical content and composition of SiO_xN_y layers can be approached by different methods. In the present study, X-ray photoelectron spectroscopy (XPS) as a non-destructive method was applied to identify the different bonds, such as Si-O and Si-N in the SiO_xN_y matrix. By combining it with a sputtering technique, we gathered information about the distributions through the SiO_xN_y thickness. The characterization was completed by using spectroscopic ellipsometry (SE).

2. Experimental details

The experiments were carried out on Cz-grown Si(100) p-type wafers with a resistivity of 5 –8 ohm cm. The first step of the processing was bombardment by atomic nitrogen (N^+) and molecular (N_2^+) ions into a shallow Si surface region by plasma-beam ion immersion (PII) implantation in a planar plasma reactor. The implantation of nitrogen ions was implemented at normal incidence at energy of 4 keV and a fluence within the 10^{16} – 10^{18} cm^{-2}interval. The second step was elimination of the possible surface contamination through a short dip in diluted HF. As the last step, the implanted Si substrates were thermally annealed in a conventional furnace at 1050 °C for 10 minutes and 20 minutes in a dry oxygen environment at atmospheric pressure. The ellipsometric measurements were performed on a variable-angle spectroscopic ellipsometer (J.A. Woollam Co.).

The depth profiles of the implanted ions were derived using the SRIM simulation code based on a Monte-Carlo method for different annealing durations taking into account the diffusion of the implanted species.

The composition of the layers was identified by XPS. The XPS measurements were carried out with PHI Quantera equipment working at the low residual pressure of 10^{-7} Pa and using monochromatic Al-Kα radiation (1486.6 eV). The charging effect was minimized by using dual beams (electrons and Ar ions) as a neutralizer. The binding energy positions were obtained with a precision of 0.2 eV.

3. Results and discussion

The thickness of the SiO_xN_y synthesized layers as obtained from the SE measurements varied between 10 nm and 40 nm with the N^+ fluence (figure 1). A reduction of the oxidation rate was observed as the N concentration in the Si substrate increased. The analysis based on the Deal-Grove model pointed to a complicated kinetics with diffusion and surface oxidation rates determined by the implantation defects and the nitrogen species [11]. The oxidation of the implanted Si was mainly determined by the oxidant diffusion through the already formed oxide layer. The thickness of the layers was determined using the ratio of the areas of the XPS peaks of the oxide to the substrate Si 2p peaks upon sputtering [12, 13]. A very good agreement with the SE thickness data was found, as illustrated in figure 1 for a fluence of 10^{18} cm^{-2}.

The XPS study yielded information on the bonding arrangements of silicon, oxygen, and nitrogen atoms and the relative elemental atomic content. Figure 2 shows photoemission spectra from SiO_xN_y layers grown in Si implanted with the highest fluence of 10^{18} cm^{-2}.

The oxygen spectra for the different fluences closely resemble each other, illustrated by the spectrum in figure 2a for 10^{18} cm^{-2} fluence. The Si2p core level spectrum in figure 2b shows two separate components in the range 98 – 107 eV. The main part located at 103.5 eV is ascribed to Si^{4+}

oxidation state in silicon oxide and/or nitride [14]. The low energy peak at 99.2 eV indicates Si-Si bonds usually taken as substrate contribution. However, the much smaller intensity of this peak has been interpreted as originating from nanosized Si clusters in the grown layer. The N1s emission shows a multicomponent behavior as evident in figure 2c. The deconvolution reveals peaks at about 398.6 eV and 400 eV characteristic for Si-O-N bonding in different configuration usually assigned to N(–SiO_3)x, O–N–Si_2.

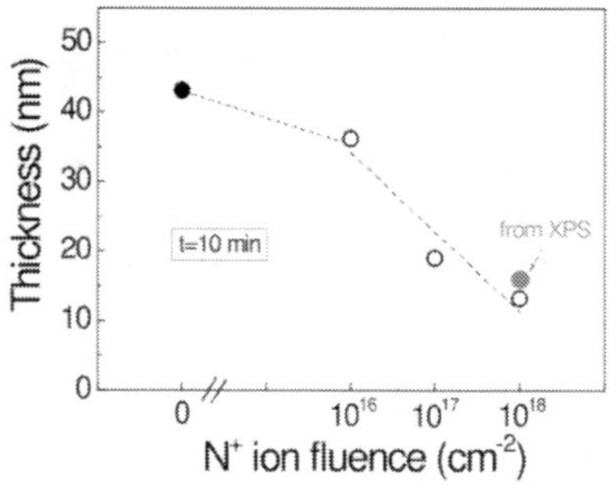

Figure 1. Thickness of the synthesized layers as a function of the N^+ PII fluence.

Analyzing the contributions of the different atoms in the SiO_xN_y composition, we can trace the evolution of the layer structure as the N^+ fluence and the annealing time are increased. The results for the atomic concentrations are presented in figure 3. It can be seen that at the annealing time of 10 minutes, the nitrogen content progressively increases at the expense of oxygen as the N^+ fluence is increased. Thus, the composition progresses from SiO_xN_y with very low N content, nearly SiO_2, to a layer with a fourfold increase of the N concentration. Nevertheless, the N concentration of 2 at. % at a fluence of 10^{18} cm^{-2} still can be estimated as low. At the longer annealling time of 20 minutes, the N concentration has a nearly constant value at a very low level so that the film composition approaches SiO_2.

A similar tendency of an increase in the N content with the ion fluence was detected by the spectroscopic ellipsometry investigation, where the volume fraction of the Si_2N_3 component was found to vary from 3 vol. % to 16.8 vol. %, in spite of the 5,5-fold increase against the 4-fold one in the atomic

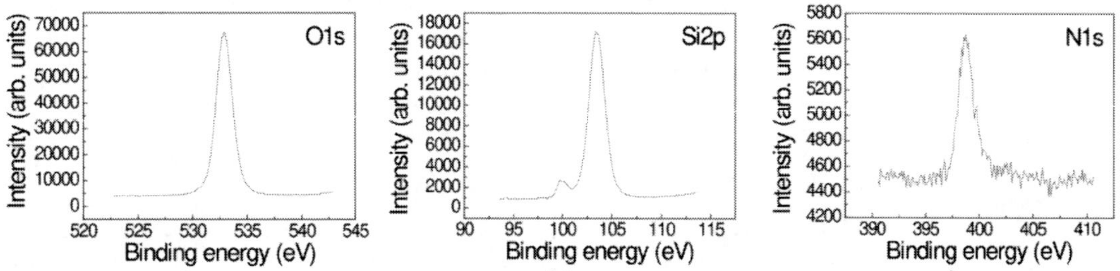

a) b) c)

Figure 2. Binding energy of N1s, Si2p and O1s core levels in layers formed into Si implanted at a N^+ fluence of 10^{18} cm^{-2}.

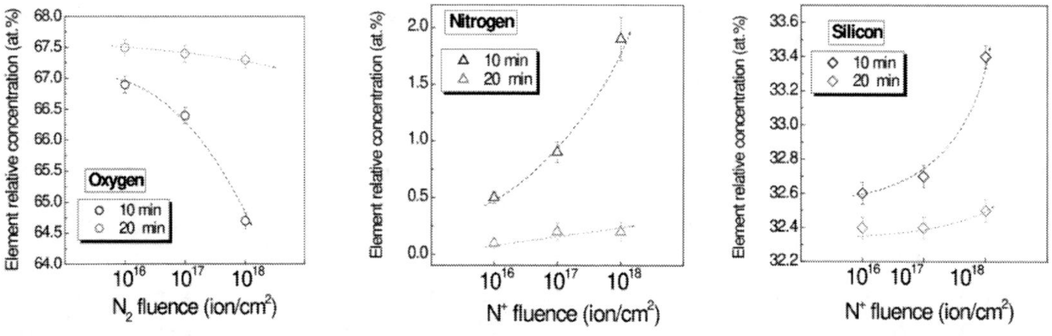

a) b) c)

Figure 3. Relative concentrations of the different atoms in the SiO_xN_y layer as a function of the N^+ fluence.

fractions above in figure 3b. However, one should bear in mind that, in the present XPS study, the photoemission originates from the ~5 nm top surface layer, while the ellipsometric data refers to the total layer volume. The implantation of the ions follows a depth profile in the Si substrate with a low concentration near the surface. Diffusion and escape of N are quite possible during the oxidation in the annealing cycle.

The SiO_xN_y layers were characterized using XPS depth profiling in a sputtering cycle with sputter rate of about 2,5 nm/min. The resulting depth distribution for a 10-nm layer is given in figure 4a. It shows a decreasing concentration mainly due to N escape from the 10 nm layer. The total concentration obtained by integrating the ion distribution in figure 4a over the whole layer amounts to 2,6 at. %.

A flattened implantation profile during annealling was inferred by modelling the implantation profile hile taking the diffusion into account, as is evident in figure 4b. It can be seen that the two profiles, the measured and the modelled, differ by the in-depth distributions.

The density of N atoms in the volume of the sputtered film studied as calculated from the profile in figure 4a amounts to $1,2\times10^{22}$ cm^{-3}. The density obtained from the simulated profile in figure 4b for the same volume is $3,1\times10^{22}$ cm^{-3}, which is of the same order of magnitude, albeit slightly higher. The differences in the calculated and the measured nitrogen distributions imply that, most probably, the diffusion is not the only process contributing to nitrogen migration during SiO_xN_y synthesis.

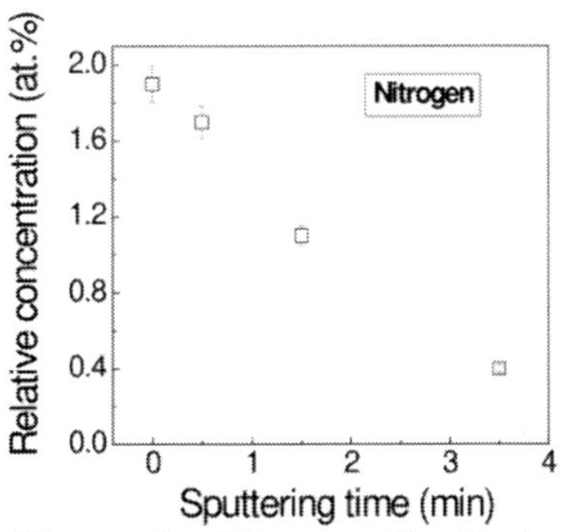

Figure 4a. Nitrogen distribution in the SiO_xN_y obtained from XPS by layer sputtering.

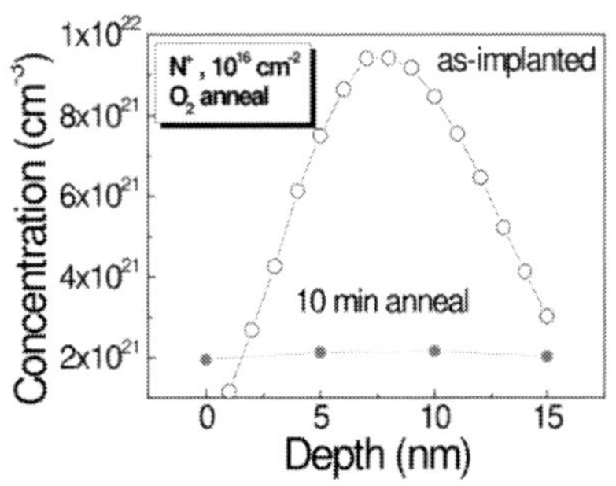

Figure 4b. SRIM-modelled nitrogen distributions before and after annealling.

4. Conclusions

XPS investigation of SiO_xN_y layers synthesized by high-temperature annealing in O_2 ambient of plasma-beam nitrogen ion implanted Si substrates revealed peaks characteristic for Si-O-N bonding in different configurations. The atomic concentrations of the Si, O and N were determined. The nitrogen distribution in the synthesized layer was obtained revealing decreasing in-depth concentrations. This was compared with the flat distribution obtained by SRIM modelling and taking into account the nitrogen diffusion during oxidizing annealling. The layers were identified as containing SiO_xN_y with a low N content of 2,6 at. %.

References

[1] Salgado T D M, Stedile F C, Krug C, Baumvol I J R and Radtke C 1999 *Nucl. Instrum. Methods Phys. Res.* B **148** 1-4 252-6

[2] Diniz J A, Sotero A P, Lujan G S, Tatsch P J and Swart J W 2000 *Nucl. Instrum. Methods Phys. Res.* B **166-167** 64-9

[3] Beck R B and Jakubowski A 2004 *J. Telecommun. Inform. Technol.* **1** 62-9

[4] Gartner M, Szekeres A, Alexandrova S, Osiceanu P, Anastasescu M, Stoica M, Marin A, Vlaikova E and Halova E 2012 *Appl. Surf. Sci.* **258** 7195– 7201

[5] Morales-Acevedo A, Carrillo-Lopez J, Aceves M, Calleja-Arriaga W, Diaz-Valdes E, Contreras G and Mejia-Garcia C 1999 *Superficies y Vacio* **9** 56-8

[6] Pandey R K, Patil L S, Bange J P, Patil D R, Mahajan A M, Patil D S and Gautam D K 2004 *Optical Mater.* **25** 1–7

[7] Chang J P, Green M L, Donnelly V M, Opila R L, Eng J, Sapjeta J, Silverman P J, Weir B, Lu C, Gustafsson T and Garfunkel E 2000 *J. Appl. Phys.* **87** 4449-55

[8] Feng S-Z, Zhang R-J, Zheng Y-X, Li J, Mao P-H, Li X-F and Chen L-Y 2007 *J. Korean Phys. Soc.* **51**/4 1593-7

[9] Canillas A, Pinyol A, Sancho-Parramon J, Ferre-Borrull J, Bertran E 2004 *Thin Solid Films* **455–456** 167–71

[10] Sahu B S, Agnihotri O P, Jain S C, Mertens R and Katok I 2000 *Semicond. Sci. Technol.* **15** L11-4

[11] Szekeres A, Alexandrova S, Vlaikova E and Halova E 2012 *Bulgarian J. Phys.* **39** 178-85

[12] Osiceanu P 2006 *Appl. Surf. Sci.* **253** 381-4

[13] Logofatu C, Negrila C, Ghita R, Ungureanu F, Cotirlan C, Ghica C, Manea A and Lazarescu M 2011 Study of SiO_2/Si Interface by Surface Techniques *Crystalline Silicon – Properties and Uses* ed S. Basu (InTech) pp 23–43

[14] Birer Ö, Sayan S, Süzer S and Aydınl A 1999 *J. Molecular Structure* **480–481** 611-4

Journal of Physics: Conference Series **514** (2014) 012036 doi:10.1088/1742-6596/514/1/012036

Surface morphology of RF plasma immersion H^+ ion implanted and oxidized Si(100) surface

M Anastasescu[1], M Stoica[1], M Gartner[1], S Bakalova[2], A Szekeres[2,4] and S Alexandrova[3]

[1]Institute of Physical Chemistry, Romanian Academy of Sciences, 202 Spl. Independentei, 060021 Bucharest, Romania
[2]Acad. G. Nadjakov Institute of Solid State Physics, Bulgarian Academy of Sciences, 72 Tsarigradsko Chaussee, 1784 Sofia, Bulgaria
[3]Technical University of Sofia, 8 Kl. Ohridski Blvd., 1797 Sofia, Bulgaria

E-mail: szekeres@issp.bas.bg

Abstract. The surface morphology of p-Si(100) wafers after RF plasma immersion (PII) H^+ ion implantation into a shallow Si surface layer and after subsequent thermal oxidation was studied by atomic-force microscopic (AFM) imaging. After PII implantation of hydrogen ions with an energy of 2 keV and fluences ranging from 10^{13} cm^{-2} to 10^{15} cm^{-2}, the Si wafers were oxidized in dry O_2 at temperatures ranging from 700 °C to 800 °C. From the analysis of the AFM images, the surface amplitude parameters were evaluated and considered in terms of the technological conditions. The amplitude parameters showed a clear dependence on the H^+ dose and the oxidation temperature, with the tendency of increasing with the increase of both the H^+ ion fluence and the oxidation temperature. The implantation causes surface roughening, changing the RMS roughness value from 0.15 nm (typical for a polished Si(100) surface) to the highest value 0.6 nm for the H^+ fluence of 10^{15} ions/cm^2. Oxidation of the H^+ implanted Si region, as the oxide is growing inward into Si, levels away the pits created by implants and results in a smoother surface, although keeping the RMS values larger than 0.2 nm.

1. Introduction

The growth of reliable thin thermal oxide on silicon in a thin-film regime is important for fabrication of smaller and faster MOS device structures. Growth of low-defect ultra-thin oxides with improved electrical and structural properties can be achieved at suitable oxidation rates and lowered oxidation temperatures by hydrogenation of silicon prior to the oxidation process. In the Si technology, hydrogenation is usually achieved by an implantation process, either beam [1,2] or plasma ion implantation (PIII) [3]. During the implantation process, due to plasma radiation and collisions of implanted ions with the surface and atoms of the substrate material, lattice defects are created, which is a shortcoming in view of preparing advanced MOS structures for the future nanoelectronics. On the other hand, ultrathin SiO_2 with easily controllable thickness, high dielectric quality and perfect interfaces is difficult to be obtained at temperatures lower than

[4] To whom any correspondence should be addressed.

900 °C. Moreover, the pre-oxidation conditions of the Si surface play a decisive role and, hence, will influence the growth kinetics of ultrathin oxides and their properties. All of the above calls for comprehensive studies of the surface, oxide and interface properties in relation with the technology of MOS device preparation.

Our recent research has been focused on the growth of high-quality ultra-thin SiO_2 layers with decreased mechanical stress at rather low oxidation temperatures through hydrogenation of the near surface region of Si wafers prior to oxidation. We have established that thin SiO_2 layers can be grown on Si and hydrogenated by either RF plasma immersion ion implantation or RF plasma treatment, thus reducing the structural strain and the concentration of defects [4,5].

In the present paper, results from the study of the surface morphology of p-Si(100) wafers created by RF plasma immersion implantation (PII) of H^+ ions into a shallow silicon surface layer followed by its thermal oxidation at temperatures below 850 °C. From the analysis of the atomic-force microscopy (AFM) results, data on the surface amplitude parameters were obtained and considered in terms of the technological conditions. The AFM observations suggest the creation of a thin, less dense, Si surface region due to the formation of voids by the hydrogen particles incorporated into the growing oxide layer. The AFM results show that, during oxidation, a surface flattening takes place probably due to faster Si oxidation facilitated by the H^+ implants along hillocks and pits leading to a more homogeneous surface morphology.

2. Experimental details

The samples studied were Cz-grown p-type Si Wacker wafers with orientation (100) and a resistivity of 5 – 8 ohm cm. All wafers were subjected to a standard RCA cleaning procedure before any technological steps. Part of the samples were loaded into a planar plasma reactor and were exposed to plasma-beam ion implantation of hydrogen with an ion energy of 2 keV and fluences varying from 10^{13} to 10^{15} cm^{-2}. No external heating was applied to the Si substrates. The low energy of the H^+ ions (2 keV) allowed to implant them in a shallow depth in the Si substrate, namely, approximately within 20 – 30 nm inwards the Si surface. After implantation, possible surface contaminations and native oxide were removed by a short dip of the samples in diluted HF.

Further, the hydrogen implanted Si wafers were oxidized in a dry O_2 atmosphere for 30 minutes at temperatures of 700 °C and 750 °C and for 20 minutes at 800 °C. In the same oxidation runs, unimplanted Si wafers were simultaneously oxidized to serve as a reference. Choosing these oxidation durations and taking into account the oxide growth kinetics at standard dry oxidation, we expected that the oxide thicknesses will be in the same range.

We should emphasize that no further thermal treatment was applied to the studied samples.

The topography measurements were performed before and after oxidation of the Si wafers under ambient conditions on a XE-100 apparatus from Park Systems (2011) in non-constant force mode. The decoupled XY-Z scanners with out-of-plane motion of less than 1 nm over a 50-micron scanned area allowed measurement of the surface roughness with an accuracy of less than 1 nm. Ultra-sharp tips (< 8 nm tip radius; PPP-NCLR type from Nanosensors™) working with a spring constant of 48 N/m and a resonance frequency of ~190 kHz were used in the measurements. The sample surfaces were scanned over an area of 1×1 μm^2 and the XEI program was used for the statistical data analysis. From the 2D and 3D AFM imaging, the amplitude parameters, namely the peak-to-peak height, S_y, the root-mean-square (RMS) roughness, S_q, the surface skewness, S_{sk} and the surface kurtosis, S_{ku}, were determined.

3. Results and discussion

The thickness of the grown SiO_2 layers, determined by ellipsometric measurements with an accuracy of ± 0.2 nm, varied within 5 – 14 nm depending on the H^+ implant fluence and the oxidation temperature. This is illustrated in figure 1, where the SiO_2 thicknesses are summarized. In comparison to standard SiO_2 grown on unimplanted Si, thicker oxides grew into H^+ implanted Si as the thickness difference increased with raising the oxidation temperature. Obviously, introduction

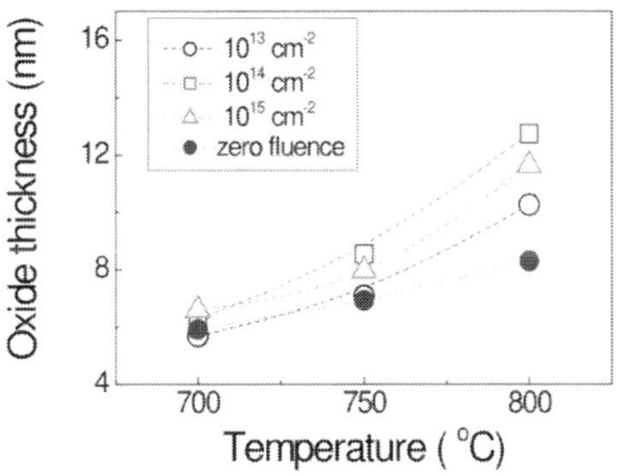

Figure 1. SiO_2 thickness versus oxidation time for (100) Si implanted with H^+ ions with different fluences (inset). For comparison purposes, the thickness of SiO_2 layers grown into unimplanted (100) Si wafers is also given. In the ellipsometric measurements, the thickness values are obtained with an accuracy of ± 0.2 nm.

of H^+ and the plasma radiation create favourable conditions for accelerated oxide growth kinetics.

The surface topographic study of bare Si wafer revealed a highly smooth good-quality surface with a RMS roughness of 0.15 nm typical for polished commercial Wacker wafers. It is well seen in figure 2a that the initial silicon surface is flat with few randomly distributed hillocks. Because the AFM measurement was made in air, the appearance of hillocks on the flat surface could be connected with native SiO_2 with a thickness of ~2 nm re-grown on Si top.

Low-energy shallow implantation of hydrogen ions into the wafers does not worsen essentially the Si surface flatness, but creates bending and a large number of deep pits. The influence of implantation becomes stronger when the H^+ fluence is higher. The overall effect is that Si surfaces subjected to PII implantation have larger amplitude parameters values with a trend to increasing as the ion fluence is increased. The AFM observations suggest the creation of a thin, less dense, Si surface region due to the formation of voids by the hydrogen particles incorporated into the near-surface Si region. The existence of a less dense region facilitates the diffusion of oxidants and, thus, a higher Si oxidation rate is only to be expected, as was observed (figure 1).

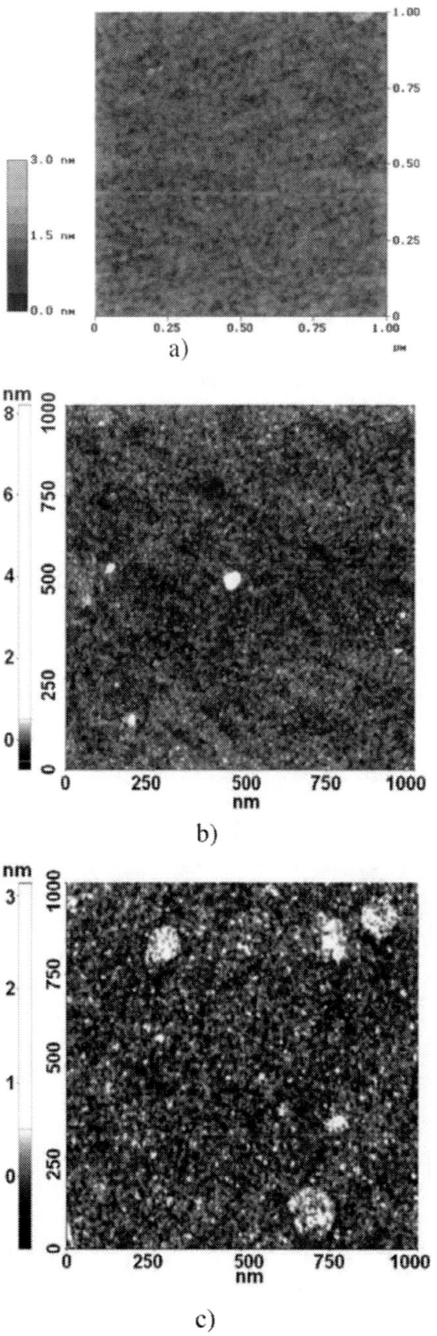

Figure 2. AFM 2D surface images of (100)Si substrate before (a) and after PII implantation of hydrogen with a fluence of 10^{14} ions/cm^2 (b) followed by thermal oxidation at 700 °C (c).

A typical AFM image taken from an implanted Si surface is presented in figure 2b for the H^+ fluence of 10^{14} ions/cm^2. A large number of pits are visualized on the considerably flat surface.

This is reflected in the slight increase of the RMS value to 0.265 nm, in the large negative skewness (S_{sk} = −14.139) and the extremely high positive kurtosis (S_{ku} = 350.591) values. The phase contrast made from the same AFM picture (not shown) pointed out that the big white spot in the middle of the image in figure 2b is dust. We would like to note that this image of dust on the surface is intentionally given for comparison purposes, because similar features appeared in many of AFM images but, as it turned out later, they could not be related to dust, as will be discussed below. The effect of oxidation on the surface morphology is illustrated in figure 2c and will also be discussed below.

The AFM amplitude parameters, calculated from the scanned area of the oxidized Si surfaces, showed a clear dependence on the H^+ fluence and the oxidation temperature. Their values are

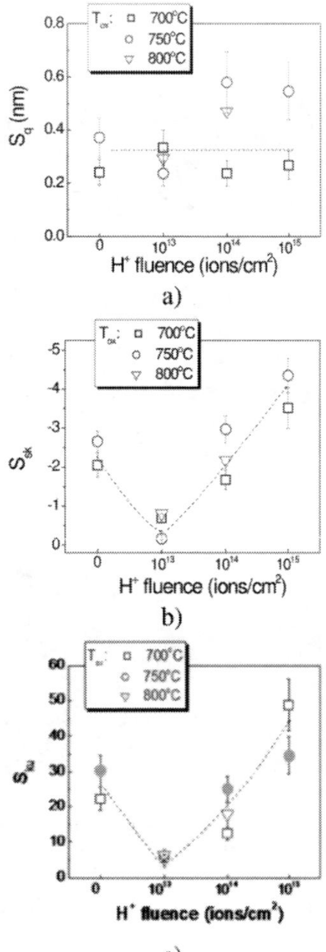

Figure 3. Dependence of the amplitude parameters, namely, root-mean-square (RMS) roughness, S_q (a), surface skewness, S_{sk} (b) and surface kurtosis, S_{ku} (c) of oxidized Si surface on the H^+ ion fluence; the oxidation temperatures are given in the inset.

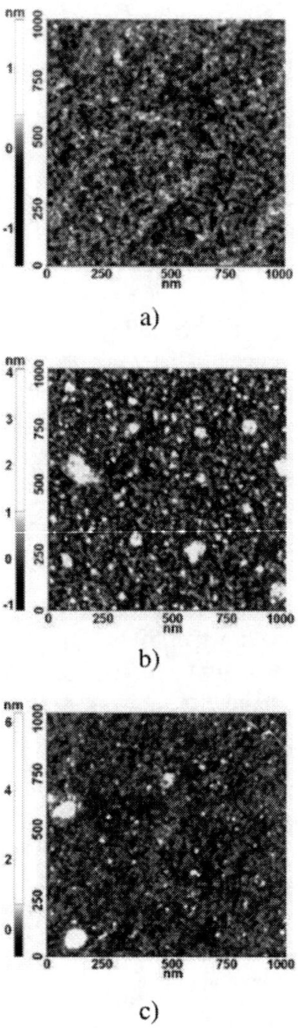

Figure 4. AFM 2D surface images of SiO_2 layers grown at 750 °C on (100)Si, implanted at different H^+ fluences: 10^{13} ions/cm^2 (a); 10^{14} ions/cm^2 (b) and 10^{15} ions/cm^2 (c).

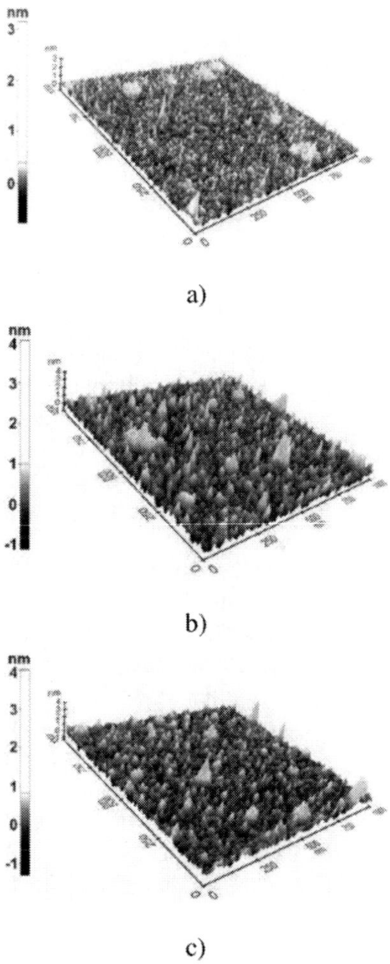

Figure 5. AFM 3D surface images from 1×1 μm^2 scanned surface area of SiO_2 layers into (100)Si, implanted with H^+ fluence of 10^{14} ions/cm^2. The oxides were grown at temperature of 700 °C (a), 750 °C (b), and 800 °C (c). The Δz scale is 3 nm (a) and 4 nm for (b) and (c).

summarized in figure 3. The surface of the SiO_2 grown on unimplanted Si wafers is considerably smooth. The large kurtosis, however, indicates a "spikiness" of the selected oxide surface.

Oxidation of the H^+ implanted Si region, as the oxide is growing inwards the Si, levels away the pits created by implants and results in a more homogeneous and smoother surface, albeit keeping the RMS values larger than those of the initial Si surface. The temperature treatment (oxidation) itself releases the structural strain caused by the implantation process, as it is suggested from the significant lowering of the 3D surface parameters (S_{sk} and S_{ku}) values. However, S_{sk} and S_{ku} still have considerably large values indicating a strong peakedness of the surface height distribution and the predominance of a valley structure comprising the surface.

For comparison purposes, figure 2c presents the 2D AFM image of a Si surface region PII implanted with a H^+ fluence of 10^{14} ions/cm^2 and oxidized at 700 °C. Randomly distributed hillocks and pits are observed on the rather flat surface (RMS = 0.26 nm). The phase contrast picture of the same AFM image revealed that the large-size spherical-like features could not be related to dust, as was the case in figure 2b, but rather to the material. They arise from the deformations of the implanted surface by thermal treatment (oxidation) and are H-induced surface blisters. Hydrogen implantation-induced surface blistering/exfoliation has been observed earlier [6] and has been further developed as an ion-cut process in direct wafer bonding techniques [7,8]. In our case, the effect of surface blistering is enhanced by increasing both the H^+ ion fluence (figure 4) and the oxidation temperature (figure 5). Although the accelerated oxidation process at elevated temperatures (figure 1) smoothens out the surface, the above-mentioned effect could lead to the minimum in the surface parameters values observed in figure 3.

As the H^+ fluence is increased, the surface roughness of the SiO_2 layer slightly increases (figures 3 and 4). The close peak-to peak (S_y) values and the more negative skewness (S_{sk}) indicate that the features on the surface are more randomly distributed. When the oxidation temperature is raised, the number of randomly distributed and sized hillocks and pits shows a tendency to increase (figure 5), leading to an increase of the surface roughness (a larger S_q) and a growing number of pits (a more negative S_{sk}).

The AFM results show that the temperature treatment (oxidation) releases the structural strains caused by the implantation process, as it is suggested from the significant lowering of the 3D surface parameters values. However, the S_{sk} and S_{ku} parameters still have considerably large values indicating a strong peakedness of the surface height distribution and the predominance of a valley structure comprising the surface. The negative values of the surface skewness are indicative for the existence of superficial voids, which may be caused by both implantation (ion bombardment) and surface reorganization (relaxation and re-ordering during thermal treatment).

4. Conclusions

The surface quality of the optimized ultra-thin oxide grown by preceding hydrogenation of Si wafers is an important milestone for further device fabrication. From the analysis of the AFM images of the samples surface, the following conclusions are drawn: (i) the surface of the RF PII implanted Si wafers have a relatively flat top and valleys; (ii) the subsequent thermal oxidation of the implanted Si wafers at temperatures of 700 °C, 750 °C and 800 °C reduces the number and size of pits on the surface but increases the average surface roughness as the temperature is raised and the valley structure comprising the surface still remains predominant. It is suggested that the surface morphology observed of H^+ implanted Si facilitates the diffusion of oxygen atoms into Si and, hence, leads to a larger oxidation rate and a lower level of oxide stress even at temperatures as low as 700 °C.

Acknowledgments

AFM measurements: the support of the EU (ERDF) and Romanian Government allowing the acquisition of the research infrastructure under POS-CCE O 2.2.1 project INFRANANOCHEM - Nr. 19/01.03.2009 is gratefully acknowledged.

Journal of Physics: Conference Series **514** (2014) 012036 doi:10.1088/1742-6596/514/1/012036

References

[1] Moutanabbir O, Terreault B and Giguère A 2009 *Phys. Status Solidi* C **6** 1958
[2] Simeonov S, Gushterov A, Szekeres A and Kafedjiiska E 2004 *Vacuum* **76** 303
[3] Chu P K 2002 *Surf. Coat. Technol.* **156** 244
[4] Alexandrova S, Szekeres A and Valcheva E 2012 *Phys. Status Solidi* C **9** 2203
[5] Alexandrova S and Szekeres A 2009 *J. Optoelectron. & Adv. Mater.* **11** 1284
[6] Kucheyev S O, Williams J S and Pearton S J 2001 *Mater. Sci. Engin.* **33** 51
[7] Bruel M 1995 *Electronics Lett.* **31**1201
[8] Christiansen S H, Singh R and Gösele U 2006 Wafer direct bonding: from advanced substrate engineering to future applications in micro/nanoelectronics *Proc. IEEE* December 2006

Journal of Physics: Conference Series **514** (2014) 012037 doi:10.1088/1742-6596/514/1/012037

Patterning of structures by e-beam lithography and ion etching for gas sensor applications

P Durina[1], A Bencurova[2], A Konecnikova[2], I Kostic[2,4], K Vutova[3], E Koleva[3], G Mladenov[3], P Kus[1] and A Plecenik[1]

[1]Department of Experimental Physics, FMFI, Commenius University, F2 Mlynská dolina, 84248 Bratislava, Slovakia
[2]Institute of Informatics, Slovak Academy of Sciences, 9 Dúbravská, 84507 Bratislava, Slovakia
[3]Acad. E. Djakov Institute of Electronics, Bulgarian Academy of Sciences, 72 Tsarigradsko Chaussee, 1784 Sofia, Bulgaria

E-mail: ivan.kostic@savba.sk

Abstract. This work deals with a method of preparation of nanometer structures for a gas detector based on e-beam lithography and ion etching of a thin TiO_2 film. The aim was the fabrication of a gas sensor with meander or comb structures of nanometer dimensions. This is of importance since both the size of the contacts and the layer thickness affect the sensor's sensitivity.

1. Introduction

In the past decade, the size of the features in ultra-large-scale integration has been continuously decreasing, leading to fabrication of nanostructures. Among all fabrication methods, the resist-based electron-beam lithography seems to be a suitable and flexible technique when preparation of nanostructures is desired, including in the development of sensors. The achievement of sub-100 nm structures using e-beam lithography is a very sensitive process determined by various factors, starting with the choice of a resist material and ending with the development process. The accurate definition of patterned structures is limited mainly by electron scattering, i.e., the well-known proximity effect [1]. To correct pattern distortions caused by electron scattering in the resist/substrate material, estimation of the lithographic parameters is required. The main factor influencing the accurate definition of patterned structures in the resist is backscattering of electrons from the substrate in the case of a low (30 keV) accelerating voltage.

Although electron backscattering can be simulated by the Monte Carlo method [2,3] or analytically [4], neither of these two methods considers the influence of the development process and equipment-related characteristics; experimental studies are, therefore, important concerning sub-100 nm structures.

2. Experimental

All experiments were conducted using the scanning electron microscope VEGA II SBH (TESCAN) equipped with a control system for nanolithography (TESCAN, Brno, Czech Republic). VEGA II

[4] To whom any correspondence should be addressed.

SBH is equipped with a thermal emission cathode, the accelerating voltage is in the range of 1 – 30 keV; the minimal probe size can be around 7 nm at a beam current of 10 pA. An accelerating voltage of 30 keV and a working distance of 5 mm were used to achieve the highest resolution. In most experiments, a spot size of 18 nm at a beam current of 40 pA was used due to the lower noise.

The other crucial factor in e-beam nanopatterning is the choice of materials sensitive to electron irradiation. Over the past decades, various types of resist materials have been investigated. Poly(methyl methacrylate) (PMMA) has been most widely used for high resolution e-beam lithography ever since the first report [5]. In brief, PMMA can be used to obtain reproducibly patterns with nm sizes. The resolution can be improved by optimizing various steps of the lithographic process, such as the writing strategy or the development process. The fabrication of very small structures (< 5 nm) utilizing this resist were reported in [6, 7].

Positive resist PMMA A2 (Microchem) with a molecular weight of 950 000 was chosen for nanostructure pattern transfer due to its high resolution, high contrast for EBL exposures, uniform resist coating, long shelf life, and good adhesion to most substrates. A thin layer of PMMA A2 resist with a thickness of 70 nm was prepared by spin coating at 3700 rpm on an Al_2O_3 substrate with a photolithographically pre-patterned platinum layer with a thickness of 20 nm. Subsequently, the resist was baked on a hot plate at 180 °C for 90 s. The exposure was carried out at 30-keV accelerating voltage and 40-pA beam current. After exposure, the resist was developed in MIBK:IPA (1:3) at 23 °C and served as a mask for ion etching of meander structures with sub-100 micrometers dimensions in a Pt layer.

Several exposure tests were performed in view of optimizing the exposure of PMMA A2 on a Pt layer. The contrast and sensitivity test consisted of large rectangles with the exposure doses being varied in the range from 10 μC/cm^2 to 120 μC/cm^2. The resist thickness in rectangles at various doses after development was measured using an Alphastep Stylus Profilometer.

The line tests allowed us to monitor the impact of the exposure dose on the linewidth of the exposed structures, to determine the optimal exposure dose needed for precise patterning and to identify the dimension limits of the patterned structures.

To determine the resistance of the PMMA resist to ion etching of the Pt layer, etch rate and etch selectivity tests were conducted. The ion etching was performed by argon ions from a PLATAR (KLAN-53M) ion source with ion energy E = 500 eV, ion current I = 20 mA, pressure of 10^{-2} Pa and 5 sccm; the etching time was 6.5 min for a 20-nm thick Pt layer.

3. Results

The sensitivity and contrast curve is shown in figure 1. The dose-to-clear D_o was 100 μC/cm^2, the contrast value γ being 4. The etch rate of PMMA A2 for etching with Ar ions is shown in figure 2.

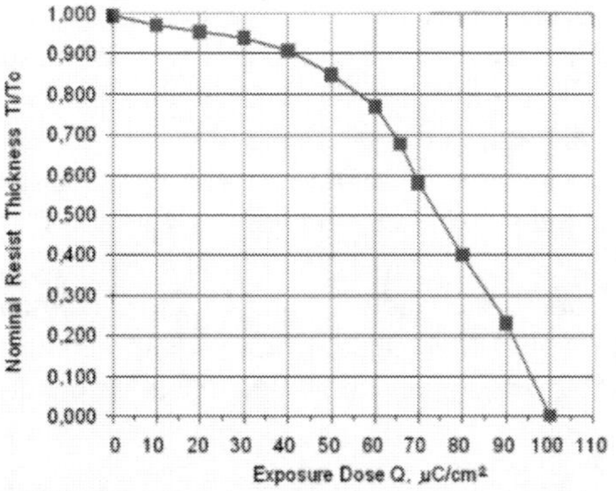

Figure 1. The sensitivity and contrast curve of the positive resist PMMA A2.

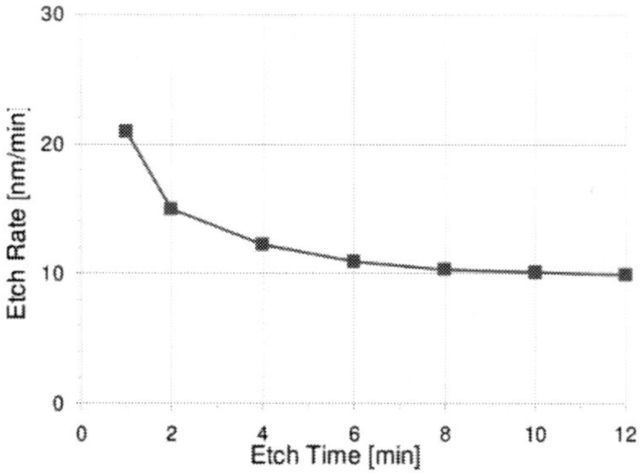

Figure 2. Etch rate of the positive resist PMMA A2 for ion etching.

The influence of the backscattered electrons in the case of a PMMA A2 positive resist is demonstrated in figure 3. The linewidth in the middle of periodical stripes is significantly narrower compared to that at the borders. In view of performing precise measurements, the stripes were made using the lift-off method in a 20-nm thick Pt layer using a PMMA A2 positive resist. The area of noticeable influence of the backscattered electrons is around 3 μm. The ends of the stripes are narrower by 20 % than in the middle.

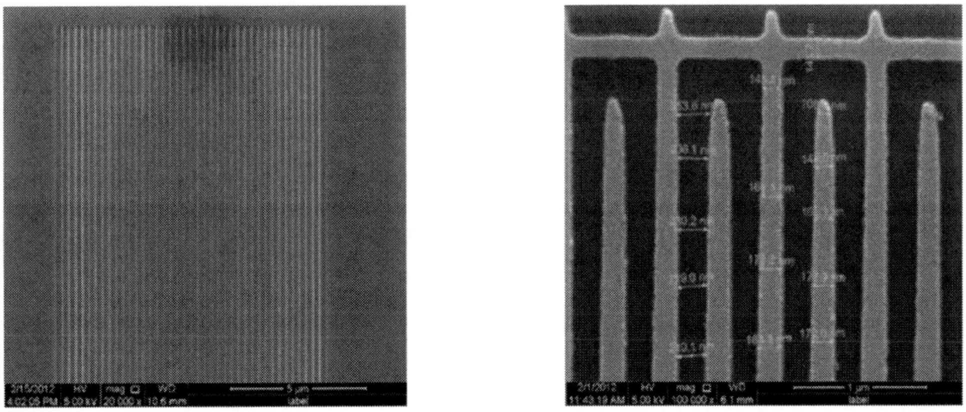

a) b)

Figure 3. Influence of backscattered electrons on the linewidth of stripe structures; a) the linewidth in the middle of periodical stripes is significantly narrower compared to that at the borders; b) stripes made using the lift-off method with a 20-nm thick Pt layer using a PMMA A2 positive resist.

The proximity effect parameters were simulated for the case of a gas sensor fabrication [8]: the thickness of the PMMA A2 resist and of the thin Pt layer on an Al_2O_3 substrate were 70 nm and 20 nm, respectively. All exposures were performed at 30-keV e-beam energy (point or Gaussian sources). Discrete data for the energy deposition function (EDF) at the interface (at a depth of 70 nm) were obtained by Monte Carlo simulation [2,3] using a point source for 10 000 particles. Analytical approximation of the discrete data obtained for the EDF was performed using a sum of two Gaussians. Afterwards, the proximity effect parameters (β_f, β_b, η_E) were determined at the interface for the point e-beam source, and for the Gaussian e-beam with δ_b=18 nm, by Monte Carlo methodology [2].

Results of the Monte Carlo simulation are presented in figure 4; the proximity effect parameters' data are summarized in table 1. A comparison of the EDF discrete data obtained (•) and its analytical fit (the curve) is shown in figure 4.

A proximity effect correction using these simulated parameters was applied to optimize the stripe and meander structures patterning in view of gas sensor development [8,9]. The minimal linewidth achieved of stripes exposed and developed in a PMMA resist was 77 nm with a space of 173 nm (figure 5a) and a 83-nm PMMA gap with a space of 177 nm at 30-keV electron energy and 18-nm probe size (figure 5b).

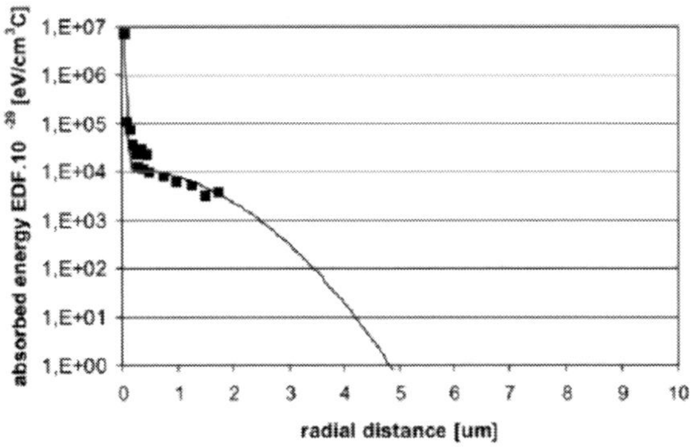

Figure 4. Comparison between the discrete data obtained (•) and its analytical fit (the curve).

Table 1. Proximity effect parameters' values.

Parameter	Point e-beam source	Gaussian e-beam source, δ_b = 18 nm
β_f	0.4216950000E-01 [μm]	0.4585048000E-01 [μm]
β_b	0.1577453000E+01 [μm]	0.1577556000E+01 [μm]
η_E	0.5816625000E+00	0.5816625000E+00

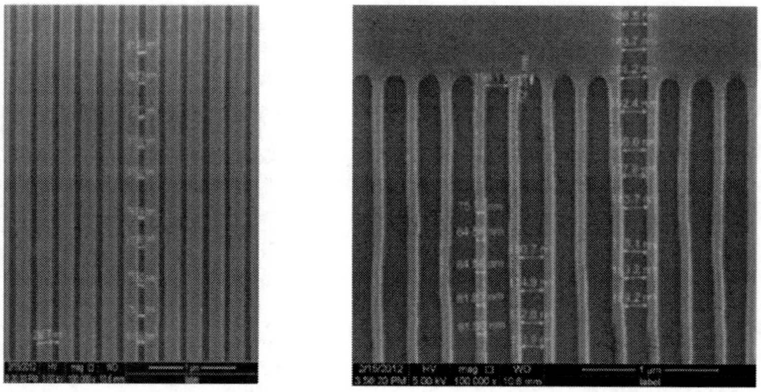

a) b)

Figure 5. a) Stripes with line/space 77 nm/173 nm in a PMMA A2 resist b) Stripes with line/space 167 nm/83 nm in a PMMA A2 resist. Exposure at 30-keV electron energy and 18-nm probe size, thermal emission.

Finally, the optimized e-beam lithography process was applied to fabricating the meander structure needed for the development of gas sensors. Meander structures of various dimension and configurations were etched into a Pt layer on an Al_2O_3 substrate. The platinum layer with a thickness of 20 nm was prepared by magnetron sputtering. Platinum bridge and contacts were prepared using photolithography and the lift-off method. The minimal linewidth of 85 nm achieved using thermal emission is shown in figure 6. The results of the measurements of the gas sensor characteristics were published in [8,9].

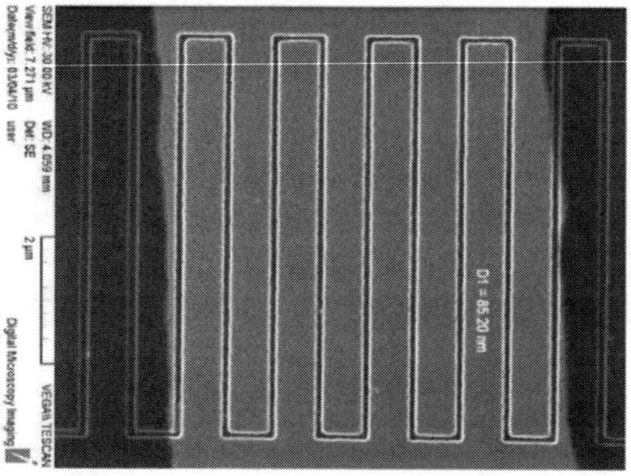

Figure 6. Meander structure patterned in thin platinum layer by electron beam lithography and ion etching using PMMA A2 (Microchem) positive resist. The channel width in platinum layer was 85 nm. The Pt layer thickness was 20 nm, prepared using magnetron sputtering.

4. Conclusions

The limitations of e-beam nanostructure patterning were investigated for a Gaussian e-beam at an electron energy of 30 keV in the case of thermal emission. The influence was evaluated of the backscattered electrons on the accurate definition of patterned structures in an e-beam resist. The proximity parameters (β_f, β_b, η_E) were simulated for a PMMA A2 positive e-beam resist on a Pt layer with a thickness of 20 nm. The smallest line/space achieved was 77 nm/173 nm.

The results of optimizing the e-beam lithography process were applied to the patterning of structures with a minimal linewidth of 85 nm in view of the development of gas sensors.

Acknowledgments
This publication was supported by the Ministry of Education of the Slovak Republic and the Slovak Research and Development Agency under contracts Nos. APVV-0520-10, APVV-0199-10, APVV-BG-SK-0037-10 and APVV-GR-SK-047-11; by the Bulgarian National Science Fund under contract DNTS/Slovakia01/10; the Research and Development Operational Program projects ITMS:26240220002, ITMS:26220220004 and ITMS:26240120026, and by the Ministry of Education of the Slovak Republic under contracts Nos. VEGA 1/0605/12 and VEGA-2/0021/12. The SEM images were taken by a SEM FEI Inspect F50 at the Institute of Electrical Engineering, Slovak Academy of Sciences in Bratislava and a Tescan VEGAII SBH at the Department of Physics, FMFI, Commenius University in Bratislava.

References

[1] Parikh M 1980 Proximity effects in electron lithography: magnitude and correction techniques *IBM J. Res. Develop.* **24** 530
[2] Vutova K and Mladenov G 1994 Modeling of exposure and development processes in electron and ion lithography *Modelling and Simulation in Mater. Sci. Engin.* **2** 239-54
[3] Vutova K and Mladenov G 2010 Computer simulation of processes at electron and ion beam lithography *Lithography* ed M. Wang (Vukovar InTech) part 1 chapter 17 319-50
[4] Raptis I, Glezos N, Valamontes N, Zervas E and Argitis P 2001 Electron beam lithography simulation for high resolution and high density patterns *Vacuum* **62** 263
[5] Haller I, Hatzakis M and Shrinivasan R 1968 High-resolution positive resists for electron-beam exposure *IBM J. Res. Develop.* **12** 251
[6] Wei C and Haroon A 1993 Fabrication of sub-100 nm structures by lift-off and by etching after electron-beam exposure of poly (methylmethacrylate) resist on solid substrates *J. Vac. Sci. Technol.* B **11** 2519
[7] Yasin S, Hasko D G and Ahmed H 2001 Fabrication of <5 nm width lines in polymethylmethacrylate resist using a water:isopropyl alcohol developer and ultrasonically-assisted development *Appl. Phys. Lett.* **78** 2760
[8] Owen G. 1990 Methods for proximity effect correction in electron lithography *J. Vac. Sci. Technol.* B **8** 1889-1892
[9] Durina P, Stefecka M, Roch T, Noskovic J, Trgala M, Pidik A, Kostic I, Konecnikova A, Matay L, Kus P and Plecenik A 2010 Patterning of nanometer structures by using direct-write e-beam lithography for the sensor development *ASDAM 2010 Proc. 8th Int. Conf. Advanced Semiconductor Devices and Microsystems* eds J. Breza, D. Donoval, E. Vavrinsky (Piscataway: IEEE) 89-92
[10] Durina P 2012 *Thesis* FMFI UK Bratislava

Journal of Physics: Conference Series **514** (2014) 012038 doi:10.1088/1742-6596/514/1/012038

New cyclometalated iridium(III) complex as a phosphorescent dopant in organic light emitting devices

P Ivanov[1,3], R Tomova[1], P Petrova[1], S Stanimirov[2] and I Petkov[2]

[1]Acad. J. Malinowski Institute of Optical Materials and Technologies, Bulgarian Academy of Sciences, Acad. G. Bonchev Str., Bl. 109, 1113 Sofia, Bulgaria
[2]Laboratory of Organic Photochemistry, Faculty of Chemistry and Pharmacy, St. Kl. Ohridski University of Sofia, 1 James Baurchier Blvd., 1164 Sofia, Bulgaria

E-mail: ivanov.petar@mail.bg

Abstract. A new cyclometalated iridium (III) bis[2-(4-chlorophenyl)benzothiazolato-N,C$^{2'}$]-acetylacetonate, $(Cl\text{-}bt)_2Ir(acac)$, was synthesized and identified by ^1H NMR and elemental analysis. The application was studied of the new compound as a dopant in the hole transporting layer (HTL) of the following organic light emitting diode (OLED) structure: HTL/EL/ETL, where HTL was 4,4'-bis(9H-carbazol-9-yl)biphenyl (CBP) or N,N'-bis(3-methylphenyl)-N,N'-diphenylbenzidine (TPD), incorporated in a poly(N-vinylcarbazole) (PVK) matrix; EL was an electroluminescent layer of bis(8-hydroxy-2-methylquinoline)-(4-phenylpheno-xy) aluminum (BAlq); and ETL was an electron-transporting layer of bis[2-(2-benzothiazoly) phenolato]zinc(II) $(Zn(btz)_2)$. We established that the electroluminescence spectra of the OLEDs at different dopant concentrations were basically the sum of the greenish-blue emission of BAlq and the yellowish-green emission of the Ir complex. It was also found that increasing the dopant concentration resulted in an increase in the relative electroluminescent intensity of the Ir complex emission, while that of BAlq decreased, thus a fine tuning of the OLED color was observed.

1. Introduction

Since Baldo et al. reported the first example of electrophosphorescence at room temperature; phosphorescent materials have attracted much attention due to their high external quantum efficiencies in OLEDs [1, 2]. Extensive investigations of this materials have been focused on heavy metal complexes of Os(II), Pt(II) and Ir(III) [3, 4]. A mixing of the singlet and triplet excited states of these d^6 complexes via spin-orbit coupling removes the spin-forbidden nature of the radiative relaxation of the triplet state, leading to high phosphorescence efficiencies [5]. In the past decade, we have witnessed an explosive growth in the interest in and application of luminescent cyclometallated iridium(III) complexes. The choice of iridium is of special interest, because: it has a quasi-octahedral geometry permitting the introduction of specific ligands in a controlled manner; iridium possesses stable and accessible oxidation and reduction states; the photophysical and electrochemical properties of iridium complexes can be tuned in a predictable way; cyclometallated iridium complexes are known for their highest triplet quantum yields (QY). The development of new iridium complexes is an area of

[3] To whom any correspondence should be addressed.

activity of many research groups. For instance, Thompson et al. synthesized a neutral emissive cyclometalated Ir(III) complexes [6] suitable as a phosphorescent dopant in OLEDs. The group of Grushin et al. reported a series of Ir(III) complexes with fluorinated 2-arylpyridines ligands [7], and Cheng et al., with substituted 2-phenylbenzothiazoles ligands [8]. Both of them showed that the complexes exhibit excellent processing and electroluminescent properties and emissive colors which can be fine-tuned via controlling the nature and position of the substituents on the aromatic rings and ligands. In this paper we present the synthesis and photophysical properties of a new benzothiazole-based iridium complex $(Cl\text{-}bt)_2Ir(acac)$ and discuss its application in phosphorescent OLEDs (PhOLEDs) as a triplet emitter that is solution processable in a poly(N-vinylcarbazole) (PVK) polymeric host.

2. Experimental

2.1. Instruments and measurements

^{1}H-NMR (600 MHz) spectra were recorded on a Bruker Avance II+ 600 device at room temperature. The spectra were referenced to the internal standard tetramethylsilane (TMS). The elemental analyses were conducted on an Elementar Vario EL III apparatus. The UV and fluorescence spectra of the complexes in solutions were recorded on a Thermo Spectronic Unicam 500 spectrophotometer and a Varian Cary Eclipse fluorescence spectrophotometer, respectively. All measurements of OLED structures were performed at room temperature and under ambient atmosphere, without any encapsulation. The electroluminescent spectra (El) and the CIE (Commission Internationale de l'Éclairage) coordinates were obtained by an Ocean Optics HR2000+ spectrometer. The current-voltage (*I-V*) curves were measured by a power supply programmable with Labview. The luminescence (*L*) was determined in a DC (direct current) mode and the light output was detected using a calibrated Hamamatsu silicon photodiode S2281-01. The electroluminescent efficiency (η_L) was calculated by equation (1) and used for quantifying the properties of the OLEDs.

$$\eta_L = L/I, \text{ cd/A}, \qquad (1)$$

where L is the luminescence (in cd/m^2) and I is the current density (in A/m^2).

*2.2. Synthesis of the iridium complex*Ligand 1 (figure 1) was obtained according to the procedure reported in [9]. The other initial reagents and solvents were purchased from Merck and used without any purification.

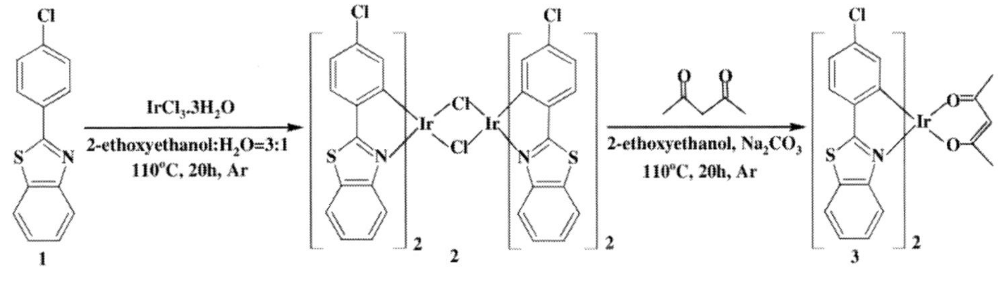

Figure 1. Synthesis of the iridium acetylacetonate complex $(Cl\text{-}bt)_2Ir(acac)$.

2.2.1. Synthesis of μ-chloro-bridged iridium dimmer (2). The cyclometalated Ir(iii) μ-chloro-bridged dimmer $(Cl\text{-}bt)_2$ $Ir(\mu\text{-}Cl)_2Ir(Cl\text{-}bt)_2$ was synthesized following a method reported by Nonoyama [10]. 2-(4-Chlorophenyl)benzo[d]thiazole (1) (2.2 equiv), iridium trichloride hydrate (1 equiv), dissolved in 2-ethoxyethanol: water = 3:1 were stirred for 20 h at 110 °C under argon. Then, the reaction mixture was cooled to room temperature and filtrated to obtain an orange-yellow precipitate. The precipitate was then washed with ethanol and hexane several times and dried at room temperature.

2.2.2. Synthesis of iridium acetylacetonate complex (3), $(Cl\text{-}bt)_2Ir(acac)$. Dimmer (2), 2,4-pentanedione (2.5 equiv) and Na_2CO_3 (5 equiv) were stirred for 20 h at 110 °C in 2-ethoxyethanol under argon. After cooling to room temperature, the product obtained was filtrated yielding an orange-yellow precipitate. Then the precipitate was washed with water, ethanol and hexane several times. Finally, the crude product was purified using liquid chromatography (silica gel, dichloromethane) and dried.

Iridium(iii)bis[2-(4-chlorophenyl)benzothiazolato-N,$C^{2'}$]-acetylacetonate, $(Cl\text{-}bt)_2Ir(acac)$: Yield: 80%. ^1H NMR (600 MHz, $CDCl_3$) δ[ppm]: 1.79 (s, 6H), 5.16 (s, 1H), 6.33 (d, J=1.8 Hz, 2H), 6.92 (dd, J_1=1.8 Hz, J_2=8.4 Hz, 2H), 7.46-7.51 (m, 4H), 7.61 (d, J=7.8 Hz, 2H), 7.94-7.95 (m, 2H), 8.04-8.06 (m, 2H); $C_{31}H_{21}Cl_2N_2S_2O_2Ir$. Elemental analysis: Calc. C 47.69%, H 2.71%, N 3.59%. Found: C 47.52%, H 2.68%, N 3.49%.

2.3. OLED fabrication

Devices with an area of 1 cm^2 were prepared on commercial polyethylene terephtalate (PET) substrates pre-coated with ITO (40 Ω/sq). The 30-nm composite films of PVK:CBP(or $TPD)_{10\ wt\%}$: $(Cl\text{-}bt)_2Ir(acac)_{X\ w\%}$ (relative to PVK) were formed by spin-coating from a 0.75% solution in CH_2Cl_2 at 2000 rpm. The films of BAlq (40 nm), $Zn(btz)_2$ (35 nm) and the Al cathode (120 nm) were thermally evaporated in vacuum better than 10^{-4} Pa at rates of 2 – 5 Å/s, as controlled by a quartz crystal microbalance.

3. Results and discussion

3.1. UV-VIS absorption and photo-luminescence

Figure 2 presents normalized UV-VIS absorption spectra of the complexes $(bt)_2Ir(acac)$ and $(Cl\text{-}bt)_2Ir(acac)$ and photoluminescence spectra (Pl) of: $(bt)_2Ir(acac)$, $(Cl\text{-}bt)_2Ir(acac)$, PVK and CBP dissolved in CH_2Cl_2. The absorption spectra of both complexes (curves 1 and 2) had almost identical shapes with peaks at 327 and 332nm (table 1), typical for this type of complexes.

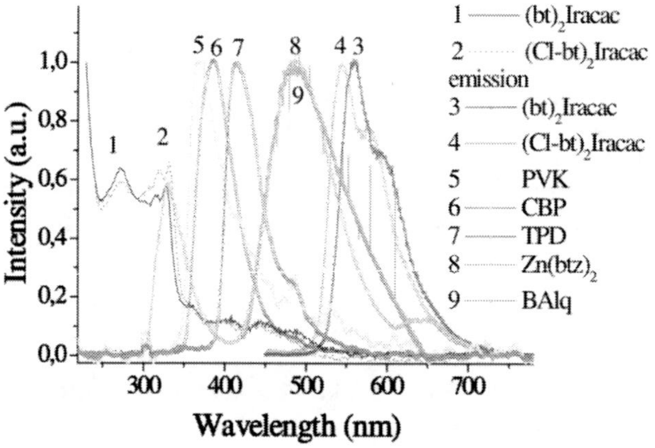

Figure 2. Normalized UV-VIS absorption spectra of: $(bt)_2Ir(acac)$ and $(Cl\text{-}bt)_2Ir(acac)$ and Pl spectra of $(bt)_2Ir(acac)$, $(Cl\text{-}bt)_2Ir(acac)$, PVK, CBP and TPD dissolved in CH_2Cl_2; and thermally evaporated films of $Zn(btz)_2$ and Balq.

The spectra can be considered as composed of two parts. The region below 360 nm contains bands of electron transitions, which leads to excitations in the bt-ligand. Most probably, the band around 330 nm is due to a transition to the lowest singlet ligand centered (^1LC) excited state. This statement is in accordance with papers discussing similar complexes published before. The shoulders above 360 nm are assigned to singlet and triplet metal-to-ligand charge transfer bands (^1MLCT, ^3MLCT), respectively. The effective spin-orbit coupling (SOC) in this complex leads to mixing of the singlet and triplet states, thus making the band of the resonant transition $^1S^0 \rightarrow {}^3MLCT$ detectable in the absorption spectrum (~480 nm). The band of the ^1MLCT transition is centered at 440 nm. Both complexes showed emission in the yellowish-green region (table 1) with peaks at 557 nm for $(bt)_2Ir(acac)$ and 543 nm for $(Cl\text{-}bt)_2Ir(acac)$ and shoulders, respectively, at 594 and 576 nm (curves 4 and 3). It is seen that the Pl spectrum of $(Cl\text{-}bt)_2Ir(acac)$ was blue-shifted with 15 nm with respect to the Pl spectrum of the unsubstituted $(bt)_2Ir(acac)$; this is typical for derivates containing electron withdrawing substituent [11, 12]. The PVK (curve 5), CBP (curve 6) and TPD (curve 7) emitted a blue

Journal of Physics: Conference Series **514** (2014) 012038 doi:10.1088/1742-6596/514/1/012038

light with peaks at 367, 385 and 414 nm respectively. This is why it could be expected that energy transfer from PVK and CBP (or TPD) to both complexes would be possible, since the ^{1}MLCT and ^{3}MLCT absorption spectra of the dopants had a broad spectral overlap with the fluorescence spectra of PVK and CBP (TPD). The Pl spectra of thermally evaporated neat $Zn(btz)_2$ and BAlq films shown in figure 2 (curves 7 and 8) were very close to each other, with comparatively broad peaks centered at $\approx$ 490 nm.

Table 1. Photophysical data of $(bt)_2Ir(acac)$ and $(Cl\text{-}bt)_2Ir(acac)$ in Ar-saturated CH_2Cl_2.

Complex	*Absorbtion λ[nm],(lgε)**	*Emission λ_{max}[nm]*	*QY*
$(bt)_2Ir(acac)$	272(3.93), 315(3.86), 327(3.88), 407(3.21), 442(3.19), 489(3.03)	558	0.50
$(Cl\text{-}bt)_2Ir(acac)$	272(3.89), 320(3.92), 331(3.94), 400(3.23), 442(3.18)	543	0.40

* ε is the molar absorption coefficient (molar absorptivity)

3.2. Electroluminescence

The electroluminescence spectra (El) of devices with two types of HTL with different concentrations of the dopant, taken at 16 V DC, are shown in figures 3 and 4. For comparison, the El spectra of devices with an undoped HTL and an EL of BAlq, $Zn(btz)_2$ and BAlq/$Zn(btz)_2$ are presented as well. It is seen that the device with the EL of BAlq emitted at λ_{max} 496 nm, while that with $Zn(btz)_2$, at 524 nm (figures 3 and 4). The shape of the El of the OLED with successively evaporated layers of BAlq/$Zn(btz)_2$ was just the same as that of the OLED with an EL of BAlq. In our opinion, it is due to the fact that in both cases the recombination zone is near the HTL/BAlq interface. As the $HOMO_{Zn(btz)2}$ level (5,05 eV) is lower than the $HOMO_{BAlq}$ (5,90 eV) level, $Zn(btz)_2$ does not stop the holes, which flow unimpeded from the $Zn(btz)_2$ layer. This is why, in the BAlq/$Zn(btz)_2$ structure, $Zn(btz)_2$ acts as an ETL only. The El spectra of the doped devices did not include any $Zn(btz)_2$, PVK and CBP (or TPD) emission and were basically the sum of the $(Cl\text{-}bt)_2Ir(acac)$ and BAlq emissions. Increasing the $(Cl\text{-}bt)_2Ir(acac)$ concentration caused a decrease in the relative intensity of the greenish-blue emission (496 nm) and an increase of the yellow emission (559 nm) of the devices. As the overlapping between the absorption spectra of the dopant and the Pl spectra of TPD (curve 7) was broader than that between the dopant and CBP (curve 6), a more intensive energy transfer to the dopant was observed in the devices with an HTL of PVK:TPD. Thus, at one and the same concentration, the ratio of the red-peak intensity (I_λ^{red}) to the blue peak intensity (I_λ^{blue}) was bigger for devices with TPD and the CIE coordinates of these devices were red-shifted compared to those with CBP (table 2).

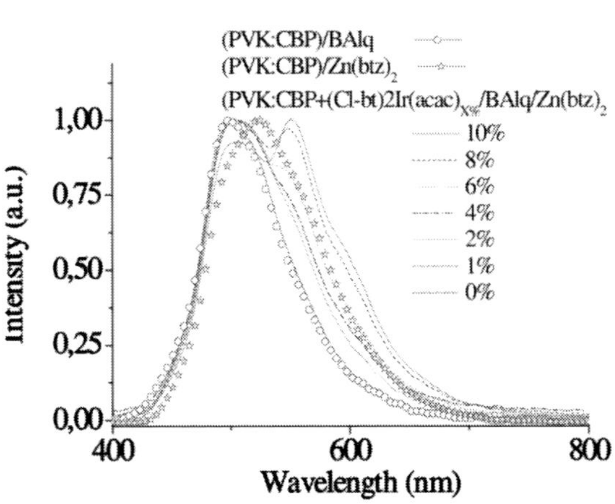

Figure 3. El spectra of devices with HTL of PVK:BCP doped with $(Cl\text{-}bt)_2Ir(acac)$ at 16 V DC.

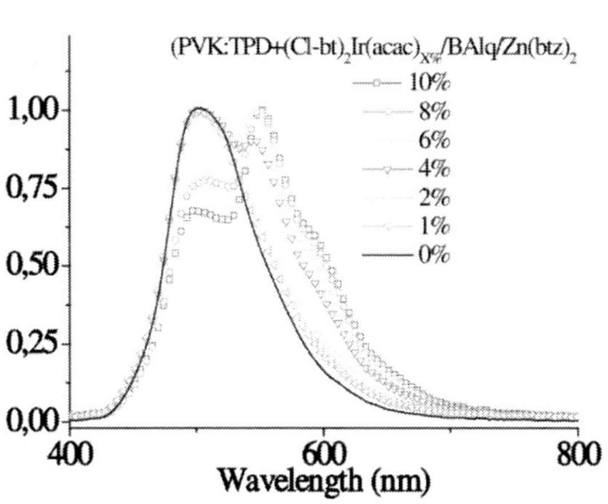

Figure 4. El spectra of devices with HTL of PVK:TPD doped with $(Cl\text{-}bt)_2Ir(acac)$ at 16 V DC.

Table 2. CIE coordinates and ratio of intensity of the red (I_λ^{red}) to the blue peak (I_λ^{blue}) for devices (PVK:CBP+(Cl-bt)$_2$Ir(acac))/BAlq/Zn(btz)$_2$ and (PVK:TPD+(Cl-bt)$_2$Ir(acac))/BAlq/Zn(btz)$_2$.

Dopant in layer, w%	CIE(x,y) PVK:CBP+(Cl-bt)$_2$Ir(acac)	$I_\lambda^{red}/I_\lambda^{blue}$ PVK:CBP+(Cl-bt)$_2$Ir(acac)	CIE PVK:TPD+(Cl-bt)$_2$Ir(acac)	$I_\lambda^{red}/I_\lambda^{blue}$ PVK:TPD+(Cl-bt)$_2$Ir(acac)
10%	(0.3590,0.4203)	1.09	(0.3689,0.4545)	1.50
8%	(0.3516,0.4013)	0.97	(0.3545,0.4492)	1.30
6%	(0.2936,0.4103)	0.83	(0.3220,0.4349)	1.02
4%	(0.2613,0.3888)	0.73	(0.3279,0.4557)	0.90
2%	(0.2552,0.3999)	0.67	(0.2632,0.4029)	0.69
1%	(0.2394,0.3824)	0.62	(0.2736,0.4348)	0.60
0	(0.2025,0.4348)	0.51	(0.2178,0.3711)	0.54

The influence of the dopant concentration on the electroluminescent performance of OLEDs with HTL of PVK:TPD and of PVK:CBP is illustrated in figure 5. We established that, at one and the same voltage, the devices containing CBP demonstrated a higher L in comparison to those with TPD. Increasing the dopant concentration from 0 to 4 w% initially decreased L and η_L of devices with CBP (respectively, from 664 to 480 cd/m^2 and from 2.32 to 2.05 cd/A); and in devices with TPD, from 440 to 410 cd/m^2 and from 2.19 to 2.08 cd/A. The further increase of the dopant concentration to 10 w% caused an increase of L and η_L to 685 cd/m^2 and 2.8 cd/A in devices with BCP, and to 575 cd/m^2 and 2.46 cd/A in devices with TPD. In addition, the El spectra and CIE of the devices with TPD, taken at one and same concentration, were very stable for a wide range of applied voltages (figure 6).

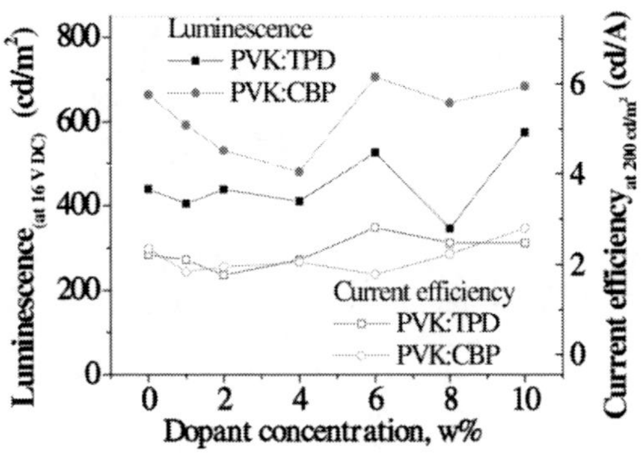

Figure 5. Luminescence (at 16 V DC) and η_L (at 200 cd/m^2) of devices with HTLs of PVK:TPD and PVK:CBP doped with different concentration of (Cl-bt)$_2$Ir(acac).

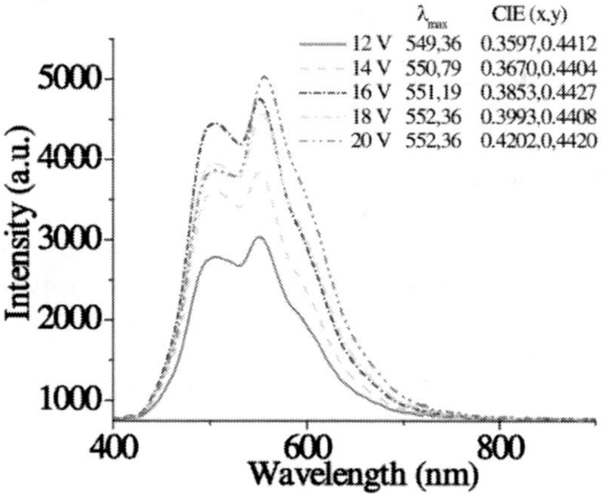

Figure 6. EL spectra at different voltages and CIE coordinates of devices doped with 6 w% (Cl-bt)$_2$Ir(acac) in HTL of PVK:TPD.

The electroluminescence in PhOLEDs originates either from the energy transfer from the host to the guest molecules, as described previously, or from charge trapping by the guest molecules, where the dopant acts as a trap site which can capture a charge carrier. As was shown by Diouf et al. [13], four main factors – exciton confinement, charge trapping, dopant concentration and molecular structure, directly affect the performance of PhOLEDs. These factors are dependent on each other and predominate individually depending on the situation. The weakly increase of the OLEDs efficiency as the dopant concentration is increased and the absence of any PVK, CBP (or TPD) emission in the El spectra of doped devices (figures 3 and 4), even at low concentrations of the dopant, are consistent with the charge trapping in the Ir complex, rather than the energy transfer, being the dominant mechanism in the OLEDs investigated.

4. Conclusions

A new iridium complex, $(Cl\text{-}bt)_2Ir(acac)$, emitting in the greenish-yellow range was synthesized and characterized as a dopant in the hole transporting layer of an OLED. It was established that a fine tuning of the emitted color from blue-greenish to orange can be achieved depending on the dopant concentration.

Including $(bt)_2Ir(acac)$ in electroluminescent devices increased the luminescence, tuned the color of emitted light in the yellow region, and raised more than twice the electroluminescent and external quantum efficiencies of the devices.

References

[1] Baldo M, O'Brian D, You Y, Shoustikov A, Sibley S, Thompson M and Forrest S 1998 *Nature* **395** 151
[2] Baldo M, Sibley S, Thompson M and Forrest S 2000 *Nature* **403** 750
[3] Xie H, Liu M, Wang O, Zhang X, Lee C, Hung L, Lee S, Teng P, Kwong H, Zheng H and Che C 2001 *Adv. Mater.* **13** 1245
[4] You Y, An G, Lee D, Kim J and Park S 2006 *J. Mater. Chem.* **16** 4706
[5] Baldo M, Lamansky S, Burrows P, Thompson M and Forrest S 1999 *Appl. Phys. Lett.* **75** 4
[6] Thompson M, Lamansky S, Djurovich P, Murphy D, Abdel-Razzaq F, Kwong R, Forrest S, Baldo M and Burrows P *US Patent 20020034656A1*
[7] Grushin V, Herron N, Lecloux D, Marshall W, Petrov V and Wang Y 2001 *Chem. Commun.* (Cambridge) 1494
[8] Chang W, Hu A, Duan J, Rayabarapu D and Cheng C 2004 *Organomet. J. Chem.* **689** 4882
[9] Deligeorgiev T, Kaloyanova S, Vasilev A and Vaquero J 2010 *Phosphorus Sulfur* **185**/11 2292
[10] Nonoyama M 1974 *Bull. Chem. Soc. Jpn.* **47** 767
[11] Chen C and Shi J 1998 *Coord. Chem. Rev.* **171** 161
[12] Wang R, Deng L, Zhang T and Li J 201 *Dalton Trans.* **41** 6833
[13] Diouf B, Jeon W, Pode R and Kwon J 2012 *Adv. Mater. Sci. Engin.* (Hindawi Publishing Corporation)

Effect of MeV electron irradiation on Si-SiO$_2$ structures

S Kaschieva[1,4], A Gushterov[1], Ch Angelov[2] and S N Dmitriev[3]

[1]Acad. G. Nadjakov Institute of Solid State Physics,
Bulgarian Academy of Sciences, 72 Tsarigradsko Chaussee, 1784 Sofia, Bulgaria
[2]Institute for Nuclear Research and Nuclear Energy,
Bulgarian Academy of Sciences, 72 Tsarigradsko Chaussee, 1784 Sofia, Bulgaria
[3]Flerov Laboratory of Nuclear Reactions, Joint Institute of Nuclear Research,
141980 Dubna, Moscow Region, Russia

E-mail: kaschiev@issp.bas.bg

Abstract: The effect of 20-MeV electron-beam irradiation on Si-SiO$_2$ structures was studied. The Si-SiO$_2$ samples were fabricated on n-type <100>-oriented Si wafers with resistivity of 4.7 Ω cm. Following a standard cleaning procedures, thermal oxidation in dry O_2 + 8 % HCl ambient was performed at 900 °C to produce oxide layer with a thickness of 20-nm. After oxidation, the samples were cooled at a rate of 1° C s^{-1} in the ambient where the oxidation was carried out. The samples were exposed to a beam of 20-MeV electrons with a flux of about 1.2×10^{15} cm^{-2}. The oxide thickness was measured by ellipsometry before and after electron irradiation, which showed that the SiO$_2$ thickness in the irradiated Si-SiO$_2$ structures increases. This result can be connected with the increase of the oxygen content at the Si-SiO$_2$ interface. If one takes into account the defect generation in the whole Si-SiO$_2$ structure, it is reasonable to expect that the oxygen motion through the SiO$_2$ oxide defects will be stimulated during MeV electron irradiation. The changes on the SiO$_2$ surface roughness induced by the high-energy electron irradiation of the Si-SiO$_2$ structures were observed by atomic force microscopy (AFM). It was seen that these changes consisted in the formation of nanocrystals at the SiO$_2$ surface. We assume that the MeV electron irradiation breaks the Si-O bonds, the free oxygen moves through the radiation defects and creates conditions whereby Si nanostructures are generated in the oxide SiO$_2$ and at the SiO$_2$ surface of the structures.

1. Introduction

The dissociation of SiO$_2$ layer in Si heterostructures using high-energy electron irradiation has been discussed by many authors [1-7]. Xi-wen Du et al. observed transformation of amorphous SiO$_2$ into crystalline Si (c-Si) by keV electron irradiation at ambient temperature [3]. The authors assumed that this transformation takes place in two steps: the first one involves the transformation of amorphous SiO$_2$ to amorphous silicon (α-Si), while the second one is the crystallization of amorphous silicon. They reported that α-Si nanodots are produced under keV electron irradiation, but c-Si nanodots are formed by heating the electron-irradiated Si heterostructure to 850 K [3]. But, the understanding of the dissociation of SiO$_2$ by high-energy electron irradiation is very limited and it is not clear whether c-Si nanodots can be obtained by electron irradiation only.

[4] To whom any correspondence should be addressed.

We present a study of the changes in amorphous SiO_2 layers during 20-MeV electron irradiation at room temperature. Using atomic force microscopy, we observed dramatic macroscopic changes on the SiO_2 after irradiation. At the same time, the oxide thickness of the samples increased (as evidenced by ellipsometry before and after MeV electron irradiation). These results correlate with our previous studies of SiO_2 layers of Si-SiO_2 structures during MeV electron irradiation [4, 5, 7]. We presumed that transformation of amorphous SiO_2 to crystalline silicon can be achieved also by MeV electron irradiation only. In this work, we investigate silicon nanostructures generation in the Si-SiO_2 structures by MeV electron irradiation. To study the effect of MeV electrons on formation of nanostructures at the SiO_2 surface, Si-SiO_2 structures were analyzed by AFM before and after electron irradiation with a flux of about 1.2×10^{15} cm^{-2}.

2. Experimental details

The studies were performed with n-type <100> oriented silicon wafers with resistivity of 4.7 Ω cm. The wafers were cleaned and then oxidized in dry O_2 + 8 % HCl at 900° C to produce an oxide layer of a thickness of 20 nm. The samples were cooled in the same oxygen ambient (in which oxidation was carried out) at a rate of 1° C s^{-1}. Then the samples were exposed to 20-MeV electron-beam irradiation perpendicular to the SiO_2 surface. This energy is sufficiently high for the electrons to penetrate through the entire SiO_2 and Si wafer and create radiation defects in the whole structure. The electron irradiation with a flux of 1.2×10^{15} cm^{-2} was carried out on the Microtron MT-25 (equipment in the Flerov Laboratory of Nuclear Reactions of the Joint Institute of Nuclear Research - FLNR, JINR, Dubna, Russia), in a vacuum chamber under pressure of about 1×10^2 Pa. The beam current was about I_e = 9 μA. The sample temperature was controlled during the entire process and kept close to room temperature. The distance between the Microtron window and the samples was 150 mm.

The oxide thickness of the samples was determined by ellipsometry before and after each electron irradiation. The morphology of the SiO_2 surfaces of the Si-SiO_2 structures before and after irradiation was observed by taking AFM images over areas of 1 μm^2 with scanning probe microscope CP-II in air at room temperature.

3. Results and discussion

The ellipsometric measurements showed that the oxide thickness of all samples increased after 20-MeV electron irradiation – from 20 nm (before) up to 21,7 nm after irradiation with a flux of 1.2×10^{15} cm^{-2}. This is in agreement with our earlier studies, when we found that this increase depends on the radiation dose and the ambient in which the irradiation is carried out [5]. A possible explanation of this result is the increase of the oxygen content at the Si-SiO_2 interface. In [6] we reported that, as a result of high-energy electron irradiation, the oxygen concentration at the Si-SiO_2 interface increases. Further, nuclear reaction analysis and X-ray emission spectroscopy demonstrated that MeV electron irradiation of Si-SiO_2 samples leads to an increase in the oxygen surface density. A possible mechanism of the radiation-stimulated oxidation may be oxygen motion through the radiation defects in the oxide layer in order to reach the silicon surface. If we take into account the defect generation in a result of MeV electron irradiation, it is reasonable to expect that the oxygen motion will be stimulated.

The microscopic morphologies of the SiO_2 surfaces of the Si-SiO_2 structures were observed by an atomic force microscope. Figure 1 is a typical AFM image of a SiO_2 film observed before irradiation. As can be seen, the sample did not exhibit any ordered surface structures.

After 20-MeV electron irradiation with the dose quoted, the AFM images revealed a roughing of the surface (shown in figure 2), with a formation of clusters, an increase in the crack density, crack healing, and, finally, a substantial modification, namely, the irradiation resulted in accumulation of nanostructures at the SiO_2 surfaces.

We suppose that the MeV electron irradiation breaks the Si-O bonds and the free oxygen moves through the radiation defects in the SiO_2 (as generated by the high-energy electrons), i.e., MeV electron irradiation creates conditions whereby Si nanostructures are generated in the oxide. This is in

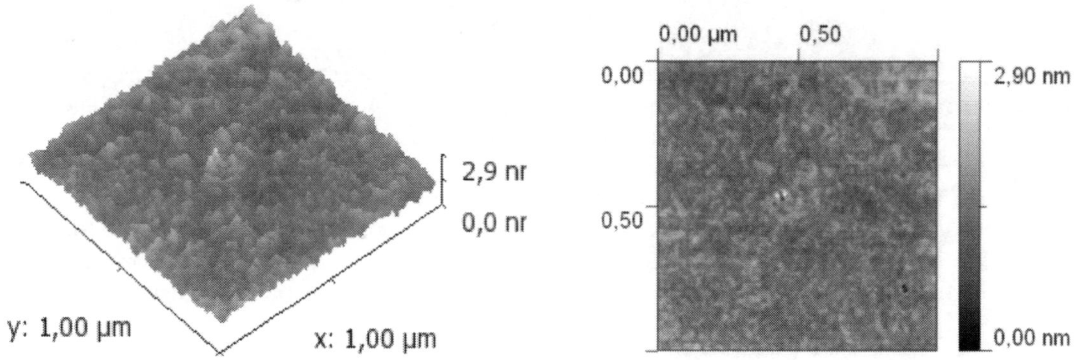

Figure 1. Typical AFM images of the Si-SiO_2 samples before electron irradiation.

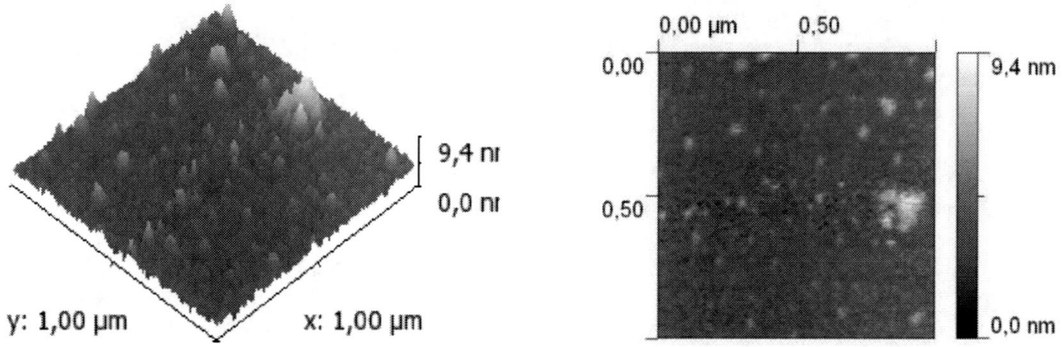

Figure 2. AFM results after 20 MeV electron irradiation with a flux of 1.2×10^{15} cm^{-2}.

good agreement with earlier results [4, 6]. We assume that the radiation defects (created by MeV electron irradiation) play also the role of sources of Si nanostructures generation in the SiO_2 [7]. Xi-wen Duet al. proposed that SiO_2 is first transformed into α-Si and then α-Si is transformed into c-Si during electron irradiation [1]. They showed the change in number and type of chemical bond during irradiation. The authors assumed that most of the α-Si islands are transformed in c-Si islands and suggested that the content of the Si-Si bond gradually increases from 0% in SiO_2 to 100% in pure Si as the time of irradiation is raised. We assume that our results (shown at figure 2) able to confirm their suggestion.

4. Conclusions

Using ellipsometry method and atomic force microscopy (AFM), it has been shown that the morphology of SiO_2 is changed as a result of the interaction between MeV electrons and the Si-SiO_2 structure. A possible reason for the observed oxide thickness increase of the irradiated samples is the increase of the oxygen concentration at the Si-SiO_2 interface. Our results demonstrate that the radiation defects (created by MeV electron irradiation) stimulate the oxygen motion through the oxide layer to reach the silicon surface, so that the oxide thickness increases and radiation-stimulated oxidation in a result of high-energy electron irradiation takes place. We assume that the radiation defects in the SiO_2 oxide act also as sources of Si nanostructures generation. MeV electron irradiation breaks the Si-O bonds, the free oxygen moves through radiation defects and creates conditions under which Si nanostructures are generated in the oxide SiO_2. Finally, we proved that nanocrystals can be created in the SiO_2 layer and at the SiO_2 surface of the Si- SiO_2 structures by MeV electron irradiation only.

Acknowledgement
This work was supported in part by Grant No. 80/20 (06.02.2012) of Joint Institute for Nuclear Research, Dubna, Russia.

References

[1] Klimenkov M, Matz W, Nepijko S A and Lehmann M *2001 NIMB* **179** 209
[2] Laha P, Banerjee I, Bajaj A, Chakraborty P, Barhai P K, Dahiwale S S, Das A K, Bhoraskar V N, Kim D and Mahaparta S K 2012 *Rad. Phys. Chem.* **81** 1600
[3] Du X-W, Takeguchi M, Tanaka M and.Furuya K 2003 *Appl. Phys. Lett.* **82**/7 1108
[4] Kaschieva S 1999 *Dr.Sci.Thesis* Dubna Russia (in Russian)
[5] Kaschieva S and Dmitriev S N 2003 *Vacuum* **69**/1-3 87
[6] Kurmaev E, Shamin S, Galakhov V, Kirilova M, Kurennykh T, Viykhodets V and Kaschieva S 1997 *J. Phys. Condens Matter.* **9** 6969
[7] Kaschieva S, Gushterov A, Angelov Ch and Dmitriev S N 2012 *J. Phys.: Conf. Series* **356** 012005

Characterization of thin MoO_3 films formed by RF and DC-magnetron reactive sputtering for gas sensor applications

R Yordanov[1], **S Boyadjiev**[1,2,4], **V Georgieva**[3] **and L Vergov**[3]

[1]Department of Microelectronics, Technical University of Sofia, 8 Kliment Ohridski Blvd., 1756 Sofia, Bulgaria
[2]Department of Inorganic and Analytical Chemistry, Budapest University of Technology and Economics, Technical Analytical Chemistry Research Group of the Hungarian Academy of Sciences, 4 Műegyetem rakpart, H-1111 Budapest, Hungary
[3]Acad. G. Nadjakov Institute of Solid State Physics, Bulgarian Academy of Sciences, 72 Tsarigradsko Chaussee, 1784 Sofia, Bulgaria

E-mail: boiajiev@gmail.com

Abstract. The present work discusses a technology for deposition and characterization of thin molybdenum oxide (MoO_x, MoO_3) films studied for gas sensor applications. The samples were produced by reactive radio-frequency (RF) and direct current (DC) magnetron sputtering. The composition and microstructure of the films were studied by XPS, XRD and Raman spectroscopy, the morphology, using high resolution SEM. The research was focused on the sensing properties of the sputtered thin MoO_3 films. Highly sensitive gas sensors were implemented by depositing films of various thicknesses on quartz resonators. Making use of the quartz crystal microbalance (QCM) method, these sensors were capable of detecting changes in the molecular range. Prototype QCM structures with thin MoO_3 films were tested for sensitivity to NH_3 and NO_2. Even in as-deposited state and without heating the substrates, these films showed good sensitivity. Moreover, no additional thermal treatment is necessary, which makes the production of such QCM gas sensors simple and cost-effective, as it is fully compatible with the technology for producing the initial resonator. The films are sensitive at room temperature and can register concentrations as low as 50 ppm. The sorption is fully reversible, the films are stable and capable of long-term measurements.

1. Introduction

MoO_3 is one of the most promising inorganic materials with a wide potential for technological applications as large-area electrochromic devices (displays and smart windows), optical switching coatings and high-density memory devices [1]. Thin MoO_x films can also be used for many other applications, such as catalysts [2], lubricants [3], solid-state microbatteries [4]. And, last but not least, MoO_3 was shown to be sensitive to NO_2 [5,6], NH_3 [7], CO [8], H_2 [9], volatile organic compounds (VOC) [10] and some other gasses. The aim of the present work was to the study the gas sensing properties of films prepared by RF and DC-magnetron sputtering. To this end, thin MoO_x and MoO_3 films were deposited on different substrates.

[4] To whom any correspondence should be addressed.

The quartz crystal microbalance (QCM) is a well-established technique for detecting physical properties of thin layers deposited on the crystal surface on the sub-nanogram level [11]. It could also be applied to monitoring the adsorption of nano-amounts of various toxic gases [10,11]. Compared with others sensors, the advantages of the QCM sensors are simple technological implementation, capability of operating at room temperature, good sensitivity and chemical process reversibility, low consumption and easy real-time monitoring.

The performance of the QCM sensors strongly depends on the properties of the sensing coating deposited on the QCM electrodes. Thin transition metal oxides films, such as SnO_2, WO_3, TiO_2 and ZnO, have been widely investigated for sensing different toxic gases. Our present research was focused on the sensing behavior of the sputtered thin MoO_3 films with respect to NH_3 and NO_2. Prototype QCM sensors with other sensitive transition metal oxides films were also fabriated by our team and tested for sensitivity to NH_3 in especially designed laboratory set-up [12,13]. Even in as-deposited state and without heating the substrates, these films, as well as the MoO_3 films investigated in the present work, show good sensitivity without an additional thermal treatment, so that manufacturing QCM gas sensors becomes simple and cost-effective, as their preparation is fully compatible with the technology for producing the initial resonator.

The thin MoO_3 films studied were characterized by X-ray photoelectron spectroscopy (XPS), X-ray diffraction (XRD) and Raman spectroscopy. The morphology of the films was studied by a high-resolution scanning electron microscopy (SEM). Thorough profile analyses on the structure were performed applying XPS, allowing interactions with the substrate and different modifications of the structure to be investigated. Laser ellipsometry was used to measure the thickness and to determine the refractive index of the films. The as-prepared gas sensors were then tested for sensing NH_3 in our laboratory-constructed system.

2. Experimental

MoO_3 films of various thicknesses (50 nm – 300 nm) were deposited on quartz resonators and silicon wafers. The deposition time varied between 10 and 60 minutes. For both RF and DC sputtering depositions, molybdenum targets were used and the sputtering was performed in the presence of oxygen as a reactive gas. Similar conditions were applied to allow for a better comparison of the methods. The films were deposited on unheated substrates and were not annealed.

The surface of the films was observed by a Hitachi S-4800 high-resolution field-emission SEM (FE-SEM) device. The structural properties of the films were characterized using a Philips X-Pert-MRD XRD system, and Raman spectroscopy was performed by an SPEX 1403 Raman double spectrometer. The composition of the films, accompanied by thorough profile analyses of the structure, was performed by an Ulvac-Phi Scanning ESCA Microprobe Quantum2000 XPS system, using a focused monochromatic Al Kα X-ray (1486,7 eV) source and a spherical section analyzer. The thickness and the refractive index of the films were measured by multi-angle four-zone null ellipsometry. Further information on the RF sputtering deposition technology and conditions, as well as on the Raman spectroscopy and laser ellipsometry equipment, can be found in our previous publications [14,15].

The sensing properties of the films were studied as the adsorption of NH_3 and NO_2 was tested. Most of the sensor structures were based on 8-mm polished AT-cut quartz plates with thermally evaporated golden electrodes (a diameter of 4 mm and a thickness of about 120 nm with a Cr underlayer) on both sides. A QCM resonance frequency around 14 MHz was thus obtained. The measurements are based on the correlation between the frequency shift and the additional mass loading of the resonator.

Sauerbrey [16] developed an empirical equation for AT–cut quartz describing the relation between the mass of a film deposited on the quartz crystal and the corresponding change in the resonance frequency, thus allowing the sorbed mass to be calculated. Other parameters of the QCM needed for the calculations, including the static capacitance, C_0, and the equivalent dynamic resistance, R_q, were measured by a selective level meter. Further, some parameters, as the dynamic capacitance, C_q, the dynamic inductance, L_q, and the quality factor, Q, were calculated [17].

The sorption properties of the MoO_3 films were determined by measuring the resonance frequency shift of the QCM structures covered with thin MoO_3 films at different NH_3 and NO_2 concentrations. This was carried out on a laboratory-constructed system, whose experimental unit, as well as the methodology of measurements, were described in detail in [12,13].

The main stages of the measurement consisted of: purging the QCM structure by dry air; creating a certain concentration of the gas measured; reaching saturation of the frequency values; purging the structure with dry air to recover the initial parameters of the sensor and prepare it for a new measurement. The NH_3 and NO_2 concentrations were controlled by mass flow controllers (MFCs) for the NH_3, NO_2 and diluting gas flows. The QCM frequency was registered by a Hameg 8123 frequency counter connected to the QCM and to a computer for data recording. The relative error in the frequency measurement was $\pm 5.25\times10^{-7}$. The experiments were carried out at 26 ± 0.2 °C.

3. Results and discussion

To observe the surface, whose quality is considered to be the most important factor determining the sorption properties of the films, we applied ultra-high resolution FE-SEM (figure 1). The films are uniform and homogeneous. They are amorphous-like and, in general, no grains with size larger than a few nanonometers are observed. The object on the surface shown in the image was only used to enable high-resolution focusing of the microscope.

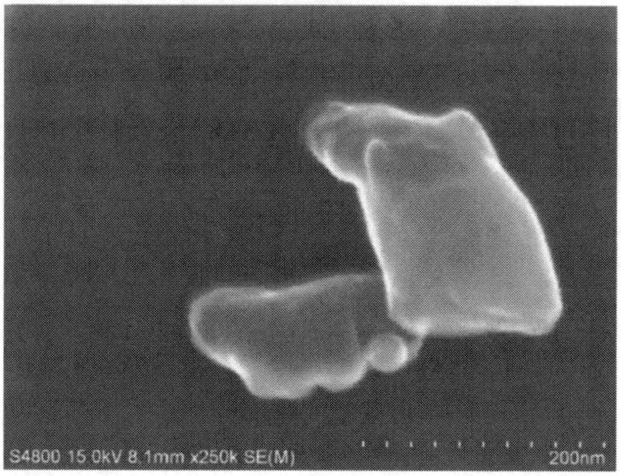

Figure 1. FE-SEM image of a DC-magnetron sputtered thin MoO_3 film at magnification 250 000×.

Our XRD research showed that both the RF and DC sputtered MoO_3 films are predominantly amorphous. These results were also supported by the Raman spectroscopy study. The results, as well as these from ellipsometry measurements, can be seen in our previous publications [14,15,18]. However, the XPS profile analyses showed that the structure changes between the surface and the bulk. In both the DC and RF sputtered film profiles, we observed close to stoichiometric MoO_2 throughout the film, almost without any significant changes in the composition. Yet, on the surface our investigations revealed a typical MoO_3 film. The Mo 3d core levels spectra are shown in figure 2. We observed sharp peaks without shoulders with shapes typical for MoO_3 [19,20]. The labels on the figure are for the $3d_{5/2}$ peaks. The doublet separation between the $3d_{3/2}$ and $3d_{5/2}$

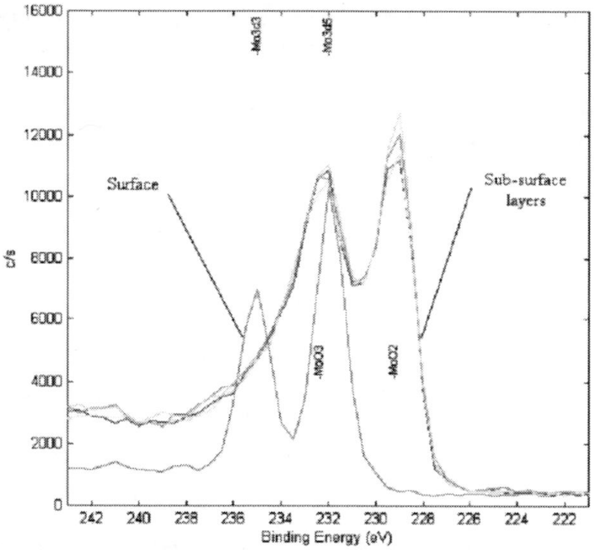

Figure 2. XPS spectra of Mo 3d core levels of a ~ 200-nm RF-sputtered thin MoO_3 film.

peaks of ~ 3,1 eV – 3,2 eV is also characteristic for MoO_3 [19-21]. The results from the compositional study of the same film are presented in figure 3. The profile was made at a sputtering energy of 2 kV with 2-minute sputtering intervals.

The Mo 3d peaks in the surface study are not positioned exactly at the most typical binding energies for the Mo^{6+} state, but at a slightly lower energy. They cannot be considered as being consistent with signals from Mo atoms in two different oxidation states, since they are sharp and without shoulders. It is more probable that the binding energy deviation of ~ 0,5 eV is caused by Mo interacting with C-impurities on the surface [19] (also detected in a considerable amount during the XPS study).

Although the sputtering during the etching of the surface could affect the structure, it is doubtful that it brought any significant changes. According to our previous experience with XPS profiles of other transition metal oxides, made at the same sputtering power, no changes are observed between the surface and the bulk [22].

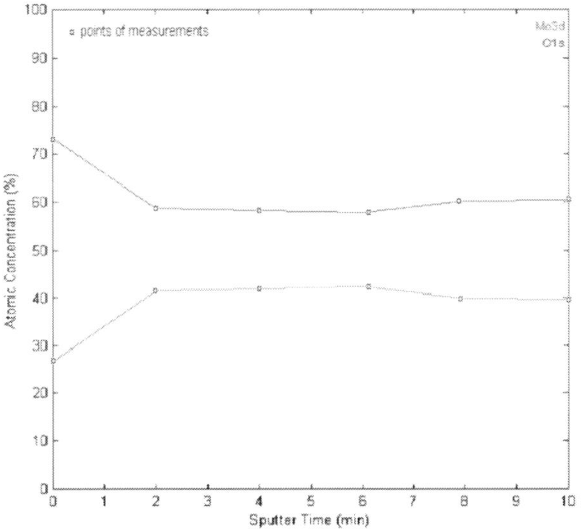

Figure 3. XPS compositional study of a ~ 200-nm RF-sputtered thin MoO_3 film.

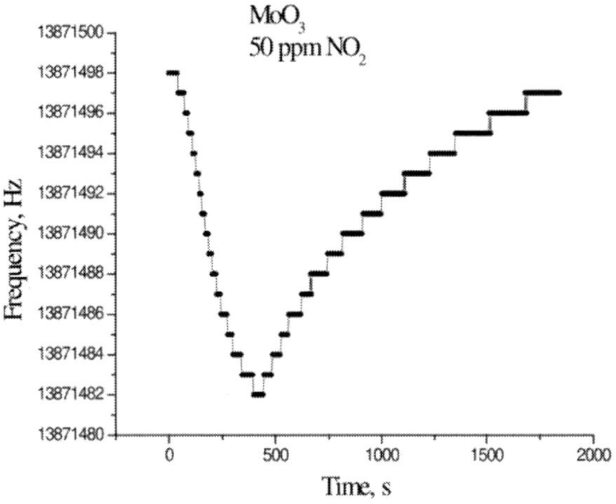

Figure 4. FTC of the MoO_3-QCM structure at 50 ppm NO_2 concentration.

Our prototype QCM sensors with MoO_3 thin films showed good sensitivity to NH_3 and NO_2. Figure 4 shows a frequency-time characteristic (FTC) of a MoO_3-QCM sensor produced by RF sputtering (together with the same MoO_3 film deposited on a referent Si wafer, whose characteristics are shown in figures 2 and 3). The FTC of the same sensor with respect to NH_3 sorption can be found in our previous publication [18]. The working frequency after loading the resonator with the MoO_3 film is about 13,87 MHz.

Both the NH_3 and NO_2 measurements reveal a very similar behavior of the sensors. When the measured gas is added to the gas flow, the sorption process starts and the frequency decreases quickly and almost linearly during the first several minutes. This process is followed by a slower decrease of the frequency until a dynamic equilibrium between the sorbed and desorbed molecules is reached, when the frequency remains constant. Turning off the measured gas flow and purging the sensor with dry air leads to a desorption process and unloading the sensor until the initial frequency is reached, thus proving the physical nature of the sorption process. This experiments show that the sorption process is reversible and the MoO_3-QCM sensor can be fully recovered without applying any additional energy.

The fact is very important that the sensors produced following the technology described exhibit a long-term parameters stability and can be used for a large number cycles. The first measurements with the RF sputtered sensor presented here were carried out more than five years ago. Its parameters and sensitivity practically have not changed during these years; we expect that it will continue operating for a long time. This fact, as well as the promising results concerning the sorption, allows these MoO_3 films to be used as gas sensors for on-line monitoring of NH_3 and NO_2 concentrations. The main advantage of the method described is that the technology for producing the sensitive films is fully compatible with the initial resonator preparation. Also, the films are able to detect even in as-deposited state and without heating the substrates. Additional thermal treatment is not necessary, which makes the manufacturing of QCM gas sensors easy and cost-effective. Further, they can be used at room temperature, which makes them suitable for portable systems for environmental control. Still, there remain problems mainly regarding the selectivity and the way to implement a large-scale manufacturing of commercial devices.

4. Conclusions

The properties of RF- and DC-magnetron sputtered thin molybdenum oxide films were studied. The films were amorphous or possibly nanocrystalline with grain size of several nanometres or less. Non-uniformity between the surface and sub-surface layers of the films was observed by XPS profiling. The films consist of stoichiometric MoO_3 with nano-size structure on the surface and amorphous nearly stoichiometric MoO_2 in the bulk.

The reactive sputtering method was found to be suitable for deposition of MoO_3 films for sensor applications. The prototype QCM sensors with MoO_3 sensitive films showed good sensitivity to NH_3 and NO_2 at room temperature and capability to register concentrations as low as 50 ppm. The sorption is fully reversible, the MoO_3 films are stable and capable of long-term measurements. The main conclusion based on the results obtained is that QCM structures covered with RF- and DC-magnetron sputtered thin MoO_3 films are suitable for NH_3 and NO_2 detection.

Acknowledgements

S. Boyadjiev is supported by the Postdoctoral Fellowship Program of the Hungarian Academy of Sciences (2013-2015).

References

[1] Granqvist C G 1995 *Handbook of Inorganic Electrochromic Materials* (Amsterdam: Elsevier)
[2] Honda K, Chen X and Zhang Z-G 2008 *Appl. Catal.* A*: Gen.* **351** 122
[3] Wang J, Rose K C and. Lieber C M 1999 *J. Phys. Chem.* B **103** 8405
[4] Julien C, Nazri G A, Guesdon J P, Gorenstein A, Khelfa A and Hussain O M 1994 *Solid State Ionics* **73** 319
[5] Ferroni M, Guidi V, Martinelli G, Sacerdoti M, Nelli P and Sberveglieri G 1998 *Sens. Actuators* B **48** 285
[6] Yamazaki T, Jin C, Nakayama A, Ito K, Yoshizawa T, Kikuta T, Nakatani N and Yamabuchi T 2005 *Jpn. J. Appl. Phys.* **44** 792
[7] Mutschall D, Holzner K and Obermeir E 1996 *Sens. Actuators* B **35-36** 320
[8] Ferroni M, Guidi V, Martinelli G, Nelli P, Sacerdoti M and Sberveglieri G 1997 *Thin Solid Films* **307** 148
[9] Hamagami J, Oh Y, Watanabe Y and Takata M 1993 *Sens. Actuators* B **13-14** 281
[10] Mirmohseni A and Rostamizadeh K 2006 *Sensors* **6** 324
[11] Cunningham A 1998 *Introduction to bioanalytical sensors* (New York: Wiley)
[12] Lazarova V, Spassov L, Gueorguiev V, Andreev S, Manolov E and Popova L 1996 *Vacuum* **47** 1423
[13] Georgieva V, Stefanov P, Spassov L, Raicheva Z, Atanassov M, Tincheva T, Manolov E and Vergov L 2009 *J. Optoelectr. Adv. Mater.* **11**/10 1363
[14] Boiadjiev S, Ivanova T, Rassovska M and Gesheva K A 2007 *J. Optoelectr. Adv. Mater.* **9**/2 264

[15] Boiadjiev S 2007 *Proc. Int. Congress on Ultrasonics* (2007 Vienna) p 1777
[16] Sauerbrey G Z 1959 *Physik* **155** 206
[17] Manolov S and Tihchev H 1982 *Generators* (Sofia Tehnika - in Bulgarian)
[18] Boyadzhiev S, Lazarova V, Rassovska M, Yordanova I and Yordanov R 2010 *Optoelectron. Adv. Mater. - Rapid Commun.* **4** 1485
[19] *Handbook of X-Ray Photoelectron Spectroscopy* 1979 ed Wagner C D, Riggs W M, Davis L E, Moulder J F and Mullenberg G E (Eden Prairie MN: Perkin-Elmer Corp.) p 104
[20] Fleisch T H and Mains G J 1982 *J. Chem. Phys.* **76**/2 780
[21] Anwar M, Hogarth C A and Bulpett R 1989 *J. Mater. Sci.* **24** 3087
[22] Boyadzhiev S I, Rassovska M M, Matsumura Y, Kotani Y, Yordanov R S and Yordanova I R 2010 *Proc. 33rd Int. Spring Seminar on Electronics Technology* (Warsawa 65)

Journal of Physics: Conference Series **514** (2014) 012041 doi:10.1088/1742-6596/514/1/012041

Deposition and characterization of thin HTS and magnetic perovskite films

B S Blagoev[1, 5], I G Gostev[1], T K Nurgaliev[1], V Strbik[2], I E Bineva[3], L Uspenskaya[4], E S Mateev[1], L Neshkov[1], E Dobročka[2] and Š Chromik[2]

[1]Acad. E. Djakov Institute of Electronics,
Bulgarian Academy of Sciences, 72 Tsarigradsko Chausse, 1784 Sofia, Bulgaria
[2]Institute of Electrical Engineering, Slovak Academy of Sciences,
9 Dúbravská cesta , 841 04 Bratislava, Slovak Republic
[3]Acad. G. Nadjakov Institute of Solid State Physics,
Bulgarian Academy of Sciences, 72 Tsarigradsko Chaussee, 1784 Sofia, Bulgaria
[4]Institute of Solid State Physics, Russian Academy of Sciences,
142432 Chernogolovka, Moscow region, Russia

E-mail: blago_sb@yahoo.com

Abstract. The *in-situ* growth and physical properties were investigated of thin films and bilayers of high-temperature superconductors (HTS) $YBa_2Cu_3O_7$ (YBCO) and ferromagnetic (FM) manganites $La_{0.7}Sr_{0.3}MnO_3$ (LSMO) and $La_{0.7}Ca_{0.3}MnO_3$ (LCMO) with thicknesses of several tens of nm obtained by magnetron sputtering. Similar twin structures were observed by an optical microscope in areas on the $LaAlO_3$ substrate surface void of films, in areas containing YBCO, in LSMO films, and in YBCO/LCMO bilayers, although the main spatial period of the twin structure seemed to be slightly different in the areas containing the LSMO film. The resistance (and its temperature dependence) of the LCMO films strongly depends on the annealing conditions. The resistance of the LSMO and LCMO films grown on Al_2O_3 substrates decreased as the temperature (T) was increased in the lower and higher temperature ranges, and increased as T was increased at medium temperatures.

1. Introduction

Perovskite materials demonstrate many interesting properties from both theoretical and applied points of view. Depending on their composition and structural properties, they can behave as dielectrics, conventional metals, high-temperature superconductors (HTS), colossal magnetoresistive materials (CMR), ferroelectric or multiferroic materials [1-3]. Some magnetic perovskites are of particular interest because of their ability to form high-quality epitaxial multilayers of HTS materials [2-7]. The compounds HTS $YBa_2Cu_3O_7$ (YBCO) and ferromagnetic (FM) $La_{0.7}Sr_{0.3}MnO_3$ (LSMO) and $La_{0.7}Ca_{0.3}MnO_3$ (LCMO) are considered as very attractive in view of technological applications in electronics, spin sensitive devices and spintronics [2, 3, 8]. The perovskites are multicomponent materials and their growth in a thin film form requires careful optimization processes. The electrical and magnetic characteristics of such thin films depend substantially on the deposition conditions and the

[5] To whom any correspondence should be addressed.

substrate properties [9-11], as lattice mismatch and defects strongly affect the film properties. The optimization of the deposition process of HTS and FM perovskite films with required characteristics is a challenging task.

The aim of this work is to study the growth conditions and the physical properties of FM and HTS perovskites in the form of both thin films and bilayers of LSMO, LCMO (manganites) and YBCO (HTS).

2. Experimental

Deposition of HTS YBCO and FM LSMO, LCMO films was performed by dual DC magnetron sputtering and single RF magnetron sputtering [12], respectively. The optimal concentration of the sputtering gas was Ar – 50 or 75 % and O_2 – 50 or 25 % for the DC and RF magnetron sputtering systems respectively. The optimal temperature of the substrate during the sputtering process was 780 °C for these materials. The sputtering process was followed by an annealing procedure of the sample in an O_2 atmosphere at T ~ 500 °C. The annealing conditions were varied in some of the experiments to find the optimal annealing conditions. To prepare HTS/FM bilayers, YBCO films were sputtered by a DC off-axis double magnetron system on top of RF-magnetron sputtered manganite films. Optical microscopy and atomic force microscopy (AFM) were used to investigate the surfaces of the films. The DC four-probe method and inductive and magneto-optical [7, 13] methods were used to study their electrical and magnetic properties.

3. Results and discussion

3.1. Characteristics of LSMO and HTS YBCO perovskite films grown on LAO substrates

Single-crystal substrates of $LaAlO_3$ (LAO) were used for growing HTS YBCO and ferromagnetic LSMO thin films. The lattice parameter of the LAO substrate (a = 0.3788 nm, pseudocubic) is close to that of LSMO (a = 0.387 nm, cubic) and YBCO (a = 0.382 nm, b = 0.388 nm, orthorhombic); this is why this substrate is suitable for the epitaxial growth of HTS YBCO and FM LSMO thin films. At room temperature, the LAO substrates exhibit a twin structure. It is formed by a ferroelastic (cubic-rhombohedral) transition occurring at ~ 813 K [2, 10] and serves as a template for the film microstructure. In spite of the twin structure, films grown under optimal conditions are characterized by a quite smooth surface (see for example the AFM scan in figure 1 of an LSMO film). Figure 2 shows an optical image (obtained in polarized light) of a film-free area of a LAO substrate (area 1) and of areas containing YBCO (area 2), LSMO (area 3) films and an YBCO/LSMO bilayer (area 4, YBCO is the top layer, LSMO is the bottom layer). It is interesting to note that nearly the same twin structure is observed in all areas of the sample, although the main period of the twin structure seems to be slightly different in the areas containing the LSMO component. Figure 3 shows a magneto-optical micrograph of a HTS YBCO film – a strip (width ~ 0.5 mm), deposited on a LAO substrate. The twin boundaries are not

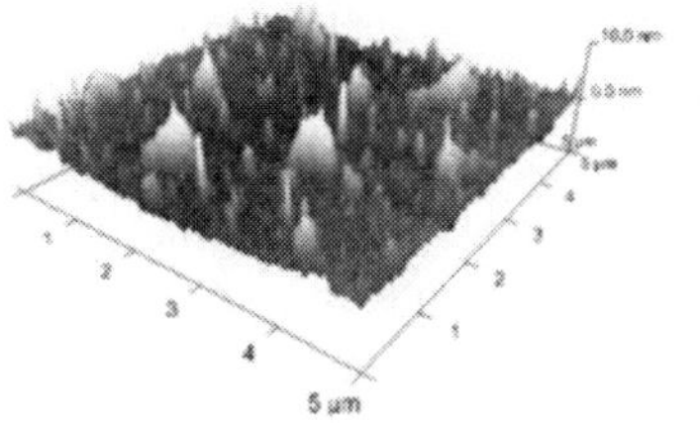

Figure 1. AFM image of a thin LSMO film deposited on a LAO substrate.

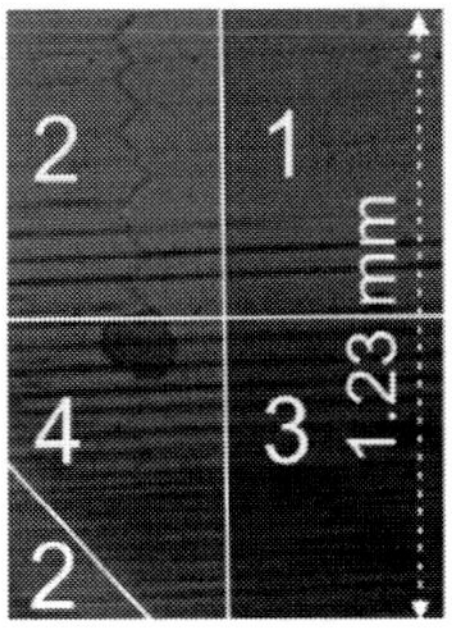

Figure 2. Polarized light optical microscopy image at T = 295 K of an YBCO / LSMO thin film deposited on a LAO substrate:
1 – film-free substrate area;
2 – YBCO film; 3 – LSMO film;
4 – area containing an YBCO/LSMO bilayer.

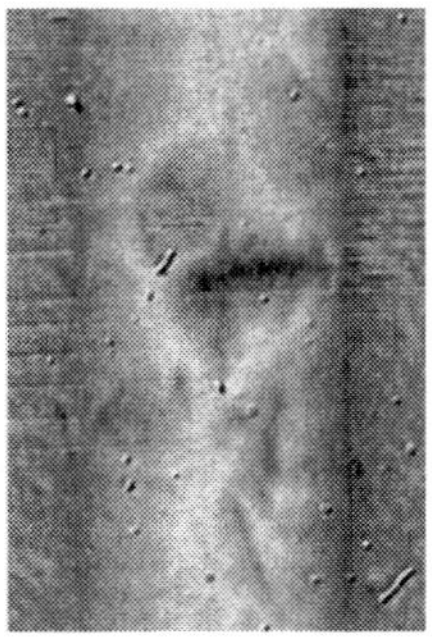

Figure 3. A trapped magnetic flux distribution pattern in the YBCO strip deposited on a LAO substrate at 77 K.

seen in the strip area (excluding the edge area of the strip and a defective area in the center of the strip). This means that the twin boundaries practically do not affect the "loops" of the electrical current (i.e. the boundaries do not "break" the paths of the electrical current) in the strip induced by the external DC magnetic field. The temperature dependence of the AC response signal, measured in a HTS YBCO thin film and in YBCO/LSMO bilayer (YBCO is the top layer and LSMO is the bottom layer) grown on a LAO substrate, are shown in figure 4. It is seen that the critical temperature of the YBCO thin film is quite high (~ 90 K), which is indicative of its good quality. The critical current density of such films is $J_c > 10^6$ A/cm^2 at $T = 77$ K. Such critical current densities are typical for high-quality HTS YBCO films of thickness of $d > 10$ nm, deposited on LAO substrates. The critical temperature of the YBCO film in the bilayer structure is somewhat lower ~ 85 K (figure 4, curve 3). This reduction is partly due to the effect of breaking the Cooper pairs in the HTS film, which leads to a deterioration of the superconducting properties of the film. The pair breaking is caused by spin-polarized charge carriers transported into the HTS layer by diffusion from the FM LSMO layer.

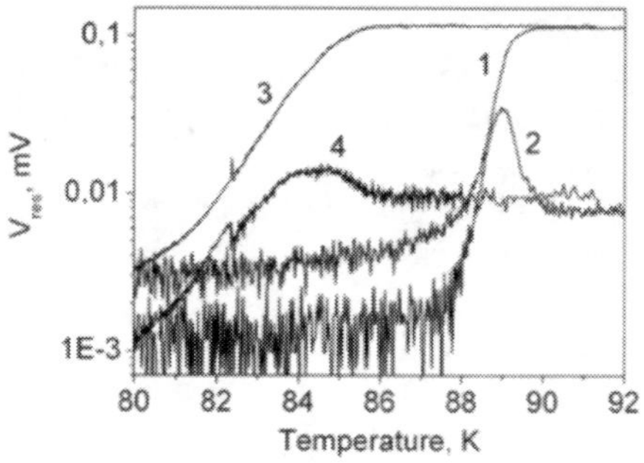

Figure 4. Temperature dependence of the real (1, 3) and imaginary (2, 4) parts of the AC response signal measured in a HTS YBCO single film (1, 2) and a YBCO/LSMO bilayer grown on LAO substrates. The thickness of the YBCO single film is ~ 60 nm, while the thicknesses of the YBCO and LSMO films in the double layer structure are ~ 40 nm and ~ 20 nm, respectively.

3.2. Annealing effect on the resistance of LCMO films grown on LAO and STO substrates

The RF magnetron sputtering process described above was used to prepare $La_{0.7}Ca_{0.3}MnO_3$ (LCMO, $a = 0.3868$ nm, $b = 0.3858$ nm, orthorhombic) films on $LaAlO_3$ and $SrTiO_3$ (STO, $a = 0.3905$ nm, cubic) substrates. The STO and LAO are the single-crystal substrates most commonly used for growing CMR manganite thin films. On the other hand, post-annealing in O_2 is very important for obtaining LCMO thin films with a good quality on such substrates. The temperature dependence of the resistance (curves 1), presented in figures 5 and 6, exhibits a rather insulating character thus indicating an oxygen deficiency in the film. An additional annealing procedure (at a temperature of about 720 °C,

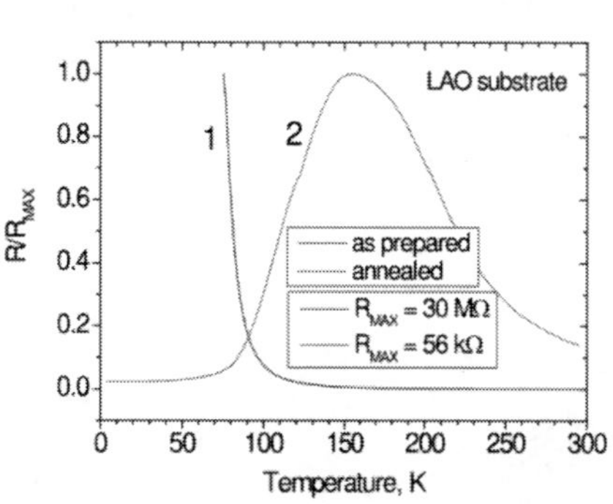

Figure 5. Temperature dependence of the resistance of a LCMO film (thickness ~ 20 nm) deposited on a LAO substrate: 1 – as deposited film; 2 – after annealing.

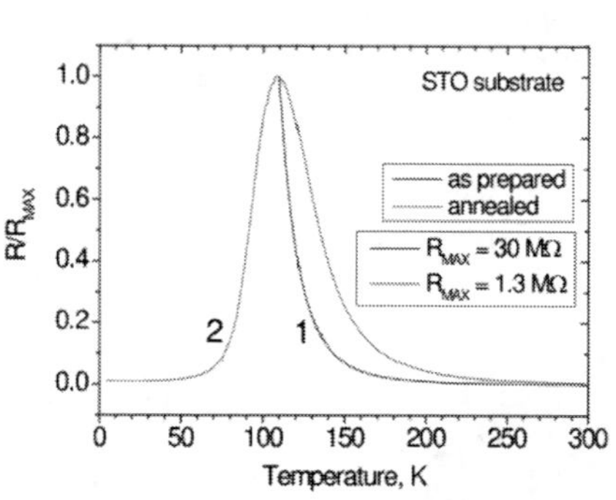

Figure 6. Temperature dependence of the resistance of a LCMO film (thickness ~ 20 nm) deposited on a STO substrate: 1 – as deposited film; 2 – after annealing.

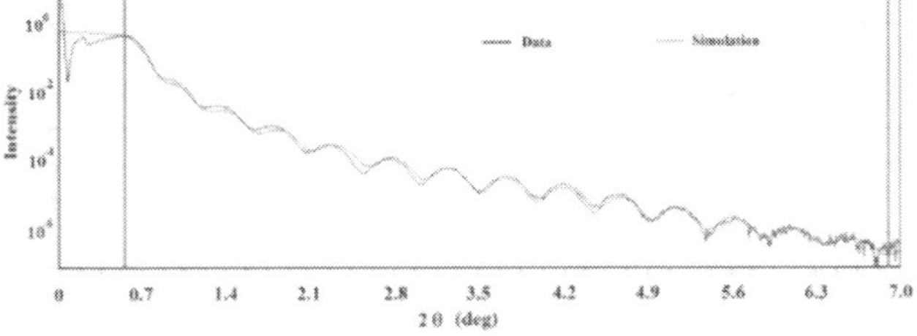

Figure 7. The X-ray reflectivity (Kiessing fringes) of a LCMO film deposited on STO with a simulation of the reflectivity.

Figure 8. AFM image of a LSMO thin film deposited on an ALO substrate.

oxygen pressure of about 35 mTorr and a duration of 10 minutes) drastically reduces the resistance and produces LCMO films with good CMR and structural properties (figures 5 and 6, curves 2 and figure 7). The X-ray reflectivity of LCMO film on STO is shown in figure 7. The oscillating dependence of the intensity indicates a high structural order, a flat surface and a uniform thickness of the film deposited. The simulation performed yielded an estimated film thickness of 18.16 nm and a surface roughness of 0.313 nm.

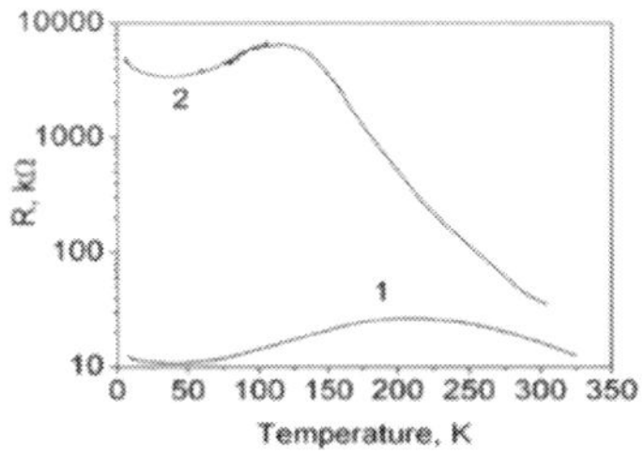

Figure 9. Temperature dependence of the resistance of LSMO (1) and LCMO (2) films deposited on ALO substrates.

3.3. LSMO and LCMO films deposited on ALO substrates

Sapphire (single crystalline Al_2O_3, ALO) is an inexpensive and readily available substrate material. On the other hand, it is not possible to grow epitaxial films of high-T_c superconductors and CMR manganites on ALO substrates because of the significant lattice mismatch and because of the possibility of certain chemical reactions to occur between the film and the substrate. The LSMO and LCMO manganite films in our experiments were directly deposited on ALO substrates by RF magnetron sputtering under the conditions described above. The films thus obtained were of polycrystalline nature and consisted of very small grains with dimensions, possibly, of several tens of nanometers (figure 8). The resistance of these films varied in a wide range as a function of the temperature T (figure 9). The resistance decreased as the temperature was increased in the lower and higher temperature ranges, and increased with the increase of T at medium temperatures (figure 9). A significant change of the resistance and its temperature dependence could be obtained by varying the annealing conditions. Such a behavior of the R vs. T dependence in LSMO and LCMO films grown on ALO substrates could be used to implement sensor-like devices.

4. Conclusions

The *in-situ* growth conditions and the physical properties were investigated of magnetron-sputtered high-temperature superconducting (HTS) $YBa_2Cu_3O_7$ (YBCO), magnetic manganite $La_{0.7}Sr_{0.3}MnO_3$ (LSMO) and $La_{0.7}Ca_{0.3}MnO_3$ (LCMO) thin films and bilayers with thicknesses of several tens of nm. Similar twin structures were observed by optical microscopy in the film-free area of the $LaAlO_3$ substrate and in the areas containing YBCO, LSMO films and YBCO/LCMO bilayers, although the main spatial period of the twin structure seemed to be somewhat different in the areas containing LSMO films. Magneto-optical images of the magnetic flux distribution in the YBCO films revealed that the twin structure modulates the critical current density, but does not interrupt the current flow. The resistance (and its temperature dependence) of the LCMO films grown on $LaAlO_3$ and $SrTiO_3$ substrates strongly depends on the annealing conditions. The resistance of the LSMO and LCMO films grown on Al_2O substrates decreased as the temperature was raised in the lower and higher temperatures ranges, and increased as T was raised in the medium temperatures range.

Acknowledgement
This work was supported by under the VEGA project 2/0173/13.

References

[1] Wu M K, Ashburn J R, Torng C J, Hor P H, Meng R L, Gao L, Huang Z J, Wang Y Q and Chu C W 1987 *Phys. Rev. Lett.* **58** 908
[2] Haghiri-Gosnet A-M and Renard J-P 2003 *J. Phys.* D: *Appl. Phys.* **36** R127
[3] Dörr K 2006 *J. Phys.* D: *Appl. Phys.* **39** R125
[4] Bozovic I, Logvenov G, Verhoeven M A J, Caputo P, Goldobin E and Geballe T H 2003 *Nature* **422** 873
[5] Sefrioui Z, Varela M, Pena V, Leon C, Santamaria J, Villegas J E, Martinez J L, Saldarriaga W and Prieto P 2002 *Appl. Phys. Let.* **81** 4568
[6] Albrecht J, Soltan S and Habermeier H-U 2005 *Phys. Rev.* B **72** 092502
[7] Gozzelino L, Laviano F, Przyslupski P, Tsarou A, Wisniewski A, Botta D, Gerbaldo R and Ghigo G 2006 *Supercond. Sci. Technol.* **19** S50
[8] Anders S, Blamire M G, Buchholz F-Im, Crete D-G, Cristiano R, Febvre P, Fritzsch L, Herr A, Il'ichev E, Kohlmann J, Kunert J, Meyer H G, Niemeyer J, Ortlep T, Rogalla H, Schurig T, Siegel M, Stolz R, Tarte E, Brake H J M, Toepfer H, Villegier J-C, Zagoskin A M and Zorin A B 2010 *Physica* C **470** 2079
[9] Chromik S, Strbík V, Dobrocka E, Dujavová A, Reiffers M, Liday J and Spanková M 2013 *Appl. Surf. Sci.* **269** 98
[10] Lehmann A G, Sanna C, Lampis N, Congin F, Concas G, Maritato L, Aruta C and Petrov A Yu 2007 *Eur. Phys. J.* B **55** 337
[11] Garcia-Hermandez M, Guinea F, de Andres A, Martinez J H L, Prieto C and Vazquez L 2000 *Phys. Rev.* B. **61** 9549
[12] Nurgaliev T, Strbik V, Miteva S, Blagoev B, Mateev E, Neshkov L, Benacka S and Chromik S 2007 *Cent. Eur. J. Phys.* **5** 6371
[13] Polyanskii A, Emergo R L S, Wu J Z, Aytug T, Christen D K, Perkins G K and Labalestier D 2005 *Phys. Rev.* B **72** 174509

Journal of Physics: Conference Series **514** (2014) 012042 doi:10.1088/1742-6596/514/1/012042

Electrical transport in epitaxial and polycrystalline thin LSMO films

V Štrbík [1, 3], B Blagoev [2], E Mateev[2], and T Nurgaliev[2]

[1]Institute of Electrical Engineering, Slovak Academy of Sciences,
9 Dúbravská cesta , 841 04 Bratislava, Slovak Republic
[2]Acad. E. Djakov Institute of Electronics,
Bulgarian Academy of Sciences, 72 Tsarigradsko Chaussee, 1784 Sofia, Bulgaria

E-mail: elekstrb@savba.sk

Abstract. We report studies on the electrical transport in $La_{0.7}Sr_{0.3}MnO_3$ (LSMO) thin films of different quality. The temperature dependence of the resistivity (ρ-T) of high-quality LSMO films ($\rho \lesssim 10$ mΩ cm) can be described by a polynomial approximation based on electron-magnon, electron-phonon, electron-electron and temperature-independent scatterings. On the other hand, the ρ-T dependence of polycrystalline LSMO films prepared under less than optimal conditions ($\rho \gtrsim 1$ Ω cm) can be described by a two-conductance-channel model. The first one is a spin-polarized tunneling between neighboring ferromagnetic grains with good contacts, while the other is implemented via a thermally-activated transport through an insulator (semiconductor) phase located mainly at the grain boundaries or between grains.

1. Introduction

In what concerns the family of manganites, a large number of studies have been focused on $La_{0.7}Sr_{0.3}MnO_3$ (LSMO) as it possesses the highest Curie temperature [1-3]. The manganites exhibit a large variety of electrical transport properties depending on their quality. At low-temperatures ($T < 10$ K), the LSMO resistivity has values from $\rho_0 \approx 60 - 80$ $\mu\Omega$ cm for high-quality epitaxial films or single crystals [4,5] or up to $\rho > 100$ Ω cm for very thin films (< 10 nm) [2], non-stoichiometric or not-optimally oxidized LSMO [5,6], or polycrystalline (granular) thin films [7]. A resistivity maximum appears in the temperature dependence of the resistivity (ρ-T) at a certain temperature T_{MI} indicating a metal-insulator transition and a simultaneous ferromagnetic (FM) – paramagnetic (PM) transition. To interpret the ρ-T dependence of high-quality LSMO thin films ($\rho \lesssim 10$ mΩ cm), the electron-magnon interaction, electron-phonon and electron-electron scattering, as well as scattering on impurities and defects, should be taken into account [4,8,9]. The ρ-T dependence of granular (polycrystalline) composite or not-optimally prepared LSMO thin films ($\rho \gtrsim 1$ Ω cm) can be described by a model of two intergrain conduction channels [10,11]. The first channel is spin-polarized tunneling between neighboring ferromagnetic grains with a good contact, the other, a thermally-activated transport through an insulator (semiconductor) phase situated at the grain boundary or between grains.

[3] To whom any correspondence should be addressed.

In this paper we present a possibility to approximate and characterize the transport mechanisms involved in the ρ-T dependences of LSMO thin films of different quality. A polynomial fit equation describes more adequately epitaxial LSMO films of high quality, while the two-channel model is better suited to the description of polycrystalline and/or less oxidized films.

2. Experimental

RF magnetron sputtering was used to deposit ferromagnetic $La_{0.7}Sr_{0.3}MnO_3$ thin films. During the sputtering in $Ar:O_2$ (1:1) atmosphere, with a total pressure of 5.3 Pa, a RF power supply of 30 W and a substrate temperature of 780°C were applied. The thickness of the LSMO layer was about 30 nm. The LSMO films deposited on $SrTiO_3$ single-crystal substrates showed high crystallographic quality, a high T_{MI} value (342 K), ferromagnetic properties at room temperature and low resistivity [12], whereas the LSMO films deposited on r-cut Al_2O_3 single-crystal substrates exhibited polycrystalline character, low T_{MI} values (below 250 K) and a high resistivity ($\rho \geq \Omega$ cm) [13].

The patterning of the basic structure was carried out by a photolithographic process and argon-ion-beam etching with cooling the substrate. After the patterning, the DC 4-point measuring method was used.

3. Results and discussions

A typical ρ-T dependence of our low-resistivity LSMO films is shown in figure 1. The values of the metal-insulator transition T_{MI} = 342 K, $\rho_{max} \leq 5$mΩ cm and the low-temperature residual resistivity $\rho_0 \approx 150$ μΩ cm are comparable with other high-quality epitaxial thin films [4,14]. We approximated the ρ-T dependence in the ferromagnetic-metallic state ($0 < T < T_{MI}$) by a polynomial fit equation $\rho(T) = \rho_0+\rho_2T^2 + \rho_N(T - T_0)^N$, where ρ_0, ρ_2, ρ_N, T_0 and N are fitting parameters [9]. ρ_0 represents the residual resistivity at low temperatures (figure 1, inset) and is determined predominantly by temperature-independent scattering on defects and impurities, i.e., ρ_0 indicates the quality of the LSMO film. The second term ($\sim T^2$) characterizes the electron-electron [5] or single-magnon scattering [15].

The first two terms of the polynomial equation describe well the ρ-T dependence up to a temperature of about 0.5 T_{MI} (figure 1, inset). The third term represents mainly the contribution from the electron-magnon interaction according to the FM double-exchange theory ($\rho \sim T^{4.5}$) [16]. This interaction (figure 1, dotted line) cannot explain the increase of resistivity at temperatures close to T_{MI} alone; thus, additional contributions from electron-phonon ($\rho \sim T^5$) [4,17] or from polaron-phonon scattering [8] should be taken into account. We fitted (figure 1, solid line) the experimental ρ-T dependence in a ferromagnetic-metallic state by the polynomial expression with the parameters ρ_0 = 0.141 mΩ cm, $\rho_2 = 1.15\times10^{-5}$ mΩ cm K^{-2}, T_0 =117 K and ρ_N = 5.97×10^{-13} mΩ cm K$^{-5.5}$ for N = 5.5.

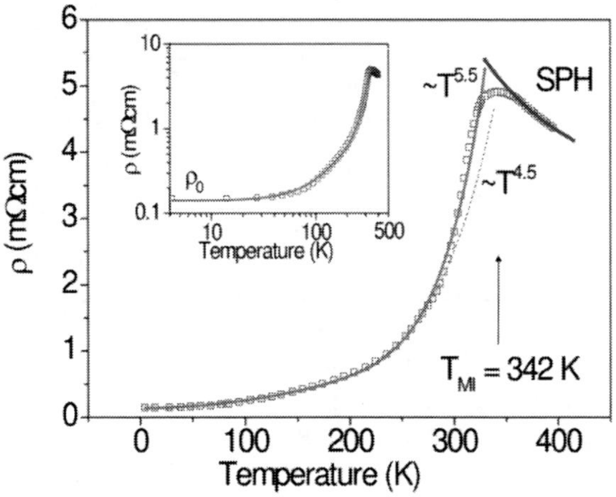

Figure 1. Approximation of the resistivity vs. temperature dependence of a high-quality LSMO thin film by a polynomial expression (solid line) in a ferromagnetic-metallic state ($0 < T < T_{MI}$) and by the small polaron hopping (SPH) model (dot-dashed line) in a paramagnetic-insulating state ($T > T_{MI}$). The dotted line indicates the contribution to resistivity by the double-exchange theory; the ρ-T dependence in a log-log scale is shown in the inset.

The electrical transport in a paramagnetic-insulating state ($T > T_{MI}$) is determined by the small polaron hopping (SPH) mechanism $\rho(T) = \rho_h T\exp(E_h/k_B T)$, where E_h is the hopping energy, k_B is Boltzmann's constant and ρ_h is a coefficient [18,9]. The SPH approximation with parameters $\rho_h = 1.538\times10^{-3}$ mΩ cm K^{-1} and $E_h = 1.076\times10^{-20}$ J is shown in Fig.1 as dot-dashed line for $T > T_{MI}$.

The ρ-T dependences of polycrystalline LSMO films can be interpreted in the framework of the phenomenological model of two intergrain conduction channels developed for granular or composite ferromagnetic materials [10,11]. The first type of channel is related to intrinsic transport properties of the system and is implemented through spin-polarized tunneling between neighboring ferromagnetic grains with good contacts. The other type is realized through thermal activation of the insulator phase arising from from the grain boundary, contaminations, crystal disorder and poor connectivity between grains. The total resistivity I is determined by the sum of the conductances of the two channels:

$$\frac{1}{\rho} = \frac{A_f}{\rho_f} + \frac{A_i}{\rho_i},$$

where A_f and A_i are effective constants and fit parameters. The dimensionless resistivity ρ_i of the insulator phase channel is modeled by a semiconductor-like resistivity

$$\rho_i = e^{W_2/(2k_B T)},$$

and the resistivity ρ_f of the channel responsible for spin-polarized tunneling, is

$$\rho_f = e^{[E_c + W_1(1-m^2)]/(2k_B T)},$$

where k_B is Boltzmann's constant, T is the absolute temperature, W_2 is the insulator-phase activation energy, E_c is the charging energy, $W_1(1\text{-}m^2)$ is the magnetic correlation function of two neighboring grains of the same volume and shape, W_1 is a temperature- and magnetic-field-independent constant and m is the magnetization normalized to the saturation value for each grain.

The numerical fit (solid line) of the experimental ρ-T dependence (dots) according to the two-channel model for three different LSMO films is shown in figure 2.

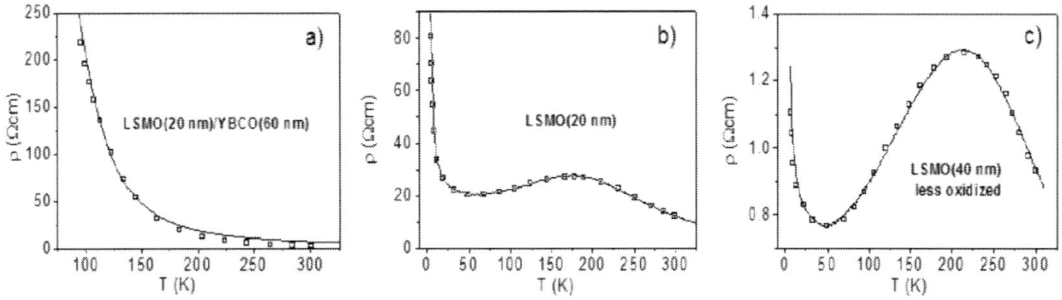

Figure 2. Numerical fit (solid line) of the experimental ρ-T dependence according to the two-channel model for three different LSMO films: a) polycrystalline LSMO film covered with non-superconducting YBCO b) thin polycrystalline film c) less-oxidized polycrystalline LSMO film.

The insulator behavior of the ρ-T dependence for the polycrystalline LSMO covered with non-superconducting YBCO (figure 2a) can be explained mainly by the transport channel with a thermal activation of carriers in an insulator phase (ρ_i) because no ferromagnetic contribution was registered. In the other two cases (Fig. 2 b, c), the ρ-T dependence is a combination of insulating and ferromagnetic phases. To approximate the experimental ρ-T dependences, it is necessary to find adequate fitting parameters for the two channels to match the experimental dependence. As fitting parameters we used A_f, A_i, W_1, W_2, E_C and T_{Curie} [10]; their values are shown in table 1.

The two-channel model is suitable for approximating the ρ-T dependences in the cases when manganite exhibits a low-temperature resistivity upturn.

Table 1. Fitting parameters of the two-channel model for the polycrystalline LSMO films in figure 2.

Sample	A_f (Ω^{-1})	A_i (Ω^{-1})	$W_1/2k_B$ (K)	$W_2/2k_B$ (K)	$E_C/2k_B$ (K)	T_{Curie} (K)
figure 2a	60000	15000000	50	600	50	340
figure 2b	8500	500000	400	1180	7	320
figure 2c	527.5	18000	442	1310	2.5	340

4. Conclusions

We presented approaches to approximating the ρ-T dependences of LSMO thin films. A polynomial fit formula describes properly the ρ-T dependence of high-quality epitaxial LSMO films ($T_{MI} > 300$ K, $\rho \leq$ mΩ cm, $\rho_0 \approx 150$ µΩ cm) in a ferromagnetic-metallic state ($T < T_{MI}$). The polynomial fit formula includes the following transport mechanisms: scattering on impurities and defects, and electron-electron, electron-magnon and electron-phonon scattering. The electrical transport in a paramagnetic-insulating state ($T > T_{MI}$) can be described by the small polaron hopping model. In the case of the LSMO films of high resistivity ($\rho \geq \Omega$ cm) and with a low-temperature resistivity upturn, the two-channel model describes well the experimental ρ-T dependences. One channel characterizes the thermally-activated transport through the insulator at the grain boundary, the other one describes the transport through spin-polarized tunneling between neighboring ferromagnetic grains with good contacts.

Acknowledgement

This work was supported by the Slovak Grant Agency VEGA under projects No. 2/0173/13 and 2/0164/11.

References

[1] Carretero-Genevrier A, Gázquez J, Idrobo J C, Oró J, Arbiol J, Varela M, Ferain E, Rodríquez-Carvajal J, Puig T, Mestres N and Obradors X 2011 *J. Am. Chem. Soc.* **133** 4053
[2] Boschker H, Kautz J, Houwman E P, Siemons W, Blank D H A, Huijben M, Koster G, Vailionis A and Rijnders G 2012 *Phys. Rev. Lett.* **109** 157207
[3] Chromik Š, Štrbík V, Dobročka E, Dujavová A, Reiffers M, Liday J and Španková M 2013 *Appl. Surf. Sci.* **269** 98
[4] Boschker H, Huijben M, Vailions A, Verbeeck J, van Aert S, Luysberg M, Bals S, van Tendeloo G, Houwman E P, Koster G, Blank D H A and Rijnders G 2011 *J. Phys. D: Appl. Phys.* **44** 205001
[5] Urushibara A, Moritomo Y, Arima T, Asamitsu A, Kido G and Tokura Y 1995 *Phys. Rev.* B **51** 14103
[6] De Léon-Guevara A, Berthet P, Berthon J, Millot F, Revcolevschi A, Anane A, Dupas C, Le Dang K, Renard J P and Veillet P 1997 *Phys. Rev.* B **56** 6031
[7] Zhu T, Yan C H, Wang Z M, Zhao H W, Sun J R and Shen B G 2001 *Solid State Comm.* **117** 471
[8] Snyder G J, Hiskes R, Di Carolis S, Beasley M R and Geballe T H 1996 *Phys. Rev.* B **53** 14434
[9] Štrbík V and Chromik Š 2012 *J. Eletrical Engin.* **63** 270
[10] Blagoev B, Nurgaliev T, Štrbík V, Mateev E, Zaleski A 2013 arXiv:1307.0302
[11] Ren G M, Yuan S L and Tian Z M 2009 *Thin Solid Films* **517** 3748

Journal of Physics: Conference Series **514** (2014) 012042 doi:10.1088/1742-6596/514/1/012042

[12] Mateev E, Blagoev B and Nurgaliev T 2012 *J. Phys.: Conf. Series* **356** 012022
[13] Blagoev B, Mateev E, Strbik V, Nurgaliev T and Uspenskaya L S 2010 *J. Phys.: Conf. Series* **223** 012015
[14] Ovsyannikov G A, Petrzhik A M, Borisenko I V, Klimov A A, Ignatov Yu A, Demidov V V and Nikitov S A 2009 *J. Experimental and Theoretical Phys.* **108** 48
[15] Jaime M, Lin P, Salomon M B and Han P D 1998 *Phys. Rev.* B **58** R5901
[16] Kubo K and Ohata N 1972 *J. Phys. Soc. Jpn.* **33** 21
[17] Millis A J, Littlewood P B and Shraiman B I 1995 *Phys. Rev. Lett.* **74** 5144
[18] Emin D and Holstein T 1969 *Annals Phys.* **53** 439

New phenomenology of gas breakdown in DC and RF fields

Zoran Lj Petrović[1,3,4], Jelena Sivoš[1], Marija Savić[1], Nikola Škoro[1], Marija Radmilović Rađenović[1], Gordana Malović[1], Saša Gocić[2] and Dragana Marić[1]

[1]Institute of Physics, University of Belgrade, POB 68, 11080 Zemun, Serbia
[2]Faculty of Natural Sciences, University of Niš, 33 Višegradska, 18000 Niš, Serbia

E-mail: Zoran@ipb.ac.rs

Abstract. This paper follows a review lecture on the new developments in the field of gas breakdown and low current discharges, usually covered by a form of Townsend's theory and phenomenology. It gives an overview of a new approach to identifying which feedback agents provide breakdown, how to model gas discharge conditions and reconcile the results with binary experiments and how to employ that knowledge in modelling gas discharges. The next step is an illustration on how to record volt-ampere characteristics and use them on one hand to obtain the breakdown voltage and, on the other, to identify the regime of operation and model the secondary electron yields. The second aspect of this section concerns understanding the different regimes, their anatomy, how those are generated and how free running oscillations occur. While temporal development is the most useful and interesting part of the new developments, the difficulty of presenting the data in a written form precludes an easy publication and discussion. Thus, we shall only mention some of the results that stem from these measurements. Most micro discharges operate in DC albeit with complex geometries. Thus, parallel plate micro discharge measurements were needed to establish that Townsend's theory, with all its recent extensions, is still valid until some very small gaps. We have shown, for example, how a long-path breakdown puts in jeopardy many experimental observations and why a flat left-hand side of the Paschen curve often does not represent good physics. We will also summarize a kinetic representation of the RF breakdown revealing a somewhat more complex picture than the standard model. Finally, we will address briefly the breakdown in radially inhomogeneous conditions and how that affects the measured properties of the discharge. This review has the goal of summarizing (rather than developing details of) the current status of the low-current DC discharges formation and operation as a discipline which, in spite of its very long history, is developing rapidly.

1. Introduction: Townsend's theory and low-current DC discharges - 100 years ago and now

With the development of basic phenomenology and theory of gas breakdown, Townsend's theory was forged some 100 years ago [1-3]. In this paper we shall give a review of how in the past 20 years the basic Townsend's theory and phenomenology have been revived, extended, revitalized and put in perspective of modelling higher current technological discharges and plasmas. This was primarily done by the groups of Art Phelps and the Gaseous Electronics Laboratory in Belgrade.

[3] Author affiliation: also at Serbian Academy of Sciences and Arts, Belgrade, Serbia.

[4] To whom any correspondence should be addressed.

Recent advances in diagnostics and modelling of complex plasma systems opened an opportunity to revisit the breakdown in gases both in DC and RF fields and also for micro gaps. We shall first discuss the experimental techniques to determine the breakdown voltage. Typical errors, such as neglecting the long-path breakdown on the left-hand side of the Paschen minimum, conditioning of the electrode and measuring properties within an unstable (oscillating) regime, will be covered briefly. In addition, we shall describe a proper methodology to establish volt-ampere (*V-I*) characteristics and how to use those with the goal of determining the breakdown voltage and the secondary electron yields.

Time-resolved imaging [4] provides us with information on the development of the anatomy of the discharge and its different modes. Using the spatial profile one may decide which of the mechanisms dominate the discharge. The Townsend regime is the low-current diffuse discharge with exponential growth towards the anode [5]. It is necessary to observe such a profile to ascertain that the discharge operates in the Townsend regime where Townsend's theory may be used to establish the condition for breakdown and effective secondary electron yield. The temporal development of the normal glow or abnormal glow following the breakdown reveals transient multi-regime operation that requires a new paradigm.

When considering volt-ampere characteristics, one may first observe the negative differential resistance in the Townsend regime which may be explained by a combination of space-charge effects and an energy-dependent secondary electrons yield. Thus, *V-I* characteristics should be used in addition to the Paschen curve to determine the secondary electrons yields. With the newly found field of discharges in and above liquids, we have also analyzed breakdown in water vapor and ethanol [6,7].

RF and microwave breakdowns have a different phenomenology as the secondary ions production of electrons at the cathode may not be necessary. Yet the RF breakdown is prone to phenomena not often observed in DC breakdowns, like S shaped (double valued) Paschen-like curves, frequency/gas number density scaling and additional mechanisms like multipactors. The principal experimental problem with the RF breakdown is the magnitude of the displacement current that thwarts the measurement of the conduction current making it difficult to ascertain the initiation of the discharge.

2. Gas breakdown, feedback mechanism, secondary electron yields and how to use them

Traditionally, the gas breakdown is characterized by Paschen curves. In figure 1 we show one such example of Paschen curves for the topical water vapor and ethanol vapor [6,7].

While most people will argue (including the present authors) that the Paschen law itself is developed with limiting assumptions

$$V_b = \frac{Bpd}{\ln(Apd) - \ln\left[\ln\left(1 + \frac{1}{\gamma}\right)\right]}, \qquad (1)$$

the breakdown equation from the Townsend's law is in principle exact if the assumptions of the primary feedback through ions are correct:

$$\gamma(E/N)\left(e^{\alpha(E/N)d} - 1\right) = 1. \qquad (2)$$

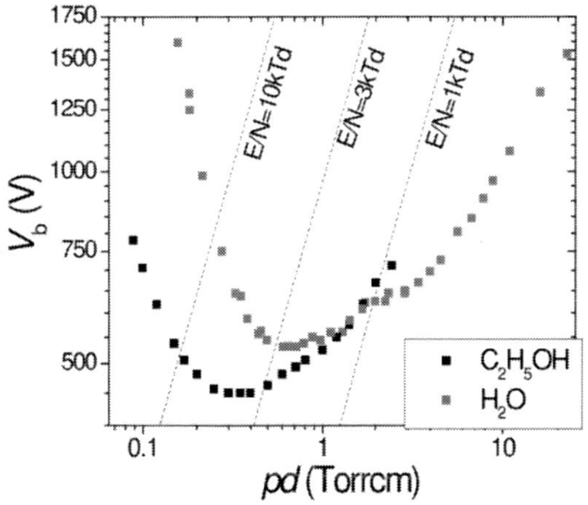

Figure 1. Paschen curves for water [6] and ethanol vapor [7].

Different forms of analytical laws have been employed for the DC breakdown in order to provide some insight [8,9], but one needs to be aware of the approximate nature of some of the basic formulae. Still, the same limitations that enter the analytical form of Paschen law may enter the implementation of equation (2). Over the years, a large amount of data on γ have been accumulated, yet seldom those have been compared and systematically collected. Even more importantly, comparisons to binary collision (beam-surface) experiments [10] have been almost avoided in the literature presumably due to poor agreement (both qualitative and quantitative). One example where both systematic collection of data and comparisons to binary experiments have been made is the paper of Phelps and Petrović [3].

In figure 2 we show a comparison of the secondary electron yields obtained from Paschen curves in argon for a range of experiments [3]. These should be compared with a constant yield of 0.08 obtained by beam experiments with an atomically clean surface [10] or to the dot-dash line obtained for the same conditions of the cathode as found in discharge experiments. Discharge results are one order of magnitude too high at higher *E/N*, one order of magnitude lower at moderate *E/N* and two orders of magnitude higher at the lowest *E/N*. Phelps and Petrović managed to reconcile the binary collision data and the discharge results by including the following:

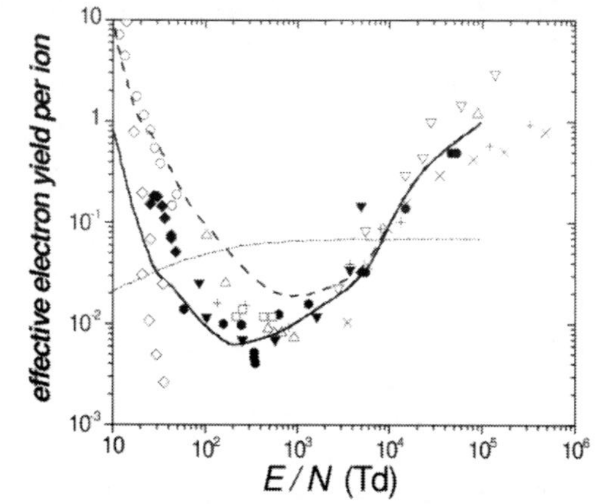

Figure 2. Effective secondary electron yield γ for argon obtained from breakdown data [3]. The different types of points indicate different experiments - see [3]. The dot-dash curve shows binary collision experiment data.

- ionization coefficient fitted in a wide range of *E/N*;
- a region close to the cathode where electrons gain energy and become in equilibrium with the local field, the simplest way of representing this being using a delay distance d_0;
- ion-induced yields at the cathode, modified to represent surfaces that are not atomically clean;
- back-diffusion, i.e. the return of newly emitted electrons back to the cathode;
- secondary electron production by metastables;
- secondary electron production by fast neutrals;
- secondary electron production by resonant radiation – the photo effect;
- trapping of resonant radiation;
- secondary electron production by fast neutrals;
- secondary production due to molecular emission.

The solid and dashed lines in figure 2 indicate the model predictions based on binary collision data (for two limits of possible contributions by molecular radiation). These two lines encompass most of the available experimental data. The good agreement with the experiments shows that all pertinent processes have been included. It also shows that the process of secondary yields modelling may be quite complex and quite challenging due to the need for a wide range of data.

The fact that Townsend's theory could associate all the yields with the flux of ions is because all fluxes are proportional to the electron flux and the system is linear in the breakdown-Townsend discharge phase. On the other hand, one cannot expect linearity to hold for higher current modes, such as glow discharges. For those, one is left with fitting the experimental data. Thus, we compared the fitting procedure with one based on the breakdown data from Phelps and Petrović [3]. We found [11] that fitting of the glow discharge is well represented by the procedure recommended for the breakdown data [3], except when *pd* is quite low and fast neutral effects become dominant. Yet, for RF discharges, for example, or for some more complex geometries, one needs to provide a clear guidance as to how the secondary electron yield may be modelled. Also, we need a considerable effort to provide the data for a number of relevant gases, as argon is the only gas covered so far by the detailed analysis.

3. Volt-ampere characteristics and spatial emission profiles

There are two ways to determine the breakdown voltage. The first option is very accurate albeit very difficult – one can extrapolate pre-breakdown currents. The second one, favored by our group, is to establish a self-sustained discharge in the low-current diffuse, i.e. Townsend, discharge and then extrapolate the measurements to zero current. All other techniques suffer from arbitrariness, either induced by the long statistical time lags or by a direct transition to the glow regime. The Townsend's regime is recognized by an exponential growth peaking at the anode and a normally broad diffusion-determined profile over the entire surface [5,12], as can be seen in figure 3.

In conducting such measurements, we (re)established that in the Townsend regime one has a negative differential resistance. As the effective resistance of the discharge is negative, sometimes, coupled with the external circuit, the overall loop resistance may become negative and oscillations may occur [13,14]. As it turns out, the space charge due to ions increases the field in front of the cathode, which increases the electron production allowing a lower field elsewhere and thus the overall voltage is reduced [14,15]. Thus, the dependence of γ on the mean energy is the reason for the negative differential resistivity, brought about by space charge induced electric field and represented as a current dependence of γ. As the discharge approaches constriction, non-linearities become important [16] and a sudden transition eventually takes place. It has also been shown that if γ were constant, the slope of the *V-I* characteristics in Townsend's regime could become positive [16]. Thus one may conclude that for the full representation of the secondary electron yield one needs to fit not only the Paschen curve but also the *V-I* characteristics. The realm of oscillations often precludes us from achieving stable operation in Townsend's regime but it is also a source of information on important processes. Thus, fitting of the induced damped or free running oscillations may reveal identity of the dominant ionic species, multiplication and may be related to basic transport properties of relevant particles.

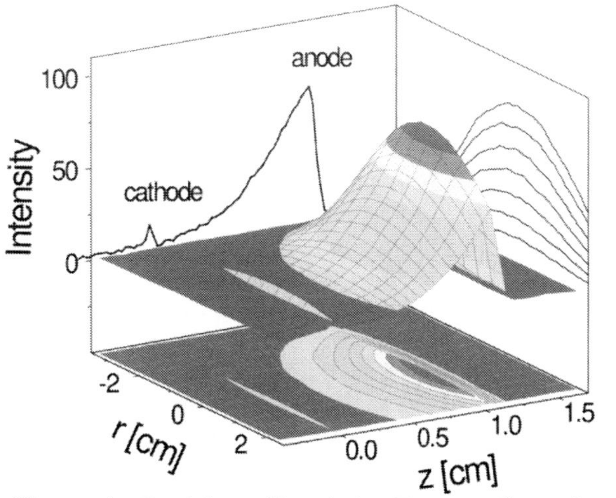

Figure 3. Spatial profile of the Townsend's regime low-current diffuse discharge [12].

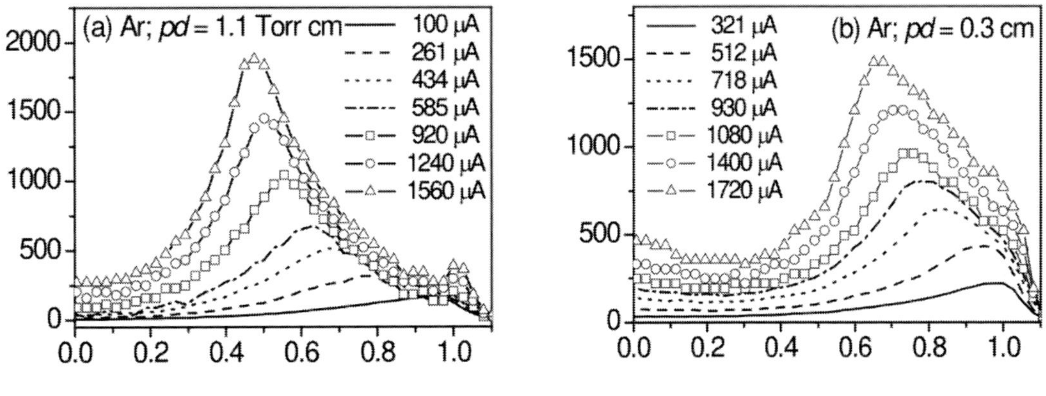

Figure 4. Axial profiles of higher current discharge regimes in argon, at (a) pd = 1.1 Torr cm (close to the Paschen minimum); (b) pd = 0.3 Torr cm (in the left-hand branch of the Paschen curve).

Another important aspect that stems from figure 3 is that spatial profiles not only give us information on the regime of operation (see figure 4 – the peak in front of the anode is for Townsend's low current discharges, the peak in the bulk corresponds to the glow discharge) but also (if put on an absolute scale) a basis to establish absolute cross sections and even profile of the field. It also shows whether and to what extent is equilibrium (with the field) developed or whether fast neutral excitation is important as recognized by the peak right in front of the cathode.

4. Scaling of the basic properties of micro discharges

Micro discharges were basically developed to take advantage of the non-equilibrium plasma that is formed around the Paschen minimum, but at a much higher pressure. Atmospheric pressure would require a 10 μm gap. On the other hand, to achieve a stable operation at high pressures one needs to use complex geometries as it proved very difficult to operate parallel plate micro discharges. Yet, many authors have assumed parallel plate geometries with narrow strips crossing at small distances and assumed that the breakdown occurs at the shortest distances. This has led to a number of papers where the left hand side of the Paschen curve showed no or little variation that could be erroneously interpreted as the onset of field emission (that was predicted to occur only for $d < 10$ μm [9]). We have made an effort to perform measurements in well defined and contained parallel plate discharges, to test the applicability of Townsend's phenomenology at small gaps [17].

Before proceeding to any modelling, we needed to test the laws of scaling, which for low pressure collision dominated discharges are E/N, pd and jd^2 (and also ω/N and B/N for time varying fields and for magnetic fields). The critical scaling is due to the current density j. Rarely are the V-I characteristics represented through j (as it should be) and even then it is not stated that j is actually determined by dividing the current by the entire area of the electrodes. In reality, however, constriction dominates in the glow regime. Even in the Townsend's and in the abnormal glow regimes the radial profile is quite different and so is the effective area. Taking advantage of transparent yet conducting materials for electrodes and also of ICCD cameras, we were able to record V-I characteristics and scale them to the real current density even in cases of complex constricted modes. First, we established that in the normal size discharges, when the current density is properly determined, it remains constant throughout the glow regime (so it is represented by a single point in V-I characteristics [18].

Secondly, we have been able to show that for discharges bordering on micro discharges (0.5 and 1 mm gaps) the jd^2 scaling works and finally we have extrapolated those findings (by using pD scaling where D is the diameter of the constricted region) to smaller gaps where we could not easily record the current profiles [17]. It was found, as can be seen in figure 5, that a Townsend type of scaling holds at those gaps until perhaps some smaller geometries where field emission really comes into play [19]. We have even shown that the spatial profiles scale well so that the ionization coefficient could be determined rather accurately from the axial emission profile of micro discharges [20].

It was also shown that the Paschen curves obtained for micro discharges (and sometimes even for the standard size discharges) that do not show a change of voltage on the left hand side beyond the minimum are due to an incorrectly assumed shortest distance for the gap when long path breakdown was allowed [21]. Reducing the chances for the long path breakdown brings back the Paschen curve to agreement with that obtained for standard dimensions/pressures.

The fact that we proved scaling and also were able to obtain accurate readings of ionization coefficients proves that it is possible to apply Townsend's phenomenology and even to some degree theory to micro discharges. The jd^2 scaling allows Townsend regime to operate at considerably higher current densities.

5. Time resolved measurements

A preliminary publication of our results on time dependent recordings of the development of the breakdown and DC discharge regimes was given in [9,22], while the majority of the data remain unpublished. The results may be summarized as follows. During the initial stage, the discharge passes through the Townsend regime and, as the current increases, glow and abnormal glow regimes are visited. During the minimum of oscillations, the discharge almost immediately returns to the

Journal of Physics: Conference Series **514** (2014) 012043 doi:10.1088/1742-6596/514/1/012043

Townsend regime and then changes occur again and again following oscillations in current. Studies such as this one performed for a parallel plate geometry are needed to provide the background for understanding developments in more complex geometries [23] and different scales.

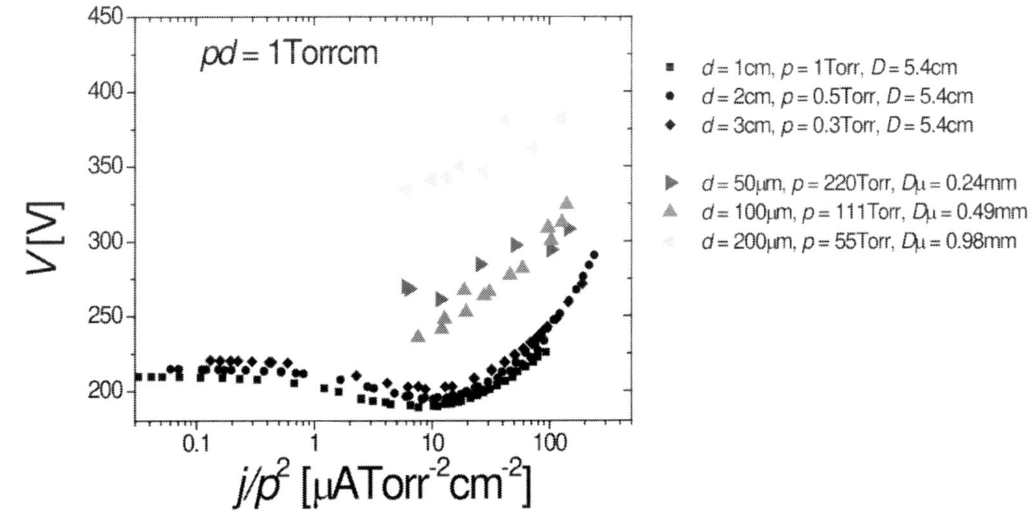

Figure 5. *V-I* characteristics for standard size and micro discharges obtained by using the jd^2 scaling and *pD* scaling for the size of the constricted regions. One should not pay attention to the actual vertical scale, normally it is taken care of by subtracting the breakdown voltages in each case, but in the case of micro discharges we could not make accurate measurements as it was not possible to achieve stable operation in the Townsend regime. The vertical scale variations are due to different conditions on the electrodes surface [17].

6. Discharges with inhomogeneous cathodes

The usual assumption in the low current limit is that the discharge is uniform over the entire surface and the radial profile is the solution to the diffusion equation. Only in the higher current (glow) regime a constriction develops and only when the field ceases to be constant along the axis. This has been questioned recently for micro discharges when it was found that the scaling laws are maintained only when an assumption of localized modes is made, the modes that have a dimension proportional to the diffusion length for the given pressure [17]. The problem is hindered further by the high pressure that reduces the diffusion length and by the fact that the ratio of gap to radius becomes very small in practical experiments [17].

We have, however, observed localized discharges in Townsend regime for standard size discharges (1 cm) and moderately low pressures [24]. In figure 6 we show one such example in nitrogen where the discharge is limited in the radial direction by the region of deposited material due to numerous pulses in the high current mode. It is clear that the deposited region has a lower secondary electron yield γ and thus the discharge cannot be self-sustained over that area for the given voltage. In order to overcome this region of reduced γ, the volt-ampere characteristics have to be quite different as compared to the one with pristine cathode. Even a positive differential resistance is observed [24].

The characteristics of such a discharge may be a prototype for the breakdown in pulsed DC discharges where numerous discharges change the properties of the cathode considerably. In addition, one may construct cathodes of different materials and design desired characteristics. For example, cathodes coated by a semiconductor [25, 26] have been often used for achieving some properties that are not easily accessible by conducting electrodes. The high resistance of the semiconductor which is in the innermost circuit assures a broader range of stable operation [14]. Combining conducting and semiconductor materials may help reduce the breakdown potential while achieving a *V-I* characteristic that is less prone to oscillations.

7. RF breakdown

When rate of change of the field is such that ions cannot complete their trajectories, then it is possible to operate in conditions when electrons may be the only particles sustaining the discharge. The feedback is provided by returning electrons in the second half of the period and thus a full circle is achieved. It has been assumed in old textbooks and papers [27] and in more recent papers [28, 29] that an optimum breakdown condition is achieved when the average electron can cross the gap in one half period. In fact, the breakdown condition has been used to obtain experimental values of the drift velocities for RF fields and convert them to DC values by assuming that in RF fields the drift velocity is a sinus function peaking with the DC drift velocity and having no delay [29]. We modelled the RF breakdown using a detailed Monte Carlo

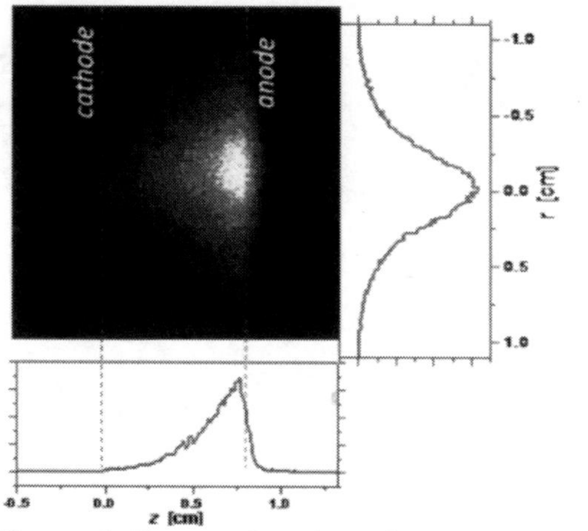

Figure 6. Townsend regime discharge in N_2 with Cu cathode and for $d = 0.8$ cm, $D = 2$ cm, $p = 2$ Torr and $V_b = 310$ V. The axial profile shows exponential growth peaking at the anode while the radial profile is very narrow covering only a small part of the diameter (D).

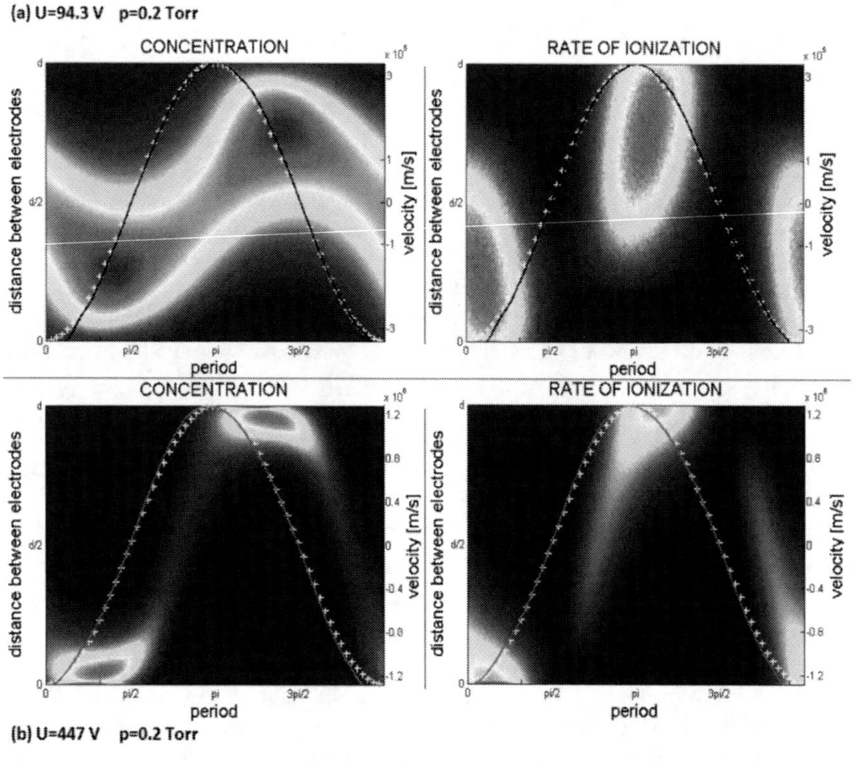

Figure 7. Development of density and ionization rate for the breakdown of RF swarms. Solid curves represent mean velocity (different colours are for better contrast), while the dotted sinusoidal lines represent inversed AC field. Both curves are for 0.2 Torr; the top curves are for the minimum breakdown voltage (93 V) while the bottom curves are for the maximum breakdown voltage (447 V) [30].

representation of all the collisions and field dependence on time. Figure 7 shows the temporal development of the concentration of electrons and of the ionization rate [30]. As the quasi-Paschen curve for the RF breakdown has double values for breakdown voltage at the same *pd* (for a range of *pd*) we show results at the lower and at the higher breakdown point. Outside these borders the discharge cannot be ignited. At the lower point, the breakdown condition is achieved by merely matching the production without much of the electrodes overlapping with the swarm. At the higher end, however, the swarm overlaps considerably with the electrodes thus representing significant losses to the electrodes. At the same time, the peaking ionization must compensate for all the losses.

We found that, while the standard explanation of the RF breakdown is very good, still some fine tuning needs to be done by a full kinetic representation (such as a Monte Carlo simulation) to be able to describe all the intricacies. In addition, one needs to extend our model by the contribution of ions and fast neutrals as well as photons. These will modify the Paschen like curve and hopefully make it more realistic but for a wide range of conditions one may find that electrons dominate in sustaining the plasma. We also have to include the secondary electrons formed by the electron impact on electrodes. This effect will lead to the so-called multipacting modes.

One should also note that numerous attempts to use analytical and semi analytical models based on simplified expressions for the basic properties have migrated from DC discharges to the RF breakdown. These results provide insight into pertinent processes [31-34] but require some form of fitting to provide quantitative agreements.

8. Conclusions

The oldest chapter in the book on plasma physics, the breakdown and low current discharges, has been changed tremendously in the past twenty years. The main agent facilitating the feedback needed to achieve DC breakdown, the ions colliding with surfaces, have been changed to include photons, fast neutrals, metastables as well as ions. By including all these processes, as well as back-diffusion [3,35] one was able to predict effective yields and Paschen curves based on the binary collision data [3].

The field of low pressure DC discharges proved to be a fertile ground for both fundamental studies and for obtaining and testing the applicability of the fundamental data that would eventually be used for modelling of more complex systems. Extensions to micro discharges and RF breakdown have been made, together with first attempts to model RF plasmas with secondary electron yields [36] as obtained in a more detailed and recent analysis.

With the new drive for benchmarking plasma modelling systems [37], we believe that the best strategy would be to start from the swarm benchmarks and then use the negative differential resistance of a DC Townsend discharge as a benchmark for space charge effects and also use some additional breakdown properties. This would provide clear and simple experimental observables that may be modelled exactly and independently and provide the next step for more complex plasma benchmarks.

Acknowledgement

This paper has been completed under the overall support of projects 171037 and III41011 of the Ministry of Education and Science of Serbia, and project 155 of the Serbian Academy of Sciences and Arts. Numerous discussion, shared research and data and ideas from the late Art V. Phelps are gratefully acknowledged.

References

[1] Townsend J S 1910 *The Theory of Ionization of Gases by Collision* (Constable, London)
[2] Townsend J S 1915 *Electricity in Gases* (Oxford: Clarendon) 313
[3] Phelps A V and Petrović Z Lj 1999 *Plasma Sources Sci. Technol.* **8** R21
[4] Marić D, Malović G and Petrović Z Lj 2009 *Plasma Sources Sci. Technol.* **18** 034009
[5] Petrović Z.Lj. and Phelps A.V. 2009 Phys. Rev.E **80** 016408
[6] Škoro N, Marić D, Malović G, Graham W G and Petrović Z Lj 2011 *Phys. Rev.* E **84** 055401R
[7] Petrović Z Lj, Sivoš J, Škoro N, Malović G and Marić D 2013 *Bul. Am. Phys. Soc. Ser. II* **58**/8

Journal of Physics: Conference Series **514** (2014) 012043 doi:10.1088/1742-6596/514/1/012043

HW1 35
[8] Kihara T 1952 *Rev. Modern Phys.* **24** 45
[9] Radmilovic-Radjenović M and Lee J K 2005 *Phys. Plasmas* **12** 063501
[10] Hagstrum H D 1954 *Phys. Rev.* **96** 336
[11] Marić D, Hartmann P, Malović G, Donko Z and Petrović Z Lj 2003 *J. Phys. D: Appl. Phys.* **36** 2639
[12] Petrović Z Lj, Donko Z, Marić D, Malović G and Živanov S 2002 *IEEE Trans. Plasma Sci.* **30** 136-7
[13] Petrović Z Lj and Phelps A V 1993 *Phys. Rev.* E **47** 2806
[14] Phelps A V, Petrović Z Lj and Jelenković B M 1993 *Phys. Rev.* E **47** 2825
[15] Živanov S, Živković J, Stefanović I, Vrhovac S and Petrović Z Lj 2000 *Eur. Phys. J.* AP **11** 59
[16] Nikolić M, Đorđević A, Stefanović I, Vrhovac S and Petrović Z Lj 2003 *IEEE Trans. Plasma Sci.* PS **31** 717
[17] Petrović Z Lj, Škoro N, Marić D, Mahony C M O, Maguire P D, Radmilović-Radenović M and Malović G 2008 *J. Phys. D: Appl. Phys.* **41** 194002
[18] Škoro N, Marić D and Petrović Z Lj 2008 *IEEE Trans. Plasma Sci.* **36** 994
[19] Radmilović-Radjenović M, Matejčik Š, Klas M and Radjenović B 20013 *J. Phys.* D:-*Appl. Phys.* **46** 015302
[20] Kuschel T, Stefanović I, Malović G, Marić D and Petrović Z Lj 2013 *Plasma Sources Sci. Technol.* **22** 045001
[21] Marić D, Škoro N, Maguire P D, Mahony C M O, Malović G and Petrović Z Lj 2012 *Plasma Sources Sci. Technol.* **21** 035016
[22] Marić D, Malović G and Petrović Z Lj 2007 *J. Phys.: Conf. Series* **86** 012009
[23] Wagenaars E, Bowden M D and Kroesen G M W 2005 *IEEE Trans. Plasma Sci.* **33** 254
[24] Gocić S R, Škoro N, Marić D and Petrović Z Lj 2010 *20th ESCAMPIG* (Novi Sad Serbia July 2010) P3.17
[25] Kurt H, Çetin S, and Salamov B G 2011 *IEEE Trans. Plasma Sci.* **39** 1086
[26] Portsel L M, Lodygin A N and Astrov Yu A 2009 *J. Phys. D: Appl. Phys.* **42** 235208
[27] Gill E B and von Engel A 1949 *Proc. R.. Soc.* A **197** 107
[28] Lisovskiy V A and Yegorenkov V D 1994 *J. Phys. D: Appl. Phys.* **27** 2340
[29] Lisovskiy V, Booth J-P, Landry K, Douai D, Cassagne V and Yegorenkov V 2006 *J. Phys. D: Appl. Phys.* **39** 660
[30] Savić M, Radmilović-Radjenović M, Šuvakov M, Marjanović S, Marić D and Petrović Z Lj 2011 *IEEE Trans. Plasma Sci.* **39** 2556
[31] Sato M and Shoji M 1997 *Jpn. J. Appl. Phys.* **36** 5729
[32] Iliev I, Gocheva-Ilieva S and Sabotinov N 2004 *Proc. SPIE* **5449** 131
[33] Iliev I, Gocheva-Ilieva S and Sabotinov N 2009 *J. Optoelectr. Adv. Mater.* **11** 1392
[34] Smith H B, Charles C and Boswell R W 2003 *Phys. Plasmas* **10** 875
[35] Radmilović M and Petrović Z Lj 2000 *Eur. J. Appl. Phys.* **11** 35
[36] Radmilović-Radjenović M and Petrović Z Lj 2009 *Eur. Phys. J.* D **54** 445
[37] Turner M M, Derzsi A, Donkó Z, Eremin D, Kelly S J, Lafleur T and Mussenbroc T 2013 *Phys.Plasmas* **20** 013507

Influence of low-pressure glow discharge on laser-induced plasma spectra

S Karatodorov[1], **V Mihailov and M Grozeva**

Acad. G. Nadjakov Institute of Solid State Physics, Bulgarian Academy of Sciences, 72 Tsarigradsko Chaussee, 1784 Sofia, Bulgaria

E-mail: skarat@issp.bas.bg

Abstract. A scheme is reported for spectrochemical analysis combining sample introduction by laser ablation and additional excitation of the ablated material by a hollow-cathode discharge. The optimal conditions allowing spatial separation of the two processes are determined. Under these conditions, enhancement of the analytical lines intensity in comparison with stand-alone laser-induced plasma is observed.

1. Introduction

Laser-induced breakdown spectroscopy (LIBS) is widely applied for spectrochemical analysis due to its advantages [1]: ability to ablate a wide variety of samples – including non-conducting materials and liquids, lack of requirements to sample preparation, and capacity for fast microprobe analysis. In spite of these features, the laser ablation spectroscopy has some drawbacks in terms of analytical performance – matrix effects, limited depth resolution due to the need of ablating sufficient mass of material for producing visible plasma, difficulties in quantitative analysis, etc [2, 3]. This is why common techniques for spectrochemical analysis have been proposed combining a direct laser sampling method with an additional source of excitation of the ablated material – inductively-coupled plasma, arc/spark, etc. [4]. Such combined techniques, e.g. laser ablation-inductively coupled plasma-mass spectrometry, exhibit excellent characteristics but are economically demanding and experimentally sophisticated. Therefore, a more economically feasible alternative is needed.

The combination of a hollow-cathode discharge and laser ablation is a possible alternative because the hollow-cathode discharge is a proven spectroscopic source with narrow lines, high signal-to-background noise ratio and rich spectra [5]. The hollow-cathode discharge provides means for enhanced electron-atom excitation of the ablated material which gives the possibility to ablate less material per pulse and achieve better depth resolution. This has already been demonstrated for the conventional glow discharge [6]. Hence, the performance of the combined technique benefits from the micro-sampling capability of laser ablation, the enhanced excitation by the hollow-cathode glow discharge and the possibility to control each of them independently.

Four parameters play a key role for the performance of the technique: the laser source, the ambient gas environment, the cathode-sample gap length and the discharge current. The laser source governs the ablation mass per pulse and the plasma plume initiation. The presence of an ambient gas and its pressure strongly influences the expansion and the emission of the laser-induced plasma. The cathode-

[1] To whom any correspondence should be addressed.

sample gap length controls the time delay between the laser pulse and the discharge excitation and, accordingly, the state in which the ablated species enter the discharge. The current governs the excitation of the ablated material and the intensity of the analytical signal.

In this work, a laser-induced plasma hollow-cathode discharge (LIP-HCD) scheme is studied. The aim is to determine the conditions under which the processes of introduction and excitation can be separated and controlled independently. This is achieved by optimization of two of the key parameters: the ambient gas pressure and the cathode-sample gap length.

2. Experimental setup, conditions and procedure

The experimental LIP-HCD setup for spectrochemical analysis, shown in figure 1, consists of three main parts – a hollow-cathode discharge, a laser source and a registration system.

The laboratory-made neon hollow cathode, with a 3-mm inner diameter and a 20-mm length, is powered by a stabilized DC power supply. The cathode is made of aluminium, as this material has a low ion sputtering rate and a poor spectrum. Anode rods are placed at the two ends of the cathode. A disc-shaped copper sample is mounted at one of the anodes. The gap between the hollow-cathode plasma and the sample can be varied. The length of the gap defines the state in which the ablated species enter the discharge.

The laser is a pulsed nanosecond Q-switched Nd:YAG laser (Quanta Ray GCR3) lasing at 1064 nm at a repetition rate of 1÷20 Hz and a pulse energy up to 400 mJ. The spectra are recorded by a PC-controlled Digikröm 480 monochromator with a CCD camera.

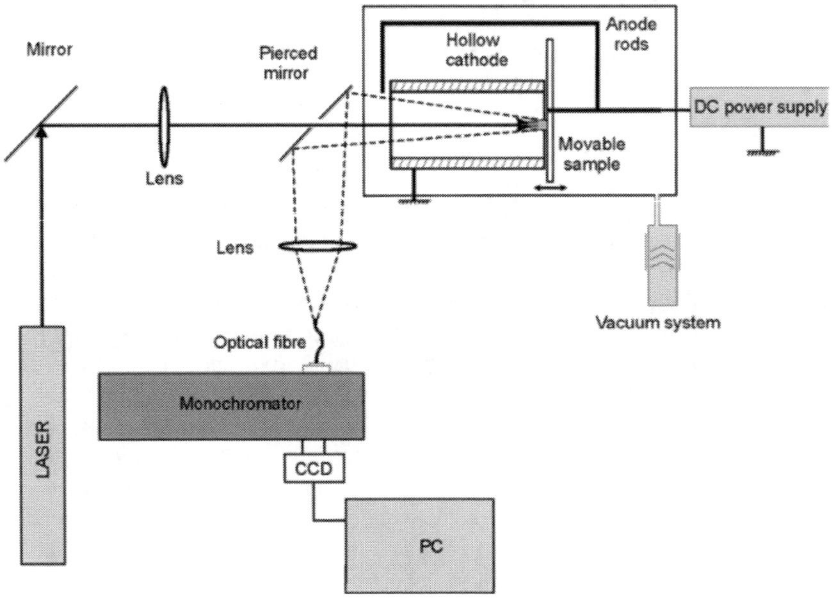

Figure 1. Experimental setup.

The laser beam is directed to the hollow cathode by a mirror and is focused by a lens at the copper sample. The focused laser beam ablates the sample surface and laser-induced plasma is formed. The plasma expands in the space above the sample and into the hollow-cathode cavity. The laser-induced plasma interacts with the hollow-cathode discharge plasma in a variety of processes, including excitation and ionization of the ablated particles. The light emitted from the hollow-cathode volume is focused at the entrance slit of the monochromator and is registered by the CCD camera.

Journal of Physics: Conference Series **514** (2014) 012044 doi:10.1088/1742-6596/514/1/012044

3. Experimental results

3.1. Influence of the ambient gas pressure on the laser-induced plasma spectra

As the spatial and energy distribution of the ablated material are very sensitive to the presence and nature of the ambient gas [7], its influence on the intensity of the emission lines is investigated. The emission spectra of laser-induced plasma in vacuum and in neon are measured. An increase of the copper lines intensity in the presence of the ambient gas is observed (figure 2).

The ablation threshold decreases significantly in the presence of the ambient gas compared to ablation in vacuum. The presence of gas next to the solid surface also increases the ablation rate due to thermal evaporation [8].

During its expansion, the laser-ablated plasma interacts with the neutral atoms of the ambient gas and excites them through different processes – electron-atom inelastic collisions, Penning ionization, etc. The atomic lines registered in the laser-induced spectra in neon, shown in figure 3, are an indication of the energy transfer from the laser-induced plasma to the ambient gas species. Hence, the expanding laser-induced plasma is dissipating its energy due to the ambient gas presence.

The influence of the ambient gas pressure on the plasma emission intensity is presented in figure 4. As it is seen, the copper lines intensity increases by an order of magnitude when the neon pressure is increased from 2 Torr to 25 Torr. This behavior of the emission is attributed to plasma contraction in a smaller volume with the pressure increase resulting in a higher species density, an increase in the collisional frequencies of plasma species, an enhancement of the electron cascade growth and stronger emission intensity [9]. A similar line intensity increase should be expected when a high repetition-rate laser is used for ablation due to the accumulation of ablated plasma above the surface of the solid target by successive pulses.

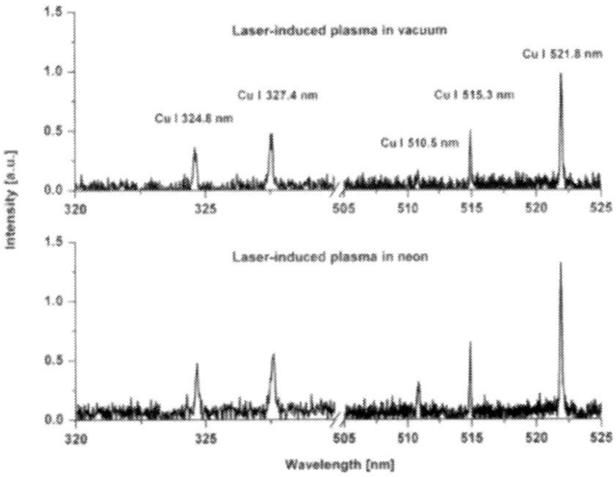

Figure 2. Laser-induced plasma emission spectra in vacuum (above) and in 4 Torr neon (below) at laser fluence 0.8 J cm^{-2}.

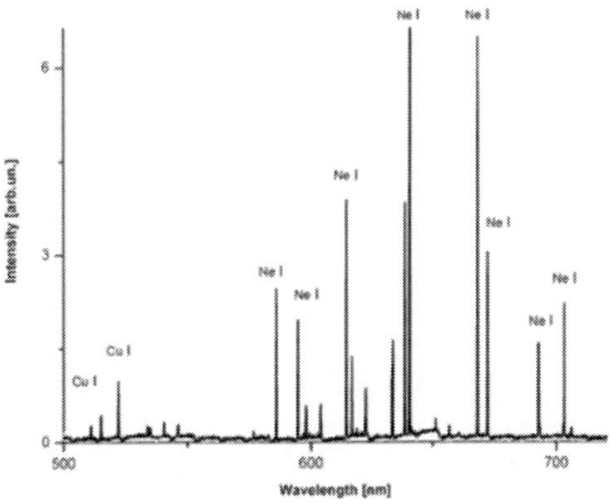

Figure 3. Laser-induced plasma spectra in neon at 4 Torr.

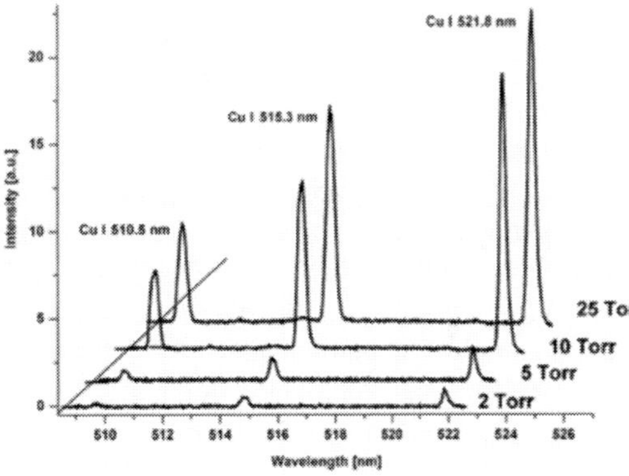

Figure 4. Laser-induced plasma emission spectra as a function of the neon pressure at a laser fluence of 0.8 J cm^{-2}.

3.2. Influence of the sample-cathode gap length on the LIP-HCD spectra

For the hollow cathode used with a 3-mm inner diameter and a 20-mm length, the hollow cathode effect is obtained at 4 Torr neon; the discharge is stable up to a 15-mA discharge current. The minimal laser fluence necessary for generation of laser-induced plasma in our experiments is 0.5 J cm^{-2}.

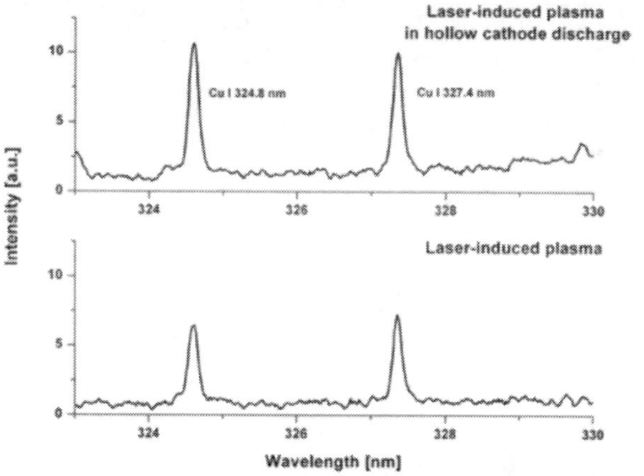

Figure 5. Copper emission spectra: from laser ablation in a hollow-cathode discharge (above) and from laser-induced plasma only (below) at 4 Torr neon, discharge current 6 mA, laser fluence 0.8 J cm^{-2}.

Figure 5 compares the analytical copper lines 324.8 nm and 327.4 nm emitted from stand-alone laser-induced plasma and from LIP-HCD. As is seen, due to the additional excitation of the ablated material in the hollow-cathode discharge plasma, an enhancement of ~30 % in the analytical signal is observed.

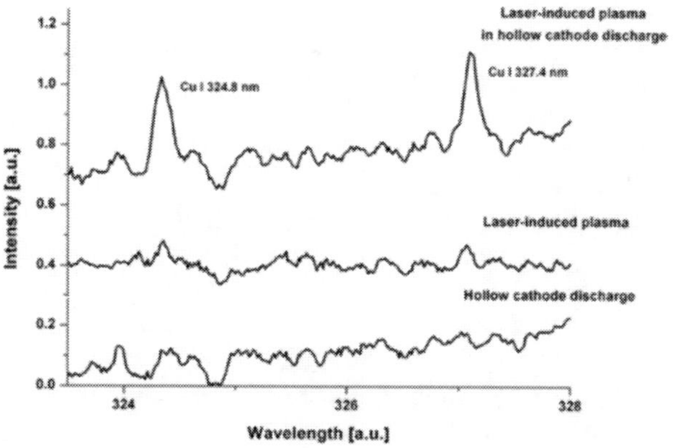

Figure 6. Segments of the emission spectra in a hollow-cathode discharge (below), laser-induced plasma (middle) and LIP-HCD (top).

When the sample is close to the cathode, due to deposition of ablated material on the inner cathode surface and this material re-entering the discharge by ion cathode sputtering, copper lines are observed in the emission spectra of the discharge long after the end of the laser ablation pulse (up to few minutes depending on the number of laser ablation pulses performed). This means that the emission comes not only from hollow-cathode excitation of the ablated material, but also by cathode sputtering. Although this can be a positive effect when detecting trace elements, if depth analysis is performed, such an effect is undesired as it complicates the depth distribution analysis of the elements in the layer. In order to decrease the deposition of material on the inner cathode surface, the sample is moved away from the cathode. The translation of the sample results in both a lower amount of deposited material and a decrease of the analytical lines intensity. The experimentally determined minimal distance at which the signal from the deposited material is negligible, i.e. comparable to the background, is 12 mm. Figure 6 presents the emission spectra excited only by a hollow-cathode discharge (below), stand-alone laser-induced plasma (middle) and LIP-HCD (top) at a sample–cathode distance of 12 mm and a discharge current of 15 mA.

It is seen that the two copper lines (324.8 nm and 327.4 nm) are well-resolved only in the LIP-HCD spectrum. They are weak, as compared to the background noise, in the spectrum of the laser-induced plasma and are completely missing in the spectrum of the hollow-cathode discharge.

We assume that at this distance the analytical signal registered results from ablation caused by the laser pulses and excitation by the hollow-cathode discharge. Thus, separation of the processes of sample introduction and excitation of the ablated material is achieved. Increasing the distance leads to a further decrease in the analytical lines emission intensity.

4. Conclusions

It is shown that the presence of an ambient gas considerably affects the analytical lines intensity. The neon pressure increase leads to an increase in the intensity of the copper lines. When expanding, the laser ablation plasma interacts with atoms of the ambient gas which results in excitation of emission lines of the gas.

By optimizing the cathode-sample gap, the presented LIP-HCD scheme allows one to separate spatially the introduction and excitation of the ablated material and, hence, to control independently each of the two processes.

References

[1] Hahn D and Omenetto N 2010 *Appl. Spectr.* **64** 335A
[2] Garcia C, Lindner H and Niemax K 2007 *Spectrochem. Acta* B **62** 13
[3] Tereszchuk K, Vadillo J and Laserna J 2007 *J. Anal. At. Spectrom.* **22** 183
[4] Marcus K 1993 *Glow Discharge Spectroscopies* (New York: Plenum Press) p 423
[5] Caroli S, Senofonte O, Violante N, Petrucci F and Alimonti A 1984 *Spectrochem. Acta* B **39** 1425
[6] Tereszchuk K, Vadillo J and Laserna J 2009 *J. Appl. Surf. Sci.* **255** 8841
[7] Gusarov A, Gnedovets A and Smurov I 2000 *J. Appl. Phys.* **88** 4352
[8] Gamaly E, Madsen N, Duering M, Rode A, Kolev V and Luther-Davies B 2005 *Phys. Review* B **71** 174405
[9] Farid N, Bashir S and Mahmood K 2012 *Phys. Scr.* **85** 015702

Journal of Physics: Conference Series **514** (2014) 012045 doi:10.1088/1742-6596/514/1/012045

Abnormal surface diffusion of particles under the action of an external time-periodic force

I G Marchenko[1,2,4] **and I I Marchenko**[3]

[1]NSC "Kharkov Institute of Physics and Technology",
1 Akademicheskaya Str., 61108 Kharkov, Ukraine
[2]Kharkov National University, 4 Svobody Sq., 61077 Kharkov, Ukraine
[3]NTU "Kharkov Polytechnic Institute", 21 Frunze, 61145 Kharkov, Ukraine

E-mail: march@kipt.kharkov.ua

Abstract. This paper delves into the abnormal surface diffusion of particles in an external time-periodic field. The relationship is studied of the surface diffusion enhancement as a function of the amplitude and frequency of external fields. It is shown that the diffusion coefficient can be significantly increased (by several orders of magnitude) by applying external time-periodic fields. The diffusion gain ratio stronglly depends on the surface temperature and on the external field amplitude and frequency.

1. Introduction

The progress in nanophysics and the intensive studies in the field of development of microsystems with prescribed properties stimulated a keen interest in the investigation of the ordered motion of particles on the surface of a solid body [1-2]. The transfer of atoms and clusters exposed to the external force caused by magnetic and electric fields, laser radiation, external voltage, etc. requires careful studies of the diffusion processes in systems with a periodic potential.

As a rule, the surface diffusion is characterized by low values of the activation barriers and a low friction coefficient [3,4]. Previous research conducted by different authors showed that special diffusion modes, the so-called subdiffusion and hyperdiffusion [5,6] can arise in systems that are characterized by a low kinetic-energy dissipation level. The ordinary diffusion is characterized by a linear relation of the root-mean-square time deviation $<x^2>\sim t$. In the case when special diffusion modes take place, this relationship is changed $<x^2>\sim t^n$. At $n<1$ we speak of subdiffusion, at $1<n<2$, of superdiffusion, and at $n>2$, of hyperdiffusion. At $n=2$, the special case of ballistic diffusion is realized. These abnormal modes are observed in a certain time domain defined by the temperature and the system's properties. The stationary state is established after a certain period of time and the spatial dispersion of the particles distribution is described by the standard expression $<x^2>\sim t$. However, the time limitation of the transport processes may result in the particle motion occurring under conditions of a strong non-equilibrium state characterized by abnormal diffusion modes [7]. Special diffusion modes can be "preserved" through the exposure of the system to periodic fields. It has been shown earlier in [8] that abnormal diffusion enhancement is possible by a

[4] To whom any correspondence should be addressed.

decrease in the ambient temperature. A possibility of enhancing the surface diffusion along a selected direction without increasing the temperature is of great practical importance, because it allows for the creation of surface structures consisting of clusters of a required shape, size and density. However, the dependence of this phenomenon on the effective force amplitude and frequency has not been studied. The purpose of this scientific paper iss to study in detail these relationships using computer simulation methods.

2. Simulation details

The surface motion of particles can be described using the following Langevin equation [9]:

$$m\ddot{x}_i = -\frac{\partial}{\partial x} U(x_1, x_2) - \gamma \dot{x}_i + F_i + \xi_i(t), \tag{1}$$

where x_i are the Cartesian coordinates components for the particle in the plane, t is the time, m is the particle mass, F_i are the components of the constant force to which the particle is exposed, $U(x_1, x_2)$ is the potential energy of the particle in the plane, and γ is the friction coefficient.
The dots denote differentiation with respect to time. The thermal fluctuations $\xi_i(t)$ represent uncorrelated white noises that obey the relationship

$$\langle \xi_i(t)\xi_j(t') \rangle = 2\gamma kT\delta(t-t')\delta_{ij}, \tag{2}$$

where k is the Boltzmann constant, T is the temperature.

Hereinafter, a "particle" will imply an adatom or a cluster consisting of many atoms that can move in the periodic potential. The potential energy of the particle on the crystal surface is defined by the surface symmetry and its atomic composition. In the simplest case of a monoatomic material and a simple rectangular surface lattice, the potential energy can be written as follows [9]:

$$U = -\frac{U_0}{2}\left[\cos\left(\frac{2\pi}{a}x_1\right) + \cos\left(\frac{2\pi}{b}x_2\right)\right], \tag{3}$$

where a and b are the lattice constants along the x_1 and x_2 axes, and U_0 is the activation barrier value.

It has been shown earlier in [10] that in this case the variables x and y are weakly related and the particle motion kinetics is defined by the component of the force acting along this direction. To simplify the computations we can refer to the case of one-dimensional particle motion in a periodic field $U(x)$.

In order to demonstrate the main physical results we studied the case of one–dimensional diffusion. Studying this case allows us to draw general conclusions that represent the main physical findings of this paper.

A moving particle is subjected to the action of the force F^{Lat} produced by a lattice:

$$F^{Lat} = -\frac{\partial U}{\partial x} = F_{cr}\sin\left(\frac{2\pi}{a}x\right). \tag{4}$$

Hereinafter the value of $F_{cr} = \frac{\pi}{a}U_0$ will be referred to as the critical force [9]. The time-periodic external field is described by the following expression:

$$F_t(t) = F_0 \sin(\omega t), \tag{5}$$

where ω is the angular frequency of the external force and F_0 is its amplitude.

The stochastic equations (1)-(2) are solved numerically using a Verlet-type algorithm with a time step of $\Delta t \approx 0.1$ fs, which is less than 1/100 of the period of oscillation. The statistic averaging is performed over an ensemble consisting of $N = 4 \cdot 10^4$ particles. To verify the model's consistency, some computations are performed at $N = 4 \cdot 10^6$. The initial conditions are specified in the following way: a particle is placed at the coordinate system origin and is randomly given a velocity with a Maxwellian temperature distribution. The system is then subjected to thermalization during 10^4 time steps. Afterwards the particle moves at the acquired velocity to the first unit cell.

The diffusion coefficient was calculated using the traditional approach:

$$D = \lim_{t\to\infty} D(t) = \lim_{t\to\infty} \frac{\langle (x(t) - \langle x(t) \rangle)^2 \rangle}{2t}, \tag{6}$$

where the angle brackets $\langle ... \rangle$ denote ensemble averaging.

While analyzing the modelling results, it is reasonable to write the equation in terms of scaled dimensionless variables, namely, time τ, distance x' and temperature T' [4]:

$$\tau = t \frac{1}{a}\sqrt{\frac{U_0}{m}}; \quad x' = \frac{x}{a}; \quad T' = \frac{kT}{U_0}. \tag{7}$$

The dimensionless friction coefficient is

$$\gamma' = \frac{\gamma a}{(mU_0)^{1/2}}, \tag{8}$$

and is set at 0.2 in the present calculations. This value corresponds to the value observed in the experiment carried out to study hydrogen diffusion on a platinum surface [11].

In order to demonstrate the suitability of using dimensionless values, we performed computations for the diffusion of atoms of different elements. Figure 1 gives the temporal dependences of the diffusion coefficients for Pt, C, H and the appropriate dimensionless coefficients $D(\tau)$ for a temperature $T' = 0.5$. It is seen that after appropriate scaling transformations the diffusion coefficients $D(t)$ fall onto one diffusion curve $D(\tau)$. In order to study the transport processes of particles exposed to an external force, it is sufficient to derive a solution for the particle of one type only. The relationships $D(t)$ for

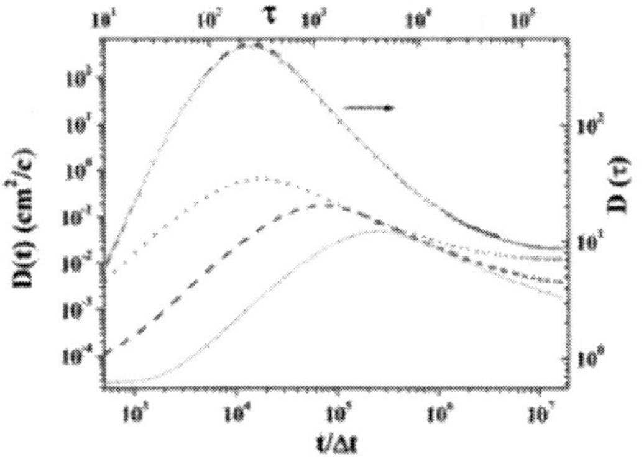

Figure 1. Dependences of the diffusion coefficient $D(t)$ for Pt, C, H on the time t and scaled dependence $D(\tau)$ on the dimensionless time τ. The dotted line denotes H; the dashed line, C; and the solid line, Pt. $T' = 0.13$, $\gamma' = 0.2$, $F = 0.15F_{cr}$.

particles of another mass can be obtained through scaling transformations. Therefore, we studied the case of hydrogen only.

The activation barrier value is 80 meV, which is a typical value for the diffusion of adatoms across the closely-packed surfaces of metals with a face-centered cubic structure and a hexagonal closely-packed structure. The lattice constant a was selected to be equal to 2 Å.

Under the action of a constant force ($\omega = 0$), different modes of particle motion (figure 2) are realized depending on the relationship between F and F_{cr} ($F' = F_0 / F_{cr}$). The results obtained are in a good agreement with the data obtained by other authors [6, 9]. The plot gives the dimensionless values of the mean-square displacements $\sigma^2 = \left(x'(t) - < x'(t) >\right)^2$ on a logarithmic scale. The modes of superdiffusion and hyperdiffusion in different time domains can be well seen in figure 2.

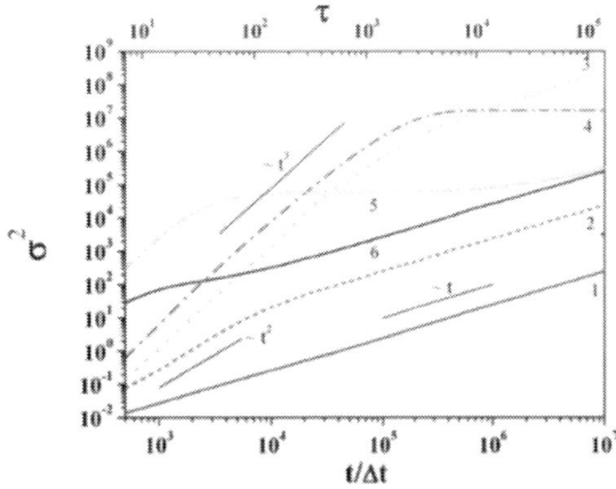

Figure 2. Mean-square particle displacement as a function of the time for different values of the constant force $F' = F / F_0$. Temperature $T = 120$ K ($T' \approx 0.13$). Curve 1 - $F' = 0$; 2 - $F' = 0.05$; 3 - $F' = 0.09$; 4 - $F' = 0.15$; 5 - $F' = 0.7$; 6 - $F' = 1.0$. The solid lines lengths show the dependences $\sim t$, $\sim t^2$ and $\sim t^3$.

3. Results and discussion

The external periodic field applied leads to a time limitation of the diffusion processes. As a result, modes of abnormal diffusion are realized that are characteristic for a time period equal to the period of the external field oscillations.

Figure 3 shows the change in the diffusion coefficient behavior as a function of the applied field frequency for different values of the force amplitude F_0 at a constant temperature. The dashed line denotes the level corresponding to the surface diffusion at $F_0' = 0$.

The figure shows that the action of an external periodic field results in a considerable diffusion enhancement in comparison with the equilibrium case. The amplification ratio depends on the frequency and force amplitude of the field applied. Increasing F_0' to the value of 0.10 (curves 1-3) is accompanied by diffusion enhancement as the force increases in the entire frequency range studied. However, the value of D' change slightly at low frequencies of the external field. As the force continues to increase (curves 4-5), the shape of the curves changes – they display a low-frequency maximum. Its value exceeds considerably the maximum at frequency $\omega = \omega_0$ corresponding to the frequency of self-oscillations of a particle. For all plots given in figure 3, a higher value of D' corresponds to a higher value of F_0' in the high-frequency range. However, at low frequencies and for $F_0' < 0.10$, a

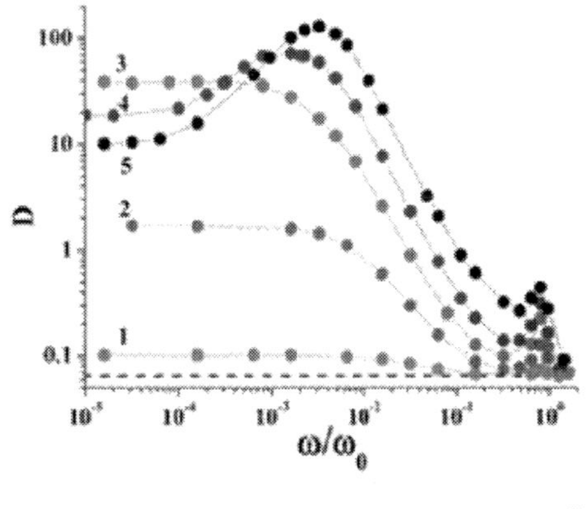

Figure 3. Diffusion coefficient D' as a function of the frequency for different amplitudes of the external field force. $T = 240$ K ($T' = 0.26$). Curve 1 - $F' = 0.02$; 2 - $F' = 0.06$; 3 - $F' = 0.1$; 4 - $F' = 0.15$; 5 - $F' = 0.25$.

more intensive diffusion corresponds to a lower force value. An increase in the force amplitude results in a shift of the maximum D'_{mx} to the higher frequencies.

Let us now consider the effect of the temperature on the diffusion frequency dependence. Figure 4 presents the plots of these relationships at a fixed value of $F'_0 = 0.15$ for different temperatures. All frequency curves have a maximum at low frequencies; the lower the temperature, the higher the D'_{mx} value. Such an abnormal behavior of the diffusion as a function of the temperature has a different nature in comparison with that of the particle diffusion exposed to a permanent force [8] discussed above. At a constant force, an exponential increase in D' with a drop in the temperature is brought about by the infinite increase in the relaxation time. However, in the case of action of a time-periodic force, we have a natural time limitation defined by the oscillation period of the external field. The abnormal temperature dependence of the enhanced diffusion is related to the existence of time-limited hyperdiffusion modes. The diffusion enhancement can reach many orders of magnitude.

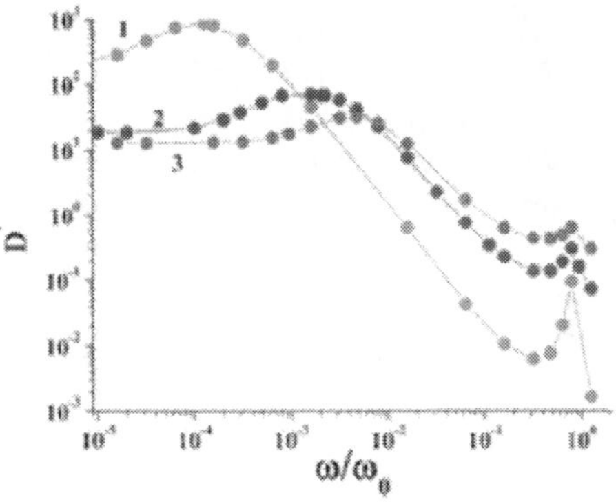

Figure 4. Coefficient D' as a function of the frequency at different temperatures. The periodic force amplitude $F' = 0.15$. 1 - $T = 120$ K ($T' = 0.13$); 2 - $T = 240$ K ($T' = 0.26$); 3 - $T = 360$ K ($T' = 0.53$); ω_0 - frequency of self-oscillations.

Figure 4 shows that at $T = 120K$ the diffusion is enhanced by more than seven orders of magnitude. The temperature drop results in a shift of the maximum to the low frequencies domain.

4. Conclusions

This paper presents studies on abnormal diffusion enhancement with the temperature drop in an external periodic field. It is shown that the diffusion enhancement depends strongly on the field amplitude and frequency.

The research reported opens new prospects for the creation of surface microstructures with required properties. This becomes especially important for low temperature methods sensitive to heat exposure. BLAG is an example of such a technique [12].

References

[1] Prakash S and Karacor M B 2009 *Surf. Sci. Rep.* **64** 233
[2] Hänggi P and Marchesoni F 2009 *Rev. Mod. Phys.* **81** 337
[3] Bruch L W, Diehl R D and Venables J A 2007 *Rev. Mod. Phys.* **79** 1381
[4] Antczak G and Ehrlich G 2007 *Surf. Sci. Rep.* **62** 39
[5] Lindenberg K, Sancho J M, Lacasta A M and Sokolov I M 2007 *Phys. Rev. Let.* **98** 020602
[6] Sancho J M and Lacasta A M 2010 *Europhys. Let.* **J187** 49
[7] Marchenko I G and Marchenko I I 2012 *JETP Let.* **95/**3 137
[8] Marchenko I G and Marchenko I I 2012 *Europhys. Let.* **100** 5005
[9] Lindenberg K, Lacasta A M, Sancho J M and Romero A H 2005 *New J. Phys.* **7** 29
[10] Lacasta A M, Sancho J M, Romero A H, Sokolov I, Mand Lindenberg K 2004 *Phys. Rev.* **E70** 051104
[11] Zheng C Z, Yeung C K, Loy M T and Xiao X 2006 *Phys. Rev. Let.* **97** 166101
[12] Huang L, Chey S J and Weaver J H 1998 *Phys. Rev. Let.* **80** 4095

Monte Carlo modelling of positron transport in real world applications

S Marjanović[1], A Banković, M Šuvakov and Z Lj Petrović

Institute of Physics, University of Belgrade, 118 Pregrevica, 11080 Belgrade, Serbia

E-mail: msrdjan@ipb.ac.rs

Abstract. Due to the unstable nature of positrons and their short lifetime, it is difficult to obtain high positron particle densities. This is why the Monte Carlo simulation technique, as a swarm method, is very suitable for modelling most of the current positron applications involving gaseous and liquid media. The ongoing work on the measurements of cross-sections for positron interactions with atoms and molecules and swarm calculations for positrons in gasses led to the establishment of good cross-section sets for positron interaction with gasses commonly used in real-world applications. Using the standard Monte Carlo technique and codes that can follow both low- (down to thermal energy) and high- (up to keV) energy particles, we are able to model different systems directly applicable to existing experimental setups and techniques. This paper reviews the results on modelling Surko-type positron buffer gas traps, application of the rotating wall technique and simulation of positron tracks in water vapor as a substitute for human tissue, and pinpoints the challenges in and advantages of applying Monte Carlo simulations to these systems.

1. Introduction

Owing to the antimatter nature of positrons, it has always been a struggle to obtain large numbers of positrons and store them for long periods of time [1]. The development of buffer gas positron traps (or Surko traps) [2,3,4,5,6] over the last two decades allowed for a constant increase in positron yields and storage times. This advancement in the experimental facilities for handling positrons led to an increased accuracy in the measurements of cross-sections for positron-matter interaction for various atoms and molecules, especially at lower energies [7,8]. In turn, the assembled cross-section data sets, if sufficiently complete (in terms of number, momentum and energy balances) [9], can be used as an input for different theories and computer simulations.

With the exception of few cases where obtaining positron plasma is the primary experimental goal [3], most positron experiments do not reach, or even aim at, such high particle densities where positron-positron interactions control the properties of the ensemble [4,5,6]. Other positron applications, like positron emission tomography (PET) scans, due to their nature, also do not produce a high enough positron density to observe positron-positron interactions. This is why most real-world applications of positrons can be analyzed using swarm calculations. The Monte Carlo technique [10], in particular, lends itself perfectly to modelling experimental conditions and geometries, as opposed to the Boltzmann codes [11], where the boundary conditions implementation can be very complicated.

[1] To whom any correspondence should be addressed.

Additionally, the Monte Carlo technique accuracy is limited only by the quality of the statistics (i.e., calculation time) and the accuracy of the cross-sections.

As many modelled applications require treatment of positrons at nearly thermal energies, it is important to include thermal effects into the calculations. Concerning the Monte Carlo technique, the problem of thermalization has only recently been adequately solved [12] providing a proper finite value of the mean energy. This technique is included in the present simulation.

2. Modelling buffer gas positron traps

The buffer gas positron trap (Surko trap) is a cylindrical device filled mostly with nitrogen gas that is used to capture, accumulate and cool a large number of positrons. Typically, the number of captured positrons ranges from $10^5 - 10^7$ e^+/s with trapping efficiencies of 10 – 25 % [2,3,4,5,6].

The main operating principle of the Surko trap is the utilization of $a^1\Pi$ electron excitation of the N_2 molecule [2]. The incoming positron energy is adjusted by a bias potential to target the cross-section for this molecule excitation just above its threshold where the process cross-section is comparable with the cross-section for positronium (Ps) formation. Through this excitation, the positrons transfer energy to buffer N_2 molecules and remain trapped. The threshold energy for this process is ~8.5 eV, which is (roughly) the energy that a positron loses in the collision. Unfortunately, the threshold for Ps formation lies very close to the threshold for $a^1\Pi$ electron excitation, so that the efficient trapping window is less then 1 eV wide [2].

We model this interaction of positrons with the N_2 background in our Monte Carlo code [13] by using the cross-section set from [9]. The fact that conditions change along the axis of the trap is taken into account by separating the simulation in three segments with different chamber diameters, gas pressures and electric potentials, which represent actual segments of the apparatus. The conditions in each segment are considered uniform.

The potential gradient existing between the segments of the trap is neglected in the model without reducing the accuracy of the simulation as this transient region is very short, much shorter than the mean-free-path of positrons under the conditions in question. Instead, as a positron reaches the end of a segment, it is immediately transferred to the next one if it has enough axial energy to overcome the potential difference. Its axial energy is increased or decreased depending on the potential difference between the segments and the direction of the positron motion. In the case when a positron's axial energy is insufficient to overcome the potential barrier, it is reflected back.

To investigate the operation of the trap, we can sample the properties of the swarm of positrons at each stage separately or, sample the entire swarm inside the trap. Through such sampling, we can observe the evolution of the positrons' energy distribution, and see how inelastic collisions with high thresholds and transitions over different potential levels first split one particle beam into several beams with well-defined discreet energies [13]. This structure is analogous to the Frank-Hertz experiment. In time, and due to collisions with smaller energy losses, these separate beams all relax to a single cloud of positrons at thermal temperature. This can be seen in figure 1, which presents the temporal development of the energy distribution function (EDF) where the magnitude is color coded while the vertical axis represents the energy. The beam enters the trap with 10-eV of energy. It bifurcates for a very brief period of time to two beams when the next stage is entered and

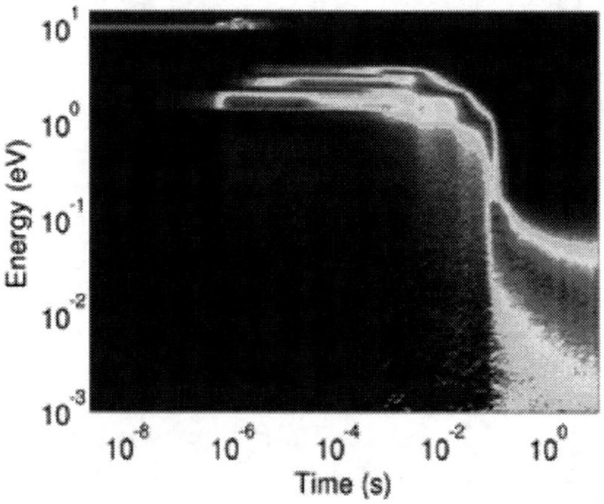

Figure 1. Evolution of the positrons' kinetic energy distribution [13].

no collisions have occurred. On the other hand, the three-pronged EDF is due to different inelastic collisions and each group represents positrons that have undergone one inelastic collision. Due to elastic and further inelastic collisions (with smaller losses), these three-beam-like distributions merge into one broad distribution that is then thermalized to the trap temperature. The thermalization time may be determined from such a simulation and compared to experiments to verify the model and the cross-sections.

Another aspect investigated by us is the loss processes that take place inside the trap and limit the number efficiency of the trap. What we see in our simulations is that at every occasion where positrons are trapped through $a^1\Pi$ electron excitation of N_2, approximately 50 % of them are lost to positronium formation. This is due to the fact that the intrinsic energy width of the modern solid neon moderators is ~1.5eV, which is relatively broad compared to the width of the trapping window between the $a^1\Pi$ electron excitation and the Ps formation channel of N_2. Since most Surko trap applications use a three- [2,3], or two-stage design [4,5,6], the maximum number efficiency that such devices can achieve is 12.5 – 25 %. Additional losses may arise due to losses to the walls, direct annihilation, and particles escaping the trap, but these losses can be reduced through experimentation and by changing the geometry and pressures inside the trap.

3. Modelling the rotating wall compression technique

Once all the positrons have been accumulated inside the Surko trap, and in order to use them efficiently in an experiment, the positron cloud needs to be compressed to be as narrow as possible. The rotating wall compression technique has long been applied to compression of single component plasmas [14]. Recently, it has been demonstrated to work even in the single particle regime, but apparently the proposed compression mechanisms differ [15,16]. As the effect occurs under the swarm conditions, our Monte Carlo code is the appropriate tool to analyze this system.

In modelling the rotating wall compression, we follow an approach similar to that of Isaac et al. [17], with the exception of the treatment of positron collisions with the background gas. Instead of adopting an effective viscous term in the equations of motion, we use our Monte Carlo procedure with the full set of interaction cross-sections. As the compression is reported to work for buffer gasses that have a strong vibrational channel [16], we run the simulations for CF_4 [9]. We can thus observe the compression for different conditions of the applied magnetic and rotating electric fields. The difficulty of employing the MC approach in modelling this particular application lies in the fact that the effect is observable at a ms timescale, while the highest characteristic motion frequency is of the order of GHz. This limits the time step of simulation to the order of several picoseconds and significantly lengthens the computation. Any practical computer code would have to be parallelized in order to achieve results at reasonable execution times. Fortunately, the Monte Carlo approach lends itself perfectly to parallelization as long as this problem is linear in nature. This is why a number of particles can be assigned to each computing unit for calculation while at predefined intervals they each report their state and entire swarm is sampled for its properties.

Figure 2 shows the profile of the positron beam at different moments [18]. The conditions simulated are: 1.5 V of applied rotating voltage at 9.63 MHz; axial magnetic field of 250 Gauss; while the axial bounce frequency of the positrons is 9.49 MHz. The pressure inside the chamber is 40 μTorr of CF_4. The radius of the positron ensemble is half the dimension of the box presented in the figure and is provided above each frame. As the rotating field is applied, the positrons are heated but lose energy preferentially near the center and remain trapped in a narrower distribution. The development is shown in frames. At the same time, the ensemble thermalizes back to a very low mean energy. Both a low energy and a small radius are required in order to optimize the production of anti-hydrogen.

4. Modelling positron interaction with tissues

The last example of positron application modelling that we address in this paper is modelling the positron interaction with a living tissue. This line of research is important for positron emission in view

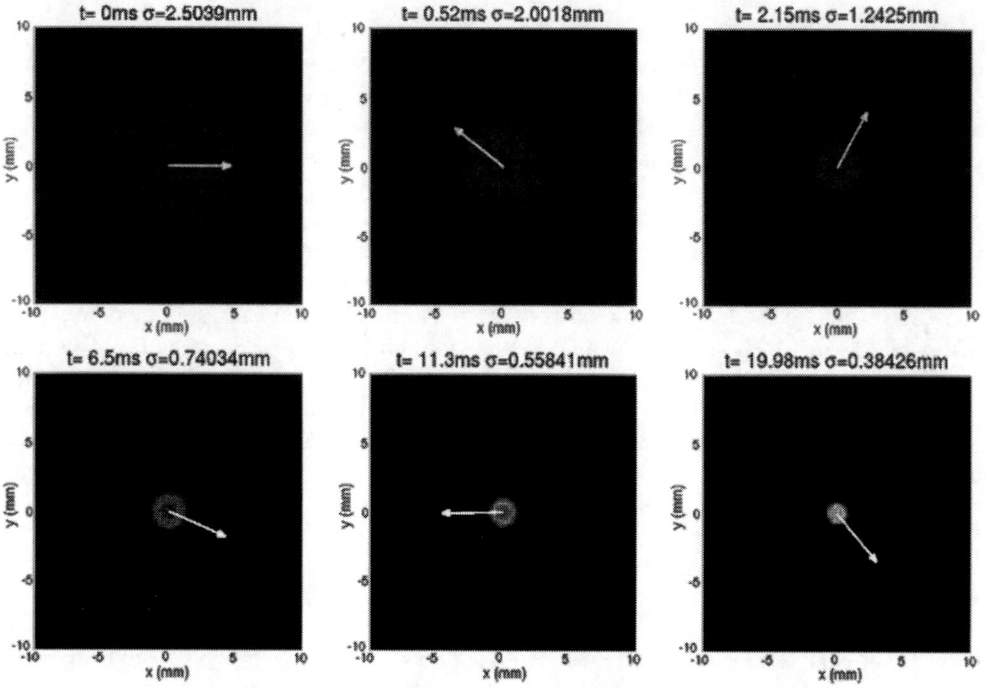

Figure 2. Profile of the positron beam sampled at different moments of compression. The white arrow shows the direction of the applied field at the moment of sampling [18].

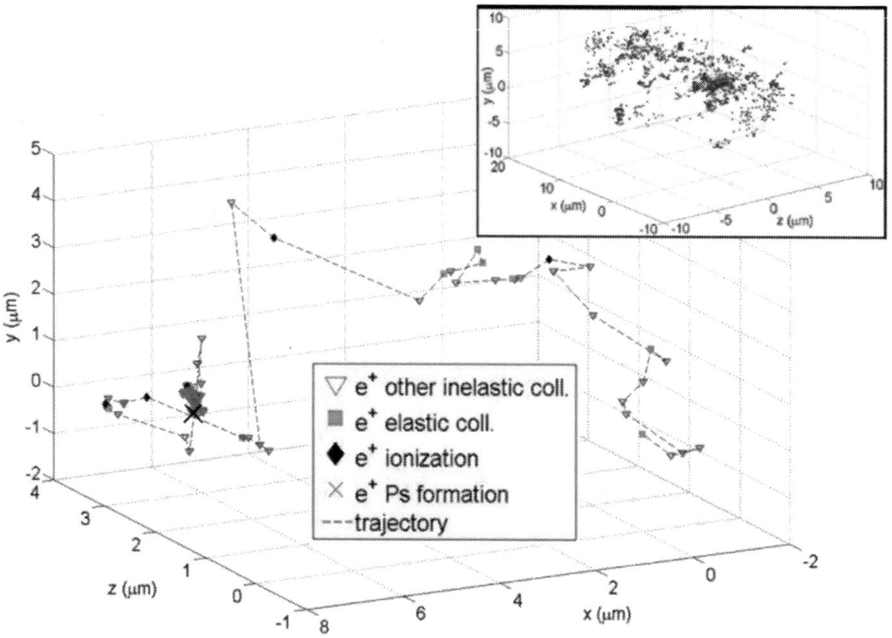

Figure 3. Example of a positron track in a water vapor background at atmospheric pressure. Inset: the same positron track with points of electron-H_2O collisions included [20].

of investigating the damage inflicted by positrons on the surrounding tissue. It also provides an insight into a possible positron radiation therapy.

As a first-order approximation for the living tissue model, we use cross-sections for water vapor at atmospheric pressure [9]. The positrons are initialized at one point and released isotropically into the medium with an energy of 1 keV. Using our MC code, we are able to follow the trajectory of positrons up until the formation of positronium, and then the newly formed positronium until annihilation. We are also able to follow any secondary electrons produced in ionization events, either by original positrons, or by other secondary electrons. The velocity of the electrons produced in an ionization event is determined by the Opal-Peterson-Beaty [19] formula, while obeying the conservation laws. As a single positron can produce several electrons, and there is no efficient loss channel for electrons, the number of electrons can become very high very fast. This has a detrimental effect on the speed of simulation. For this reason, when an electron thermalizes (in effect reaches energy less than 100 meV) it is considered harmless and is removed from the simulation.

Figure 3 shows an example of a visualized track that a positron makes. Included in the inset of the figure 3 as blue dots laid over the positron's track are the points of collisions of the secondary electrons. This shows that the range of electrons can be much larger than that of positrons, collisions are much more frequent and that electrons could actually play a more direct role in tissue damage. This can be seen more accurately in the energy deposition to the medium by both electrons and positrons reported in [20].

5. Conclusions

A large part of positron applications lies in the domain of swarm physics, where Coulomb interactions do not play a significant role. Monte Carlo simulation, as a well-developed swarm tool, allows us to investigate these systems in detail, and proceed from such fundamental properties as cross-section data, through sampling of averaged observables, to estimating the macroscopic qualities of large-scale, applicable, devices. The flexibility of this numerical method allows us to adapt it quickly to many different, collision dominated, gas-filled systems.

In this paper we show how the simple Monte Carlo technique may be used in combination with accurate available data for cross-sections in much the same way as it is applied to electron kinetics in ionized gases and plasmas to describe the transition from beams to swarms of charged particles, to manipulate those ensembles and to describe their major applications. This paper gives an overview of our technique, while a more detailed description of particular applications will be given in focused technical papers.

References

[1] Mills Jr A P and Gullikson E M 1986 *Appl. Phys. Lett.* **49** 1121
[2] Murphy T J and Surko C M 1992 *Phys. Rev.* A **49** 5696-5705
[3] Surko C M and Greaves R G 2004 *Phys. Plasmas* **11** 2333-48
[4] Sullivan J P, Jones A, Caradonna P, Makochekanwa C and Buckman S J 2008 *Rev. Sci. Instrum.* **79** 113105
[5] Clarke J, van der Werf D P, Griffiths B, Beddows D C S, Charlton M, Telle H H and Watkeys P R 2006 *Rev. Sci. Instrum.* **77** 063302
[6] Cassidy D B, Deng S H M, Greaves R G and Mills A P 2006 *Rev. Sci. Instrum.* **77** 073106
[7] Marler J P and Surko C M 2005 *Phys. Rev.* A **72** 062702
[8] Makochekanwa C, Bankovic A, Tattersall W, Jones A, Caradonna P, Slaughter D S, Nixon K, Brunger M J, Petrovic Z, Sullivan J P and Buckman S J 2009 *New J. Phys.* **11** 103036
[9] Petrović Z Lj, Banković A, Dujko S, Marjanović S, Malović G, Sullivan J P and Buckman S J 2013 *AIP Conf. Proc.* **115** 1545
[10] Dujko S, White R D, Petrović Z Lj and Robson R E 2010 *Phys. Rev.* E **81** 046403
[11] White R D, Robson R E, Dujko S, Nicoletopoulos P and Li B 2009 *J. Phys.* D: *Appl. Phys.* **42** 194001

[12] Ristivojevic Z and Petrović Z Lj 2012 *Plasma Sources Sci. Technol.* **21** 035001
[13] Marjanović S, Šuvakov M, Banković A, Savić M, Malović G, Buckman S J and Petrović Z Lj 2011 *IEEE Trans. Plasma Sci.* **39** 2614
[14] Danielson J R, Surko C M and O'Neil T M 2007 *Phys. Rev. Lett.* **99** 135005
[15] Cassidy D B, Deng S H M, Greaves R G and Mills A P 2006 *Rev. Sci. Instrum.* **77** 073106
[16] Greaves R G and Moxom J M 2008 *Phys. Plasmas* **15** 072304
[17] Isaac C A, Baker C J, Mortensen T, van der Werf D P and Charlton M 2011 *Phys. Rev. Lett.* **107** 033201
[18] Marjanović S, Banković A, Šuvakov M and Petrović Z Lj, 2013 *Bull. Am. Phys. Soc.* **58** 90-91
[19] Opal C B, Peterson W K and Beaty E C 1971 *J. Chem. Phys.* **55** 4100-06
[20] Petrović Z Lj, Marjanović S, Dujko S, Banković A, Malović G, Buckman S, Garcia G, White R and Brunger M 2013 *Appl. Radiat. Isotopes* Available: 10.1016/j.apradiso.2013.01.010

Journal of Physics: Conference Series **514** (2014) 012047 doi:10.1088/1742-6596/514/1/012047

Experimental and numerical investigation of the refinement of Hf by EBM

Veliko Donchev[1], Katia Vutova and Vania Vassileva

Acad. E. Djakov Institute of Electronics, Bulgarian Academy of Sciences,
72 Tsarigradsko Chaussee, 1784 Sofia, Bulgaria

E-mail: velikod@gmail.com

Abstract. Electron beam melting (EBM) in vacuum is one of the most promising technologies for refining and recycling of metals that react with oxygen when heated. Hf is such a metal. Pure Hf (with a small content of gas and metal impurities) is needed for a variety of applications in the aerospace industry and metallurgy, in the production of components of nuclear reactors, microprocessors, optical components etc. We conducted experiments with the ELIT-60 equipment on Hf ingots at electron beam powers of 12, 15, 17 kW and obtained data about the concentration of impurities by the ICP-MS method. For further understanding and optimizing the Hf refining processes, a non-stationary heat model was applied for numerical simulation of the heat transfer processes. Simulation data about the liquid pool variation during the e-beam treatment was thus obtained. The flatness of the crystallization front shape, which is connected to the structure quality, was investigated by optimization criteria related to the curvature of the liquid/solid boundary curve. We also describe an algorithm for calculation of the criteria. One of the criteria was applied to EBM of Hf for different electron beam powers; the results obtained were confirmed by the experimental data. Combining experimental, theoretical and simulation results, a proper technological regime is proposed for better Hf refining.

1. Introduction

Electron beam melting and refining (EBMR) is a method applied in the special electrometallurgy for production of pure metals and alloys and of new materials [1-6]. The electron beam technologies and, in particular, the technologies for EBMR of metals and alloys have been recognized as a competitive, if not the only, method for obtaining new materials for uses in various fields: nuclear industry, medicine, electronics, instrument engineering, transport, etc. The principle and a description of the EBMR process have been presented in [1, 2, 3, 6]. Computational modeling gives a possibility for a better understanding and investigating the heat transfer mechanisms and can be used for optimizing the electron beam process for obtaining new materials with improved characteristics. In [7], a stationary heat model was described and applied to studying the EBMR process. This mathematical model was extended to a time-dependent heat model with a Pismen-Rekford numerical scheme that is absolutely stable with respect to the time [8].

[1] To whom any correspondence should be addressed.

A mathematical optimization technique, including a proposition for analytical optimization criteria, their discretization according to the numerical scheme and methods for solving the discrete optimization problems, were proposed and applied as well [8]. In this paper, an optimization problem for achieving flatness of the molten pool shape (liquid/solid metal boundary) is proposed. The criterion is based on calculating the curvature of the liquid/solid boundary which describes the geometry of the molten pool contour and is very important for studying and optimizing the quality of the pure metal obtained after EBMR. The flatness of the liquid/solid boundary is directly connected to the quality of the structure of the metal obtained [3, 6, 8].

2. An optimization problem for EBMR of metals

2.1. Formulation of the optimization problem

The temperature fields during EMBR of metals and alloys in cylindrical ingots are investigated by applying a non-stationary heat model [8]. By $T(r;\ z;\ t)$ we denote the temperature at a distance r, height z and heating time t, due to the assumed radial symmetry [3, 8].

The flat molten pool shape permits the formation of a vertical dendrite structure and a uniform impurities' displacement toward the ingot top surface. At a fixed moment of heating t_f, the liquid/solid contour can be defined mathematically as a curve:

$$\Gamma(t_f)=\{(r,z)\,|\,T(r,z,t_f)=T_{melt}\}, \tag{1}$$

where T_{melt} is the melting temperature of the investigated metal or alloy.

The following optimization criteria for minimization of the curvature of the crystallization front $\Gamma(t)$ are defined:

$$\mathrm{K}_{1,max}(P_b,r_b,V;t_f)=\max_{s\in[S_1,S_2]}\left\|\frac{d\Gamma(t_f)}{ds}\big(r(s),z(s)\big)\right\|\to min, \tag{2}$$

$$\mathrm{K}_{max}(P_b,r_b,V)=\int_{t_1}^{t_2}\mu_1(t)\mathrm{K}_{1,max}(P_b,r_b,V;t)dt\to min, \tag{3}$$

where $[t_1,t_2]$ is a fixed heating time interval, μ_1 is a weight function, s is a natural parameter for $\Gamma(t)$ and t_f is a fixed moment of heating. The temperature field T is the solution of equation (4) subject to the initial condition (5) and boundary conditions (6-9) for metal samples with height H and diameter $D = 2R$, for electron beam radius r_b and heating time F.

$$\frac{1}{r}\frac{\partial}{\partial r}\left(r\frac{\partial T}{\partial r}\right)+\frac{\partial^2 T}{\partial z^2}+\frac{V}{a}\frac{\partial T}{\partial z}=\frac{\rho C_p}{\lambda}\frac{\partial T}{\partial t},\quad x\in\Omega=(0,R)\times(0,H)\times(0,F], \tag{4}$$

$$T(x)=T_0(r,z),\quad x\in\Omega_0=\{x\in\overline{\Omega}:t=0\}, \tag{5}$$

$$\lambda\frac{\partial T}{\partial z}(x)=\left|P_s(r,t)-\alpha\sigma\left(T^4(x)-T^4_{room}\right)-C_p.W_v.T(x)\right|,\quad x\in\{x\in\overline{\Omega}:z=H,t>0\} \tag{6}$$

$$\lambda\frac{\partial T}{\partial r}(x)=\lambda_2\frac{\partial T^1}{\partial r}(x),\quad x\in\{x\in\overline{\Omega}:H-Q\le z\le H,r=R,\ t>0\}, \tag{7}$$

$$\lambda\frac{\partial T}{\partial r}(x)=-\alpha\sigma\left(T^4(x)-T^4_{room}\right),\quad x\in\{x\in\overline{\Omega}:0<z<H-Q,r=R,\ t>0\}, \tag{8}$$

$$\lambda\frac{\partial T}{\partial z}(x)=\lambda_2\frac{\partial T^2}{\partial z}(x),\quad x\in\{x\in\overline{\Omega}:z=0,\ t>0\}. \tag{9}$$

The control variables (on which the criteria depend) are the beam power P_b, the beam radius r_b and the casting velocity V. P_b is connected with the e-beam energy distribution P_s. For a fixed heating time, t_f, the beam energy distribution $P_s(r, t_f)$ is a Gaussian-like function and:

$$\iint\limits_{[0,R]\times[0,2\pi]} rP_s(r,t_f)drd\varphi = P_b \qquad (10)$$

In (4)-(9) ρ, [g/m^3], is the density of the metal; C_p, [W s/g K], is the heat capacity; λ, [W/m K], is the heat conductivity; a, [m^2/s], is the heat diffusivity, $a = \frac{\lambda}{\rho C_p}$. The term $\frac{V}{a}\frac{\partial T}{\partial z}$ characterizes the casting, i.e. the heat added by the poured molten metal; α is the metal's emissivity, $\sigma = 5.6704.10^{-8}[J/s.m^2K^4]$ is the Stefan-Boltzmann constant; $C_p.W_v.T$ describes the evaporation losses, T_{room} is the room temperature. More details about the heat model (4)-(9) are presented in [8].

2.2. Discretization of the optimization problem and numerical solution methods

The discretization of the problem (4-9) is made on the net:

$$W_{h_1,h_2,\tau} = \{(r_i, z_j, t_n) \mid r_i = ih_1, z_j = jh_2, t_n = n\tau; i = \overline{0,N}, j = \overline{0,M}, n = \overline{0,P}\} \qquad (11)$$

by a modifed Pismen-Rekfort scheme which is absolutely stable in τ. In the numerical scheme, the approximation of the differential operators in eq.(4) is made via the difference operators $\Lambda_1 T_i = (1/r_i)(r_{i-1/2}T_{\bar{r},i})_{r,i}, \Lambda_2 T_j = T_{\bar{z}z,j} + (V/a)T_{\overset{\circ}{z},j}$. For input functions P_b and V, the discrete temperature field $\{T^n_{i,j}\}^{n=\overline{1,P}}_{i=\overline{1,N},j=\overline{1,M}}$ is calculated applying the non-stationary heat model [8]. Hence, if the criteria (2) and (3) are approximated over the net (11), the defined analytical problem (2, 4-9) or (3-9) is transformed to a discrete one.

For numerical calculation of the criteria (2) and (3), the curve $\Gamma(t)$ can be approximated by a discrete set of points $G(t_f)$ at a time t_f (figure 1):

$$G(t_f) = \left\{(r_i, z_j) : \left|T^{n_f}_{i,j} - T_{melt}\right| < \varepsilon\right\} \qquad (12)$$

for an apriori fixed $\varepsilon > 0$. The choice of ε depends on the steps h_1, h_2. Then $\Gamma(t_f)$ can be interpolated using a linear regression method by an analytical function $\Gamma^l(t_f)$ (figure 1). After parameterizing the curve $\Gamma^l(t_f)$, its curvature $k^l(r;t_f) = \left\| \frac{d\Gamma^l(t_f)}{ds}\big(r(s), z(s)\big) \right\|$ (figure 1-b) can be easily calculated by an explicit formula in each point of the curve. Usually, the parameter is r and the curvature is calculated via the formula:

$$k^l(r;t_f) = \frac{|f''(r;t_f)|}{\sqrt{((f'(r;t_f))^2 + 1)^3}}. \qquad (13)$$

For criterion (3), the calculation (13) must be repeated for all points in a regular net of points over the area $D=[r_0,r_{max}]\times[t_1,t_2]$, where: $r_{max}=max\{r\in[r_0;R]\,|\,T(r,H,t_f)>T_{melt}\}$, i.e. the radius of the molten pool and $0<r_0<r_{max}$. Then the integral (3) must be calculated by a quadrature or numerically.

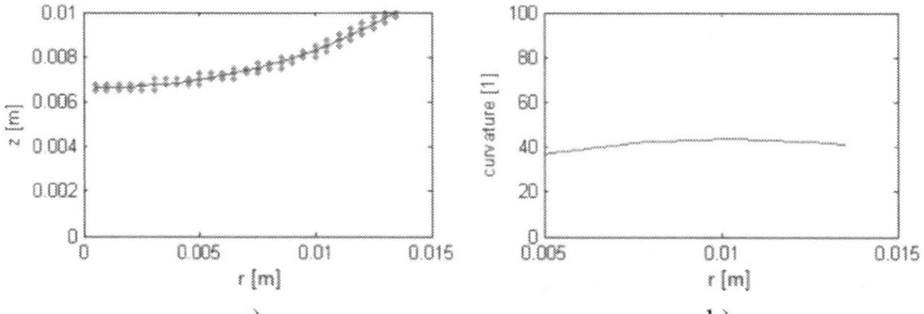

Figure 1. (a) An interpolation of the liquid/solid metal contour via a linear regression method; (b) the curvature of the approximated curve, $\|\frac{d\Gamma^{l}(t_f)}{ds}(r(s),z(s))\|$. The curve is approximated as a polynomial of degree 3 for ε = 100*K*. The numerical experiment is for Hf ingot with *H* = 10*mm; R* = 30*mm*; P_b = 12*kW*; r_b = 10*mm; V* = 0*mm/min*; t_f = 300*s*. 1 - the set $G(t_f)$, 2 - the curve $\Gamma^{l}(t_f)$.

After discretization of the model and implementation of the developed method for calculation of the criterion, a discrete optimization problem has to be solved. For a fixed P_b, r_b and *V*, the discrete temperature field $\{T^n_{i,j}\}^{n=\overline{1,P}}_{i=\overline{1,N},j=\overline{1,M}}$ and the criterion (2)-(3) are calculated. Optimal values of the control parameters are sought for which the criteria (2)-(3) are minimized. For solving the discrete optimization problem obtained, heuristic methods can be applied because the dependence of the criteria on the control variables is implicit [8].

3. Results and discussion

Electron beam melting and refining of hafnium was performed using 60 kW equipment (ELIT-60 in the IE-BAS). The feeding material – hafnium chips, were degreased and compacted in the form of disks with a diameter of 60 mm and a height of 10 mm. To investigate the refining processes in Hf recycling, different technological regimes (conditions) were implemented. The ranges studied of the process parameters (electron beam power P_b and refining time *F*) were: P_b = 12, 15, 17 kW and *F* in the range 2-5 min. The beam radius was r_b = 10*mm*. Using the ICP-MS method, results on the removal of impurities were obtained, namely, the highest purification of Hf ingot was achieved at a beam power of 12 kW and a short processing time [9].

A simulation of the EBM of Hf was made for the above technological regimes. The thermal fields were calculated for all beam powers and the criterion (2) was calculated for t_f = 1, 2, 3, 4, 5 min. The liquid/solid contour $\Gamma(t)$ was approximated by a linear regression with a polynomial of degree 3.

The function Γ^{l} was parametrized by the radial distance *r*. Thus, the criterion used was:

$$K_{1,max}(P_b,r_b,V;t_f)=\max_{r\in(0,r_1]}\frac{|f''(r;t_f)|}{\sqrt{((f'(r;t_f))^2+1)^3}}, \tag{14}$$

where $\{(r, f(r;t_f)) : 0 \le r \le r_{max}\}$ is the parametrization of the approximation Γ^l of the molten pool contour at the moment t_f, calculated for r_b = 10 mm; V = 0 mm/min, r_1 = 12 mm and for $P_b \in \{12kW, 15kW, 17kW\}$. The data obtained for $K_{1,max}(P_b, r_b, V, t_f)$ (denoted as $K_{1,m}$ in Table 1), the depth h_m of the liquid metal pool, the ratio S_m/V_m of the melted top surface areas S_m and melted volume V_m is shown in table 1.

Table 1. Calculated parameter values at EBM of Hf.

P_b, kW	$K_{1,m}$	h_m, [mm]	S_m/V_m, [mm^{-1}]	$K_{1,m}$	h_m, [mm]	S_m/V_m, [mm^{-1}]	$K_{1,m}$	h_m, [mm]	S_m/V_m, [mm^{-1}]	$K_{1,m}$	h_m, [mm]	S_m/V_m, [mm^{-1}]
		t=2min			t=3min			t=4min			t=5min	
12	49.2	3.67	47.60	49.7	3.67	46.64	48.4	3.67	46.48	43.7	3.67	46.48
15	49.2	5.8	41.87	52.3	6.00	44.29	52.1	6.00	44.29	52	6.0	44.29
17	45.8	4.61	41.25	46.8	4.61	43.44	47.4	4.61	43.44	47.4	4.61	43.44

These results allow one to recommend values of the technological parameters in order to optimize the EBM process. The minimal curvature, i.e. the maximal flatness of the molten/solid metal contour, permits the formation of a better structure in the cast ingot; the lowest value of $K_{1,max}$ is observed at a heating time t_f = 5 min and electron beam power P_b = 12 kW.

The minimal value of the liquid pool depth and high values of the ratio S_m/V_m and of S_m ensuring more efficient refining processes (higher purification) are achieved at the same parameters values (P_b = 12 kW; t_f = 5 min) - table 1. These recommended regime conditions for obtaining a good quality (structure and composition) of the refined material correspond to the experimental result in [9]. The approximated liquid/solid contour and the graph of the curvature in Fig. 1 correspond to the numerical experiment for the lowest value of $K_{1,max}(P_b, 10, 0, t_f)$ achieved at P_b = 12 kW and t_f = 5 min. An important advantage of the optimization scheme proposed is that experimental data from a chemical analysis of the impurities' concentrations for different technological regimes are not needed.

4. Conclusions

In this paper, the EBMR process of Hf is investigated experimentally and theoretically (by a non-stationary heat model with a corresponding optimization technique) with the purpose of improving the quality (structure and composition) of the ingots treated. The optimization problem proposed aims to obtain maximum flatness of the liquid/solid metal contour that permits formation of a better structure of the cast metal (vertical dendrite structure and an uniform impurities' displacement toward the top surface of the cast ingot) during EBMR. The maximal curvature of this boundary in terms of the beam power, beam radius and casting velocity is minimized analytically. Simulation results concerning the liquid pool geometry are presented and recommendations are given for appropriate process conditions ensuring better refining processes (higher purification). The technological parameters recommended for obtaining good-quality refined materials correspond to the experimental results showing that the minimal impurities concentrations (higher purification of Hf) are obtained at 12 kW beam power and short heating time for the ranges investigated of the process conditions. The mathematical heat model and the corresponding optimization scheme for achieving maximum flatness of the liquid/solid boundary can be applied to investigating and optimizing the EBMR regimes for different metals without the need of experimental results for the impurities' concentrations.

Acknowledgments

The authors express gratitude to Mr. R. Nikolov and Mr. A. Stoimenov for the technical support. This research was funded by the National Science Fund at the Ministry of Education, Youth and Science of Republic of Bulgaria under contracts DO 02-127 (BIn- 5/2009).

Journal of Physics: Conference Series **514** (2014) 012047 doi:10.1088/1742-6596/514/1/012047

References

[1] Mitchel A and Wang T 2000 *Proc. Conf. Electron Beam Melting and Refining. State of the Art 2000* (October 2000 USA) p 2

[2] Mladenov G 2009 *Electron and Ion Technologies* (Sofia Academic Publ House) in Bulgarian

[3] Mladenov G, Koleva E, Vutova K and Vassileva V 2011 Experimental and theoretical studies of electron beam melting and refining In *Practical Aspects and Applications of Electron Beam Irradiation* ed M Nemtanu and M Brasoveanu (Research Signpost) p 43

[4] Bakish R 2006 *Proc. 7th Int. EBT Conf.* (June 2006 Bulgaria) p 233

[5] Murr L, Gaytan S, Ramirez D, Martinez E, Hernandez J, Amato K, Shindo P, Medina F and Wicker R 2012 *J. Mater. Sci. Technol.* **28**/1 1

[6] Choudhury A and Hengsberg E 1992 *ISIJ Int.* **32**/5 673

[7] Vutova K, Donchev V, Vassileva V and Mladenov G 2013 *J. Metal Sci. and Heat Treat.* **11**/701 53

[8] Vutova K and Donchev V 2013 *Materials* **6** 4626

[9] Vutova K, Vassileva V, Mladenov G, Koleva E, Prakash T and Munirathnam N 2011 *Proc. Int. Conf. TMS* (April 2011 USA) vol 3 p 35

Investigation of the corona current in a vacuum bulb

Vasyl Chyhin

Lviv Academy, 32 Hvardiyska, Lviv, Ukraine

E-mail: vchygin@gmail.com

Abstract. The dependence of the negative corona current on the gas pressure was studied experimentally and theoretically in view of designing a gas-pressure sensor to be applied in the production of light bulbs. The gas pressure was varied in the range 1×10^{-2} Torr – 7.4×10^{2} Torr. The dependence of the current on the gas pressure is characterized by a strong heterogeneity. This allowed the implementation of a prototype of a high-speed sensor for a wide range of gas pressures. A mathematical model was developed of the negative corona current behavior by taking into account the ionization of the gas molecules, the attachment and detachment of electrons, the charge drift and the surface ion-electron emission. The results of the numerical simulations describe satisfactorily the experimental dependences.

1. Introduction

Studying of the negative corona current dependence on the gas pressure is of importance, particularly in what concerns the problem of implementing sensors for fast measurement of the gas pressure (air, nitrogen, argon, etc.) in the manufacture of light bulbs. Such measurements must be performed in the pressure range of 1×10^{-2} Torr – 1×10^{3} Torr within a time interval not exceeding 0.1 s – 0.2 s. This requirement has to do with the fact that the time of pumping and filling the light bulb gas at one position on the factory carousel is in the order of 1.8 s. At present, the gas pressure in light bulbs on a production line is not measured under dynamic conditions due to the large inertia of the known industrial sensors and their narrow operating range. For example, the thermal sensors of the VT-3 and FAB-1 type [1] have a time constant in the order of several seconds and cover a low gas-pressure range (in the order of 0.01 Torr – 30 Torr). The Pirani thermal-conductivity sensor [2] has the same inertia characteristics. Low-inertia sensors, such as the PMDH-1 deformation-discharge type [3] and the Baratron membrane-capacitive type [2] do not cover the pressure range required. Choosing the negative corona as a gas-pressure sensor is related to its characteristic property of undergoing pulsations and to the strong dependence of its parameters on the concentration of electronegative gases [4-7]. In particular, the corona pulses charge and current significantly decrease as the oxygen concentration in argon and nitrogen is increased.

2. Experimental setup

The measurements were carried out on the setup presented schematically in figure 1. It consists of a corona-discharge unit, a high-voltage power supply U, a measuring unit, intermediate memory and an inverter. The discharge unit is a metal bulb whose volume is equal to the volume of a typical light bulb. The cathode is a thin needle with a radius of 20 μm – 100 μm made of refractory metal. The anode is a flat metal disk with a diameter of 35 mm. The length, d, of the discharge gap is in the order of 0.6 cm. Vacuum-tight leads allow the high voltage to be fed in (through the limiting resistance R_o) and the output signal to be fed out (through the measuring resistance R_m). The high-voltage power supply provides a maximum discharge current of 70 mA and a minimum current of 0.1 mA with an accuracy of 0.3%. The measurement unit and intermediate memory consist of an analog-digital converter (ADC) type ADS1286 and a PIC16F876 processor-controller. The corona current detected, which can be as low as a few μA at low gas pressures, is amplified and digitized; the data obtained is stored in the intermediate memory and red by the inverter, which converts them into gas-pressure readings and displays them. The bulb is connected to a vacuum system equipped with a high-precision manometer and a PMT-2 type vacuum sensor. The system is evacuated by a 2NVR-5DM type vacuum pump; the air inflow is controlled precisely by a needle valve.

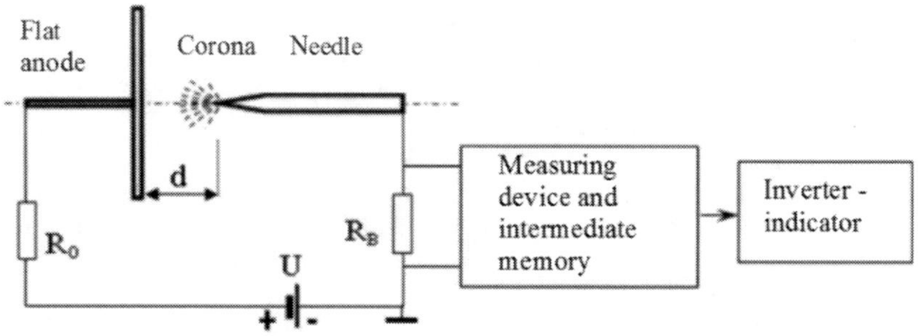

Figure 1. Schematic of the experimental setup.

3. Measurement results

Periodical pulsations were observed when a constant high voltage in the range 1500 V – 2000 V was applied to the anode-cathode gap. The average current reachs values in the order of 10^{-6} – 10^{-5} A. The fluctuations in the pulse rate and current amplitude were minimized by selecting optimal values of the voltage and the limiting resistance. Under such conditions, we recorded the pulse shape and duration and the peak and average currents. The shape and the other parameters of the negative-corona current pulses, as well as their dependence on the concentration of oxygen, differed substantially in the different gases. Thus, the current amplitude and the pulse duration and charge in N_2 were considerably greater than in air, whith the pulse rate being lower.

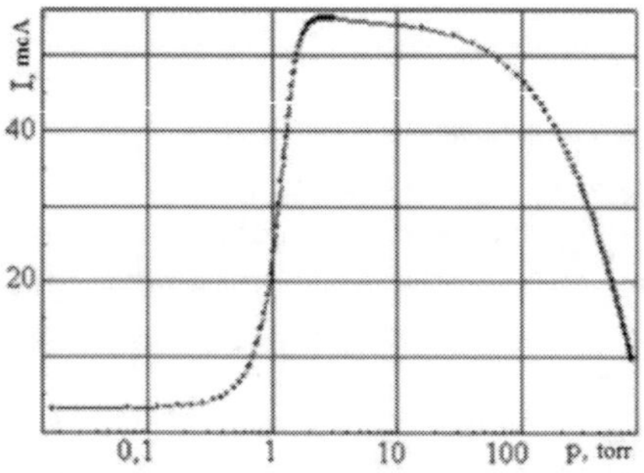

Figure 2. Typical measured dependence of the corona current on the air pressure.

Figure 2 presents a typical dependence of the corona current on the air pressure measured in the range 10^{-2} – 740 Torr. As one can see, the shape of curve $I(p)$ is complex. As the pressure is raised, the current rises sharply from a few to several tens of μA around an air pressure of 1 Torr. Then it forms a slightly inclined plateau in the range of 1 – 10 Torr, to fall gradually to a level in the order of 10 μA at an air pressure of 740 Torr.

Figure 3 shows a typical time dependence of the corona sensor current at one position on the factory lamp carousel. It is recorded every 0.2 seconds during the lamp being purged on the gas-vacuum line. As can be seen, within the time limits of the bulb residing at one position on the carousel (1.8 s), the current reaches maximal and minimal values and an exponential "saturation" of the current is observed. This is indicative of the possibility of applying this technique of fast gas pressure measurements to industrial production of light bulbs.

4. Modelling the corona discharge

Below we describe a simplified model of the negative corona current pulsation in a sharply non-uniform electric field. The following elementary kinetic processes were taken into account: ionization, excitation of gas molecules by electron collisions, ionization by photons, attachment of electrons to oxigen moleculs, their detachment from the O_2^- ion due to collisions, charge drift and surface photo- and ion-electron emission. The numerical computation was based on the differential continuity equations for fluxes of positive and negative ions and electrons, supplemented by Poisson's equation for an electrical field in a quasi-one-dimensional space:

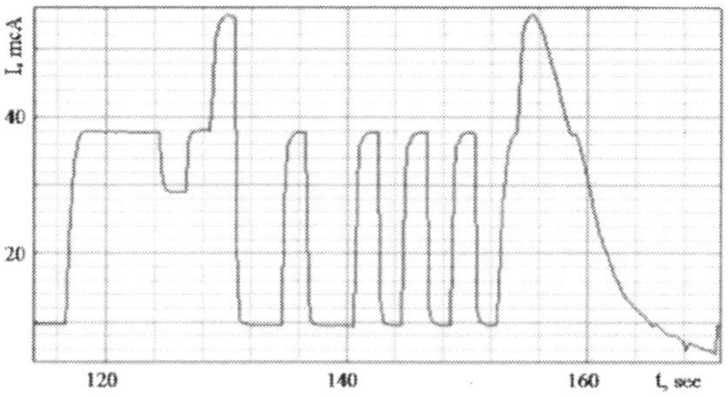

Figure 3. Typical plot of measured corona current in gas sensor on time of its recording.

$$\frac{\partial n_e}{\partial t} + div(w_e n_e) = \alpha n_e w_e - \eta n_e w_e + k_d n_n n_0 \quad , \tag{1}$$

$$\frac{\partial n_p}{\partial t} - div(w_p n_p) = \alpha n_e w_e \, , \tag{2}$$

$$\frac{\partial n_n}{\partial t} + div(w_n n_n) = \eta n_e w_e - k_d n_n n_0 \, , \tag{3}$$

$$divE = 4\pi e(n_p - n_e - n_n) \, . \tag{4}$$

Here n_e, n_p, n_n are the densities of electrons and positive and negative ions and w_e, w_p, w_n are their drift velocities; α, η, k_d are the ionization, attachment and detachment coefficients of the main gas molecules of density n_0. The boundary conditions for positive and negative ions are self-evident: their number density is equal to zero at the anode and the cathode. The boundary condition for electrons at the cathode ($x = 0$) is formulated in terms of the secondary ion emission coefficient, γ_i:

$$j_e(0,t) = \gamma_i \cdot j_p(0,t) \, , \tag{5}$$

where $j_e = n_e \cdot w_e$ and $j_p = n_p \cdot w_p$ are the electron and the positive ion current densities.

Because of the very low current density of the negative corona, it was assumed that, during the pulsation period, the gas temperature is constant at all discharge points. Under this condition, the coefficients of kinetic processes involving neutral particles are constant in time, while those related to ionized particles depend on the local field intensity only. The shape of the current channel was set by choosing two discharge gap regions: the cathode (generating) one as cylindrical, and the drift one, as parabolic. The dependence $F \sim n_p^{2/3}$ was assumed for the current channel cross-section on the positive ion density n_p.

Taking into account that the main current pulse characteristics in a negative corona are determined by the cathode generating region [4-7], for this region we used formulas only for the positive ion flow current $I_p = j_p \cdot S$ and the displacement current $I_{dc} = {dE}/{dt}$, where S is the square of the cathode surface and E is the electrical field strength on the cathode surface. Thus, the cathode current I_c at any point in time is defined as $I_c(t) = I_p(t) + I_{dc}(t)$. The total corona current was obtained by the

Journal of Physics: Conference Series **514** (2014) 012048 doi:10.1088/1742-6596/514/1/012048

summation $I_{total} = \Sigma I_c(t)\Delta t$ for several tens of current pulses. Bearing in mind the characteristic times of the various processes, the time step Δt for the numerical solution of the equations was set within the range 10^{-11} s – 10^{-9} s. The differentiations were carried out on a non-uniform spatial grid with 160 cells and a minimal step of 10^{-4} cm on the cathode surface.

5. Numerical modelling results

The numerical simulation yielded the dependence of the average current on the air pressure in the discharge gap. The corona parameters used were set close to the experimental ones: d = 0.6 cm, r_0 = 0.004 cm, U = 2300 V, $R = 10^6\ \Omega$. Figure 4 shows typical dependence of the current on the air pressure in the range of 200 Torr– 740 Torr. In case 2, the cathode tip radius and the current-channel radius in the cathode zone increase in an inverse proportion to the gas pressure, in contrast with case 1, where they are constant.

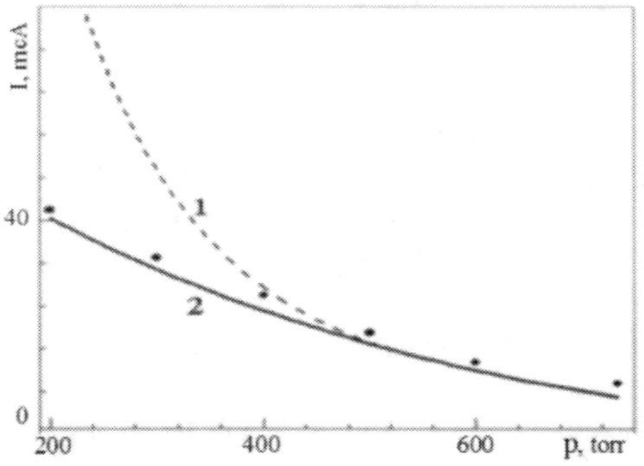

Figure 4. Corona discharge current dependence on air pressure: dots - measured values, 1, 2 - the data of numerical simulations.

As it can be seen, the current values calculated (curve 2) are sufficiently close to the measured ones within the entire range of air pressures. At lower gas pressure (240 – 400 Torr), a better agreement between the simulation and the experimental results was achieved by increasing the working area of tip radius and the cathode-current channel to values in the order of 0.05 cm.

6. Conclusions

The complex dependence of the negative corona current on the air pressure was measured and its nature was partly explained. The technique of measuring the gas pressure based on the corona current is suitable for use in the production of light bulbs. A more detailed description of the corona-current dependence on the gas pressure would require a detailed analysis of the dynamics of the variation of the working cathode surface size during the discharge transition to a glow mode.

References

[1] Rozanov L 1982 *Vacuum technique* (Moscow: Higher School)
[2] *Alcatel vacuum technology for industry and research. Product catalogue* 2002 (Alcatel-France)
[3] Manometer converter of gas discharge-deformation type PMDH-1 1980 *Technical Specifications № 2832001-TU* 1989 (Ryazan Russia)
[4] Chyhin V 2002 Physical mechanisms of negative corona current pulse with secondary oscillation *Conden. Matter Phys.* **5** 429-48
[5] Chyhin V 2007 *Physical mechanisms of negative corona pulsation* (DrSc Thesis Kharkiv Ucraine)
[6] Chyhin V and Karpyak S 2007 Numerical modelling of low-temperature non-equilibrium plasma of pulsing corona and breakdown *Conden. Matter Phys.* **10** 209-17
[7] Chyhin V 2011 Experimental investigation and numerical modeling of corona discharge plasma *Phys. Collection of Shevchenko Scientific Society* **8** 257-85

Evaluation of the scrape-off-layer plasma parameters by a horizontal reciprocating Langmuir probe in the COMPASS tokamak

M Dimitrova[1,3,4], Tsv K Popov[2], P Ivanova[3], E Vasileva[3], E Hasan[2,3], J Horáček[1], P Vondráček[1], R Dejarnac[1], J Stöckel[1], V Weinzettl[1], J Havlicek[1], F Janky[1] and R Panek[1]

[1]Institute of Plasma Physics, Academy of Sciences of the Czech Republic v.v.i., 3 Za Slovankou, 182 00 Prague 8, Czech Republic
[2]Faculty of Physics, St. Kl. Ohridski University of Sofia, 5 J. Bourchier Blvd., 1164 Sofia, Bulgaria
[3]Acad. E. Djakov Institute of Electronics, Bulgarian Academy of Sciences, 72 Tsarigradsko Chaussee, 1784 Sofia, Bulgaria

E-mail: dimitrova@ipp.cas.cz

Abstract. The scrape-off-layer (SOL) parameters in the COMPASS tokamak are studied by using a Langmuir probe mounted on a horizontal reciprocating manipulator. The radial profiles of the plasma potential, the electron energy distribution function and the electron densities are derived from the measured current-voltage probe characteristics by applying the first-derivative probe technique (FDPT). It is shown that close to the tokamak wall the electron energy distribution function is Maxwellian, while in the SOL, in the vicinity of the last closed flux surface and inside the confined plasma, the electron energy distribution function is bi-Maxwellian with a low-energy electron fraction dominating over a higher energy one. The radial profiles of the electron pressure and the parallel electron power flux density in COMPASS are also presented.

1. Introduction

The electric probes are simple plasma diagnostic tools allowing one to evaluate quickly and reliably the edge plasma parameters. In magnetized plasma, the interpretation of the electron part of the current-voltage (*IV*) characteristics above the floating potential still remains difficult [1] because the electron part of the *IV* characteristics is distorted due to the influence of the magnetic field.

In tokamaks, the assumption for a Maxwellian electron energy distribution function (EEDF) is generally valid. However, experimental evidence does exist suggesting non-Maxwellian distributions in tokamak SOL plasmas [2]. It is clear that the knowledge of the real EEDF is of great importance in understanding the underlying physics in SOL plasma. This fact notwithstanding, only a few experimental works have so far been reported devoted to the EEDF direct measurement. Our investigations on the CASTOR tokamak aimed at evaluating the real EEDF indicated a bi-Maxwellian

[4] To whom any correspondence should be addressed.

Journal of Physics: Conference Series **514** (2014) 012049 doi:10.1088/1742-6596/514/1/012049

one in the edge plasma [3]. Recent probe measurements in the liquid lithium divertor area of the NSTX also showed a bi-Maxwellian EEDF [4].

In this paper we report results of measurements in the SOL made by a horizontal reciprocating probe in the COMPASS tokamak [5] with D-shaped plasma. The radial distribution of the plasma potential, the electron temperatures and densities are presented and discussed. The radial distributions of the electron pressure and the parallel electron power flux density in COMPASS are also presented.

2. The first-derivative Langmuir probe technique for evaluating the EEDF in tokamak edge plasma

The FDPT for evaluating the plasma parameters in tokamak edge plasma was published and discussed in detail in [3]. It was shown there that the electron current flowing to a cylindrical probe negatively biased by potential U is given by:

$$I_e(U) = -\frac{8\pi e S}{3m^2\gamma}\int_{eU}^{\infty}\frac{(W-eU)f(W)dW}{\left[1+\frac{(W-eU)}{W}\psi(W)\right]}, \tag{1}$$

where W is the electron energy; e, m and n are the electron charge, mass and density; S is the probe area; U is the probe potential with respect to the plasma potential U_{pl} ($U = U_p - U_{pl}$). The geometric factor γ assumes values in the range of $0.71 \leq \gamma \leq 4/3$.

Here $f(W)$ is the isotropic EEDF normalized by:

$$\frac{4\pi\sqrt{2}}{m^{3/2}}\int_0^{\infty} f(W)\sqrt{W}dW = \int_0^{\infty} f(\varepsilon)\sqrt{\varepsilon}d\varepsilon = n. \tag{2}$$

In the presence of a magnetic field B at low gas pressures, the diffusion parameter $\psi = \psi(W,B)$ depends on the Larmor radius $R_L(W,B)$, as well as on probe size and orientation with respect to the magnetic field. As it was shown in [3], for cylindrical probes oriented perpendicular with respect to the magnetic field the diffusion parameter can be written as:

$$\psi_{\perp}(W,B) = \frac{R\ln\left(\frac{\pi L'}{4R}\right)}{16\gamma R_L(W,B)}, \tag{3}$$

where L' is the characteristic cross-section size of the turbulent structures (blobs), R is the radius of the probe.

As the magnetic field intensity increases, so does the value of the diffusion parameter. When $\psi(W,B) >> 1$ **(high value of the magnetic field *B*)**, the EEDF is represented by the first derivative of the electron probe current, as was shown in [3,6,7]:

$$f(\varepsilon) = \frac{3\sqrt{2m}\gamma}{2e^3 S}\frac{\psi}{U}\frac{dI_e}{dU}. \tag{4}$$

Note that in equation (4) the diffusion parameter is inversely proportional to the square root of the energy. This will cause the parameter to diverge at small energies and can cause an artifact in the resulting distribution function. This artifact in the range from zero to the value of the electron temperature (on energy scale) is due to the mathematical approach rather than to physical phenomena [1,3].

3. Langmuir probe measurements in the COMPASS tokamak edge plasma

To diagnose the edge plasma and the SOL in the horizontal direction in the COMPASS tokamak, the *IV* characteristics were measured by using a reciprocating Langmuir probe. The graphite cylindrical probe tip (length 1.5×10^{-3} m and diameter 8×10^{-4} m) was placed perpendicular to the magnetic field lines. Before the shot, the probe was moved to a chosen initial position with respect to the center of the tokamak chamber and, in synchronization with the shot onset, started a fast reciprocation towards the last closed flux surface (LCFS) and back. The maximum extension of the reciprocating probe motion was 0.06 m.

A series of reproducible ohmic discharges in hydrogen with D-shape cross-section plasma were performed. We present below results from shot #3908, which is typical for the series. The toroidal magnetic field was 1.15 T. The measurements were performed during the steady-state phase of the discharge at a low loop-voltage (about 1.5 V), a constant plasma current of 170 kA and a line average electron density of 8.3×10^{19} m^{-3}. The position of the horizontal reciprocating probe (HRP) versus time is presented in figure 1. The red line indicates the probe position when the probe signal is collected (probe extension 0.04 m). In the same figure, the dashed line shows the position of the LCFS in the plasma, derived from the magnetic reconstruction.

The probe was biased with respect to the tokamak chamber wall by a triangular voltage $U_p(t)$ at a frequency of 1 kHz supplied by a KEPCO 100-4M power supply. The ramp-up/ramp-down phases lasted 0.5 ms, so that the probe could be assumed stationary in space during each sweep cycle. The probe potential and the probe current versus time were recorded by the tokamak Data Acquisition System. Thus, we could construct the *IV* characteristics at different probe positions and, using the FDPT, evaluate the real EEDF. In the far SOL (probe positions in the range 0.771 – 0.751 m) it appears to be Maxwellian. When going deeper into the plasma (probe positions in the range 0.751 – 0.731 m), the nature of the EEDFs measured changes: Besides the predominant low-energy fraction, a fraction of electrons with higher energy appears in the electron energy distribution function. In this case, the non-Maxwellian EEDF can be approximated by a bi-Maxwellian one – i.e., by a sum of two Maxwellian EEDF for two different temperatures. An example of a bi-Maxwellian EEDF measured at

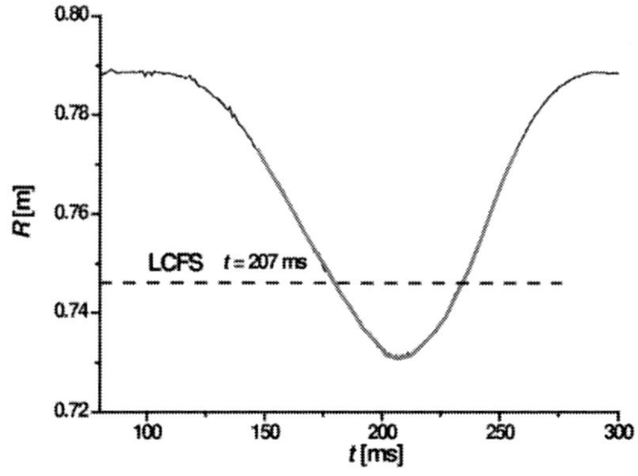

Figure 1. Position *R* of the HRP tip with respect to the tokamak chamber center versus time. The red line indicates the probe position when the probe signal is collected. The LCFS is indicated by a dashed line.

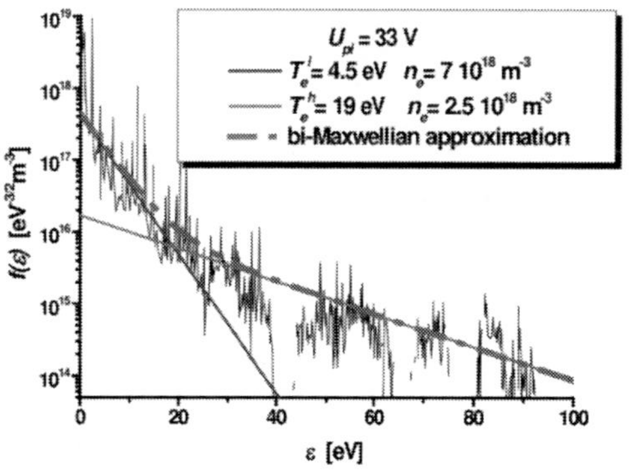

Figure 2. Experimental EEDF (black line) and the sum of the model ones for two electron temperatures (dashed dotted line). Shot #3908, probe position 0.731 m from the tokamak center.

the deepest probe position (0.731 m from the tokamak center) is presented in figure 2. The low-temperature electron group (blue line, $T_e^l = 4.5$ eV) has a density higher than that of the high-temperature one (red line, $T_e^h = 19$ eV). The electron densities were calculated by using equation (2).

4. Results and discussion

Figure 3 shows the radial distribution (with respect to the position of the LCFS) of the electron temperature – the triangles represent the temperature of the more populated low-energy electron fraction, while the squares, the temperature of the suprathermal electrons in the bi-Maxwellian EEDF. The dots indicate the temperature of the Maxwellian EEDF. The position of the LCFS evaluated by EFIT is indicated in all figures by a dashed line.

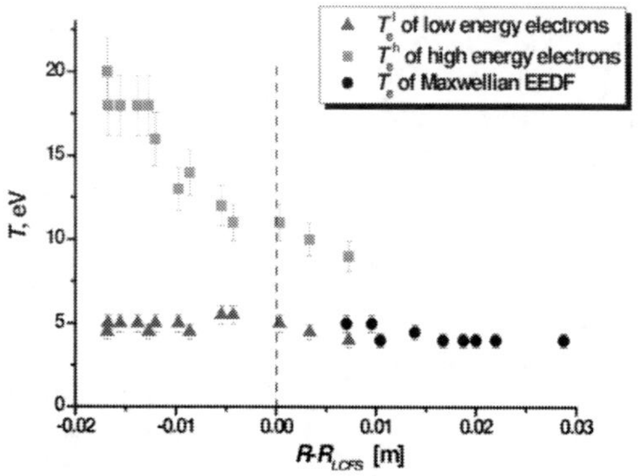

Figure 3. Radial distribution of the electron temperature T_e for shot #3908.

We should point out that more detailed theoretical and experimental investigations must be performed to clarify the origin of the low-energy fraction in the bi-Maxwellian EEDF. In [4], a "heuristic model" accounting for the inelastic collision effects (i.e. excitation and ionization of neutral hydrogen) is proposed to explain this EEDF feature. Indeed, the energy balance of the reaction $\mathrm{H}+\mathrm{e} \rightarrow \mathrm{H}^{+}+2\mathrm{e}$ for electrons with energy higher by 10 – 15 eV than the energy of ionization (13.6 eV) is in agreement with the energy of the low-temperature electrons registered. The values of the rate coefficient [8] for electron temperatures in the range 15 – 20 eV are close to the maximum. Performing a detailed energy balance necessitates that ionization through neutral hydrogen excited states be taken into account as well. On the other hand, the energy of the electrons in the far SOL is ~5 eV and the most probable reactions are dissociation with threshold of 4.5 eV and excitation. The behavior of the electron temperature spatial distribution at the vicinity of the separatrix can be explained by the plasma turbulence and non-local kinetic effects [9,10,11].

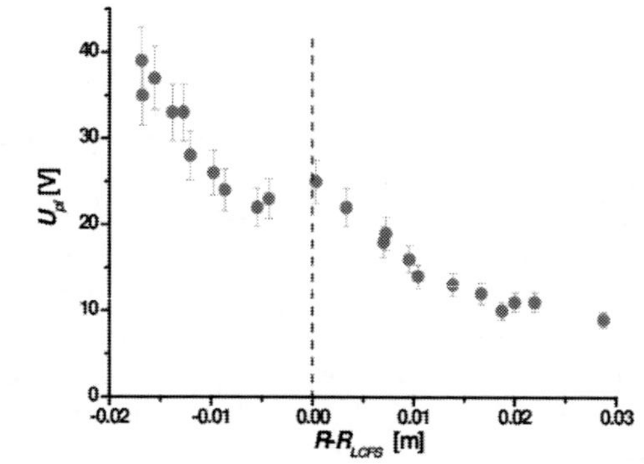

Figure 4. Radial distribution of the plasma potential for shot #3908.

The radial distribution of the plasma potential is presented in figure 4. The dots show the values evaluated by the FDPT.

In figure 5, the electron densities evaluated are represented by the same symbols as used in figure 3. Figure 6 shows the ratio between the electron densities of the two electron groups. The solid symbols represent the data for the reciprocating probe motion from its initial position to its maximum extension. The empty symbols illustrate the data measured during the backward probe motion. It is seen that the density of the high-energy electron group increases from 15% to 43% within a few millimeters after the LCFS.

Using the results for the electron temperatures and densities obtained by probe measurements, we can also calculate other plasma parameters. Figure 7 presents the radial distribution of the electron pressure p_e, namely, $p_e = n_l kT_e^l + n_h kT_e^h$, where the EEDF is by-Maxwellian, and $p_e = n_M kT_e^M$ for

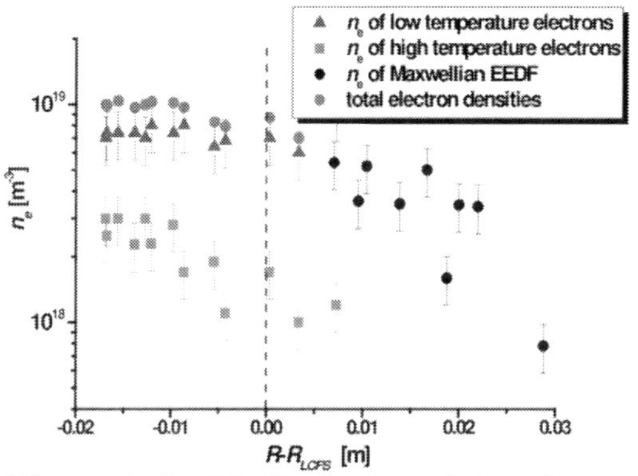

Figure 5. Radial distribution of the electron densities n_e for shot # 3908.

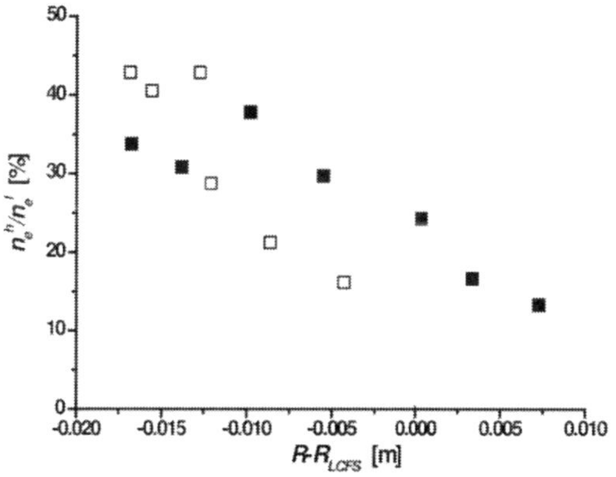

Figure 6. Ratio between densities of the high and the low temperature electron groups.

a Maxwellian case. The values of T in these equations are expressed in Kelvin and k is the Boltzmann constant. It is seen that, within the experimental error, the electron pressure radial distribution has an exponential behavior versus the radial position.

Figure 8 presents the results from the calculations of the parallel electron power flux density [8] $Q_{\|}^{e} = \frac{7}{2} c_s p_e$. In the SOL, the power flux density decreases exponentially as $Q_{\|}^{e} = Q_0 \exp\left(-\left(R - R_{LCFS}\right)/\lambda_q\right)$, where the decay length $\lambda_q = 0.015$ m and the power flux density at the LCFS $Q_0 = 0.9$ MW/m^2 (the solid line in figure 8).

Such a profile shape is regularly observed when other probe techniques are used and in most of tokamaks [12]. This is usually interpreted in terms of turbulent structures (blobs) transporting plasma radially away from the LCFS with a roughly constant speed, with the plasma loosing both its density and temperature in parallel with the magnetic field lines by acoustic streaming (determined by the ion sound velocity, or the parallel pressure gradient).

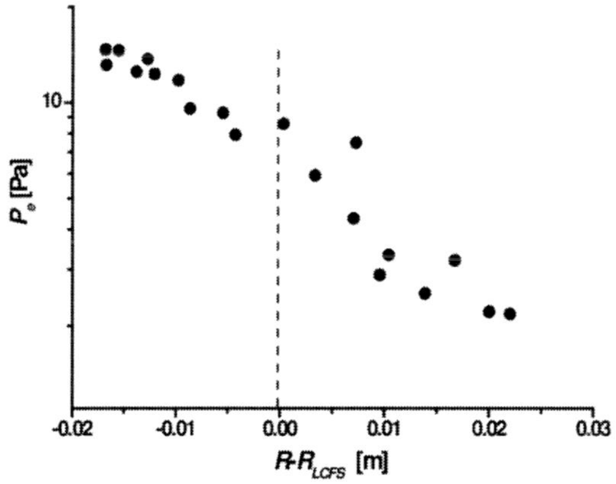

Figure 7. Radial distribution of the electron pressure for shot # 3908.

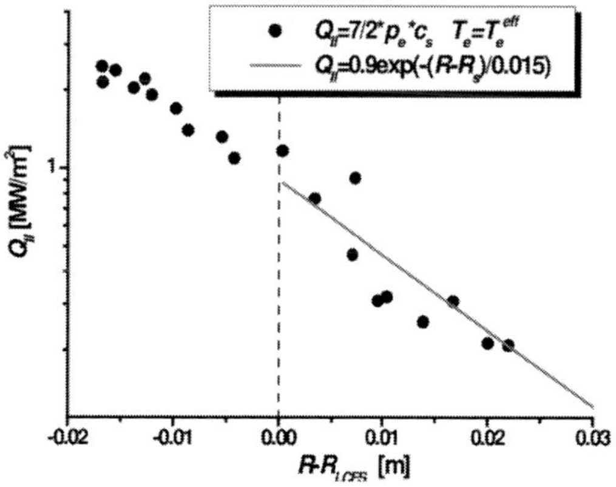

Figure 8. Radial distribution of the parallel electron power flux density for shot # 3908.

5. Conclusions

The SOL parameters in the COMPASS tokamak are studied by using a horizontal reciprocating Langmuir probe. Data for the radial distribution of the plasma potential, the electron energy distribution function and the electron densities are derived from the measured current-voltage probe characteristics by applying the first-derivative probe technique. It is shown that close to the wall of the tokamak chamber the energy distribution function of the electrons is Maxwellian, while in SOL, around the last closed flux surface and inside the confined plasma, the electron energy distribution function is bi-Maxwellian with a low energy electron fraction dominating over the higher energy one. The radial profiles of the electron pressure and the parallel electron power flux density in the COMPASS tokamak are also presented.

Acknowledgements

This research has been supported by the European Community under the contract of Association between EURATOM/IPP.CZ and EURATOM/INRNE.BG; Bulgarian Science Fund through the Association EURATOM-INRNE, International Atomic Energy Agency (IAEA) Research Contract No 17125/R0 as a part of the IAEA CRP F13014 on "Utilisation of a Network of Small Magnetic Confinement Fusion Devices for Mainstream Fusion Research" and 5th IAEA Joint Experiment, by the JOINT RESEARCH PROJECT between Institute of Plasma Physics v.v.i., AS CR and Institute of Electronics BAS BG, the grant project GA CR P205/12/2327 and by MSMT Project # LM2011021.

References

[1] Popov Tsv K, Ivanova P, Dimitrova M, Kovačič J, Gyergyek T and Čerček M 2012 *Plasma Sources Sci. Technol.* **21** 025004 A
[2] Dejarnac R, Gunn J P, Stöckel J, Adámek J, Brotánková J and Ionita C 2007 *Plasma Phys. Control. Fusion* **49** 1791-1808
[3] Popov Tsv K, Ivanova P I, Stockel J and Dejarnac R 2009 *Plasma Phys. Control. Fusion* **51** 065014
[4] Jaworski M A, Bell M G, Gray. T K, Kaita R, Kallman J, Kugel H W, LeBlanc B, McLean A G, Sabbagh S A, Soukhanovskii V A, Stotler D P and Surla V 2012 *Fusion Engin. Design* **87** 1711-18
[5] Pánek R, Bilyková P, Fuchs V, Hron M, Chráska P, Pavlo P, Stockel J, Urban J, Weinzettl V, Zajac J and Zacek F 2006 *Czechoslovak J. Phys.* **56**/2 B125–137
[6] Arslanbekov R R, Khromov N A and Kudryavtsev A A 1994 *Plasma Sources Sci. Technol.* **3** 528-38
[7] Demidov V I, Ratynskaia S V and Rypdal K 2002 *Rev. Sci. Instrum.* **73** 3409-39
[8] Stangeby P C 2000 *The Plasma Boundary of Magnetic Fusion Devices (Series in Plasma Physics)* (IOP)
[9] Tskhakaya D, Jachmich S, Eich T, Fundamenski W and JET EFDA Contributors 2011 *J. Nuclear Mater.* **415** S860-4
[10] Chodura R 1992 *Contr. Plasma Phys.* **32** 219-30
[11] Batishchev O V, Krasheninnikov S I, Catto P J, Batishchev A A, Sigmar D J, Xu X Q, Byers J A, Rognlien T D, Cohen R H, Shoucri M M and Shkarofskii I P 1997 *Phys. Plasmas* **4** 1672-80
[12] Horacek J, Pitts R A, Vondracek P, Panek R, Goldston R, Stangeby P C, Kocan M, Janky F, Havlicek J and Arnoux G 2013 Heat flux deposition on ITER-like HFS limiter on tokamak COMPASS *Presented at 18th ITER ITPA Div-SOL meeting* (March.2013 Hefei China)

Journal of Physics: Conference Series **514** (2014) 012050 doi:10.1088/1742-6596/514/1/012050

Determination of the electron energy distribution function in the ISTTOK tokamak

M Dimitrova[1,4,5], C Silva[2], H Fernandes[2], P Duarte[2], Tsv K Popov[3], J Stockel[1] and R Panek[1]

[1]Institute of Plasma Physics, Academy of Sciences of the Czech Republic v.v.i., 3 Za Slovankou, 182 00 Prague 8, Czech Republic
[2]Associação Euratom/IST, Instituto de Plasmas e Fusão Nuclear, Instituto Superior Técnico, Universidade Técnica de Lisboa, 1 Av. Rovisco Pais, 1049-001 Lisboa, Portugal
[3]Faculty of Physics, St. Kl. Ohridski University of Sofia, 5 J. Bourchier Blvd., 1164 Sofia, Bulgaria
[4]Acad. E. Djakov Institute of Electronics, Bulgarian Academy of Sciences, 72, Tzarigradsko Chaussee, 1784 Sofia, Bulgaria

E-mail: dimitrova@ipp.cas.cz

Abstract. The first derivative probe technique was applied to study the ISTTOK tokamak plasma. This technique employs the electron part of the Langmuir probe current-voltage (*IV*) characteristic and yields information on the plasma potential and the electron energy distribution function (EEDF). The *IV* characteristic was measured with new electrical probes mounted on a horizontal manipulator, one oriented in parallel and the other perpendicularly to the magnetic field lines. Using the first-derivative probe technique, the plasma potential and the EEDF at different radial positions were acquired. We show that, in the vicinity of the last close flux surface (LCFS), the EEDF is non-Maxwellian and can be approximated by a bi-Maxwellian one with a dominant cold electron population and a minority group of hot electrons. In the limiter shadow, the EEDF obtained is Maxwellian.

The comparison of the plasma parameters evaluated using the first-derivative probe technique for both probes shows a satisfactory agreement.

1. Introduction

In non-magnetized, low-density plasmas, Langmuir probes (LP) allow local measurements of the plasma potential, the charged particles density and the electron energy distribution function, $f(\varepsilon)$ (EEDF). In magnetized plasmas, the interpretation of the electron part of the current-voltage (*IV*) characteristic above the floating potential is difficult – the electron part of the *IV* characteristic is distorted due to the influence of the magnetic field. This is why a new, advanced method to solve the problem has to be applied.

The recently-developed [1,2] first-derivative probe technique (FDPT) makes use of the electron part of the Langmuir probe *IV* characteristic and provides information on the plasma potential and the

[5] To whom any correspondence should be addressed.

electron energy distribution function. Dedicated investigations aiming at evaluating the EEDF employing this method have revealed a bi-Maxwellian EEDF in the CASTOR edge plasma [2], in the COMPASS tokamak and, more recently in the liquid lithium divertor area of NSTX [3]. This technique was applied to the investigation of ISTTOK tokamak plasma.

In the ISTTOK tokamak, we acquired data on the plasma potential and the EEDFs at different radial positions using electrical probes, one oriented parallel and the other perpendicular to the magnetic field. The results demonstrated that, in the vicinity of the LCFS plasma, the EEDF is not Maxwellian, but may be approximated by a bi-Maxwellian with a dominant cold electron population and a group of hot electrons. In the limiter shadow, the EEDF obtained is Maxwellian.

2. The first-derivative probe technique for evaluation of the plasma parameters in tokamak plasma

The FDPT for evaluating the plasma potential, electron temperature and density in tokamak edge plasma was published and discussed in detail in [2]. It was shown there that the electron current flowing to a cylindrical probe negatively biased by potential U_p is given by:

$$I_e(U) = -\frac{8\pi e S}{3m^2\gamma}\int_{eU}^{\infty}\frac{(W-eU)f(W)dW}{\left[1+\frac{(W-eU)}{W}\psi(W)\right]}, \qquad (1)$$

where W is the electron energy; e and m are the electron charge and mass; S is the probe area; U is the probe potential with respect to the plasma potential U_{pl} ($U = U_p - U_{pl}$). The geometric factor γ assumes values in the range $0.71 \le \gamma \le 4/3$.

The important parameter in equation (1) is the diffusion parameter $\psi = \psi(W)$. In the presence of a magnetic field B at low gas pressures, ψ depends on the Larmor radius $R_L(W,B)$, as well as on the shape, size and orientation of the probe with respect to the magnetic field. As it was shown [2] in strongly turbulent tokamak plasma for probes oriented perpendicular to the magnetic field lines ($\psi_\perp$) and for probes oriented along magnetic field lines ($\psi_\parallel$), the diffusion parameter can be expressed as:

$$\psi_\perp(W,B) = \frac{R\ln\left(\pi L'/4R\right)}{16\gamma R_L(W,B)} \quad \text{and} \quad \psi_\parallel(W,B) = \frac{\pi L'}{64\gamma R_L(W,B)}. \qquad (2)$$

Here L' and $R_L(W,B)$ are the characteristic cross-section size of the turbulent structures (blobs) and the electron Larmor radius. As the number of collisions in the probe sheath increases, so does the value of the diffusion parameter. When $\psi(W,B) >> 1$ (high value of the magnetic field B), the EEDF is represented by the first derivative, instead of the second derivative, of the electron probe current, as was shown in [1,2,4]:

$$f(\varepsilon) = \frac{3\sqrt{2m}\gamma}{2e^3 S}\frac{\psi}{U}\frac{dI_e}{dU}. \qquad (3)$$

3. Langmuir probe measurements in the ISTTOK tokamak

Radial measurements were performed of the current-voltage characteristics with new electrical probes, one oriented parallel and the other perpendicular to the magnetic field in the ISTTOK tokamak and mounted on a horizontal manipulator. Figure 1 shows the cylindrical Langmuir probe tips with a length of 3 mm and a diameter of 0.75 mm. The position of the manipulator was changed from shot-to-shot; during the different shots the plasma parameters were reproducible at a toroidal magnetic field $B = 0.45$ T. The probes were biased with respect to the tokamak chamber wall by a triangular voltage $U_p(t)$ with a frequency of 1 kHz [5].

Figure 1. Cylindrical perpendicular and parallel Langmuir probes.

An example of the results obtained for shot #34810 (working gas H_2, plasma current 4 kA and average electron density 3.5×10^{18} m^{-3}) are presented in figure 2. The experimental *IV* characteristics are measured by both probes during the steady-state plateau of the discharge at a position of 80 mm from the tokamak chamber center. Figure 3 presents the first derivative of the smoothed *IV* curves and the fit with the model curve (first derivative of equation (1)). Using this comparison, one can evaluate the plasma potential.

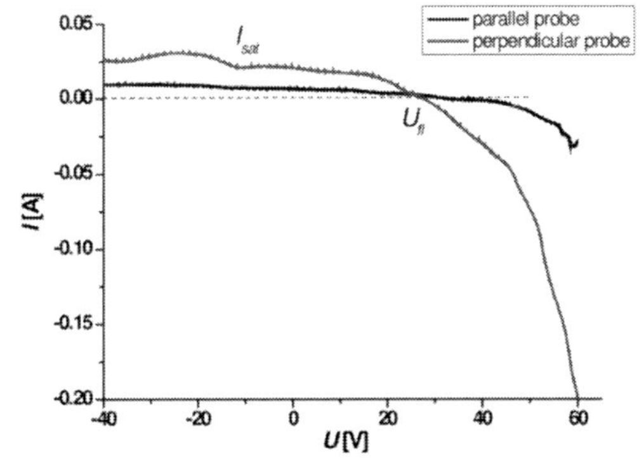

Figure 2. Current-voltage characteristics for perpendicular and parallel probes at position 80 mm from the tokamak center.

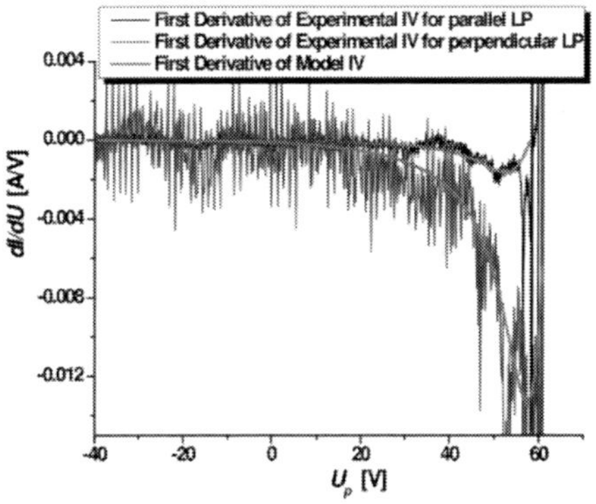

Figure 3. First derivatives of the smoothed experimental *IV* characteristics. Model curves – red lines.

Figures 4a and 4b present the EEDFs obtained by parallel and perpendicular probes and applying equation (3). It is seen that the EEDFs can be approximated by a bi-Maxwellian – i.e., by a sum of two Maxwellian EEDFs with two different electron temperatures and densities. In this case, we have the

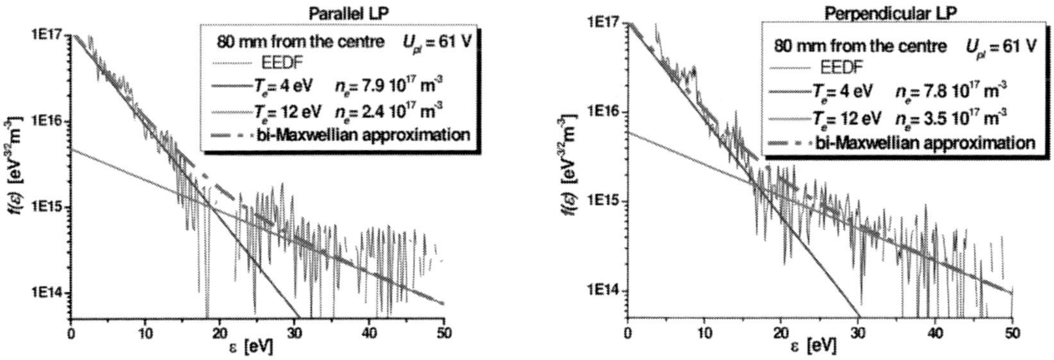

a) parallel probe **b)** perpendicular probe

Figure 4. Experimental EEDF for perpendicular and parallel probes (black curves) at probe position 80 mm from the tokamak chamber center. The blue line represents the distribution of the low-energy electron population; the red line is the high energy one. The dash-dot line is the bi-Maxwellian approximation (a sum of the blue and the red ones).

same temperatures on both probes, namely, 4 eV and 12 eV. In the limiter shadow (88 mm from the center) the EEDF is Maxwellian (figure 5) with electron temperature $T = 7.5 \pm 0.7$ eV and electron density $n_e = (4.8 \pm 1)10^{17}$ m^{-3}.

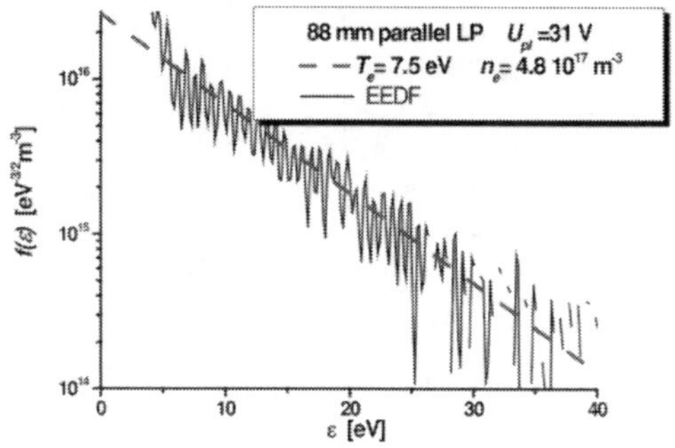

Figure 5. The EEDF obtained at 88 mm probe position from the tokamak chamber center.

4. Results and discussion

The radial distributions of the plasma potential, ion saturation current density, electron temperatures and densities are presented in figure 6 a-d. The results refer to both the perpendicular and the parallel probe. Figures 6a presents the radial distribution of the ion saturation current density. Notwithstanding of different orientation of the probes, the current densities values evaluated by both of the probes practically coincide and thus we can compare other plasma parameters measured by them. Figure 6b shows the radial distribution of the plasma potential. The position of the poloidal limiter is indicated at 85 mm. Based on the position of the maximum of the plasma potential radial distribution, we concluded that the position of the LCFS is at 75 mm from the tokamak chamber center.

The accuracy of the plasma potential evaluation is ~5 V. The discrepancy in the values obtained in the confined plasma can be explained by the slightly different radial positions of the probes where the gradient of the plasma potential is large.

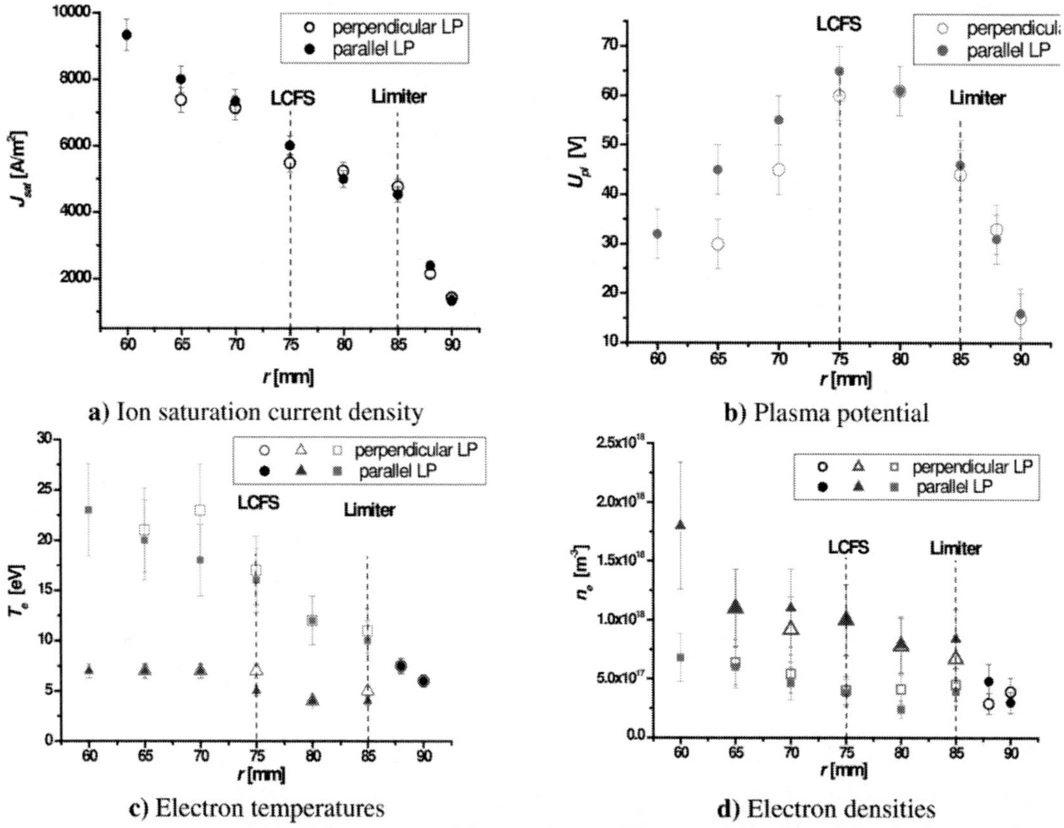

a) Ion saturation current density

b) Plasma potential

c) Electron temperatures

d) Electron densities

Figure 6. Plasma parameters at different radial positions.

Figure 6c and 6d present the radial distribution of the electron temperatures and densities. The triangles correspond to the values of the more populated low-energy electron fraction, while the squares indicate the values of the suprathermal electrons from the bi-Maxwellian EEDF. The dots indicate the temperature of the Maxwellian EEDF. It is seen that the density of the hot-electron group is about 40-50% of the low-temperature one. Taking into account the uncertainties in all values measured, the temperatures of the low-energy electron group are evaluated by an accuracy of about 10%, while that for the suprathermal electrons is about 20%. The uncertainties of the electron densities are within 30%. Thus, the comparison of the parameters evaluated using the FDPT for both probes shows a satisfactory agreement.

The behavior of the electron temperature spatial distribution in the vicinity of the LCFS can be explained by the plasma turbulence and non-local kinetic effects [6]. To explain the bi-Maxwellian EEDF behaviour, a "heuristic model" accounting for the ionization of neutral hydrogen is proposed in [3]. It has to be noted that, in the limiter shadow, the EEDFs established are Maxwellian with a temperatures of about 6-7 eV which are not enough for ionization. The most probable reactions in this energetic diapason are dissociation with threshold of 4.5 eV and excitation.

4. Conclusions

The first-derivative probe technique was used to evaluate the radial distribution of the main plasma parameters (plasma potential, the electron energy distribution function and the electron densities) in the strongly magnetized turbulent ISTTOK tokamak plasmas. The measurements are performed from probe position from 60 to 90 mm in regards to the tokamak center by new electrical probes mounted on a horizontal manipulator, one oriented in parallel and the other perpendicularly to the magnetic field lines. The position of the LCFS at 75 mm from the tokamak chamber center was evaluated by the maximum of the radial profile of the plasma potential.

Radial evaluations of the EEDF show that, at the vicinity of the LCFS the EEDF is non-Maxwellian and can be approximated by a bi-Maxwellian one with a dominant cold electron fraction and a minority group of hot electrons. In the limiter shadow the EEDFs obtained are Maxwellian.

Acknowledgements

This research has been supported by the European Community under the contract of Associations between EURATOM/IPP.CZ and EURATOM/INRNE.BG; Bulgarian Science Fund through the Association EURATOM-INRNE, International Atomic Energy Agency (IAEA) Research Contract No 17125/R0 as a part of the IAEA CRP F13014 on "Utilisation of a Network of Small magnetic Confinement Fusion Devices for Mainstream Fusion Research" and MSMT Project # LM2011021.

References

[1] Popov Tsv K, Ivanova P, Dimitrova M, Kovačič J, Gyergyek T and Čerček M 2012 *Plasma Sources Sci. Technol.* **21** 025004 A
[2] Popov Tsv K, Ivanova P I, Stockel J and Dejarnac R 2009 *Plasma Phys. Control. Fusion* **51** 065014
[3] Jaworski M A, Bell M G, Gray T K, Kaita R, Kallman J, Kugel H W, LeBlanc B, McLean A G, Sabbagh S A, Soukhanovskii V A, Stotler D P and Surla V 2012 *Fusion Engineering and Design* **87** 1711-18
[4] Arslanbekov R R, Khromov N A and Kudryavtsev A A 1994 *Plasma Sources Sci. Technol.* **3** 528
[5] Mitov M, Bankova A, Dimitrova M, Ivanova P, Tutulkov K, Djermanova N, Popov Tsv K, Stöckel J and Dejarnac R 2012 *J. Phys.: Conf. Series* **356** 012008
[6] Tskhakaya D, Jachmich S, Eich T, Fundamenski W and JET EFDA Contributors 2011 *J. Nuclear Mater.* **415** S860-4

Journal of Physics: Conference Series **514** (2014) 012051 doi:10.1088/1742-6596/514/1/012051

PIG charged particle source with hydrogen supply from a metal-hydride cathode

Ie V Borgun, D L Ryabchikov, I N Sereda[1] and A F Tseluyko

V N Karazin Kharkiv National University, Kharkiv, Ukraine

E-mail: igorsereda@karazin.ua

Abstract. We present the results of an experimental investigation of a Penning-type charged-particles source with a metal-hydride cathode. The main characteristic of the experiment is internal hydrogen supply from the metal-hydride cathode under the conditions of ion-stimulated desorption; we studied its influence on the source's emissive characteristics. An additional mode of source operation was observed involving axial electron emission; the decisive effect was revealed of the desorbed hydrogen on the axial electron emission. The ion energy distribution function was measured and its dependence on the external discharge parameters was determined.

1. Introduction

The use of metal-hydride cathodes (MH) based on Zr-V alloys in hydrogen plasma devices is of interest for modern science and technology. These materials are capable of storing hydrogen isotopes followed by desorption through heating with a uniform gas release in a wide range of working temperatures. In the case of using MH as a cathode material, desorption is caused by both heating and the discharge current. A distinctive characteristic of these compounds is a decrease of the ionization potential of desorbed hydrogen by 0.3 eV – 0.5 eV due to recombination of hydrogen atoms at the metal-hydride surface followed by desorption of the molecules in an excited and thermodynamically non-equilibrium state. This results in an increase of the hydrogen ionization cross-section by a factor of 1.3 to 1.5 as compared to common molecular hydrogen [1]. Additional advantages of using a MH-cathode are its compactness and safe hydrogen storage, as well as the possibility to realize local gas supply where necessary in the plasma device.

The early experiments with a Penning-type ion source using a MH-cathode revealed several particularities. One of them was an additional mode at a high discharge voltage and a low pressure [2]. In this mode, electrons were detected in an output axial flux from the cathode-reflector side (the side opposite to the MH-cathode), whereas an ion flux was only maintained on the MH-cathode side. The negative current increased with the discharge voltage and at some point exceeded the ion current. Another effect was the strong dependence of the hydrogen pressure on the MH-cathode temperature, which made it difficult to stabilize the discharge mode.

This is why the work reported here was devoted to solving the problem of stabilizing the pressure and revealing the factors responsible for the axial electron emission from the discharge.

[1] To whom any correspondence should be addressed.

2. Experimental setup

The experiments were carried out using a Penning-configuration discharge cell in a longitudinal magnetic field (figure 1). Three types of cathodes were used. The first one was an ordinary MH-cathode; the second type was the same MH-cathode with forced cooling, and the third one was a copper cathode with hydrogen supplied in order to simulate hydrogen desorption (figure 2).

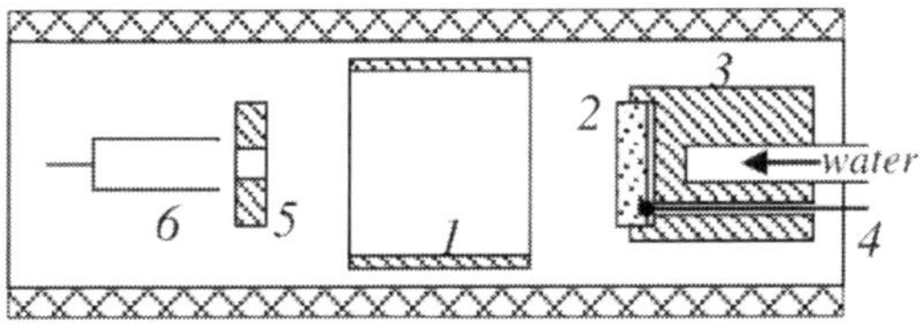

Figure 1. Schematic of the discharge cell: 1 – anode, 2 – MH-cathode, 3 – cathode holder, 4 – thermocouple, 5 – cathode reflector, 6 – collector.

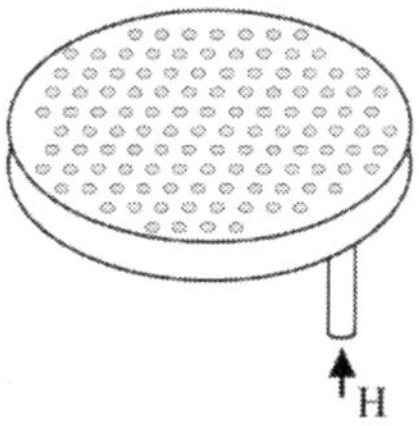

Figure 2. Schematic of the copper cathode used for hydrogen desorption simulation.

The MH-cathode (2) was a disk with a diameter of 2 cm and a thickness of 0.5 cm. It was pressed from a powder mixture of $Zr_{50}V_{50}H_x$ alloy and copper saturated with hydrogen. The initial saturation of the MH-cathode with hydrogen was about 900 cm^3 under normal conditions. In order to stabilize the MH-cathode temperature, it could be placed in a water-cooled copper cathode holder (3) with a diameter of 2.5 cm. To ensure a good thermal contact of the MH-cathode with the cathode holder, its surface was covered by heat-conducting grease. The MH-cathode temperature was controlled by a thermocouple (4). The copper cathode reflector (5) had a diameter of 2 cm and a thickness of 0.5 cm with a 0.5-cm diameter hole at the center. A collector (6) was placed behind the hole with the purpose of measuring the axial charged-particles current. An electrostatic energy-analyzer was used to study the energy spectrum of the axial charged-particles flux (it could mounted instead of the collector). To separate the electrons or ions in the flux, a potential of +3 kV or –0.2 kV was applied.

The cylindrical anode had a diameter of 3.7 cm with a length of 3 cm. The cathode- anode distance was 1 cm. The entire electrode system was fixed inside a quartz cylinder to prevent the ignition of a parasitic discharge in the vacuum chamber. The cathode was kept at ground potential, while a positive potential was supplied to the anode.

In the simulation experiments, a copper cathode of special design was used (figure 2). Hydrogen was fed from a balloon through small holes on the cathode's working surface, thus simulating hydrogen desorption. The ratio between the local and the additional fluxes of hydrogen, as well as the intensity of the external magnetic field, were the same as in [2], where an axial electrons output was registered in a discharge with a MH-cathode. Solid copper cathodes were used in the check experiments.

The residual pressure in the vacuum chamber did not exceed 5×10^{-6} Torr. The investigations were carried out at pressures within 10^{-6} Torr – 10^{-4} Torr. The working pressure was varied via the initial supply of hydrogen from a balloon into the vacuum chamber.

3. Results and discussion

Figure 3 shows typical dependences on the discharge voltage of the current through the collector placed behind the central hole in the cathode reflector for all cathode types used.

One can see three working modes taking place in succession as the discharge voltage is raised. The first one ($U_d \leq 1.6$ kV) with axial electrons output and the second one ($U_d \geq 1.6$ kV) with ions output have been known for a long time and have been observed irrespective of the electrode material and the working gas type [2]. The third mode ($U_d > 3$ kV) only arose when a MH-cathode saturated with hydrogen was used. In the case of the MH-cathode without cooling, the third mode started at 3.0 – 3.5 kV (curve 1 in figure 3), and at approximately 3.0 kV or less (curve 2 in figure 3), in the case of the water-cooled MH-cathode. The higher voltage needed for the onset of the third mode in the case of the MH-cathode is obviously due to the ion contribution to the total output current. In the simulation experiment, the current through the collector diminished in the same way (curve 3 in figure 3). But if the current reduction in the case of MH-cathodes was caused by the electron contribution to the output flux, than measurements in the simulation experiments by the retarding field method showed that there were no electrons in the flux. Thus, no third mode took place in the simulation and in the check experiments. This fact is indicative of the decisive role played by desorbed hydrogen in the processes of axial electron emission.

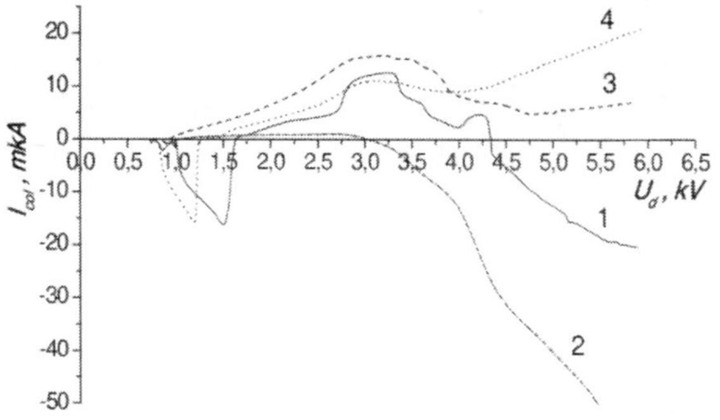

Figure 3. Dependence of the current through the collector behind the cathode reflector on the discharge voltage at $P = 3\times10^{-5}$ Torr, $H = 0.1$ T. 1 – discharge with a MH cathode, 2 – discharge with a water-cooled MH cathode, 3 – simulation experiment, 4 – check experiment.

Moving in an axial direction against an electric field, the electrons can take up energy from a HF-field in a HF-instability of a diocotron type. As was shown in [3], the transition to the third mode is followed by a step-wise increase of the HF-oscillation frequency and amplitude. Taking into account the well-known ratio for the frequency of diocotron oscillations $f \sim E_r / H$ [4], the rise in the oscillation frequency is obviously due to the increase of the radial electric field (E_r) related to a redistribution of the axial (E_z) and radial (E_r) electric fields leading to an extension of E_r (the potential on the axis of the system drops). The distributions of the radial (E_r) and axial (E_z) electric fields are thus established in such a way that the primary ionization occurs in the anode layer. The comparison between the MH-cathode experiments and the simulation experiment reveals that the additional hydrogen flux from the cathode affects the axial electric field (E_z) distribution in the gap and the formation of the E_z gradient [3].

Thus, the decisive factor for the appearance of the axial electron flux emission is a HF-instability developed under the conditions of a non-equilibrium hydrogen desorption from the MH-cathode. The electron output from one side only is obviously caused by the E_z gradient that appears due to the drop in the potential barrier induced by the cathode-reflector [3]. When the hydrogen stored in the MH-cathode is exhausted, the third mode vanishes and the discharge behaves as in the case of ordinary cathodes. Thus, activated hydrogen desorbed from MH-cathode not only brings about the appearance of an axial electron flux, but also causes a redistribution of the electrical field in the gap.

However, another problem in implementing a working hydrogen source is the strong dependence of the desorbed hydrogen flux on the MH-cathode temperature, which makes it difficult to stabilize the discharge mode. As was shown in [5], forced MH-cathode cooling stabilizes the working discharge pressure and eliminates the hydrogen kick due to uncontrolled thermal decomposition of hydride phases. A low MH-cathode temperature (lower than that leading to decomposition of hydride phases) ensures hydrogen desorption by ion-stimulated processes only and would thus allow one to control the hydrogen desorption rate by controlling the discharge current. While in previous experiments with a MH-cathode in a Penning discharge, additional hydrogen was provided from an external source (a balloon), it is reasonable to investigate the behavior of a discharge that works only on hydrogen desorbed from the MH-cathode, thus eliminating the necessity of an external plasma-forming gas supply.

Figure 4 shows typical dependences of the collector current on the discharge voltage in the case of a water- cooled MH-cathode. Curve 1 registered at $P = 5\times10^{-6}$ Torr corresponds to the initial residual pressure in the vacuum chamber without hydrogen supply with the discharge working only on desorbed hydrogen. The other curves refer to an initial hydrogen supply in the vacuum chamber. One can see that there are no sufficient differences in the discharge behavior as the pressure is varied.

The source is characterized by an axial electron emission both when hydrogen is added from a balloon (curves 2 and 3) and when hydrogen is desorbed from the MH-cathode only (curve 1). Therefore, how hydrogen is supplied does not affect the discharge behavior. This gives one an opportunity to work with hydrogen desorbed from the MH-cathode only, with water-cooling of the MH cathode used to stabilize the pressure via its influence on the discharge current [4].

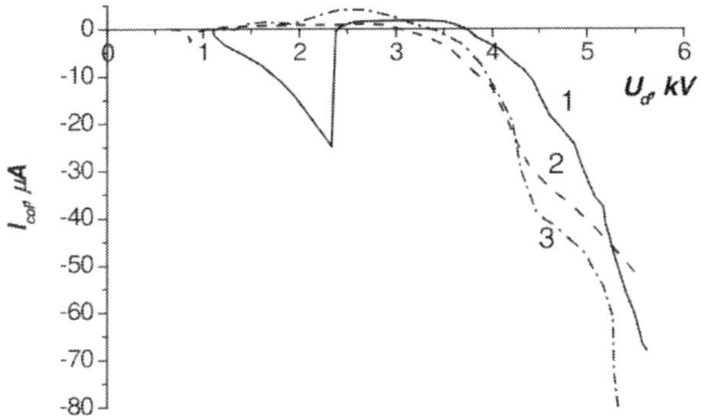

Figure 4. Dependence of the collector current on the discharge voltage at $H = 0.1$ T and different pressures. 1 – $P = 5\times10^{-6}$ Torr; 2 – $P = 3\times10^{-5}$ Torr; 3 – $P = 5\times10^{-5}$ Torr.

The analysis of the energy spectra of axially-moving charged particles revealed a number of particularities. Figure 5 shows the ion energy distribution function (IEDF) for all cathodes used in the discharge gap. Figure 5(a) refers to $U_d = 2.5$ kV corresponding to the second discharge mode with an axial ion flux output; figure 5(b), to $U_d = 3$ kV corresponding to the transition to the third mode (electrons in the output flux start appearing); and figure 5(c), to the third discharge mode (the output axial current has a negative sign).

One can see that, as the discharge undergoes a transition to the third mode, the distribution function shifts and widens towards the lower energy values. This is caused by the widening of the intensive ionization field from the anode layer to the discharge axis at lower values of the space potential. This is observed most distinctly when a MH-cathode without water-cooling is used (curve 3), when intensive hydrogen desorption in a non-equilibrium state occurs under the conditions of thermal hydride phases decomposition btought about by the discharge current.

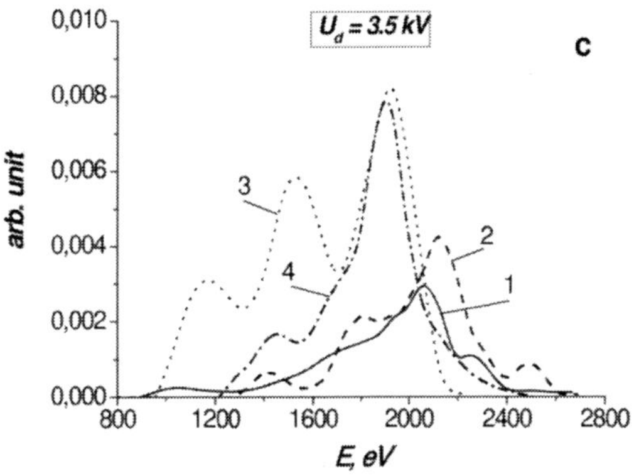

Figure 5. Ion energy distribution function at $P = 3\times10^{-5}$ Torr, $H = 0.1$ T for different discharge voltages U_d and cathode types. 1 – check experiment, 2 – simulation experiment, 3 – MH-cathode, 4 – water-cooled MH-cathode.

The intensive desorbed hydrogen flux leads to intensification of the ionization by the MH-cathode so that and ions are detected generated along the low-potential. Therefore, the widening of the IEDF is caused by hydrogen desorption followed by ionization intensification along the axis. In the case of hydrogen desorption due only to ion-stimulated processes (curve 4), the situation is the same but the flux of desorbed hydrogen is is reduced and the number of ions generated along the axis drops.

The differences of the simulation experiments results (curve 2) as compared with the other cases studied are due to the constant ratio between the local and the additional flux of hydrogen in the whole range of discharge voltages. The addition balloon hydrogen flux was chosen the same as in [2], when axial electrons output was registered corresponding to $U_d \approx 3.5$ kV. At lower U_d, the desorbed hydrogen flux was considerably lower than in the simulation experiment.

The studies on the electron energy distribution function showed that the average electron energy was about 20 eV and did not depend substantially on the hydrogen desorption.

4. Conclusions

The possibility is shown of a discharge to operate with hydrogen desorbed from a MH-cathode due to ion-stimulated processes only. This hydrogen does not influence substantially the discharge emissive characteristics, which gives one the opportunity to use data reported in previous experiments.

The decisive factor determining the axial electron flux emission is a HF-instability developing under the conditions of non-equilibrium hydrogen desorption from the MH-cathode. The axial electron current from one side only is due to the additional neutral hydrogen flux from the cathode side. When the hydrogen stored in the MH-cathode is exhausted, the third mode of operation disappears and the discharge behaves as in the case of copper cathodes.

References

[1] Shmal'ko Yu F, Klochko Ye V and Lototsky N V 1996 *Int. J. Hydrogen Energy* **21** 1057
[2] Klochko Ye V, Ryabchikov D L, Sereda I N and Tseluyko A F 2010 *Probl. Atomic Sci. Tech. Series: Plasma electronics and new acceleration methods* **4** 226
[3] Borgun I V, Ryabchikov D L, Sereda I N and Tseluyko A F 2013 *Probl. Atomic Sci. Tech. Series: Plasma Phys.* **1** 228
[4] Khauer W 1966 *J. Appl. Phys.* **37** 602
[5] Agarkov A V, Ryabchikov D L, Sereda I N and Tseluyko A F 2013 *Probl. Atomic Sci. Tech. Series: Plasma Electronics and new Acceleration Methods* **4** 301

Journal of Physics: Conference Series **514** (2014) 012052 doi:10.1088/1742-6596/514/1/012052

Energy spectra of Penning electrons in non-local plasma at middle and high pressures

M Stefanova[1,3], P Pramatarov[1], A Kudryavtsev[2] and R Peyeva[1]

[1]Acad. G. Nadjakov Institute of Solid State Physics,
Bulgarian Academy of Sciences, 72 Tsarigradsko Chaussee, 1784 Sofia, Bulgaria
[2]St. Petersburg State University,
7-9 Universitetskaya Nab., 199034 St. Petersburg, Russia

E-mail: mstef@issp.bas.bg

Abstract. A recently-developed collisional electron spectroscopy (CES) method enabled us to measure the energy spectra of groups of fast non-local electrons in a collisional mode at high pressures, where no collisional energy relaxation of electrons in the different groups takes place in the volume, and the different groups of electrons behave independently of each other. We recorded the energy spectra of groups of fast electrons created via Penning ionization of Ar and N_2 impurities by metastable He atoms at He pressures of 30 and 200 Torr. The experiments were conducted in the non-local negative glow plasma of a short dc microdischarge. The Penning electrons' energy spectra were recorded by means of an additional electrode – a sensor located at the boundary of the discharge volume, in contrast with the classical Langmuir probe. The spectra are characterized by the appearance of maxima at characteristic energies corresponding to the energy of the electrons released via Penning reactions. Using the Penning electrons' energy spectra, one can detect and identify the presence of different atomic and molecular admixtures in He at high pressures.

1. Introduction

The plasma electron spectroscopy (PLES) method, developed in 1994 by Kolokolov et al. [1], represents a new approach to studying elementary plasma processes and atomic constants. It makes use of the relation between the electron energy distribution function (EEDF) and the characteristics of the different elementary processes. On the basis of the PLES method, a new method for identification of gas impurities was proposed by Kudryavtsev [2], whereby gas impurities are identified by recording groups of fast non-local electrons released via Penning ionization of the impurity particles by metastable He atoms. A further development of the PLES method for identifying unknown species in a gas phase was reported in [3]. In [1-3], standard Langmuir probes were used for EEDF measurements in the positive-column afterglow plasma at low pressures (0.5 – 3.5 Torr). The main disadvantages of this method are a low sensitivity due to the small probe size and a large plasma volume due to the low working pressures. Moreover, the measurements in the afterglow plasma require a temporal resolution of the recording system, which further complicates the measurements.

[3] To whom any correspondence should be addressed.

In this work we used the new collisional electron spectroscopy (CES) method, whose basic principles were presented in [4-6]. The CES method is based on measuring the energy spectra $R(\varepsilon)$ of fast non-local electrons released in Penning ionization of the impurities atoms or molecules by the metastable atoms of the main gas:

$$A^* + B \rightarrow A + B^+ + e\,\{\varepsilon_p\}. \quad (1)$$

The Penning electrons' energy is $\varepsilon_p = \varepsilon_m - \varepsilon_i$, where ε_m is the excitation energy of the metastable particles A^* and ε_i is the ionization energy of the particles B. The use of helium as a main gas is convenient, since the metastable He levels have high excitation energies (19.8 eV and 20.6 eV).

The CES method enables one to measure the energy spectra of groups of fast electrons in a collisional mode at high pressures, where the energy relaxation of the electrons in the different groups does not occur by collisions in the volume, and the different groups of electrons behave independently from each other. This has to do with the fact that the electron loses only a small portion $\delta < 10^{-4}$ of its initial kinetic energy ε_p in one elastic collision with a He atom. As a result, the electron energy relaxation length, λ_ε, exceeds considerably the electron mean free path λ. In the case of He, the ratio is $\lambda_\varepsilon / \lambda \approx 70$, and the parameter $\lambda_\varepsilon p$ is $\approx$ 5 cm Torr, where p is the gas pressure of the main gas.

When λ_ε is greater than the specific plasma length L

$$\lambda_\varepsilon > L, \quad (2)$$

the EEDF is formed in a non-local mode and the electrons move in the restricted plasma volume with conservation of their full energy $\varepsilon = w + e\varphi(r)$ (kinetic plus potential) [7]. If the Penning electrons' energy is $\varepsilon_p > e\Phi_w$ (where Φ_w is the potential between the discharge axis and the wall), these electrons reach the wall in a free diffusion. In the absence of an electric field, or when it has a small value (e.g., negative glow plasma, afterglow plasma, where the electron temperature T_e, and hence, $e\Phi_w$, are small), the Penning electrons' energy spectrum exhibits sharp peaks near the ε_p energies, reproducing the spectrum $R(\varepsilon)$ of reactions of type (1) [1].

The inequality (2) is known as the condition for non-locality. Over the specific plasma length L, the EEDF depends weakly on the local plasma parameters and is the same at any point of the plasma. As a result, the non-local plasma has an important feature: by measuring the EEDF at the plasma boundary, one acquires information on the whole plasma volume. In non-local plasma, each group of Penning electrons reaches the plasma boundary with its initial energy and, hence, the EEDF can be recorded by means of an additional electrode – a sensor located at the boundary of the plasma volume. Thus, in contrast to the classical Langmuir probe, a sensor with a large collecting area can be used in order to enhance significantly the sensitivity of Penning electrons measurements.

The negative glow plasma of a short dc microdischarge is the most suitable medium for non-local formation of the EEDF. The important characteristics of the non-local negative glow plasma are: high metastable atom density, high rate of Penning ionization, equipotentiality, small dimensions, stable operation at high pressures, low operating voltage, simple design and simple power supply. Temporal resolution of the recording system is not required, which further simplifies the device.

2. Experimental

The experiments were carried out in the non-local negative glow plasma of a short (without a positive column) dc glow discharge. A schematic view of the discharge device is shown in figure 1. It consists of a plane disk-shaped (4 mm in diameter) cathode and an anode placed at a distance of 1 or 2 mm. A ring-shaped additional electrode, the sensor, is located coaxially between the anode and the cathode. The sensor is made of 0.4-mm molybdenum wire, 4 mm in diameter. It has a large collecting area, comparable to that of the anode. The plasma volume is reduced

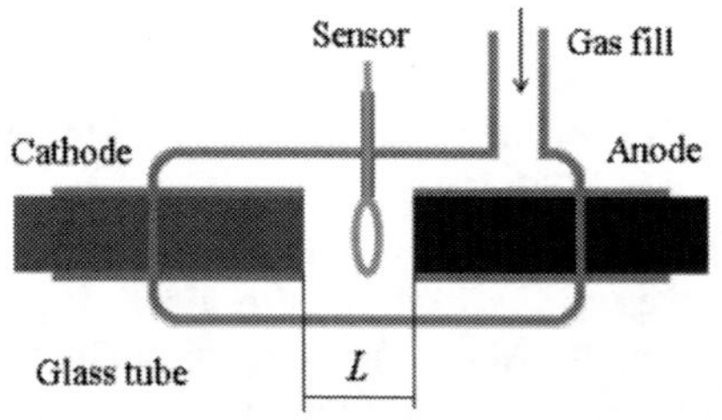

Figure 1. Schematic view of the microplasma discharge device.

dramatically; it is about 0.01 cm^3, which is more than two orders of magnitude less then the one used in previous works [1-3]. The small size of the plasma volume allows a significant increase in the operating pressures.

The energy spectra of the Penning electrons were obtained by measuring the second derivative of the probe *VA*-characteristic (d^2I/dV^2) with respect to the scanning voltage applied, which, according to the Druyvesteyn's relation, is proportional to the EEDF [8]. The so-called "modulation method" [8] was used. The recording system is typical for the probe measurements [5].

The Penning electrons' energy spectra were obtained in He as the main gas with small admixtures of Ar and N_2. The total gas pressure ranged from 30 Torr to 200 Torr, with the discharge current being varied from 0.5 mA to 12 mA.

2. Results and discussion

A typical shape of the sensor's *VA*-characteristic is shown in figure 2. It should be noted that, despite the high pressure (200 Torr), the essential parts of a Langmuir probe characteristic are well seen. The second derivative of the sensor current with respect to the scanning voltage applied, i.e. the EEDF, was obtained in the voltage range −20 V – 5V.

The energy spectra of the Penning electrons recorded at high pressures of the main gas are similar to those obtained in our earlier papers [5, 6] at intermediate pressures. Maxima appear in the EEDF at characteristic energies corresponding to Penning ionization of Ar or N_2 admixtures by metastable He atoms.

The electron energy spectrum measured when a small amount (0.2 %) of Ar is added to 200 Torr He is illustrated in figure 3. Two well-expressed characteristic Ar maxima are seen at about 4 eV and 4.8 eV. The slow part of the EEDF (reduced 50 times) is shown in figure 3 as well. The fine structure of the Ar maximum is the result of the existence of two metastable He atom states (triplet $He(2^3S_1)$ and singlet $He(2^1S_0)$), having an energy difference of 0.79 eV. Fast electrons with energies of 4 eV and 4.8 eV are released in Penning reactions of Ar atoms with triplet and singlet metastable He atoms (reactions (3) and (4)), which form the two Ar peaks in the electron energy spectrum:

$$He(2^3S_1) + Ar(0) \rightarrow He(0) + Ar^+(^2P_{3/2}) + e\ \{4.06\ eV\}, \quad (3)$$
$$He(2^1S_0) + Ar(0) \rightarrow He(0) + Ar^+(^2P_{3/2}) + e\ \{4.85\ eV\}. \quad (4)$$

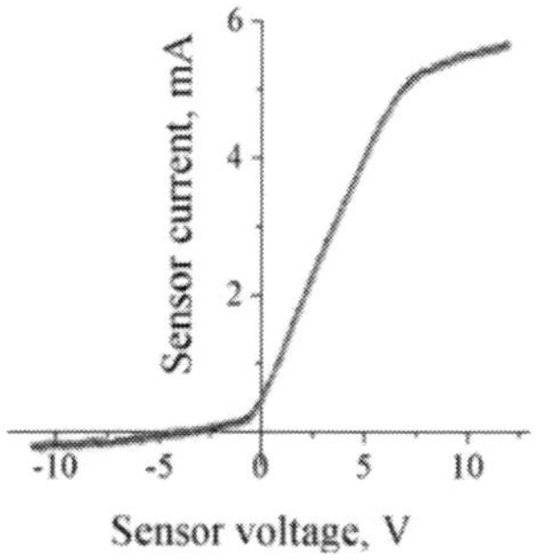

Figure 2. Sensor *VA*-characteristic at 200 Torr He pressure and 7 mA discharge current.

The Ar peaks in figure 3 have comparable amplitudes. As the cross sections for Penning ionization of Ar with singlet and triplet metastable He atoms (16.4×10^{-16} cm^2 and 5.3×10^{-16} cm^2, respectively [9]), differ by about a factor of three, one can conclude that, at the particular discharge conditions, the density of singlet metastable He atoms is about three times as low as the density of triplet metastable He atoms.

The characteristic Ar maxima were easily obtained due to the large sensor surface.

Thus, the measured energy, ε_p, at which the recorded Ar maxima appear in the EEDF allows one to detect and identify the presence of Ar atoms in the plasma at high pressures.

Except Penning reactions, fast electrons are produced by supper-elastic collisions of slow electrons with triplet

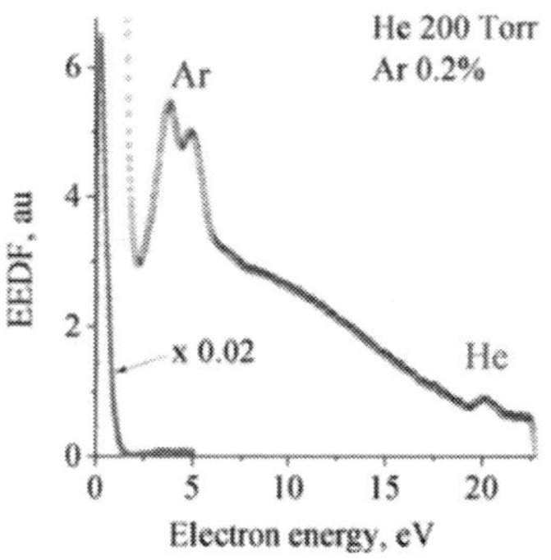

Figure 3. Electron energy spectra in He-Ar mixture at 11 mA discharge current, 0.9 V modulating voltage and 1 mm cathode - anode gap.

and singlet metastable He atoms, which in non-local plasma form maxima in the EEDF as well. The peak at about 20 eV in figure 3 corresponds to electrons arising from such processes:

$$He(2\,^3S_1) + e \rightarrow He + e\ \{19.82\ eV\}, \tag{5}$$
$$He(2\,^1S_0) + e \rightarrow He + e\ \{20.61\ eV\}. \tag{6}$$

At high He pressures, it can be observed at currents exceeding 10 mA only. This peak can be used for absolute calibration of the energy scale and for determination of the metastable He atoms density.

The Penning electron energy spectrum obtained in 200-Torr pure He is shown in figure 4. A maximum appears at about 4.2 eV due to a Penning reaction between N_2 molecules present in the plasma as residual air impurities and triplet metastable He atoms leading to the formation of molecular nitrogen ions in ground state [1, 5, 6]:

$$He(2^3S_1) + N_2(0) \rightarrow He(0) + N_2^+(X\,^2\Sigma) + e\ \{4.2\ eV\}. \tag{7}$$

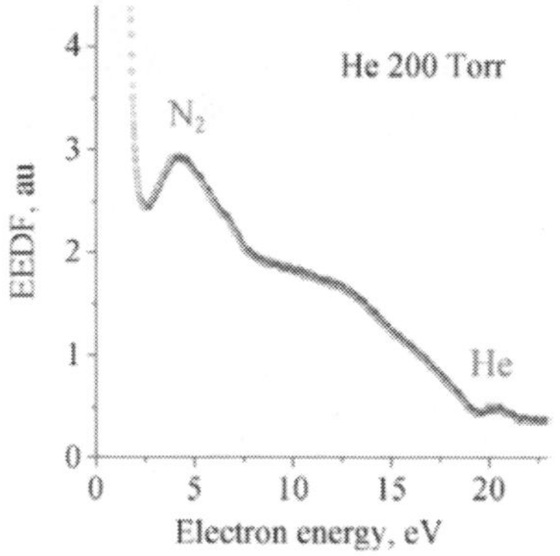

Figure 4. Electron energy spectrum in pure He at 11 mA discharge current, 0.5 V modulating voltage and 1 mm cathode – anode gap.

The characteristic He maximum at about 20 eV (reactions (5) and (6)) is seen in figure 4 as well. It is weak and appears only at discharge currents over 8 mA.

The characteristic N_2 maximum recorded in pure He illustrates the capability of the CES method to identify the presence of very low densities of N_2 impurities by measuring the Penning electron energy spectrum.

Figures 5 illustrates the Penning electron energy spectra at 34 Torr He with admixture of 1 % N_2. Well-expressed characteristic N_2 maxima at 1.8 eV (figure 5a) and at 4.2 eV (figure 5b) are observed. The N_2 maxima correspond to Penning reactions of N_2 molecules with triplet and singlet metastable He atoms (reactions (7) and (8), respectively):

$$He(2^1S_0) + N_2(0) \rightarrow He(0) + N_2^+(B\,^2\Sigma) + e\ \{1.8\ eV\}. \tag{8}$$

The maximum at 1.8 eV is superimposed over the steep rise of the tail of the bulk electrons distribution. It appears at a low discharge current (1 mA) and low modulating voltages. The peak at 4.2 eV is well expressed at discharge currents above 2 mA, while the peak at 1.8 eV is immersed in the slow part of the EEDF at this current. The rate constant for reaction (8) with singlet metastable He

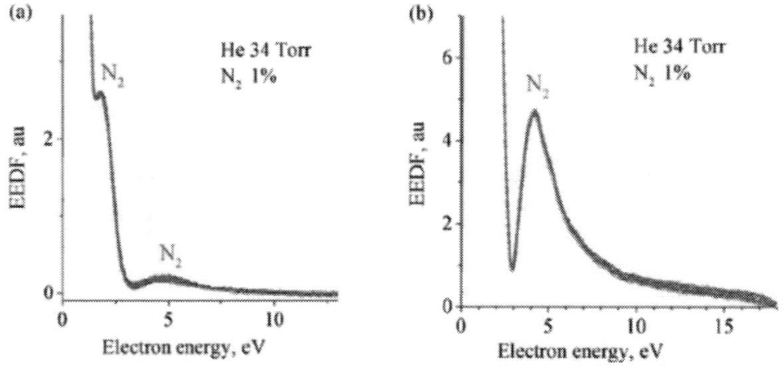

Figure 5. Electron energy spectra in a He-N_2 mixture at a 2-mm cathode-anode gap and: (a) 1 mA discharge current, 0.6 V modulating voltage; (b) 2 mA discharge current, 0.8 V modulating voltage.

atoms is about twice as high as than that for reaction (7) with triplet metastable He atoms [1], while the density of singlet metastable He atoms is lower than that of the triplet metastable He atoms.

4. Conclusions

The recorded energy spectra of Penning electrons at He pressures of 30 and 200 Torr, illustrated above, are characterized by the appearance of maxima at characteristic energies corresponding to the energy of the electrons released in Penning reactions involving Ar and N_2 impurities. Well-expressed maxima in the electron energy spectra are easily obtained as a result of the high number of Penning electrons collected by the large sensor surface. The non-local negative glow plasma of a short dc microdischarge is a suitable medium for recording electron energy spectra and for detection and identification of gas impurities.

To the best of our knowledge, no observation has so far been reported of maxima in the EEDF at high pressures due to electrons produced in Penning reactions of metastable He atoms with impurity atoms or molecules.

The maxima observed in the EEDF demonstrate the possibility to detect and identify the presence of different atomic and molecular admixtures in He at high pressures.

Acknowledgments

The authors wish to thank Mrs. Ts. Rankova for the preparation of the microplasma discharge device.

References

[1] Kolokolov N B, Kudryavtsev A A and Blagoev A B 1994 *Physica Scripta* **50** 371
[2] Kudryavtsev A A 1999 *Int. Workshop Results of Fundamental Res. for Investment; Russian Technol. for Industrial Appl.* (May 1999 St. Petersburg Russia) Abstract Book p 94
[3] Sheverev V A, Khromov N A and Kojiro D R 2002 *Anal. Chem.* **74** 5556
[4] Kudryavtsev A A and Tsyganov A B 2007 *US Patent* 7309992; Kudryavtsev A A, Tsyganov A B and Chirtsov A S 2011 *Patent of Russian Federation* 2422812 (in Russian)
[5] Kudryavtsev A, Pramatarov P, Stefanova M and Khromov N *J Instrumentation IOP* 2012 **7** PO7002
[6] Kudryavtsev A A, Stefanova M, Pramatarov P, Khromov N, Peyeva R and Patrikov T 2012 *Contributed Papers 7th Int. Conf. Plasma Phys. Plasma Technol.* vol 2 pp 628–31 (Sept. 2012 Minsk Belarus)
[7] Tsendin L D 1995 *Plasma Sources Sci. Technol.* **4** 200
[8] Godyak V A and Demidov V I 2011 *J. Phys. D: Appl. Phys.* **44** 233001
[9] Schmeltekopf A L and Fesenfeld F C 1970 *J. Chem. Phys.* **53** 3173

Journal of Physics: Conference Series **514** (2014) 012053 doi:10.1088/1742-6596/514/1/012053

Galvanic manifestation of coherent degenerate Zeeman sublevels in a gas discharge

V Steflekova and D Zhechev[1]

Acad. G. Nadjakov Institute of Solid State Physics, Bulgarian Academy of Sciences, 72 Tsarigradsko Chaussee, 1784 Sofia, Bulgaria

E-mail: spectron@issp.bas.bg

Abstract. The degenerate Zeeman sublevels were found to contribute to the ionization in a gas discharge in accordance with their coherent state. The voltage across the discharge was monitored while the same ensemble of excited atoms was alternately self-aligned ↔ nonaligned or oriented ↔ aligned. In the latter case, the coherence was optically induced; the corresponding opto-galvanic signals, i.e. amplitude- and time-resolved ones, were compared. Each state of the ensemble of degenerate Zeeman sublevels, i.e. (self-) aligned, oriented or disordered was characterized by its own rate of ionization. Various hollow cathode discharge (HCD) media were studied, namely, Ne/As, Ne/Cu, Ne/Ni, Ne/Cd, Ne/Li and Ne/Si in the corresponding commercial HCD spectral lamps.

1. Introduction

Hanle was the first to align an excited atomic state, i.e. to prepare a coherent superposition of degenerate magnetic states *m* such that $\Delta m = 0, \pm 2$ [1] using linearly polarized resonant illumination for alignment. Later, the same coherence was found to be an attribute of gas discharge plasma [2]. It arises without any external reason (self-alignment), i.e. due to the space anisotropy of some internal process of excitation. In fact, all real gas discharge sources generate self-alignment. Another Δm- combination, i.e. $\Delta m = \pm 1$ is known as orientation [3]. It arises due to absorption of circularly polarized light. Both (self-) alignment and orientation manifest themselves optically, i.e., as linearly- and circularly-polarized spontaneous emission from interfering *m*- states.

In a real gas discharge medium, either (self-) alignment or orientation mean that the corresponding $\Delta m = 0, \pm 2$ or $\Delta m = \pm 1$ state dominate in number over the rest of the *m*- states permitted. Therefore, a certain definitive *m*-states disposition dominates along any arbitrary space axes. On the other hand, in low-temperature plasma the process of ionization is anisotropic. Thus, the yield of charged particles should also depend, generally, on the magnetic states *m*. This nonenergetic contribution to the ionization (and conductivity, respectively) is present implicitly in some earlier investigations [4, 5].

In this study, the excited ensemble of atoms is investigated galvanically at the two Δm- transitions, i.e. i) self-aligned→ non-aligned and ii) aligned ↔oriented.

A hollow cathode discharge (HCD) was used. Besides its spectroscopic properties, HCD is also known as a medium where the excited atoms are self-aligned by a characteristic beam, such as fast electrons [6, 7]. Since the self-alignment is an attributive property of HCD, the comparison in the

[1] To whom any correspondence should be addressed.

ionization between self-aligned and nonaligned atoms is based on the magnetic destruction of this coherence (case i). As for the comparison aligned↔oriented atoms, these coherences are light induced (case ii).

2. Experimental set-up

The ionization was compared by measuring the voltage change ΔU across the discharge. Figure 1 illustrates the experimental set-up used. Commercial HCD spectral lamps Ne/As ("Pye Unicam"), Ne/Cu, Ne/Ni, Ne/Cd, Ne/Li and Ne/Si ("Narva") were studied in a dc mode of operation The measurement of ΔU in a self-aligned → nonaligned ensemble (case i above) was based on an external scanning (by means of a stepper motor) magnetic field B (Helmholtz coils) destroying the self-alignment. The values of $\Delta U(B)$ were measured by using a modulating (179 Hz) magnetic field B_m. In order to improve the S/N ratio, a phase-sensitive detection was applied. A lock-in amplifier type 232B (Unipan) provided the $\Delta U(B)$ dependence whose shape is close to that of the first derivative $\partial U(B)/\partial B$. The experimental set-up is presented schematically in figure 1.

The measurement of an ensemble of Ne atoms being aligned ↔ oriented (case ii above) was performed by introducing light-induced alignment/orientation. Two lasers, He-Ne (λ = 632.8 nm) and He-Cd (λ = 325.0 nm), and a polarizer (linear or achromatic waveplate $\lambda/4$, 320-800 nm) were used. In essence, the set-up used was an opto-galvanic (OG) (in this case, a polarization OG) spectroscopy set-up [8]. The intensity of the illuminating light in the different polarizations was adjusted and controlled.

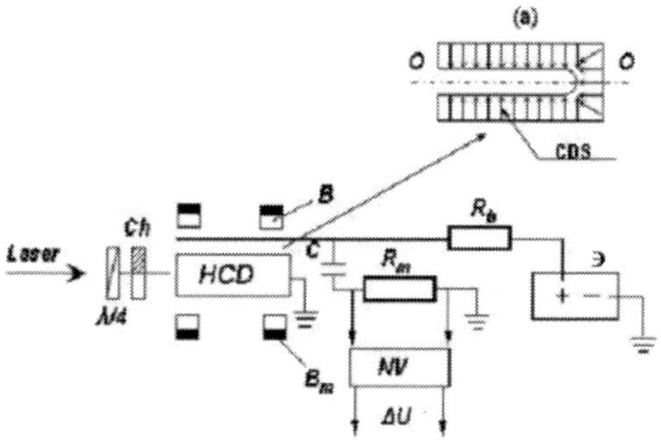

Figure 1. Experimental set-up: $\lambda/4$ – polarizer (linear or achromatic wave-plate $\lambda/4$, 320-800 nm), Ch – mechanical chopper, HCD – hollow cathode discharge, CDS – cathode dark space, B – magnetic field, B_m – modulating magnetic field, C – decoupling capacitor, R_m – measuring resistor, NV – nanovoltmeter (selective type 237 or lock-in type 232B), R_b – ballast resistor, э – power supply.

3. Experimental results and discussion

3.1. Magnetic-field-induced transitions self-aligned→nonaligned ensemble of atoms

The weak scanning magnetic field B applied was found to generate a peak-like galvanic change in the discharge. Figure 2a illustrates the galvanic behavior of Ne/As HCL as a function of B. The signals are close in shape to the first derivative $\partial U(B)/\partial B$. The maximum of the primitive function $\Delta U(B)$ (at $B = 0$) corresponds to the maximal self- alignment. The experimental circuit *HCD in* $(B_0 \pm B_m)$ - CR_m - *NV lock–in* (figure 1) reduces the degree of self-alignment at any value of $B \neq 0$. Thus, the field B reveals directly the difference in the conductivity at the transition self-aligned →nonaligned ensemble.

The magnetic field was directed along the geometric axis OO of the cathode (figure 1). Thus, the field B destroys the self-alignment along the radius $R \perp B$. This self-alignment arises due to the characteristic beam-like electrons along any radius R. In this geometry, resonances of magnetic depolarization (Hanle-signals) $\partial I_\lambda(B)/\partial B$ (I_λ being the spectral line intensity in a given polarization) were observed earlier in the spontaneous emission from self-aligned levels in a HCD [7].

We should emphasize the fact that the behavior of the signals $\partial U(B)/\partial B$ as a function of the direction of B is identical to that of the Hanle-signals $\partial I_\lambda(B)/\partial B$ observed earlier [7], i.e. a reduction of the signal when B deviates from either OO or **R**. The two kinds of signals are also of identical behavior with respect to the discharge current *i*, i.e., their width rises with *i*. However, the manifestation of a Hanle-signal as a light emission polarization assigns it to a concrete quantum state. On the contrary, many excited (self-aligned) levels contribute to the plasma conductivity and the signal measured is an integral characteristic.

The resonances $\partial U(B)/\partial B$ were measured to be of a higher amplitude near the operating inflection *I-V* points, where $\partial U/\partial I < 0$.

Simultaneously, a lower signal-to-noise ratio was observed for $\partial U/\partial I < 0$. Earlier, this was found to be typical for HCD due to Penning ionization of the sputtered atoms [9]. Here, the external galvanic perturbation manifests itself as both a higher amplitude and higher noise.

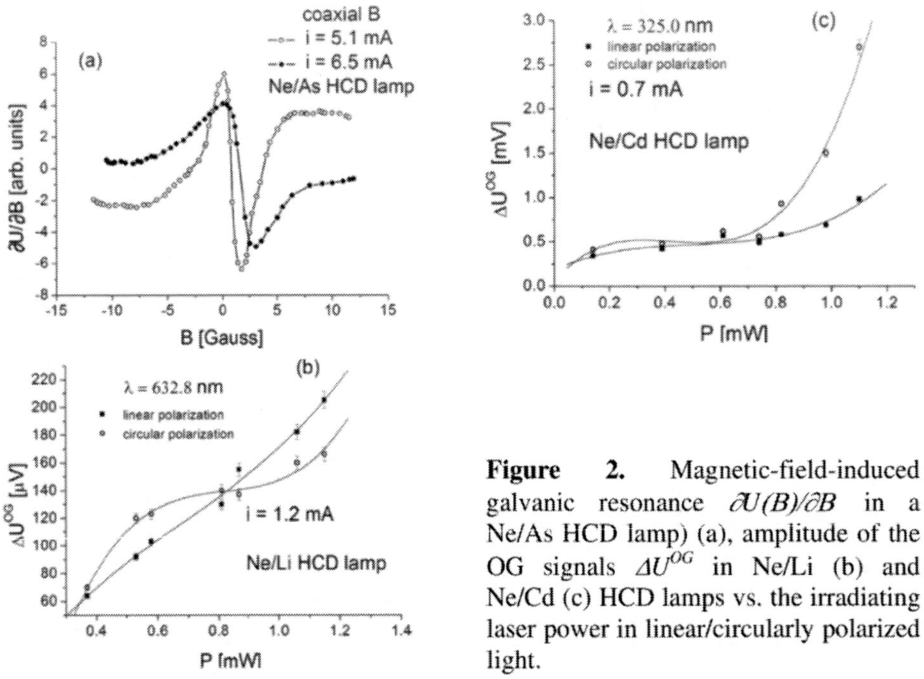

Figure 2. Magnetic-field-induced galvanic resonance $\partial U(B)/\partial B$ in a Ne/As HCD lamp) (a), amplitude of the OG signals ΔU^{OG} in Ne/Li (b) and Ne/Cd (c) HCD lamps vs. the irradiating laser power in linear/circularly polarized light.

3.2. Aligned/oriented ensemble of atoms

The set-up *laser- Ch – HCD - CR_m - selective NV* (figure 1) generates and detects two types of amplitude OG (AOG) signals ΔU^{OG}, i.e., due to illumination by either of linearly- or circularly-polarized light (figure 2(b, c)). The values of ΔU^{OG} were detected as a function of the laser power *P*; they characterize the conductivity of the same ensemble of atoms that are successively aligned ↔ oriented. The comparison of the values $\Delta U^{OG}(P)$ reveals a general tendency, i.e. $\Delta U^{OG} \sim P$. However, within this trend, a different steepness of the dependence $\Delta U^{OG}(P)$ characterizes the aligned ↔ oriented atoms. These results correlate with earlier measurements of time-resolved OG signals.

The signals ΔU^{OG} measured may be summarized as $\Delta U^{OG} \propto \pm |\Delta U(\lambda_{21}) \pm \Delta U'(\Delta m = \pm 1, \pm 2)|$, where the dominating energy term $\Delta U(\lambda_{21})$ describes the conventional OG efficiency of the light-induced population transfer. The second term $\Delta U'(\Delta m)$ describes the galvanic contribution of the coherent magnetic sub-states.

The different values of ΔU^{OG} for linearly- and circularly-polarized irradiating light reveal a dependence of the cross-sections of excitation and ionization on the type of coherence $\Delta m = \pm 1, \pm 2$ of the degenerate state.

4. Conclusions

The galvanic signals $\partial U(B)/\partial B$ and $\Delta U^{OG}(P)$ measured are manifestations of both different type and degree of ordering of the atomic magnetic state *m*. The signals reveal implicitly different rates of ionization in the ensemble of atoms self-aligned ($\Delta m = 0, \pm 2$) → nonaligned and aligned ↔ oriented ($\Delta m = \pm 1$).

The magnetic disordering of the self-alignment induces additional conductivity due to the contribution of all self-aligned levels. This effect may be of opposite signs.

Two axes of self-alignment are found in the Ne/Cu HCD lamp.

Since the self-alignment is an a priori coherence, any real gas discharge medium generates the corresponding additional conductivity. The latter is sensitive to a weak magnetic field that destroys the coherence. Maximal $\partial U(B)/\partial B$ signals are observed in vicinity of the inflection *I/V* point.

The difference in the amplitude distinguishes the signals $\Delta U^{OG}(P)$ as arising from aligned ↔ oriented ensemble of atoms having different contribution to the conductivity.

The different galvanic signals measured in both self-aligned → nonaligned and aligned ↔ oriented ensembles of atoms are manifestation of the galvanic contribution of the coherent degenerate Zeeman sublevels in a gas discharge.

References

[1] Hanle W 1924 *Zeitschrift für Physik* **30** 93-105
[2] Aleksandrov E B, Chaika M P and Khvostenko G I 1993 Interference of atomic states *Springer Series on Atoms and Plasmas* **7** (Springer Berlin Heidelberg New York)
[3] Chaika M P 1975 *Interference of degenerate atomic states* (Leningrad State University Press) p 33
[4] Series G W 1981 Comments *At. Mol. Phys.* **10** 199-201
[5] Julien L and Pinard M 1982 *J. Phys.* B: *At. Mol. Phys.* **15** 2881-98
[6] Fang D and Marcus R K 1993 in: *Glow discharge spectroscopies* ed R Marcus (Plenum Press New York) p 45
[7] Zhechev D 2001 Coherent, opto-galvanic and coherent-galvanic properties of hollow cathode discharge Doctoral Thesis, Bulgarian Academy of Sciences Sofia p 14
[8] Barbieri B and Beverini N 1990 *Rev. Modern. Phys.* **62** 603-44
[9] Zhechev D and Atanassova S 1998 *Opt. Commun.* **156** 400-408

Influence of the structural inhomogeneity on the luminescent properties of nitride multiple quantum wells grown by MOCVD

I S Asenova[1], E P Valcheva, K Kirilov and A Poturlyan

Faculty of Physics, St. Kliment Ohridski University of Sofia,
5 J. Bourchier Blvd, 1164 Sofia, Bulgaria

E-mail: iasenova@phys.uni-sofia.bg

Abstract. We investigated GaN/InGaN multiple quantum wells grown by MOCVD for LED application in the blue and green regions. The sample considered was characterized by a significant number of defects on the interfaces between the layers. We examined the heterostructure by means of cathodoluminescence. Due to the composition and the layer's thickness fluctuations on a small and a large scale, we observed peak splitting and broadening. In order to justify our assumption, we compared the experimental results with our theoretical calculations. The theoretical model used is based on the T-matrix formalism.

1. Introduction

The III-nitride layered heterostructures have acquired widespread applications as building blocks in manufacturing of devices for optical communication, high-electron-mobility transistors, ultra-violet lasers and resonant-tunneling-based structures for THz applications [1-3]. Along with the attractive characteristics [4], they also possess very specific ones, such as well pronounced piezoelectric and spontaneous polarizations along the low-symmetry axis. These features should always be taken into account when considering thin III-nitride layers, especially in superlattices and multiple quantum wells. The optical properties of such structures are greatly affected by the presence of macroscopic polarization in the layers. Another important factor that influences the optical properties is the interface roughness, which is manifested as wells with fluctuations [5-7].

In this study, we examined by means of cathodoluminescence an InGaN/GaN multiple quantum well (MQW) grown by MOCVD on a GaN substrate. All measurements were conducted at room temperature. Further, we compared the experimental results with our theoretical calculations based on the model briefly described in [8]. Our main goal was to demonstrate that, despite the idealizations and the simplifications we have made, our calculations are in a satisfactory agreement with the experimental results. We conducted our calculations in the framework of the effective-mass approximation and the T-matrix formalism. To ensure the accuracy desired, the Airy function formalism was used to solve the one-dimensional Schrödinger's equation for the MQW potential.

[1] To whom any correspondence should be addressed.

Journal of Physics: Conference Series **514** (2014) 012054 doi:10.1088/1742-6596/514/1/012054

2. Theoretical framework and calculation details

The multi-layered structure examined was composed of three layers of $In_xGa_{1-x}N$ (forming the potential wells) alternating with three layers of GaN (forming the potential barriers), grown on a GaN substrate. In order to calculate the levels of the electrons and the holes and, therefore, the optical transition energy, we examined the one-dimensional potential of the MQW. Due to the spontaneous and piezoelectric polarization in the layers, the profiles of the wells and the barriers assume a trapezoidal shape. The magnitudes of the corresponding electric fields were calculated to be [9]:

$$E_{\mathrm{w,b}} = \frac{\left(P_{\mathrm{b,w}} - P_{\mathrm{w,b}}\right)L_{\mathrm{b,w}}}{\left(\varepsilon_{\mathrm{b,w}} P_{\mathrm{w,b}} - \varepsilon_{\mathrm{w,b}} P_{\mathrm{b,w}}\right)\varepsilon_0} \tag{2.1}$$

in the wells and barriers regions, respectively. Here $P_{\mathrm{w,b}}$ represents the total polarization in a layer, $L_{\mathrm{w,b}}$, its thickness, $\varepsilon_{\mathrm{w,b}}$, the relative permittivity of the materials, and ε_0, the permittivity of vacuum. Having determined the potential across the MQW, we found the electron and hole states in the wells (using the effective-mass approximation) and, thereupon, the interband transition energies. In order to perform the calculations, one needs a set of material parameters, which are systematized in table 1. All the parameters of GaN and InN were extracted from [10] The parameters of $In_xGa_{1-x}N$ were calculated following [9] by setting $x = 0.2$:

$$\begin{aligned} E_{\mathrm{g}} &= 3.39 - 2.65x \\ \chi &= 4.26 + 1.67x - 0.67x^2 \\ m_{\mathrm{e}}^* &= 0.2 - 0.08x \\ \varepsilon &= 9.8 + 5.2x \end{aligned}, \tag{2.2}$$

where E_{g} is the energy gap in bulk $In_{0.2}Ga_{0.8}N$ in eV, χ is the electron affinity of the material in eV, m_{e}^* is the electron effective mass in units of m_0 (i.e. free electron mass), and ε is the dielectric constant in units of ε_0.

Proceeding from the above parameters and considering (2.1), we calculated the band offsets and the potential profile of the structure. The alternating layers of GaN and $In_{0.2}Ga_{0.8}N$ had thicknesses of $L_{\mathrm{b}} = 10\,\mathrm{nm}$ and $L_{\mathrm{w}} = 4\,\mathrm{nm}$, respectively. The offset in the conduction band was calculated to be $\Delta E_{\mathrm{C}} = 0.3\,\mathrm{eV}$ and the offset in the valence band, $\Delta E_{\mathrm{V}} = 0.26\,\mathrm{eV}$. A plot of the potential profile of the MQW examined, as well as the electron (E1) and heavy hole (HH1) ground states, are shown in figure 1.

The transition $\mathrm{E1} \rightarrow \mathrm{HH1}$ as calculated to have an energy of 2.81 eV, which corresponds to a wavelength of $441\,\mathrm{nm}$. In order to estimate the influence of the structural inhomogeneity of the $In_{0.2}Ga_{0.8}N$ layers (the wells), we performed the same calculations for MQW structures with different

Table 1. Parameters of GaN and $In_{0.2}Ga_{0.8}N$.

Parameter	GaN	$In_{0.2}Ga_{0.8}N$
Bandgap energy E_{g} (eV)	3.39	2.86
Electron affinity χ (eV)	4.26	4.56
Effective mass (m_0)		
electron m_{e}^*	0.2	0.18
heavy hole m_{hh}^*	1.4	1.55
Dielectric constant ε (ε_0)	9.8	10.84
Total polarization P ($\mathrm{C\,m^{-2}}$)	-0.03	-0.062
Lattice constant c (Å)	5.19	5.6

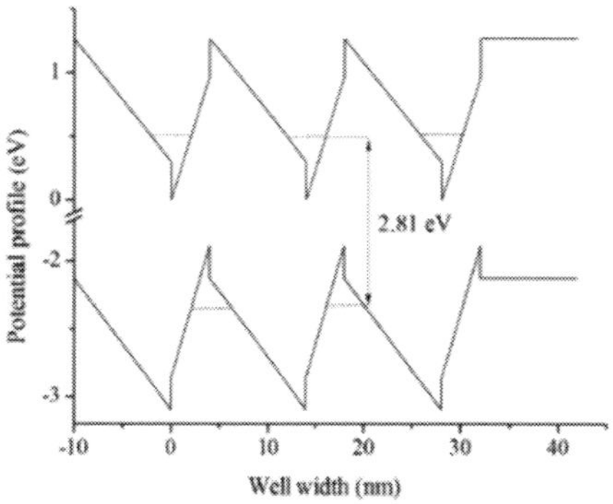

Figure 1. Potential profile across the growth axis of three-period $In_{0.2}Ga_{0.8}N$/GaN MQW. The ground electron and hole levels, along with the transition energy, are also depicted.

well widths. We considered well-width fluctuations of ± 2 monolayers (ML), i.e. $\pm c_{\mathrm{InGaN}}$, and kept all other parameters the same throughout the calculations.

3. Results and discussion

The luminescent spectrum shown in figure 2 was obtained from the sample considered above. Two distinct peaks are clearly observed. The leftmost peak on figure 2(a) is observed at $\lambda = 364\,\mathrm{nm}$. This wavelength corresponds to the bandgap of GaN, hence the peak is due to transitions between the band edges in the GaN substrate. The second peak is wider and split at $\lambda = 425\,\mathrm{nm}$ and $\lambda = 441\,\mathrm{nm}$. It is attributed to the transition $\mathrm{E1} \rightarrow \mathrm{HH1}$ in the MWQ. The calculation results indicate a transition energy of $E_{E1\rightarrow HH1} = 2.81\,\mathrm{eV}$ or $\lambda = 437\,\mathrm{nm}$ for the quantum wells with a nominal width ($L_w = 40\,\mathrm{nm}$). For the quantum wells whose width is increased by 2 ML, the transition energy is calculated to be $E^{+2}_{E1\rightarrow HH1} = 2.72\,\mathrm{eV}$ or $\lambda = 456\,\mathrm{nm}$, while for wells whose width is decreased by 2 ML, the results are $E^{-2}_{E1\rightarrow HH1} = 2.95\,\mathrm{eV}$ or $\lambda = 420\,\mathrm{nm}$. All results are plotted over the measured luminescent spectrum on figure 2(b).

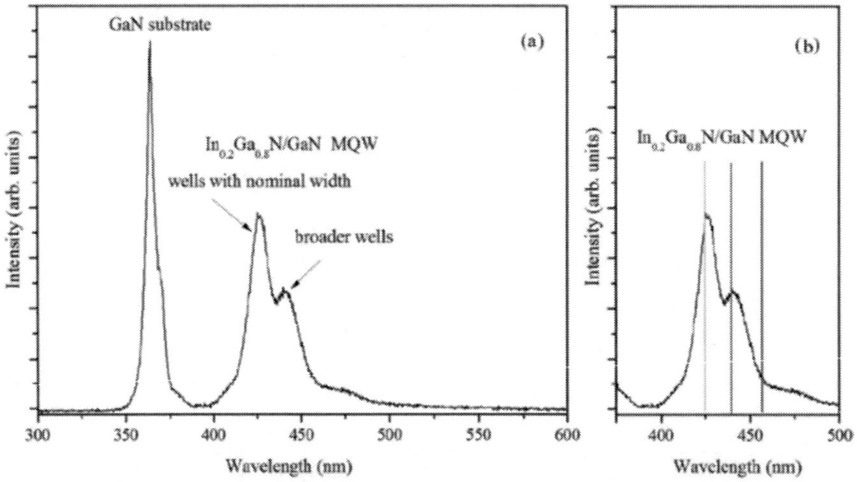

Figure 2. (a) Room temperature cathodoluminescent spectrum of $In_{0.2}Ga_{0.8}N$/GaN MQW. The first peak is attributed to the transitions between the band edges in the GaN substrate. The second peak is due to the optical transitions in the MQW. The split in the second peak is attributed to fluctuations in the InGaN layer; (b) $\mathrm{E1} \rightarrow \mathrm{HH1}$ transition wavelength, calculated for: nominal width of the wells (red line); wider wells (blue line); narrower wells (green line). The corresponding wavelengths are depicted over the spectrum of the MQW.

The splitting observed in the luminescent spectrum is most likely due to fluctuations in the InGaN well width, which inevitably occur during the growth process. This assumption is supported by the fact that the difference of $110\,\mathrm{meV}$ between the peaks (in the measured spectrum) is very close to the difference of $90\,\mathrm{meV}$ between $E_{E1\rightarrow HH1}$ and $E^{+2}_{E1\rightarrow HH1}$, or to the difference of $130\,\mathrm{meV}$ between $E_{E1\rightarrow HH1}$ and $E^{-2}_{E1\rightarrow HH1}$. In our theoretical results, we observed a certain tendency toward the lower transition energies, which is most likely due to a composition ratio that differs from the nominal during the growth process, or to the chosen set of band parameters. For example, a higher value of the InGaN bandgap energy would have resulted in higher transition energies. Apart from this, our calculations are in a good agreement with the experimental results.

4. Conclusions

We examined both theoretically and experimentally the transition energies in InGaN/GaN MQW grown by MOCVD. We calculated the optical transition energy taking account of the well-width fluctuations. The splitting of the peak that corresponds to the E1 → HH1 transition was explained by the well-width fluctuations. The luminescence spectrum observed supported our theoretical results, although the latter showed a certain shift toward the lower energies. This might be corrected in further calculations by choosing another set of parameters or by changing the composition ratio.

Acknowledgements

This work was supported by the Sofia University Fund for Scientific Research under contract No.76/2013. Pierre Ruterana is acknowledged for providing the sample.

References

[1] Beeler M, Trichas E and Monroy E 2013 *Semicond. Sci. Technol.* **24** 074022-48
[2] Cho A, Sivco D, Ng H, Gmachl C, Tredicucci A, Hutchinson A, Chu S and Capassco F 2001 *J. Crystal Growth* **227-228** 1-7
[3] Camacho A S, Gutiérrez R M and Ardilla G 2005 *Microelectr. J.* **36** 922-7
[4] Monemar B 1999 *J. Mater. Sci.: Mater. Electr.* **10** 227-54
[5] DeCuir A, Fred E, Passamore B, Muddasani A, Manasrech M, Xie J, Morkoç H, Ware M and Salamo G 2006 *Appl. Phys. Lett.* **89** 151112-14
[6] Dimitrov S, Valcheva E and Donchev D 2007 *Nanosci. & Nanotechnol.* **7** 115-8
[7] Paskov P, Bergman J, Darakchieva V, Paskova T, Monemar B, Iwaya M, Kamiyana S, Amano H and Akasaki I 2005 *Phys. Stat. Sol.* C **2** 2345-8
[8] Asenova I and Valcheva E 2012 *J. Phys.: Conf. Series* **398** 012014
[9] Piprek J 2007 *Nitride Semiconductor Devices* (Weincheim: Willey-VCH) p 354
[10] Vurgaftman I and Meyer J 2003 *J. Appl. Phys.* **94** 3675-96

Temperature-dependent interaction potential between NF_3 molecules and thermophysical properties of gaseous NF_3

M Damyanova[1,3], E Balabanova[1] and U Hohm[2]

[1]Acad. E. Djakov Institute of Electronics,
Bulgarian Academy of Sciences, 72 Tsarigradsko Chaussee, 1784 Sofia, Bulgaria
[2]Institut für Physikalische und Theoretische Chemie der TU Braunschweig,
10 Hans-Sommer-Str., D-38106 Braunschweig, Germany

E-mail: milenadam@gmail.com

Abstract A temperature-dependent effective intermolecular interaction potential is applied to describe the interaction between two nitrogen fluoride (NF_3) molecules in gas phase. To this end, a spherically-symmetric (*n*-6) Lennard-Jones temperature-dependent potential (LJTDP) is used. The (*n*-6) LJTDP takes into account the influence of vibrational excitation of the molecules on the potential parameters, namely, the equilibrium distance r_m and the potential well depth ε. The potential parameters at $T = 0$ K were obtained from the very small amount of existing thermophysical equilibrium and transport properties of low-density NF_3 gas. Fitting formulae are tabulated for a fast and reliable prediction of the thermophysical properties and potential parameters in the temperature range between 200 K and 1200 K. A comparison is also presented between our estimates for some thermophysical properties of the NF_3 gas with the available experimental and calculated data.

1. Introduction

Nitrogen trifluoride (NF_3) is a gas primarily used in the semiconductor industry. Its usage has increased during the last years as it is still not included in the list of gases controlled by the Kyoto protocol. Because of its high percentage conversion to fluorine, NF_3 is used as a substitute to other fluorinated species, such as perfluorocarbons and SF_6 for plasma etching and CVD chambers cleaning. However, there is a lack of sufficient thermophysical data on gaseous NF_3 over a large temperature range, which would allow for a reliable prediction of its thermophysical behavior. This is probably be the reason for the relatively poor knowledge of its intermolecular interaction energy as well.

Oghaz et al. [1] have determined an effective interaction potential function by inversion of the reduced viscosity collision integrals at zero density. In their later work [2], a new approach has been proposed to the estimation of viscosity in dilute and dense pure NF_3. They have also calculated the viscosity in the zero-density limit between 224 K and 349 K by using different potential energy functions, namely Lennard-Jones (12-6), Mie (14-7), Hartree-Fock-Dispersion and the Maitland-Smith extended [*n*,*m*] Lennard-Jones potential.

[3] To whom any correspondence should be addressed.

Journal of Physics: Conference Series **514** (2014) 012055 doi:10.1088/1742-6596/514/1/012055

The main goals of our study were:

i) to obtain an effective interaction potential energy function of two NF_3 molecules by using simultaneously the available equilibrium (second *pVT*-virial coefficient, *B*, and acoustic second virial coefficient, (β) and transport properties (viscosity, η) of gaseous NF_3;

ii) to predict the temperature dependence of $B(T)$, $\eta(T)$ and $\rho D(T)$ in the large temperature interval between 200 K and 1200 K.

2. Theoretical background

The (n-6) LJTDP was introduced and discussed in full detail in our earlier work, see e.g. [3]. Briefly, the intermolecular interaction potential energy $U(r,T)$ is modelled as

$$U(r,T)=\frac{\varepsilon(T)}{n-6}\left[6\left(\frac{r_m(T)}{r}\right)^n-n\left(\frac{r_m(T)}{r}\right)^6\right]. \tag{1}$$

Here r and T are the center-of-mass distance and temperature, respectively, $r_m(T)$ is the equilibrium distance, $\varepsilon(T)$, the potential energy well depth, and n, the repulsive parameter, the latter being independent of the temperature. The temperature dependence of $r_m(T)$ and $\varepsilon(T)$ is the result of a Boltzmann-type averaging over the occupation of the molecule's higher vibrational levels. It is described by:

$$r_m(T)=r_m(T=0\,\mathrm{K})+\delta f(T) \tag{2}$$

and

$$\varepsilon(T)=\varepsilon(T=0\,\mathrm{K})\left(\frac{r_m(T=0\,\mathrm{K})}{r_m(T)}\right)^6, \tag{3}$$

where δ is a temperature-independent fit parameter, whereas the function $f(T)$ indicates the influence of the vibrational excitation of the molecule on its geometrical radius. The function $f(T)$ can be calculated at any T if the vibrational frequencies of the molecule are known.

3. Results and discussion

The potential parameters at T = 0 K were obtained by minimizing the sum F (equation 4) of the squared deviations between the experimental and calculated thermophysical properties, normalized to their relative experimental errors $a_{\exp}$. Here M is the number of experimental data points.

$$F=\sum_1^{M_B}\left[\frac{\ln\left(\frac{B_{\exp}}{B_{calc}}\right)}{a_{\exp B}}\right]^2+\sum_1^{M_\beta}\left[\frac{\ln\left(\frac{\beta_{\exp}}{\beta_{calc}}\right)}{a_{\exp \beta}}\right]^2+\sum_1^{M_\eta}\left[\frac{\ln\left(\frac{\eta_{\exp}}{\eta_{calc}}\right)}{a_{\exp \eta}}\right]^2 \tag{4}$$

The available experimental data for the thermophysical properties of NF_3 are very scarce. In the case of the second *pVT*-virial coefficient *B*, we used the data of Lemmon and McLinden [4]. The accepted $a_{\exp}$ of these data is in the range between 2 % and 5 %. For the second acoustic virial coefficient β, only the experimental results of Hurly [5] are available in the literature. Although measured in the small temperature range between 200 K and 425 K, these measurements have a large impact on the resulting potential parameters due to their high relative accuracy (error about 0.2 –1 %). Hurly's experimental data for the viscosity [5, 6] are also substituted into our minimization procedure. Their accuracy is in the range between 0.7 % and 3 %. Our survey of the literature showed that no experimental data are available for the self-diffusion coefficients of NF_3.

The minimization procedure was carried out with all available data. The results obtained for the potential parameters of the (*n*-6) LJTDP at $T = 0$ K are: $\varepsilon/k_B = (273.89 \pm 3.44)$ K, $r_m = (4.254 \pm 0.014)\times10^{-10}$ m, $n = 24.13 \pm 0.62$ and $\delta = (0.0045 \pm 0.0001)\times10^{-10}$ m with a root-mean-square deviation of $RMS = \sqrt{F/M} = 0.786\ a_{exp}$. The quality of our (*n*-6) LJTDP can be seen in figure 1. In this figure, the calculated thermophysical properties P_{calc} are compared to the experimentally obtained input data P_{exp}. It can be seen that the relative deviations $(P_{exp} - P_{calc})/a_{exp}$ lie mostly in the range between $\pm 1 \cdot a_{exp}$. This means that the (*n*-6) LJTDP is able to reproduce the experimental data within their error limits. Additionally, the data points shown in figure 1 are distributed more or less symmetrically towards the zero line defined by our calculated data.

In figures 2-4, our results obtained with the (*n*-6) LJTDP are compared to other correlations which are given in the literature. We also included the self-diffusion coefficient ρD in this comparison.

In figure 2, our calculations of the second *pVT*-virial coefficient, *B*, are compared to the correlations of Tsonopoulos [7]. The maximum deviation is –4.07 cm^3/mol. This small deviation shows that both models do coincide well within their error bars. A comparison of the viscosities obtained from different models is shown in figure 3. It can be seen that the relative deviation between our calculations and the results of other authors are mostly in the range between ±3 %.

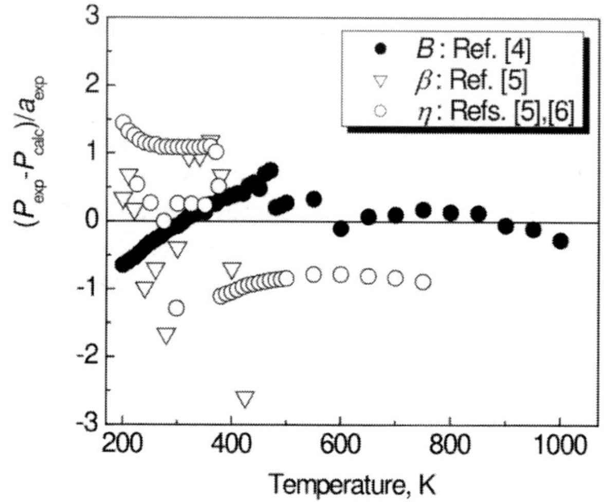

Figure 1. Relative deviations between the (*n*-6) LJTDP solution (zero line) and the experimental input data of NF_3.

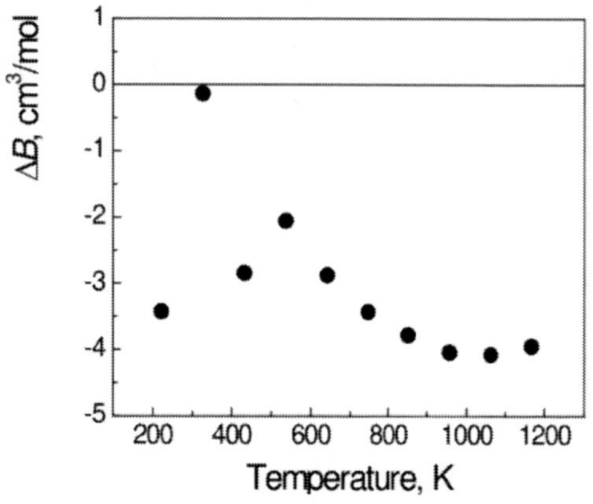

Figure 2. Absolute deviations $\Delta B = B_{[7]} - B_{LJTDP}$ of the second *pVT*-virial coefficients between the results of [7] and the values obtained in this work.

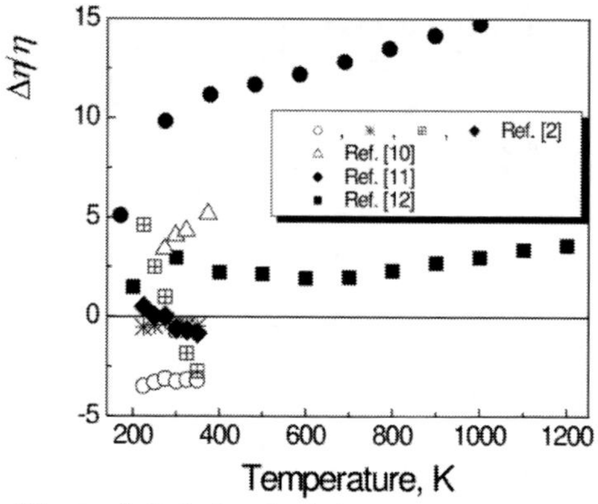

Figure 3. Relative deviations $\Delta\eta/\eta = 100(\eta_{Lit} - \eta_{LJTDP})/\eta_{LJTDP}$ of the viscosity calculated in [2] (various potential energy functions) and [10]-[12] and the values obtained in this work.

In figure 4, the relative deviations between the binary self-diffusion coefficient ρD of Liley [12] and our results are shown. In table 1, we present our recommended values for the potential parameters r_m and ε as well as for the thermophysical properties B, η, and ρD.

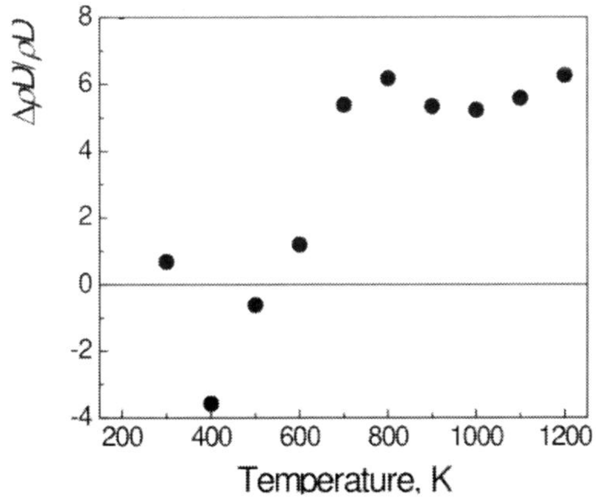

Figure 4. Relative diviations $\Delta\rho D/\rho D = 100(\rho D_{Li} - \rho D_{LJTDP})/\rho D_{LJTDP}$ of the binary self-diffusion coefficient between the results calculated in Ref. [12] and the values obtained in this work.

Table 1[a]. Recommended data for NF_3.

T, K	r_m, 10^{-10} m	ε/k_B, K	B, cm^3/mol	η, μPa·s	ρD, 10^{-3}·g/(m·s)
200	4.259855	271.63	-197.72	12.22	16.00
250	4.264275	269.95	-125.50	14.99	19.55
300	4.269267	268.06	-83.38	17.68	22.93
350	4.274497	266.10	-55.85	20.25	26.15
400	4.279821	264.12	-36.47	22.70	29.18
450	4.28513	262.16	-22.11	25.02	32.08
500	4.290367	260.25	-11.05	27.22	34.84
550	4.295491	258.39	-2.27	29.31	37.44
600	4.300466	256.60	4.85	31.30	39.92
650	4.305244	254.90	10.74	33.19	42.30
700	4.309771	253.30	15.68	35.00	44.59
750	4.313991	251.81	19.87	36.74	46.79
800	4.317862	250.46	23.46	38.41	48.91
850	4.32136	249.25	26.57	40.02	50.97
900	4.324483	248.17	29.27	41.58	52.96
950	4.327245	247.22	31.63	43.10	54.88
1000	4.329675	246.39	33.72	44.57	56.76
1050	4.331804	245.66	35.56	46.00	58.59
1100	4.333671	245.03	37.21	47.40	60.37
1150	4.335303	244.48	38.68	48.77	62.11
1200	4.336738	243.99	40.01	50.10	63.81

[a] The parameters and properties given in the table are fitted by the following formulae:
$r_m(T)/(10^{-10}\text{m}) = 4.254+0.17062\cdot\exp(-902.14/T^*)+2.76691\cdot10^{-5}T^*-2.13554\cdot10^{-8}T^{*2}$,
$B(T)/(\text{cm}^3/\text{mol}) = 70.63-33411.7/T^*-3.43064\times10^6/T^{*2}-1.23330\times10^8/T^{*3}$,
$\eta(T)/(\mu\text{Pa}\cdot\text{s}) = -0.66622+0.07090957T^*-3.50206\cdot10^{-5}T^{*2}+9.33423\times10^{-9}T^{*3}$,
$\rho D(T)/(10^{-3}\cdot\text{g}/(\text{m}\cdot\text{s})) = -0.013169+0.088446T^*-4.35150\times10^{-5}T^{*2}+1.18078\cdot10^{-8}T^{*3}$,
with $T^*=T/\text{K}$.

Journal of Physics: Conference Series **514** (2014) 012055 doi:10.1088/1742-6596/514/1/012055

4. Conclusions

The (n-6) LJTDP proposed is capable of reproducing different thermophysical equilibrium and transport properties of low-density gaseous NF_3. The values recommended of the potential parameters, second virial coefficient B, viscosity η and self-diffusion coefficient ρD allow for a reliable interpolation of different properties in the temperature-range between 200 K and 1200 K; they can be used for further calculation necessary in some engineering and scientific projects.

References

[1] Oghaz N M, Shahidi D, Ghiamati E and Haghighi B 2010 *Chem. Phys.* **369** 59-65

[2] Oghaz N M, Ghiamati E, Haghighi B, Nasrabad A E, Papari M M and Bamdad M 2012 *J. Mol. Liq.* **165** 55-62

[3] Zarkova L, Hohm U and Damyanova M 2006 *J. Phys. Chem. Ref. Data* **35** 1331-64

[4] Lemmon E W and McLinden M O 2001 NIST Reference fluid thermodynamic and transport properties *NIST Standard Reference Data Program Version 7.0* (Gaithersburg)

[5] Hurly J J 2002 *Int. J. Thermophys.* **23** 667-96

[6] Hurly J J 2004 *Int. J. Thermophys.* **25** 625-41

[7] Tsonopoulos C 1978 *AIChE J.* **24** 1112-15 as cited in: BYU 2008 *DIPPR 801 Chemical Database*

[8] Tommasini F, Levi A C, Scoles G, De Groot J J, van den Broeke J W, van den Meijdenberg C J N and Beenakker J J M 1970 *Physica* **49** 299-341

[9] Svehla R A 1962 *NASA Technical Report R-132* as cited in: van Ditzhuyzen P G, Thusse B J, van der Meij L K, Hermans L J F and Knaap H F P 1977 *Physica* A **88** 53-87

[10] *Matheson Gas Data Book* 1974 unabridged ed (Matheson Company Inc. East Rutherford New Jersey) as cited in: BYU 2008 *DIPPR 801 Chemical Database*

[11] Cowling T G and Chapman S 1952 *The Mathematical Theory of Non-Uniform Gases* (Cambridge University Press Cambridge) as cited in: BYU 2008 *DIPPR 801 Chemical Database*

[12] Liley P E 1975 *Transport properties of selected elements and compounds in the gaseous state* CINDAS-TPRC Report **29**

Journal of Physics: Conference Series **514** (2014) 012056 doi:10.1088/1742-6596/514/1/012056

Current status of the development of the problem-oriented software package GYREOSS

M Damyanova[1,3], S Sabchevski[1], I Zhelyazkov[2], E Vasileva[1], E. Balabanova[1], P Dankov[2] and P Malinov[2]

[1]Acad. E. Djakov Institute of Electronics, Bulgarian Academy of Sciences, Association EURATOM-INRNE, 72 Tsarigradsko Chaussee, 1784 Sofia, Bulgaria
[2]Faculty of Physics, St. Kl. Ohridski University of Sofia, Association EURATOM-INRNE, 5 James Bourchier Blvd., 1164 Sofia, Bulgaria

E-mail: milena@ie.bas.bg

Abstract. A great number of problem-oriented software packages (e.g., DAPHNE, ESRAY, GYROSIM, just to name a few) are used for computer-aided design (CAD) and optimization of high-power gyrotrons for fusion research (most notably for ECRH and ECCD). In them, adequate self-consistent physical models formulated in a two-dimensional space (2D) are implemented. Most of the problems encountered in recent years in the development of megawatt-class gyrotrons, however, can be attributed to physical factors and phenomena that are both non-stationary (for instance various electron beam instabilities, dynamics of the trapped particles) and inherently three-dimensional by nature (e.g., non-uniformity of the emission, misalignment of the electrodes and magnetic coils, etc.). In order to address these problems and to take into account most of the physical factors that have so far been neglected, a work on the development of a novel problem-oriented software package called GYREOSS (which stands for Gyrotron Electron-Optical System Simulation) was initiated recently. Since then, GYREOSS has evolved into a test bed for experimenting with different numerical methods, algorithms, and programming techniques for ray tracing (trajectory analysis) and particle-in-cell (PIC) simulations. In this paper, we outline the current status of GYREOSS and present both its physical model formulated in 2D and 3D and its program realization. We illustrate the capabilities of the latest version of the code, which includes a novel 3D field solver and an efficient relativistic particle pusher. Finally, we present an outlook for a further development of GYREOSS.

1. Introduction

The gyrotrons are fast-wave vacuum tubes operating on a physical principle known as electron-cyclotron maser instability, which is caused by the dependence of the relativistic cyclotron frequency of accelerated electrons gyrating in a strong magnetic field on their energy. The interaction of a helical beam in which most of the kinetic energy of its electrons is associated with their rotational motion (gyro-motion) with the transverse components of the electric field of an electromagnetic wave having a phase velocity greater than the speed of light in vacuum takes place in a resonant structure (cavity resonator) with characteristic dimensions that are much larger than the wavelength of the radiation.

[3] To whom any correspondence should be addressed.

Such operational principle, which lacks the disadvantages and the limitations of the conventional (slow-wave) vacuum tubes makes the gyrotrons the most powerful sources of coherent radiation at millimetre and sub-millimetre wavelengths (sub-THz and THz frequencies) in a CW mode. Definitely, among numerous applications of gyrotrons in the physical research and the technologies, the most prominent one is their use as sources of high-power radiation for electron cyclotron resonant heating (ECRH) and electron cyclotron current drive (ECCD) of magnetically confined plasmas in fusion devices (e.g., tokamaks and stellarators). The current state-of-the-art of these devices is well represented by a series of megawatt-class tubes in the frequency range 140 – 170 GHz [1]. For their computer-aided design (CAD), optimization and analysis, many problem-oriented software packages are available that are specialized to one or another subsystem of the tube, e.g., EOS (DAPHNE, ESRAR, GUN-MIG/CUSP), electrodynamical system (CAVITY, SELFT) or the whole device (GYROSIM). In all of them, adequate physical models formulated in a two–dimensional (2D) space are realized. Many physical factors and phenomena (for instance non-uniformity of the emission, misalignment of the electrodes and magnetic coils, various instabilities, etc.), however, are inherently three-dimensional (3D) and non-stationary by nature and are therefore neglected in the 2D simulations. In order to take them into account, the development of GYREOSS package (which stands for Gyrotron Electron-Optical System Simulation) was initiated recently. It follows a concept presented elsewhere [2, 3] and is based on a physical model formulated in both 2D and 3D and is being developed as a problem-oriented platform for particle in cell (PIC) simulations and trajectory analysis (ray tracing). In this paper we present both the current status of GYREOSS and an outlook for its further development.

2. Novel field solver and relativistic particle pusher of the GYREOSS package

Starting from an initial version of the package [4], which was based on the *gmsh* (a 3D mesh generator with a build-in CAD system with pre-, post-processing, and visualization capabilities [5]) and GetDP (a General Environment for Treatment of Discrete Problems [6]) in recent years GYREOSS has evolved into a test bed for experimenting with different numerical methods, algorithms, and programming techniques for ray tracing (trajectory analysis) and particle-in-cell (PIC) simulations. Its current version is implemented as a library of scripts using FreeFEM++ [7] and its integrated development environment (IDE) FreeFEM++-cs. The most characteristic for the current status of GYREOSS are two of its recently developed computational modules, namely a novel 3D (2D) field solver and a relativistic particle pusher. They will be presented briefly below.

The distribution of the electrostatic potential u in the electron-optical system (EOS) is governed by the Poisson's equation and the electric field $\mathbf{E} = -\nabla u$ is a gradient of that potential. More specifically, we consider a bounded computational domain $\Omega \subset \mathrm{R}^d$ of dimension $d = 3$ (3D) or $d = 2$ (2D) with a boundary $\partial\Omega = \bigcup_{i=1}^{N} \partial\Omega_i$, which is composed of N surfaces $\partial\Omega_i$ in the case of 3D or N contours in 2D. The electrostatic potential is a solution of the following boundary value problem for the Poisson equation

$$\Delta u = -f \text{ in } \Omega, \tag{1}$$

with mixed boundary conditions, i.e., Dirichlet boundary conditions (DBCs) on the electrodes (designated by the indices $l = 1, \ldots, L$)

$$u = U_l \text{ on } \partial\Omega_{Dl}, \tag{2}$$

and Neumann boundary conditions (NBCs)

$$\frac{\partial u}{\partial n} = 0 \text{ on } \partial\Omega_{Nm}, \tag{3}$$

on the appropriately specified closing boundaries (e.g., symmetry planes), designated by the indices $m = 1, \ldots, M$. The regions with Dirichlet $\partial\Omega_D$ and Neumann $\partial\Omega_N$ boundary conditions do not overlap, i.e., $\partial\Omega = \partial\Omega_D \cup \partial\Omega_N$ and $\partial\Omega_D \cap \partial\Omega_N = \varnothing$, where $\partial\Omega_D = \bigcup_{l=1}^{L} \partial\Omega_{Dl}$, $\partial\Omega_N = \bigcup_{m=1}^{M} \partial\Omega_{Dm}$ and $L+M = N$. It is well known that with such mixed boundary conditions the Poisson equation has a unique solution. Here the right-hand side of the Poisson equation $f = \rho / \varepsilon_0$ is defined by the space charge distribution ρ and the permittivity of vacuum ε_0.

For the solution of the above boundary value problem the finite element method (FEM) and the tools of the FreeFEM++ software package are used. The corresponding weak variation form of the Poisson equation reads: Find $u \in H^1$ such that

$$\int_\Omega \nabla u . \nabla u_\tau = \int_\Omega f u_\tau , \tag{4}$$

for all $u_\tau \in H_0^1$, where H^1 is the Hilbert space and H_0^1 consists of continuously differentiable functions u_τ in Ω and vanishing on the boundary. Here and below we use the traditional notations for the Sobolev function spaces $H^m(\Omega)$ that consist of all functions on Ω that are square integrable together with their derivatives up to the order m, i.e.

$$H^m(\Omega) = \left\{ u \in L_2(\Omega) : \frac{\partial^j u}{dx^j} \in L_2(\Omega), 1 \le j \le m \right\}, \tag{5}$$

where $L_2(\Omega)$ denotes the space of square integrable functions on Ω. In this paper, for convenience we denote both the unknown (trail) functions and the test functions by one and the same letter but mark the latter functions by a subscript τ.

The above formulation of the boundary value problem of the field solver is implemented using the finite elements and the tools provided by the FreeFEM++ [7] package for solving partial differential equations by the FEM. Two pieces of code with its program realization for the 2D and 3D cases respectively are shown in figure 1.

```
// GYREOSS ** GYREOSS ** GYREOSS ** GYREOSS ** GYREOSS //
func rhsLap2d=0.0;
func g2d=Uanode;
func gn2d=0.0;

//* Finite Element Spaces: *//

fespace Ph2(Th2,P2);

//Function spaces
// u,ut - trial and test functions for the potential

Ph2 u2,ut2;

macro Grad2(u2) [dx(u2),dy(u2)] // EOM

//* Conventional formulation for the Poisson's equation *//

problem Poisson2d(u2,ut2,solver=GMRES)=
int2d(Th2)((Grad2(ut2)' *Grad2(u2))*y)
-int2d(Th2)(-rhsLap2d*ut2*y)
+on(1,u2=g2d)
+on(3,u2=g2d)
+on(5,u2=gn2d);

//* Solution of the BVP *//

Poisson2d;
```

```
// GYREOSS ** GYREOSS ** GYREOSS ** GYREOSS ** GYREOSS //
func rhsLap=0.0;
func g=Uanode;
func gn=0.0;

//* Finite Element Spaces: *//

fespace Ph3(Th3,P23d);

//Function spaces
// u,ut - trial and test functions for the potential

Ph3 u,ut;

macro Grad3(u)  [dx(u),dy(u),dz(u)] // EOM

//* Conventional formulation for the Poisson's equation *//

problem Poisson3d(u,ut,solver=GMRES)=
int3d(Th3)(Grad3(ut)' *Grad3(u))
-int3d(Th3)(-rhsLap*ut)
+on(1,u=g)
+on(3,u=g)
+on(5,u=gn);

// Solution of the BVP

Poisson3d;
```

Figure 1. Excerpts from the codes of the field solvers in 2D (left) and 3D (right).

The computational domain Ω is discretized by a mesh T_h consisting of cells K, which are triangles in 2D and tetrahedrons in 3D. Both externally generated meshes (using gmsh and TetGen) as in the previous version of GYREOSS and grids generated by the embedded mesh tools of FreeFEM++.

An example which shows the surface meshes that represent the electrodes of the gun and the closing boundaries of the computational domain as well as the 3D tetrahedral mesh used in the simulations of a coaxial EOS are presented in figure 2.

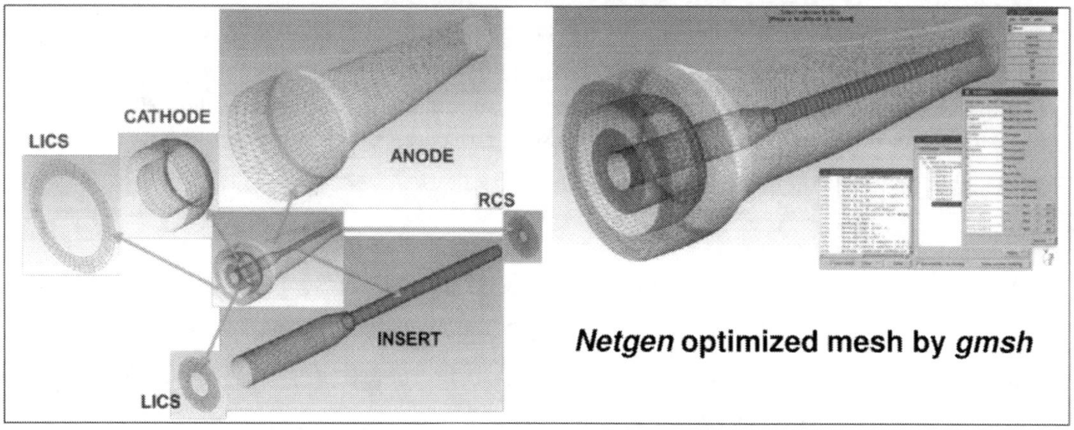

Figure 2. Surface meshes forming the boundary of the computational domain (left) and the 3D tetrahedral mesh (right) of a coaxial EOS for a 2 MW/170 GHz gyrotron.

From a physical point of view, the DBC correspond to the potentials applied to the electrodes that form the EOS and are also known as geometric boundary conditions. The electrodes are considered as perfectly conducting equipotential surfaces, where the tangential component of the electric field vanishes. The NBC are usually specified on the closing contours (surfaces) between the electrodes, and the condition $\frac{\partial u}{\partial n} = 0$ implies simply that the equipotential lines cross these boundaries at a right angle. It should be noted that the physical boundary conditions (DBCs or NBCs) are treaded by the finite element method as either *natural* or *essential*. For example, the DBCs are essential boundary conditions (EBCs) while the NBCs are natural boundary conditions. The main difference between them is the following. The natural BCs are already included in the integration in the finite element equations and there is no need to impose them additionally. On the other hand, although the EBCs are satisfied automatically in the finite-element formulation, they still need to be implemented (using one or another method, e.g., direct application; Lagrangian multiplier or Penalty Method). In another words, the essential boundary conditions are built into the solution space, while the natural boundary conditions are built into the weak form.

The particle in cell (PIC) method implies representing the electron beam by an ensemble of N macro particles and tracing their motion in a self-consistent electromagnetic field whose electric $\mathbf{E}$ and magnetic $\mathbf{B}_i$ components include not only the applied external fields but the self-fields of the space charge as well [8, 9]. According to the PIC algorithm, this is being performed integrating numerically the following relativistic Newton–Lorentz equations of motion

$$\frac{d\mathbf{w}_i}{dt} = \frac{q}{m_0}\left(\mathbf{E}_i + \frac{\mathbf{w}_i}{\gamma_i} \times \mathbf{B}_i\right), \quad \frac{d\mathbf{r}_i}{dt} = \mathbf{v}_i, \quad \mathbf{w}_i = \gamma_i \mathbf{v}_i \tag{6}$$

for all electrons (i = 1, 2,..., N), where q/m_0 is the charge to mass ratio of an electron at rest, $\mathbf{E}_i$, $\mathbf{B}_i$ being the electric and the magnetic field at the current particle position $\mathbf{r}_i$, and the relativistic Lorentz factor of the i-th particle is defined by its velocity $\mathbf{v}_i$ and the speed of light in vacuum c as $\gamma = [1 - \frac{|\mathbf{v}|^2}{c^2}]^{-1/2} = \sqrt{1 + \frac{\mathbf{w}^2}{c^2}}$. Among the numerous explicit and implicit numerical methods for integration of the equations of motion (e.g., schemes of Boris, Boris–Bunneman, Tajima, Runge–Kutta method, Verlet's method, and predictor–corrector method) evaluating all important factors (accuracy, stability, CPU time etc.) finally we have selected the leapfrog method of the

Journal of Physics: Conference Series **514** (2014) 012056 doi:10.1088/1742-6596/514/1/012056

relativistic Boris–Bunneman scheme. The leap-frog method [8, 9] reduces the relativistic equations of motion (6) to the following time–centered implicit relations

$$\frac{\mathbf{w}_i^{n+1/2} - \mathbf{w}_i^{n-1/2}}{\Delta t} = \frac{q}{m_0}\left(\mathbf{E}^n + \frac{\mathbf{w}_i^{n+1/2} - \mathbf{w}_i^{n-1/2}}{2\gamma_i^n}\right); \quad \frac{\mathbf{r}_i^{n+1} - \mathbf{r}_i^n}{\Delta t} = \mathbf{v}_i^{n+1/2} = \frac{\mathbf{w}_i^{n+1/2}}{\gamma_i^{n+1/2}}, \quad (7)$$

where the advancement in time $t^n = n\Delta t$ is performed with a step Δt. Integration of (7) is performed following the classical 3-step method by Boris [10], which involves a complete separation of the fields $\mathbf{E}_i$, $\mathbf{B}_i$ (introducing two intermediate velocities and using the scalings $\mathbf{E} \leftarrow \frac{q\Delta t}{m}\mathbf{E}$, $\mathbf{B} \leftarrow \frac{q\Delta t}{m}\mathbf{B}$), followed by a half-step electrical acceleration ($\mathbf{w} \leftarrow \mathbf{w} + \mathbf{E}$), then a Larmor rotation (gyromotion) of the velocity vector ($\mathbf{w} \leftarrow \frac{2}{1+\mathbf{B}^2}(\mathbf{w} + \mathbf{w} \times \mathbf{B}) \times \mathbf{B}$) and, finally, another half-step acceleration ($\mathbf{w} \leftarrow \mathbf{w} + \mathbf{E}$). This stepping scheme introduces an error which is second order with respect to the ratio of the time step to the cyclotron period. By satisfying the condition $\omega_c \Delta t < 0.35$ (ω_c being the electron cyclotron frequency) the relative error of the rotation angle of the Boris scheme is kept to values less than 1%. This stepping technique has two important physical, namely it is time-reversible and energy-conserving. Since the leap-frog method is not self-starting initially a half step backwards from the initial conditions for the particles is performed in order to kickstart the process. The method described above is coded in the novel relativistic particle pusher (aka mover), which is used to update (advance) the positions and the velocities of the traced particles on each time step of the PIC algorithm as well as for calculation of the electron trajectories if the code is used as a ray-tracing tool for trajectory (electron–optical) analysis. As an example, figure 3 shows the electron trajectories in the coaxial gyrotron, which was presented in figure 2.

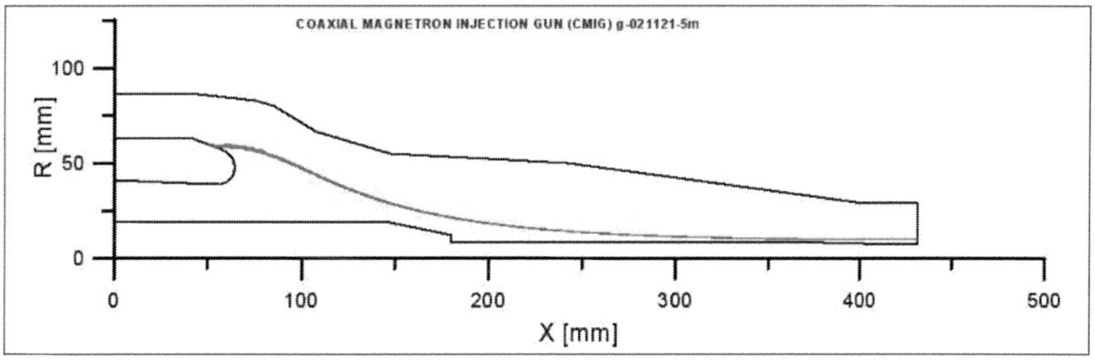

Figure 3. Electron trajectories in the coaxial gyrotron shown in figure 2.

3. Conclusions and outlook

The problem-oriented software package GYREOSS is being developed as a versatile tool for numerical studies and CAD of EOS of gyrotrons. Its development follows a well-defined concept formulated earlier, which sets the following guiding requirements to the code: (i) based on adequate (self-consistent) physical models; (ii) utilizing state-of-the-art computational platforms, numerical libraries and programing techniques; (iii) extensibility; (iv) portability; (v) efficiency. This paper presents a novel electrostatic field solver of the GYREOSS software package. It is parameterized in such a way as to provide convenient data structures of the electromagnetic fields at the current particle positions in both 2D and 3D which are used by the newly developed second-order relativistic particle pusher in which the leap-frog method and the Boris–Buneman scheme are implemented. Although the current version is sequential, both the data structures and the algorithms are coded in a way that would allow an easy parallelization of the code in the future. Currently we are studying and comparing various schemes for allocation of the currents and charges to the computational grid. The aim is to develop efficient modules that not only realize the above mentioned step ("scattering") of the PIC algorithm but also take full advantage of both the advanced interpolator and the mesh adaptation (adaptive mesh refinement) provided by FreeFEM++.

Acknowledgements
This work was carried out in the framework of Task 2.1.2 of the scientific program of the EURATOM-INRNE Association.

References

[1] Thumm M 2013 State-of-the-art of high power gyro-devices and free electron masers 2012 *KIT Sci. Reports* 7641 (KIT Scientific Publishing, 2013) 1–144

[2] Sabchevski S, Zhelyazkov I, Thumm M, Piosczyk B, Tran T M, Hogge J-Ph and Pagonakis J Gr 2007 Recent evolution of the simulation tools for computer aided design of electron-optical systems for powerful gyrotrons *Computer Modelling in Engin. & Sci.* **20** 203–20

[3] Sabchevski S, Illy S, Piosczyk B, Borie E and Zhelyazkov I 2008 Towards the formulation of a realistic 3D model for simulation of magnetron injection guns for gyrotrons (A Preliminary study) *Wissenschaftliche Berichte* FZKA 7409 1–34

[4] Damyanova M, Sabchevski S and Zhelyazkov I 2010 Pre- and post-processing of data for simulation of gyrotrons by the GYREOSS software package, *J. Phys.: Conf. Series* **207** 012032

[5] Geuzaine Ch and Remacle J-F 2013 *Gmsh: a three-dimensional finite element mesh generator with built-in pre- and post-processing facilities* (version 2.8.2 July 16 2013) http://geuz.org/gmsh/

[6] Dular P and Geuzaine Ch 2013 *GetDP: a general environment for the treatment of discrete problems* (version 2.4.1 July 16 2013) http://geuz.org/getdp/

[7] Hecht F 2013 *FreeFem++* 3rd edition version 3.23 (Laboratorie Jacques-Louis Lions, Université Pierre et Marie Curie, Paris) 1–392 http://www.freefem.org/ff++/ftp/freefem++doc.pdf

[8] Hockney R W and Eastwood J W 1981 *Computer simulation using particles* (McGraw-Hill New York)

[9] Birdsall C and Langdon A 2004 *Plasma physics via computer simulation* (Series in Plasma Phys.) (Taylor & Francis London)

[10] Buneman O 1976 The advance from 2D electrostatic to 3D electromagnetic particle simulation *Computer Phys. Communic.* **12** 21–31

Journal of Physics: Conference Series **514** (2014) 012057 doi:10.1088/1742-6596/514/1/012057

Formation of copper clusters on the surface of a xenon buffer layer

I G Marchenko[1,2,4] and I I Marchenko[3]

[1]NSC "Kharkov Institute of Physics and Technology",
1 Akademicheskaya, 61108 Kharkov, Ukraine
[2]Kharkov National University, 4 Svobody Sq., 61077 Kharkov, Ukraine
[3]NTU "Kharkov Polytechnic Institute", 21 Frunze Str, 61145 Kharkov, Ukraine

E-mail: march@kipt.kharkov.ua

Abstract. We applied the molecular dynamics technique to the study of the formation of three-dimensional nanoclusters on a xenon surface. The main physical mechanisms were established of the processes of formation of their structure. Equilibrium configurations of copper clusters on the solid xenon layer were determined and the energies of their formation were calculated.

1. Introduction

Studying the formation processes of three-dimensional nanoclusters on different solid surfaces is of great importance for the development of new technologies. One of the most promising technologies used for the development of such nanocluster structures is the buffer-layer-assisted growth (BLAG) [1]. During the BLAG process atoms, (or molecules) are deposited from vapor phase onto a solid inert gas multilayer (as Xe) that prevents their immediate contact with the substrate. Experiments have shown that these species are mobile and form 3D nanostructures on the Xe layer. Thermal desorption of the rare gas then delivers them to the substrate.

The buffer-layer-assisted growth allows one to form nanostructures of single elements or compounds on any substrate, potentially with control over the size and spatial distribution of the nanostructures. These nanostructures can be incorporated to form composites by a sequential growth of nanostructures and thin film deposition.

The efficient use of the BLAG method requires information on the energy profile of the cluster creation and diffusion on the buffer layer of an inert gas. So far, data are lacking on the nucleation of such structures and the physical mechanisms of their growth. No studies have been carried out of the mobility of both adatoms and clusters of different sizes on the surface of a solid phase of inert gas, making it difficult to develop adequate theoretical models describing the aggregation of clusters. Computer simulation methods are widely used for studying the properties of complex systems. The method of molecular dynamics is considered to be an efficient tool that allows one to study different defects in crystals. The essence of this method consists in the computer solution of the equations of motion for each atom of the system with a specified Hamiltonian system interaction.

[4] To whom any correspondence should be addressed.

The goal of the present study was to use the molecular dynamics method for the investigation of the physical processes of the formation of three-dimensional copper clusters on the surface of solid xenon.

2. Simulation details

The molecular dynamics method was used to study the mobility of copper clusters on the surface of solid xenon. The interaction of the metal atoms was described by the embedded atom model [2]. The interaction energy of the copper atoms was calculated as the sum of pair interactions and the energy of "embedding":

$$E = \sum_i F_i(\rho_i) + \frac{1}{2}\sum_{j\neq i}\phi_{ij}(R_{ij}), \tag{1}$$

where ϕ_{ij} is the pair interaction potential, and $F_i(\rho_i)$ is the energy necessary for embedding atom i in an electron gas at a point with electron density ρ_i . The electron density is found as

$$\rho_i = \sum_{j\neq i}\rho_j^0(R_{ji}), \tag{2}$$

where $\rho_j^0(R_{ij})$ is the electron density from an atom of type j at a distance R_{ij}. The functions describing the interatomic pair interaction, the atomic electron density, and the embedding function in the electron gas for copper atoms were obtained with the use of experimental data and quantum-mechanical calculations of the various crystallographic structures and the Cu dimer [3]. We could then use those results for calculating the characteristics of small copper clusters.

The Xe–Xe and Cu–Xe interactions are described by Lennard-Jones potentials with an auxiliary function that changes the potential in such a way that the potential value and its first derivative become equal to zero at a distance equal to the cutoff radius r_c [4]:

$$\phi(r) = 4\varepsilon\left[\left\{\left(\frac{\sigma}{r}\right)^{12} - \left(\frac{\sigma}{r}\right)^{6}\right\} + \left(6r_c^{-12} - 3r_c^{-6}\right)\left(\frac{r}{r_c}\right)^2 - \left(7r_c^{-12} - 4r_c^{-6}\right)\right]. \tag{3}$$

The value of r_c was chosen equal to $r_c = 0.995$ nm for both potentials. The values of σ and ε for the Xe–Xe interaction were taken from [5]: $\varepsilon = 0.01997$ eV, $\sigma = 0.398$ nm.

The Lennard-Jones potential has been used previously by other authors for studying the interaction of metals with a rare gas surface [6]. In analogy with [6], the parameters of the potential describing the Cu–Xe interaction were chosen consistent with the experimentally observed value of the binding energy of the xenon atom to the (111) surface of copper (0.20 eV) and the value of the equilibrium distance of a xenon adatom from the copper surface (0.36 nm) [7]. The values obtained for the parameters of the potential are $\varepsilon = 0.0212$ eV and $\sigma = 0.369$ nm.

The computer simulation of the formation and diffusion of metallic clusters was carried out by the molecular dynamics method. The calculated cell is a parallelepiped with dimensions of 5.26, 4.55, and 5.37 nm along the X, Y, and Z axes, respectively. The X axis had the crystallographic direction $\left[10\bar{1}\right]$, the Y axis $\left[1\bar{2}1\right]$, and the Z axis $[111]$. Periodic boundary conditions were used along the X and Y directions. The diffusion of copper clusters occurred on a substrate represented by 13 atomic layers of closely-packed (111) Xe planes. The first three layers were fixed in positions of an ideal fcc lattice.

3. Results and discussion

The copper deposition on the buffer xenon layer at 50 K was calculated using the method of molecular dynamics. Figure 1 gives an example of the computational results for a structure formed on the Xe-surface during the deposition of 0.1 copper monolayer (a typical deposition value for BLAG).

Figure 1 shows two large metal clusters embedded into a solid xenon layer. These clusters result from the merging of copper adatoms and clusters of a smaller size. The large metal clusters are surrounded by xenon atoms forced out of the substrate. The implanted copper clusters are immobile nuclei of three-dimensional metal clusters. The figure also shows clusters of a smaller size formed via merging of individual surface-migrating copper adatoms. Their diffusion was studied earlier [8].

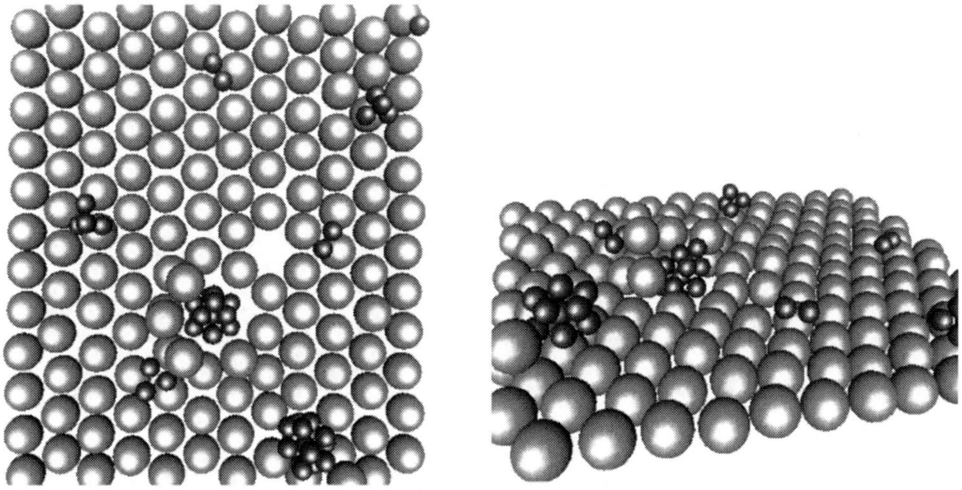

Figure 1. Different projections of the atom surface structure obtained through simulation of the deposition of 0.1 copper monolayer onto the xenon substrate (111), T = 50 K. The balls of a larger radius are xenon atoms, the copper atoms have a dark color.

The results allow us to represent the growth of three-dimensional nanoclusters in the following way. On reaching a critical size, the metal cluster is embedded into the xenon layer and forms a cluster with a surface vacancy. A displaced atom of inert gas is bound to the cluster and remains next to it. Figure 2 gives an example of a change in the Xe-surface structure observed during the investigation of

a) 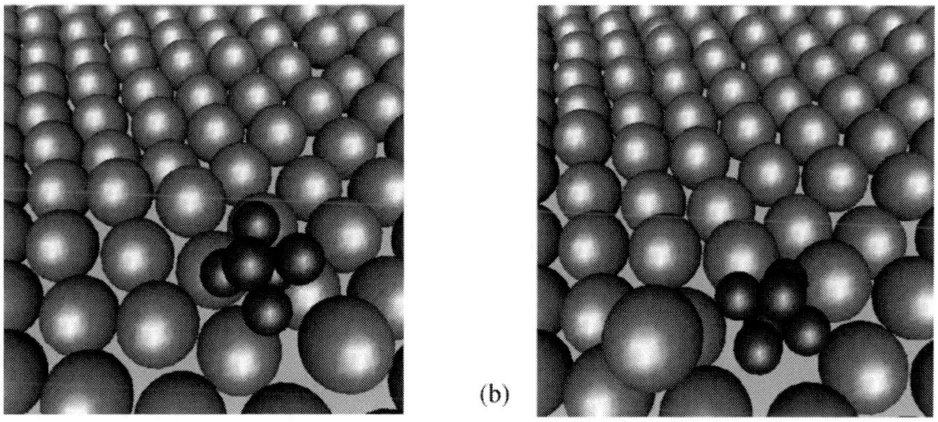 (b)

Figure 2. Atomic configuration of copper clusters (balls of a smaller size) before and after the displacement of xenon atom from the surface layer, T = 50 K. The bright balls of larger radius are xenon atoms; the copper atoms have a smaller radius and a dark color.

the migration of the cluster. The time interval for a change in the structure as shown in figures 2a and 2b is less than 0.1 ns. The figures show that after the displacement of xenon atoms the cluster embeds itself into the layer of inert gas. This cluster can grow by other approaching small-size metal clusters that can freely migrate on the buffer surface layer. Subsequently, the displaced inert gas atoms wrap up the cluster from the top forming a xenon "coat".

To determine the reasons for the initial embedment of metal clusters into the inert gas layer, we calculated the formation energy of the complex of copper clusters with the surface vacancy. The formation energy was calculated as the difference of energies between two systems. One system was represented by the $Cu^{(n)}$ cluster located on the atomically smooth surface of xenon. Figure 3 shows the atom configurations of $Cu^{(n)}$ clusters located on the atomically smooth surface of xenon that have a minimum formation energy. The other system was represented by the relaxed configuration of the $Cu^{(n)}$ cluster located in the surface vacancy and the xenon adatom. A vacancy was formed next to the cluster on the Xe-surface.

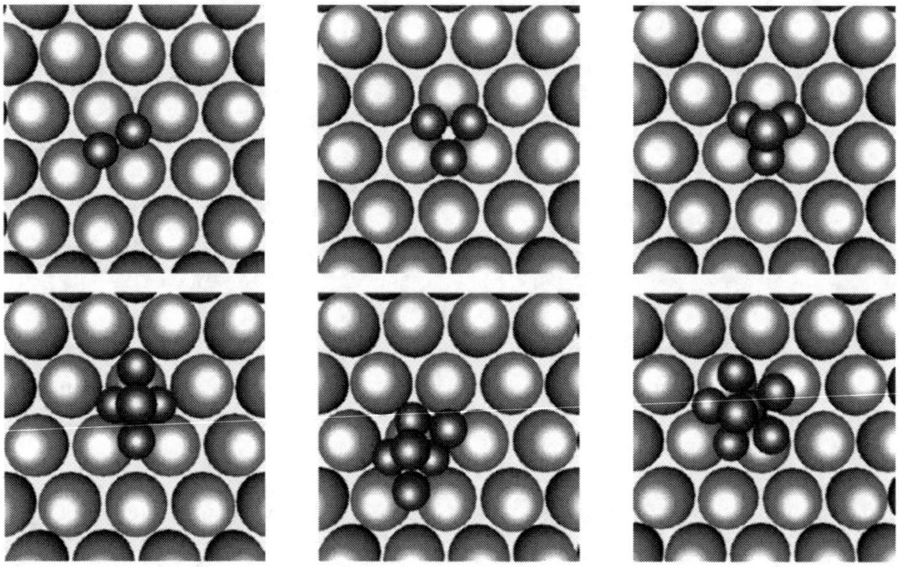

Figure 3. Atomic configuration of copper clusters on the xenon surface. The brighter balls of a larger radius are xenon atoms; the balls of dark color and a smaller size are copper atoms.

The xenon adatom was located at such a distance from $Cu^{(n)}$ that the elastic interaction between the defects could be neglected. The computations were performed in the following way: the initial defective configuration was kept at T = 50 K for a period of time sufficient for the system thermalization. Afterwards, the energy was diverted from the system by switching on the friction mode until T = 0 was reached.

The computational results are given in figure 4. The figure shows that it is energy-advantageous for the copper clusters to be implanted into the surface with the displacement of the xenon atom. However, the formation of the $Cu^{(6)} - V$ cluster is the most

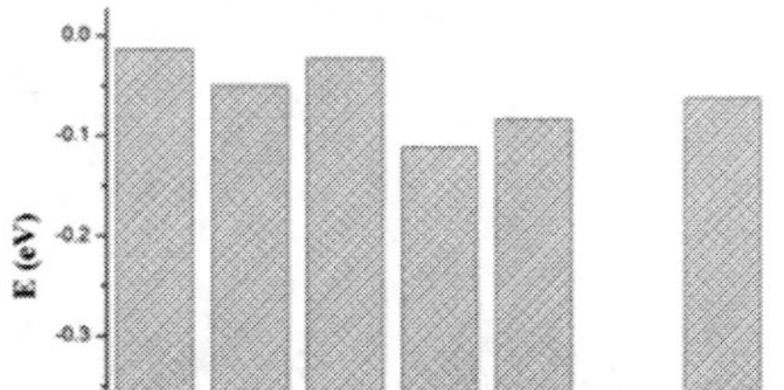

Figure 4. Formation energy of the complexes of copper cluster consisting of N copper atoms with a xenon surface vacancy on the {111} surface.

energy–efficient process. A decrease and an increase in the cluster size result in the decay of the probability of a cluster formation process. The energy gain is 0.4 eV. Presently, the mechanisms that produce such an essential change in energy are not clear. However, the interesting fact is that the volume of the $Cu^{(6)}$ cluster is approximately equal to that of the displaced xenon atom. The data presented show that a change in the bonding energy of vacancies and metal clusters can depend strongly on their sizes and influence significantly the kinetics of nucleation and growth of three-dimensional clusters on the buffer layer.

4. Conclusions

The paper describes an application of the method of molecular dynamics to the study of the formation of three-dimensional nanoclusters on a xenon surface. The main physical mechanisms of the formation processes of their structure were established.

Equilibrium configurations of copper clusters on the solid xenon layer were determined; the energies of their formation were calculated. The method of computer simulation was used for the computation of the formation energy of defects consisting of surface vacancies and copper nanoclusters.

It was established that copper clusters are embedded into the xenon substrate by displacing its atoms to the surface. This allows one to explain the available experimental data related to a change in the spectra of the X-ray photoemission with the deposition time [9]. The results obtained show that the change in the bond energy of surface vacancies and metal clusters of different sizes can significantly influence the kinetics of nucleation and growth of three-dimensional clusters on the buffer layer.

References

[1] Weaver J H and Waddill G D 1991 *Science* **251** 1444
[2] Daw M S and Baskes M J 1984 *Phys. Rev.* B **29** 6443
[3] Mishin Y, Mehl M J, Papaconstantinopolous D A, Voter A F and Kress J D 2001 *Phys. Rev.* B **63** 224106
[4] Stoddarrd D and Ford J 1973 *Phys. Rev.* A **8** 1504
[5] Minkowycz W J and Sparrow E M 2000 *Advances in Numerical Heat Transfer* (Taylor & Francis New York) vol 2
[6] Tomassone M S, Sokoloff J B, Widom A and Krim J 1997 *Phys. Rev. Lett.* **72** 4798
[7] Seyller Th, Caragiu M, Diehl R D, Kaukosina P and Lindroos M 1998 *Chem. Phys. Lett.* **291** 567
[8] Marchenko I G and Neklyudov I M 2006 *Low Temp. Phys.* **32** 957
[9] Ohno T R, Patrin J C, Ayyala U S and Weawer J H 1991 *Phys. Rev.* B **44** 1891

Journal of Physics: Conference Series **514** (2014) 012058 doi:10.1088/1742-6596/514/1/012058

Computer modelling of nanoscale diffusion phenomena at epitaxial interfaces

M Michailov[1] and B Ranguelov

Institute of Physical Chemistry, Bulgarian Academy of Sciences,
Acad. G. Bontchev Str., Block 11, 1113 Sofia, Bulgaria

E-mail: mike@ipc.bas.bg

Abstract. The present study outlines an important area in the application of computer modelling to interface phenomena. Being relevant to the fundamental physical problem of competing atomic interactions in systems with reduced dimensionality, these phenomena attract special academic attention. On the other hand, from a technological point of view, detailed knowledge of the fine atomic structure of surfaces and interfaces correlates with a large number of practical problems in materials science. Typical examples are formation of nanoscale surface patterns, two-dimensional superlattices, atomic intermixing at an epitaxial interface, atomic transport phenomena, structure and stability of quantum wires on surfaces. We discuss here a variety of diffusion mechanisms that control surface-confined atomic exchange, formation of alloyed atomic stripes and islands, relaxation of pure and alloyed atomic terraces, diffusion of clusters and their stability in an external field. The computational model refines important details of diffusion of adatoms and clusters accounting for the energy barriers at specific atomic sites: smooth domains, terraces, steps and kinks. The diffusion kinetics, integrity and decomposition of atomic islands in an external field are considered in detail and assigned to specific energy regions depending on the cluster stability in mass transport processes. The presented ensemble of diffusion scenarios opens a way for nanoscale surface design towards regular atomic interface patterns with exotic physical features.

1. Introduction

The space- and size-dimensional effects in nanoscale surface physics are a cutting-edge problem in materials science [1-4]. This extended concern is motivated by a huge number of technological applications where detailed knowledge of the fine atomic structure of interfaces is essential. Typical examples are self-organization phenomena leading to formation of regular nanoscale objects and superlattices with exotic physical characteristics. The atomic-scale ordering of interfaces is considered as a complex superposition of ensemble of interactions in the system. In general, depending on the range they act, the interactions could be classified as two types being simultaneously in subtle competition. The first, short-range direct electronic interactions commonly relate to localized chemical bond between particles, while the second are due to long-range acting forces resulting from charge density waves, electrostatic dipole-dipole interactions, surface-state Friedel oscillations, elastic deformation of substrate lattice, etc. [3]. In the range of sub-monolayer adatom concentration, the interplay between these competing interactions generates a diversity of temperature-dependent

[1] To whom any correspondence should be addressed.

two-dimensional atomic structures. The interface ordering evolves to the final equilibrium state via a complex diffusion scenario where single atoms, atomic clusters and vacancies are involved [1,5,6]. Although a large number of theoretical and experimental studies have already revealed the general features of mass transport in such low-dimensional systems, fine details of diffusion processes causing the specific ordering in the overlayers are still deficient [3,5]. A general reason is that in complex systems with reduced size, confined space dimensionality and broken symmetry as surfaces and interfaces, the simultaneously acting physical effects could not be distinguished and, therefore, their impact could not be evaluated separately. In these cases, the experimental results are strongly influenced by a number of side effects. Thus, the interpretation of advanced experiments conducted by powerful techniques as scanning tunnelling microscopy (STM), thermal energy atom scattering (TEAS) and high resolution spot profile analysis low energy electron diffraction (HR SPA LEED) is often problematic. The presence of imperfections on real surfaces, including atomic terraces, steps, kinks, vacancies, defects, grain boundaries, finite domain size, domain walls, charge density waves, surface-state Friedel oscillations, etc. makes the general scenario of atomic ordering at epitaxial interface generally not amenable to pure theoretical treatment. Overcoming these difficulties, the atomistic simulations treat successfully a number of problems of these complicated structures. Employing a reliable computational model, the computer simulations reveal important details of the interactions and the ordering resulting from them. Hence, they can be implemented to predict unexpected phenomena and to conduct and extend the physical experiment in a right direction [7]. Here we demonstrate an application of the Monte Carlo simulations pointed to two important surface diffusion problems in materials science: formation of regular surface patterns on epitaxial interface and diffusion of atomic clusters in an external field. We also discuss the equilibrium structure of the interface organized upon network of interactions and selective cluster size-dependent diffusion that could initiate formation of atomic surface patterns.

2. Computational and physical models

The computational model in the present study implies classical canonical Monte Carlo (MC) sampling with tight binding second moment approximation of electron density of states (TB SMA) many-body potential between interacting atoms. This semi-empirical TB SMA potential has been extensively applied in contemporary atomistic simulations of transition metals and intermetallic alloys [7-10]. Its advantage relates to the ability of reproducing real physical properties of surfaces and interfaces including point defects, dislocations, vacancies, grain boundaries [8,9]. Depending on the specific diffusion problem, we also apply an external field to direct the cluster migration in a specific direction and to evaluate the cluster stability. The physical model is a three-dimensional continuum space Pb/Cu(111) interface with system size of $L_x = 76$, $L_y = 44$ and $L_z = 5$ lattice units. The simulations were performed on a crystal surface with large atomic terraces. The model enables one to gain insight into the diffusion of adsorbed atoms both directly into the outermost surface layer and through the surface steps with different anisotropy and, therefore, different free energy and relaxation ability. Full lattice dynamics for all atoms is applied to allow a complete spatial relaxation of the system. The system equilibrium is realized after an average Monte Carlo time of 10^6 steps per atom. All simulation details are described elsewhere [3,10,11].

3. Diffusion barriers and nanoscale surface patterns on fcc(111) interface

The atomic-scale surface morphology provides a variety of energy barriers for surface diffusion [3, 9]. For volume-immiscible systems, the migration of a single atom on the surface evolves in two diffusion scenarios. The first relates to a simple random Einstein walk on top of the interface resulting in aggregation of atoms in large clusters and decoration of steps of substrate atomic terraces. The second scenario is more complex and involves diffusion inside the first atomic layer of the substrate. This process is governed by the subtle competition between the elastic strain energy of the substrate and the energy gained by mixing. On an atomically smooth interface, this scenario leads to formation of a stable two-dimensional (2D) surface alloy and can be realized if the energy of migrating atoms is higher than the energy barrier for incorporation inside the substrate [4]. Refining the classical diffusion model, we showed that on vicinal surfaces, the intermixing could take place even in the case when the adatom energy is lower than that for direct incorporation inside the first atomic layer [3,9]. This effect, known as incomplete alloying, is realized exclusively via surface diffusion through the steps of atomic terraces. The simulation results reveal a 30-% lower activation energy for diffusion via step-terrace mechanism compared to that of direct incorporation. The temperature evolution of the surface-confined intermixing is demonstrated in figure 1. At a low temperature T, figure 1a, the adatoms only attach to atomic steps and decorate the terrace without interatomic exchange. More interesting is the diffusion at high temperatures. Apparently, on large terraces or flat substrate areas, the direct atomic exchange between adsorbed and substrate atoms is hampered because of energy reasons. In contrast, near the step edges, a very distinct diffusion of Pb atoms within the Cu matrix is realized, figure 1b. This atomic exchange is limited inside the terrace plane to a specific length and generates alloyed stripes in the vicinity of steps. The stripes, having a temperature-dependent characteristic width, are followed by a pure non-alloyed interface (middle part of the terrace).

a) **b)** **c)**

Figure 1. Snapshot of random, close-to-equilibrium atomic configuration of 0.1 monolayers Pb (light gray balls) on Cu(111) stepped surface (green balls) with initial Cu terrace of 16 atomic rows. **a)** T = 200 K, low-temperature step decoration. **b)** T = 300 K, formation of 2D stripes due to Pb diffusion across the atomic steps. No surface alloying takes place on a smooth, step-free surface area. **c)** T = 300 K, interface configuration for initial Cu terrace of 6 atomic rows. Completely alloyed Pb-Cu islands are formed on top of the pure non-alloyed substrate [9].

The observed dramatic diffusion slowing down towards the terrace center is a result of the interplay between elastic strain and energy gain of mixing. Two time-scale diffusion mechanisms of direct atomic exchange and vacancy-mediated diffusion take place in this complex scenario [12]. A difference is also seen in the stripe width L_S^A and L_S^B on upper "A" and lower "B" terrace step, figure 2. It originates from a step anisotropy on (111) interface orientation and a different step free energy. Our simulation data show $L_S^A \approx 1.4 L_S^B$ at 300 K which is in line with the extended relaxation ability of "A" step on the terrace. The decrease of terrace width to $L_T < 3L_S$ makes the terrace unstable and completely transparent for adsorbed atoms. The instability initiates terrace rupture and formation of 2D alloyed islands, as it is shown in figure 1c. In this case, note that the surface intermixing takes place exclusively at the terrace level so that entirely alloyed islands patch the pure non-alloyed substrate. Hence, in a specific temperature gap, fine tuning of the energy of the adsorbed atoms via temperature variation can be applied to atomic-scale surface design. It is essential to point out that the final configuration seen in figure 1b represents a close-to-equilibrium state and does not change during the subsequent time evolution of the interface. The results presented reveal a classification order of surface-confined intermixing (blocked, incomplete and complete, see figure 3) and open a way for the formation of a variety of nanoscale surface patterns via fine tuning of the system temperature [9].

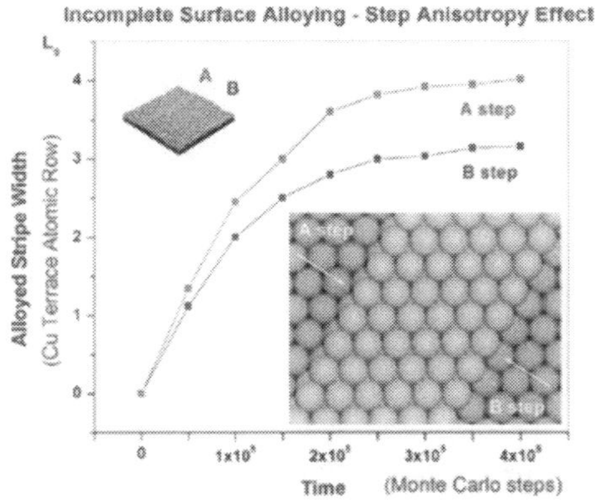

Figure 2. Step anisotropy effect at incomplete surface alloying. Snapshot of a random MC atomic configuration of 0.1 ML Pb/Cu(111) at 300 K and initial substrate terrace width of 16 atomic rows. The larger alloyed stripe width at step A is a result of the eased atomic relaxation of the terrace in that direction. (from [9] Copyright © 2009 The American Physical Society).

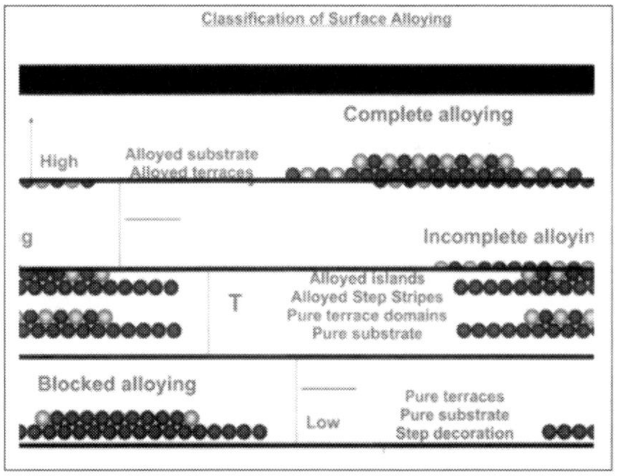

Figure 3. Classification of surface alloying with corresponding atomic configurations at the epitaxial interface (light balls indicate adsorbed atoms and dark balls indicate substrate atoms). The presented order opens up a way for fine modification of the atomic surface structure and formation of regular patterns. (from [9] Copyright © 2009 The American Physical Society).

4. Diffusion of atomic clusters in external field

In contrast to diffusion of single atoms, the migration of atomic clusters on an epitaxial interface is a tricky problem [1,2,6]. The complexity originates from a variety of diffusion mechanisms which control the island migration. For a single atom, the diffusion coefficient follows an exponential decay as $D \approx \exp\left(-\frac{E_d}{k_B T}\right)$ where E_d is the energy barrier for migration to a randomly chosen site on the surface. In the case of clusters, $E_d = \sum_i E_{d_i}$ and E_d corresponds to an effective energy barrier which originates from the barriers E_{d_i} of competing specific diffusion mechanisms. For instance, the diffusion of 2D cluster is tightly dependent on the specific mass transport mechanism and scales with its size as $D_{2D} \approx N^{-\alpha}$ where $\alpha = 3/2$ for periphery diffusion, $\alpha = 1$ for terrace diffusion and $\alpha = 1/2$ for evaporation-condensation diffusion [5]. In the case of compact islands, shape effects of open or closed crystallographic forms, soliton propagation, misfit dislocation and commensurability with the substrate lattice also have a substantial impact on the island migration. All these effects have been studied in detail both experimentally and theoretically [1,2,13]. An essential problem in cluster diffusion phenomena is the atomic islands integrity throughout the surface migration. Loss of integrity leads to a variety of size, shape and lattice-misfit effects and, therefore, can dramatically influence the mass transport process. Here we discuss the impact of an external field on the stability and migration of homoepitaxial atomic islands on an epitaxial interface.

The stability of an island depends on ensemble of competitively acting forces due to lateral, normal and external field interactions. Evidently, stronger lateral and normal interactions hamper the island decomposition into small clusters. The simulation results reveal critical values of the imposed external field that cause disintegration of the islands into morphologically different atomic complexes. This effect is demonstrated in figure 4 a, b, c. Three energy regions with respect to the cluster stability can be specified. At a weak external field along the (111) direction (1.5 % of the cluster interactions), only single atoms are detached from the cluster periphery, while the total cluster integrity is preserved, figure 4a. Increasing the external force to 2 % of normal and lateral bonding, the island disintegrates into several atomic clusters, all being compact and smaller in size, figure 4b. This effect of decomposition changes the surface diffusion mechanisms and assists dramatically the mass transport process on the interface. Let us point out that the small clusters are more stable than the large ones with respect to the same external force. They are less strongly bonded to the substrate and, therefore, they move more easily and quickly on the interface. In this case, the diffusion behavior of entire cluster gliding takes place in contrast to a periphery diffusion mechanism. Finally, attaining the specific critical value of 5 %, the external field destroys totally the atomic cluster to an ensemble of single atoms, which is seen in figure 4c. An important conclusion following from the simulation results is that the transition between the regions with a specific cluster morphology (single atoms, small compact clusters and total disintegration) is not smooth. Hence, the impact of the external filed on the cluster stability and related cluster disintegration demonstrates a staircase-like behavior. This effect can be applied to a selective cluster size-dependent mass transport in nanoscale pattern formation on epitaxial interfaces.

Figure 4. Diffusion of anatomic cluster in an external field. Depending on the field intensity E_{ext} related to the cluster atom interactions, three modes of disintegration are observed: **a)** E_{ext} = 1.5 % – detachment of single peripheral atoms only, **b)** E_{ext} = 2 % – fragmentation into small compact clusters, **c)** E_{ext} = 5 % – total atomic disintegration.

5. Conclusions

The present study reveals important details of atomic diffusion mechanisms on stepped surfaces in bulk immiscible systems. On epitaxial interface with atomic terraces, the computational model reveals an energy region of formation of 2D-alloyed stripes or 2D-alloyed atomic islands. The stripes formed at the terrace edges are followed by a flat, non-alloyed, area, thus forming regular patterns on the surface. The stripe width is step-anisotropy dependent and correlates with the relaxation ability of the terraces in the specific direction. Atomic terraces having a width smaller than $3L_S$ are unstable and completely transparent for the adsorbed foreign atoms. This process leads to formation of 2D alloyed islands on top of a pure non-alloyed substrate. The scenarios presented demonstrate nanoscale patterning at epitaxial interfaces through a fine diffusion control of adatoms on the crystal surfaces. By a simple variation of the temperature, a terrace width distribution or a lattice overlayer/substrate misfit, the interface generates a variety of regular nanoscale ensembles, including alloyed stripes, alloyed islands or entirely alloyed terraces. In the presence of an external field, depending on its intensity, the atomic clusters migrating on the epitaxial interface disintegrate in three different ways: detachment of single peripheral atoms solely, fragmentation to small compact clusters and total atomic disintegration. The present report also demonstrates the extended capability of atomistic models in computer simulations to unravel simultaneously acting effects, to distinguish between them, and finally to evaluate their specific contribution to experimentally-observed complex physical phenomena.

References

[1] Brune H 1998 *Surf. Sci. Rep.* **31** 121
[2] Einax M, Dieterich W and Maass P 2013 *Rev. Modern Phys.* **85**/3
[3] Michailov M 2011 in: *Nanophenomena at Surfaces: Fundamentals of Exotic Condensed Matter Properties* ed M Michailov (Springer Series in Surf. Sci.) vol 47 chapter 6
[4] Kellogg G 2002 in: *Surface Alloy and Alloyed Surfaces* ed. D P Woodruff (Elsevier) p 152
[5] Giesen M 2001 *Prog. Surf. Sci.* **68**/1
[6] Tsong T 2005 *Physica* A **357** 250
[7] Landau D P and Binder K 2000 *A Guide to Monte Carlo Simulations in Statistical Physics* (Cambridge University Press)
[8] Cleri F and Rosato V 1993 *Phys. Rev.* B **48** 22
[9] Michailov M 2009 *Phys. Rev.* B **80** 035425
[10] Georgiev N and Michailov M 1998 *Phys. Rev.* B **58** 13895
[11] Michailov M 2004 in: *Computer Simulations Studies in Condensed Matter Physics XVI* ed D. P. Landau, S. Lewis and H. Schuettler (Springer Series Proc. in Phys.) vol 95 pp 226-49
[12] Michailov M 2010 *Solid State Phenomena* **159** 121
[13] Michailov M and Vladkov M 2007 *Surf. Sci.* **601** 3912
[14] Terssoff J 1995 *Phys. Rev. Lett.* **74**/3

Mobility of positive ions in CF_4

V Stojanović[1,4], Z Raspopović[1], J V Jovanović[2], J de Urquijo[3] and Z Lj Petrović[1]

[1]Institute of Physics, University of Belgrade, POB 68, 11080 Belgrade, Serbia
[2]Faculty of Mechanical Engineering, University of Belgrade, 16 Kraljice Marije, 11000 Belgrade, Serbia
[3]Centro de Ciencias Físicas, Universidad Nacional Autónoma de México, P.O. Box 48-3, 62251 Cuernavaca Mor., México

E-mail: stoyanov@ipb.ac.rs

Abstract. Cross-section sets for transport of positive ions in CF_4 that fit the available experimental data for mobility are assessed by normalizing the available experimental and theoretical cross-sections within the framework of the swarm method. Transport parameters for positive ions in CF_4 in DC fields at a gas temperature of $T = 300$ K are calculated as a function of the reduced electric fields E/N (N being the gas density) by using Monte Carlo simulation.

1. Introduction

Control of reactive ion plasmas is necessary in applications related to semiconductor modifications. A cornerstone in the modelling of such plasmas is precise information on the transport of ions, as the reactive ions' flux and energy control the surface processes. In plasmas in collision-dominated regimes, if the negative ions dominate over the electrons, the plasmas become electronegative with properties distinctly different from the electropositive plasmas, especially in the sheath region. In modelling and analyzing the plasma chemistry, the cross-sections are needed for all relevant species and not only concerning the momentum transfer. We focused our attention on the transport of positive ions originating from CF_4 by using the swarm method [1] and attempted also to compile a comprehensive cross-section set based on the available thermochemical data and on a simple procedure. This paper is a preliminary publication of such a set.

The CF_3^+ ion is the most abundant positive ion in pure CF_4 plasmas and its mixtures. It is also known that F^+ ions are produced due to the F atom's significant ionization potential with respect to that of other radicals. In (intensive) exothermic charge-transfer collisions of F^+ ions with CF_4, CF_3^+ ions are mainly produced [2]. No tested cross-section sets for these two ions in a CF_4 gas can be found in the literature.

Many authors [3, 4, 5, 6] have used collisional models for ions in CF_4 or its mixtures taking into account the long-range effects of ion-induced dipole interaction. Although based on a simple theory, a very wide range of possible reactions of ions with a CF_4 target have been considered [4, 5] in a consistent way. Taking into account a large number of reactions may lead to a selection of numerous processes with a scattering probability higher [4, 5] than that observed in the existing crossed-beam experiments [7].

[4] To whom any correspondence should be addressed.

The aim of this study was to compile cross-section sets for F^+ and CF_3^+ ions and use Monte Carlo (MC) simulation to test them by comparing the calculated mobility data with the available experimental results. Input cross-sections were prepared over a wide collision-energy range, from 0.01 eV up to 500 eV. The results for the higher energies in this interval were obtained by extrapolation. However, the energy range of swarm studies used to normalize them is quite narrow and close to the thermal energies.

2. Cross section sets

2.1. CF_3^+ on CF_4

The cross-section set for scattering of CF_3^+ on the CF_4 molecule has been calculated recently by using available data and a simple procedure known as Nanbu's theory [4, 5]. It groups reactions scaled by their scattering probability according to reaction rate coefficients allowing separation of elastic from reactive endothermic collisions. The thermodynamic threshold energy and the branching ratio are accounted for following the Rice-Ramsperger–Kassel (RRK) theory [5]. A wide range of ion-molecule reactions were included in the cross-section set presented in [4] covering endothermic reactions only.

In this work, as a basis of the calculation of cross-section sets for F^+ and CF_3^+ ions in CF_4, we focused our efforts only on the processes observed in the experiment conducted by Peko *et al.* [7] rather than taking into account all processes listed by Nanbu [4, 5]. These reactions, with the thermodynamic threshold energies based on the existing thermochemical data [9], are shown in table 1.

Table 1. List of reactions and the corresponding thermodynamic threshold energies Δ [7] for CF_3^+ (a) and F^+ (b). Endothermicities are in eV and the underlined products have the same laboratory velocity as the reaction ion in the model.

a)			b)		
No	*reaction* $CF_3^+ + CF_4 \rightarrow$	Δ (eV)	*No*	*reaction* $F^+ + CF_4 \rightarrow$	Δ (eV)
R1	$\mathbf{CF_3^+} + F + \underline{CF_3}$ (DCT)[a]	-5.7	P1	$\mathbf{CF_3^+} + F + \underline{F}$ (DCT)[a]	+2.7
R2	$\mathbf{CF_2^+} + F_2 + \underline{CF_3}$ (DCT)[a]	-10.3	P2	$\mathbf{CF_2^+} + F_2 + \underline{F}$ (DCT)[a]	-1.9
R3	$\mathbf{CF^+} + F + F_2 + \underline{CF_3}$ (DCT)[a]	-13.1	P3	$\mathbf{CF^+} + F + F_2 + \underline{F}$ (DCT)[a]	-4.7
R4	$C^+ + 2F_2 + \underline{CF_3}$ (DCT)[a]	-19.2	P4	$C^+ + 2F_2 + \underline{F}$ (DCT)[a]	-10.79
R5	$\underline{\mathbf{CF_2^+}} + F + CF_4$ (CID)[b]	-6.2	P5	$\mathbf{F^+} + CF_3 + \underline{F}$ (CID)[b]	-10.8
R6	$\underline{\mathbf{CF^+}} + \underline{F_2} + CF_4$ (CID)[b]	-7.4	P6	$\mathbf{CF_3^{++}} + F + \underline{F}$ (DCT)[a]	-21
R7	$\underline{\mathbf{F^+}} + \underline{CF_2} + CF_4$ (CID)[b]	-12.3	P7	$\mathbf{CF_2^{++}} + F_2 + \underline{F}$ (DCT)[a]	-21.1
R8	$\underline{C^+} + \underline{F} + \underline{F_2} + CF_4$ (CID)[b]	-15.1			

[a] Dissociative charge transfer; [b] Collision-induced dissociation.

We used the polarizability value of 3.86×10^{-30} m^3 [10] for the CF_4 molecule. When this value is substituted in Langevin's formula [11] for the zero-field mobility of CF_3^+ ions in CF_4, the value of 1.13375 $cm^2\ V^{-1}\ s^{-1}$ is obtained, which is close to the low-field mobility measured of 0.96 $cm^2\ V^{-1}\ s^{-1}$ [8].

The ionization potentials of CF_3 and CF_4 and the bond values between ions/atoms were taken from [4, 9]. We than applied a simple semi-analytic momentum transfer theory (MTT) to obtain the integral momentum-transfer cross-section. The cross-sections obtained were modified in such a way that the MC results agree perfectly with the experimental results for the reduced mobility (see figure 2 (a)). A more detailed explanation of the procedure applied can be found in our earlier papers [12, 13]. The lack of reactivity at low collision energies has also been found by other authors [2].

The inelastic and reactive cross-sections obtained using the Nanbu theory were scaled in the threshold region in order to fit the experimental cross-sections obtained by Peko *et al.* [7]. Finally, in order to provide data for modelling plasma sheaths and plasma etching applications, the cross-sections were extrapolated up to 500 eV following the trends in the experimental measurements. A complete set of cross-sections is shown in figure 1 (a).

2.2. F^+ on CF_4
Because of the small endothermicity (2.7 eV), the exothermic reaction of F^+ with CF_4 produces mainly CF_3^+ ions at lower energies (reaction P1 in table 1(b)). The careful extrapolation of the measurements [7] towards the lower energies for reaction P1 leads to a low-energy capture process curve that is positioned approximately at β *tot (β = 0.12), which is the low energy part of the exothermic cross-section for CF_3^+ production. The total cross-section is the sum of the elastic and capture (exothermic reaction) cross-sections; these are shown in figure 1(b).

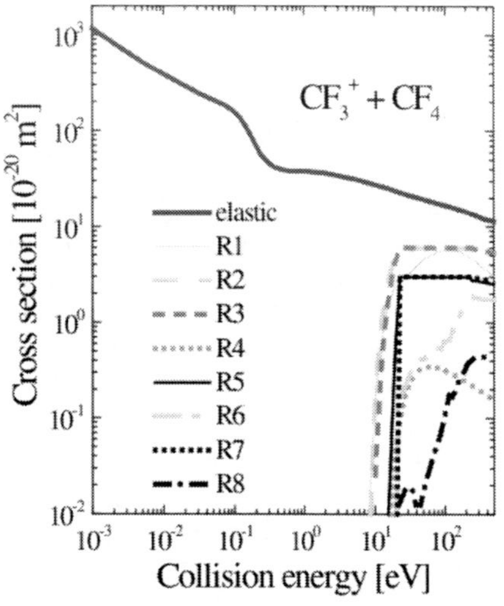

Figure 1(a). Cross-section set for CF_3^+ ions in CF_4 as a function of the collision energy.

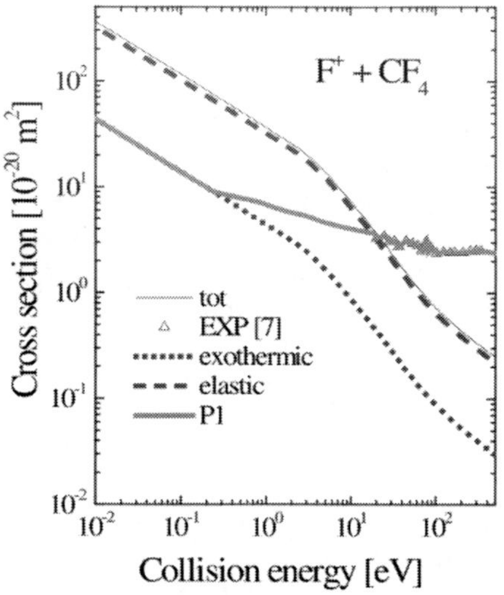

Figure 1(b). Cross-sections for elastic scattering and exothermic reaction for F^+ scattering on CF_4.

3. Results and discussion
The Monte Carlo technique was applied to calculations of the transport parameters in DC electric fields. In this work we used a Monte Carlo code that takes properly into account the thermal collisions [12, 14]. The code has passed all the tests and the benchmarks that were covered in our earlier studies [12, 15].

Figure 2 presents the reduced mobility of CF_3^+ ions (a) and F^+ ions (b) colliding with CF_4 as a function of the reduced electric field at room gas temperature (T = 300 K). The low-field mobility measured (0.96 $cm^2 V^{-1} s^{-1}$) is in a good agreement with the value measured for the CF_3^- ion in CF_4 of 0.99 $cm^2 V^{-1} s^{-1}$ [16]. The mobility curve of CF_3^+ in CF_4 shows a well-defined maximum for 200 Td $< E/N <$ 300 Td (1 Townsend = 10^{-17} V cm^2), followed by a decrease due to the predominance of the repulsive part of the interaction potential. The agreement of the experimental values for the reduced mobility of CF_3^+ ions in CF_4 with the MC results in the polarization limit also helped in determining the F^+ mobility at low electric fields. The presence of exothermic CF_3^+ formation of unknown intensity complicates the experimental observation of the F^+ mobility in CF_4 at low electric fields. Thus lack of experimental values for the reduced mobility of F^+ ions in CF_4 forced us to look for an agreement between the MC results and the experimental data in [16] for the mobility of F^- ions drifting through CF_4. The difference of around 15 % is not as good as that in CF_3^+,but is of the same order as the negative ion scatterring data; we thus can claim that a consistency of the MC results and the experimental values in [16] was achieved. More experimental data are needed to establish the effects of the exothermic process observed in the transport of F^+ ions in CF_4 so that the cross-section sets could be improved further.

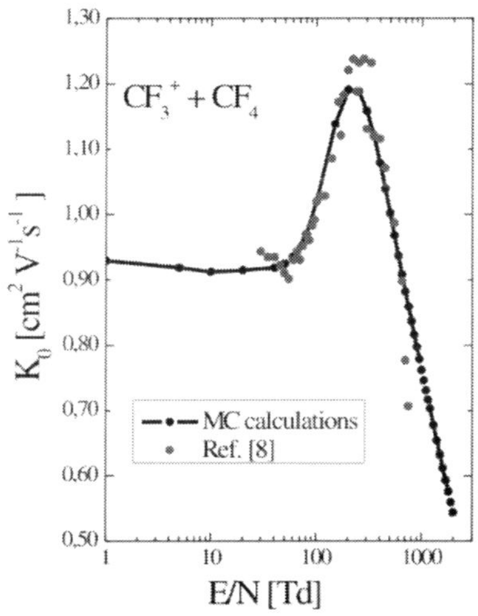

Figure 2(a). Mobility of CF_3^+ ions in CF_4 as a function of *E/N*.

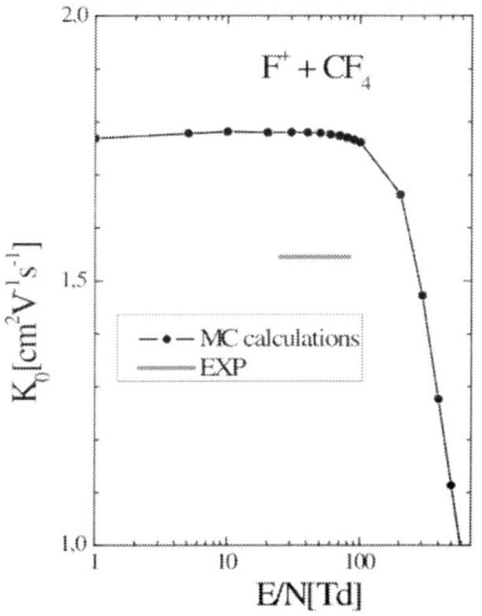

Figure 2(b). Mobility of F^+ ions in CF_4, as a function of *E/N*. EXP denotes averaged results of [12].

4. Conclusions

We present preliminary reults on the complete cross-section sets for the F^+ and CF_3^+ ions in CF_4 gas tested in a wide energy range. These sets were compiled by using simple theories covering both endothermic and exothermic reactions and the existing theoretical and experimental data for both binary collisions and swarms. The data for CF_3^+ were determined with sufficient accuracy thanks mainly to the available swarm data, while the data for F^+ are of lesser quality but still allowing proper plasma modelling.

Acknowledgments

The results were obtained in the Laboratory of Gaseous Electronics, Institute of Physics at the University of Belgrade with the support of the Ministry of Education and Science, Projects Nos. 171037 and 410011.

References

[1] Petrović Z Lj, Raspopović Z M, Stojanović V D, Jovanović J V, Malović G, Makabe T and De Urquijo J 2007 *Appl. Surf. Sci.* **253** 6619
[2] Morris R A, Viggiano A A, van Doren J M and Paulson J F 1992 *J. Phys. Chem.* **96** 2597
[3] Georgieva V, Bogaerts A and Gijbels R 2004 *Phys. Rev.* E **69** 026406
[4] Georgieva V, Bogaerts A and Gijbels R 2003 *J. Appl. Phys.* **93** 2369
[5] Denpoh K and Nanbu K 1998 *J. Vac. Sci. Technol.* A **16** 1201
[6] Donko Z and Petrović Z Lj 2006 *Jpn. J. Appl.Phys.* **45** 8151
[7] Peko B L, Dyakov I V, Champion R L, Rao M V V S and Olthoff J K 1999 *Phys. Rev.* E **60** 7449
[8] Basurto E and de Urquijo J 2002 *J. Appl. Phys.* **91** 36
[9] Fisher E R, Weber M E and Armentrout P B 1990 *J. Chem. Phys.* **92** 2296
[10] Tipton A B, Deam A P and Boggs J E 1964 *J. Chem. Phys.* **40** 1144
[11] Mason E A and McDaniel E W 1988 *Transport properties of ions in gases* (New York: Wiley)
[12] Petrović Z Lj, Jovanović J V, Stojanović V, Raspopović Z M and Ristivojević Z 2008 *Eur. Phys. J.* D **48** 87
[13] Vrhovac S B and Petrović Z Lj 1996 *Phys. Rev* E **53** 4012; Vrhovac S B, Petrović Z Lj,

Viehland L A and Santhanam T S 1999 *J. Chem. Phys.* **110** 2423; Jovanović J V, Vrhovac S B and Petrović Z Lj 2002 *Eur. Phys. J.* D **21** 335
[14] Ristivojević Z and Petrović Z Lj 2012 *Plasma Sources Sci. Technol.* **21** 035001
[15] Petrović Z Lj and Stojanović V D 1998 *J. Vac. Sci. Technol.* A **16** 329; Stojanović V D, Jelenković B M and Petrović Z Lj 1997 *J. Appl. Phys.* **81** 1601
[16] Dutton J, Goddings A, Loucas A K and Williams A W 1987 *Proc. XVIII Int. Conf. Phys. Ionized Gases* ed W T Williams (Swansea) p 20

2D numerical modelling of the gas temperature in a high-temperature high-power strontium atom laser excited by nanosecond pulsed longitudinal discharge in a He-$SrBr_2$ mixture

T P Chernogorova[1], K A Temelkov[2,3], N K Koleva[2] and N K Vuchkov[2]

[1]Faculty of Mathematics and Informatics, St. Kl. Ohridski University of Sofia, 5 James Bourchier Blvd., 1164 Sofia, Bulgaria
[2]Metal Vapor Lasers Laboratory, Acad. G. Nadjakov Institute of Solid State Physics, Bulgarian Academy of Sciences, 72 Tsarigradsko Chaussee, 1784 Sofia, Bulgaria

E-mail: temelkov@issp.bas.bg

Abstract. Assuming axial symmetry and a uniform power input, a 2D model (*r*, *z*) is developed numerically for determination of the gas temperature in the case of a nanosecond pulsed longitudinal discharge in He-$SrBr_2$ formed in a newly-designed large-volume high-temperature discharge tube with additional incompact ZrO_2 insulation in the discharge-free zone, in order to find the optimal thermal mode for achievement of maximal output laser parameters. The model determines the gas temperature of a nanosecond pulsed longitudinal discharge in helium with small additives of strontium and bromine.

1. Introduction

It is well-known that gas the temperature is one of the basic gas-discharge plasma parameters, which determines the heavy-particles interaction, the heavy-particle concentration, the gas discharge stability, etc. Therefore, the experimental and theoretical determination of the gas temperature is obviously of significant and fundamental importance; it is performed in various fields, such as gas-discharge laser physics, gaseous discharges, plasma technologies, gas-discharge mass spectroscopy, absorption and emission spectroscopy, and plasma in general. It is also well-known that the techniques that are widely used for gas temperature measurement based on measurements of the Doppler broadening of spectral lines and of the focal distance of a thermal lens are definitely imprecise, i. e., characterized by an inadmissible experimental error.

Assuming that the gas temperature varies only in the radial direction and considering uniform and non-uniform power inputs, gas temperature distributions have been calculated through analytical solution of the steady-state heat-conduction equation in the case of a nanosecond pulsed longitudinal discharge in a series of gas-discharge tube constructions [1, 2]. Unfortunately, despite the tremendous efforts, the analytical solution of the abovementioned steady-state heat conduction equation for the 2D (*r*, *z*) and 3D (*r*, φ, *z*) cases encountered some insuperable obstacles.

[3] To whom any correspondence should be addressed.

2. Experimental setup

A schematic diagram of a new high-temperature large-volume discharge tube is shown in figure 1. The basic tube with an inner diameter of 71.5 mm and an outer diameter of 76-mm is made of fused quartz. A ceramic insert, confining the discharge zone, with a 30.5-mm inner diameter, a 38.5-mm outer diameter and a length of 98 cm is sleeved coaxially in the basic tube. The zone between the ceramic insert, the basic quartz tube and the holders of the ceramic insert is incompactly filled with ZrO_2 fibrous insulation. The discharge operates in a self-heating regime. The temperature at the quartz tube surface is measured by a thermocouple.

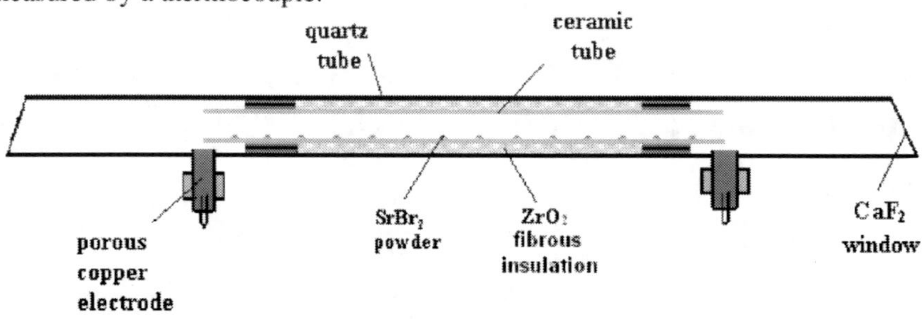

Figure 1. Discharge tube design.

The electrodes of the discharge tube are made of porous copper and are of a special design. CaF_2 windows are glued to the ends of the discharge tubes. The nanosecond pulsed longitudinal discharges investigated are excited by an electrical circuit based on a high-voltage rectifier and a high-voltage pulsed excitation circuit with interacting circuits (IC scheme).

3. Results and discussion

A 2D numerical solution of the steady-state heat conduction equation is found for the case of one discharge zone with radius R_1, and three discharge-free zones, namely, a ceramic tube within $R_1 \leq r \leq R_2$, a discharge-free zone within $R_2 \leq r \leq R_3$ incompactly filled with ZrO_2 fibrous insulation, and a basic tube made of quartz within $R_3 \leq r \leq R_4$. The dependence of the thermal conductivity k has the form $k = B\,T_g^{\,a}$, where B and a are constants (within a certain temperature range), which are specific for each gaseous or solid medium. The constants B and a, which determine the thermal conductivity, can be obtained through fitting the existing experimental data taken from [3]. The thermal conductivities needed are presented in table 1.

Table 1. $k = B\,T_g^{\,a}$ – thermal conductivity in W m^{-1} K^{-1}.

	He	Al_2O_3	ZrO_2	He-ZrO_2	quartz
B	34.9×10^{-4}	44323.1	7326.2×10^{-4}	655.9×10^{-4}	705.9×10^{-4}
a	0.670	-1.227	0.130	0.366	0.487

For the four zones considered, the2D heat conduction equations have the following form:

$$\frac{1}{r}\frac{\partial}{\partial r}\left(rk_i(T^{(i)})\frac{\partial T^{(i)}}{\partial r}\right)+\frac{\partial}{\partial z}\left(k_i(T^{(i)})\frac{\partial T^{(i)}}{\partial z}\right)+q_v=0,\quad R_{i-1}<r<R_i,\quad 0<z<L,\ R_0=0, \tag{1}$$

where q_v = 1.9 W cm^{-3} is the power deposited into the discharge zone per unit volume and i = 1, 2, 3, 4.

The equations are solved at the following boundary conditions:

$$\left.\frac{\partial T^{(1)}}{\partial r}\right|_{r=0}=0,\quad \left.T^{(4)}\right|_{r=R_4}=T_w,\quad 0\leq z\leq L$$

and for i = 1, 2, 3

$$T^{(i)}\left(R_i-0,z\right)=T^{(i+1)}\left(R_i+0,z\right),\quad 0\le z\le L,$$

$$k_i(T^{(i)})\frac{\partial T^{(i)}}{\partial r}\bigg|_{R_i-0}=k_{i+1}(T^{(i+1)})\frac{\partial T^{(i+1)}}{\partial r}\bigg|_{R_i+0},\quad 0\le z\le L,$$

$$T^{(i)}\Big|_{z=0}=T^{(i)}\Big|_{z=L}=T_{w,i},\quad R_{i-1}\le r<R_i,$$

$$T^{(i)}\Big|_{z=0}=T^{(i)}\Big|_{z=L}=0.5\left(T_{w,i}+T_{w,i+1}\right),\quad r=R_i,$$

$$T^{(4)}\Big|_{z=0}=T^{(4)}\Big|_{z=L}=T_{w,4},\quad R_3<r\le R_4,$$

The problem formulated above is solved by using homogeneous difference schemes and an iteration procedure for solving the non-linear set of algebraic equation obtained.

The gas-discharge temperature distribution in axial (a) and radial (b) directions are shown for a nanosecond pulsed longitudinal discharges in He (45 Torr) in figure 2 and in He (45 Torr) with small admixtures of strontium (0.6 Torr) and bromine (1.2 Torr) in figure 3 and 4, respectively.

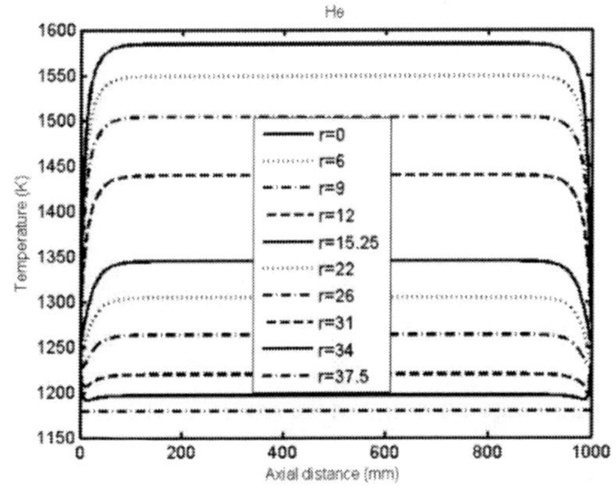

a)

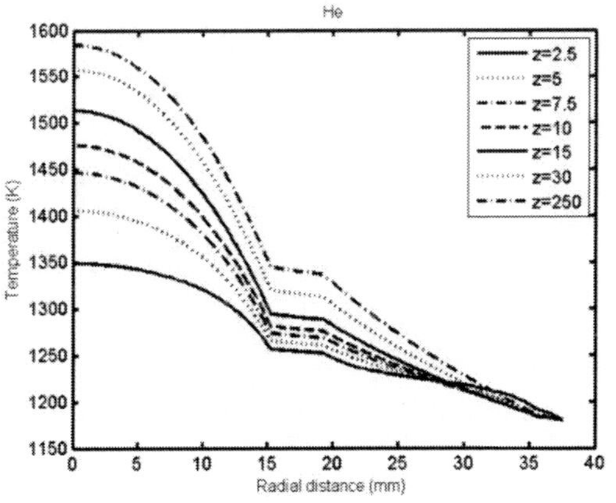

b)

Figure 2 (a) and (b). Gas-discharge temperature distribution in axial (a) and radial (b) directionss for nanosecond pulsed longitudinal discharges in He (45 Torr).

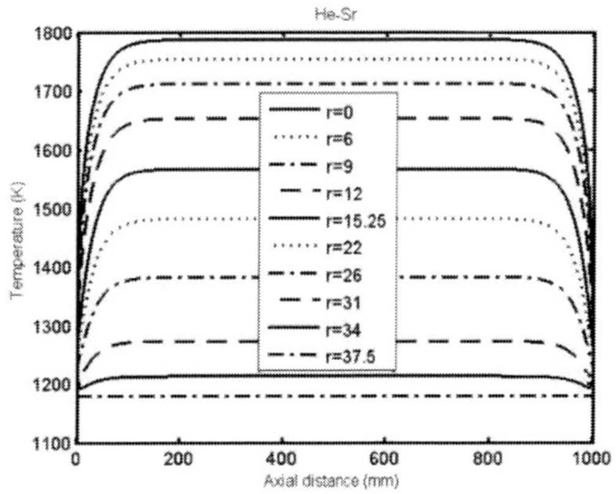

a)

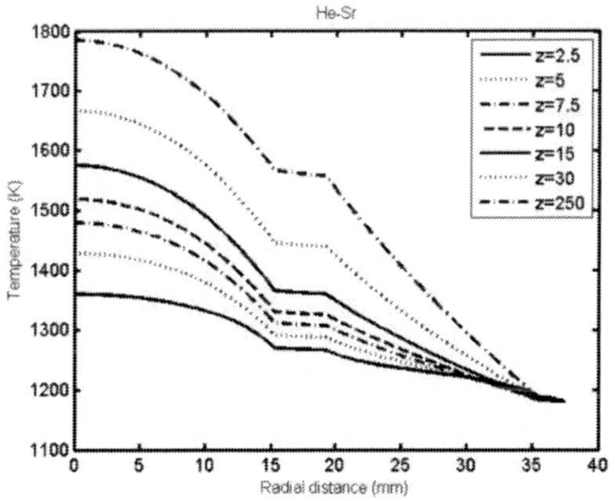

b)

Figure 3 (a) and (b). Gas-discharge temperature distribution in axial (a) and radial (b) directions for nanosecond pulsed longitudinal discharges in Sr-He (0.6 Torr - 45 Torr) mixture.

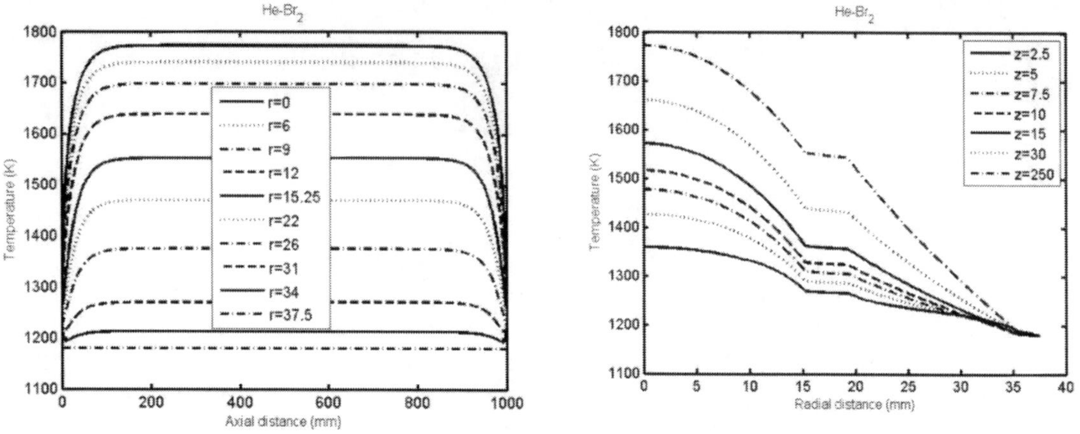

a) b)

Figure 4 (a) and (b). Gas-discharge temperature distribution in axial (a) and radial (b) directions for nanosecond pulsed longitudinal discharges in Br-He (1.2 Torr - 45 Torr) mixture.

4. Conclusions

A new discharge tube design was developed with incompact ZrO_2 active-volume insulation of the discharge-free zone, in order to increase additionally the operating temperature; its experimental investigation is in progress. Assuming axial symmetry and a uniform power input, results from 2D numerical modeling (r, z) of the gas temperature in nanosecond pulsed longitudinal discharges in various mixtures are presented for the newly-designed large-volume high-temperature gas discharge tube. The development of a 2D numerical model for determining the gas temperature in this gas discharge is a further, more complex, step toward approaching the real experimental conditions in the discharge studied.

Acknowledgement

This work was supported by Bulgarian Science Fund for under Grant No 106/2013 (University of Sofia) and contract BK-03-13 (ISSP, Bulgarian Academy of Sciences).

References

[1] Temelkov K A, Vuchkov N K, Ekov R P and Sabotinov N V 2008 *J. Phys.* D: *Appl. Phys.* **41** 105203
[2] Temelkov K A, Vuchkov N K, Freijo-Martin I and Ekov R P 2010 *J. Phys.* D: *Appl. Phys.* **43** 075206
[3] Handbook 1976 *Tables of physical constants*, ed I. K. Kikoin (Atomizdat Moscow) pp 257-61

AUTHOR INDEX

AUTHOR INDEX

AUTHOR INDEX

Institute of Physics
Dirac House, Temple Back
Bristol BS1 6BE UK

CURRAN ASSOCIATES INC
proceedings.com